The Penguin Dictionary of
ELECTRONICS

The Penguin Dictionary of
ELECTRONICS

Editor: Valerie Illingworth
for Market House Books

THIRD EDITION

PENGUIN BOOKS

PENGUIN BOOKS

Published by the Penguin Group
Penguin Books Ltd, 27 Wrights Lane, London W8 5TZ, England
Penguin Putnam Inc., 375 Hudson Street, New York, New York 10014, USA
Penguin Books Australia Ltd, Ringwood, Victoria, Australia
Penguin Books Canada Ltd, 10 Alcorn Avenue, Toronto, Ontario, Canada M4V 3B2
Penguin Books (NZ) Ltd, Private Bag 102902, NSMC Auckland, New Zealand

Penguin Books Ltd, Registered Offices: Harmondsworth, Middlesex, England

First edition 1979
Second edition 1988
Third edition 1998
10 9 8 7 6 5 4 3 2 1

Copyright © Market House Books Ltd, 1979, 1988, 1998
All rights reserved

Typeset in 8.5/10.5 pt Tmes New Roman PS
Printed in England by Clays Ltd, St Ives plc

PREFACE

The Penguin Dictionary of Electronics is primarily concerned with the words, terms, and abbreviations used in electronic research and industry and in solid-state theory. This third edition of the dictionary has substantially revised the second edition and contains many definitions of terminology in the related fields of computing, communications, control, electrical engineering, and music technology together with items of historical interest. There are now over 4800 entries in the dictionary, illustrated by almost 400 diagrams and tables.

We hope that the dictionary will be of use not only to students and teachers of electronics, physics, and related subjects but also to researchers, technicians, and technologists working in electronics or in the emerging fields of music and audio technology, who use electronic equipment in their work.

York 1998

This edition is dedicated to Carol Young, who died in 1997. Carol will be remembered not only as the author of the first two editions of the highly successful *Penguin Dictionary of Electronics* but as a woman who established a national reputation for her work, in the voluntary sector, for the community.

Living in Harlow, she made major contributions to the establishment of a successful Women's Refuge, a Well Women Centre, and a Playbus. She went on to chair the National Playbus Association, and later, as her family grew up, she became a professional manager heading the Rural Unit of the National Council for Voluntary Organisations.

CONTRIBUTORS

Editor:
Valerie Illingworth BSc, MPhil

Contributors for the Third Edition:
Andy M. Tyrell BSc, PhD, CEng, MIEE, Senior MIEEE
David M. Howard BSc, PhD, CEng, FIOA, MAES
Paul E. Garner BSc
Stuart J. Porter BSc, DPhil, AMIEE, MIEEE
Tony E. Ward BSc, MBA, CEng, MIEE, MIEEE
John Wood BSc, PHD, CEng, MIEE, MIEEE
Department of Electronics, University of York

For the First and Second Editions:
Carol Young BSc (author)
John Young BSc, MSc (consultant)

GUIDE TO THE DICTIONARY

The terms defined in the dictionary – the headwords – are listed in strict alphabetical order, ignoring spaces, punctuation, and numbers, and are printed in bold type.

An abbreviation or variant form is given in brackets immediately after the headword.

An alternative name is given after the headword, preceded by the label *Syn.* (short for synonym).

The meaning of each headword is explained in one or more definitions at the most commonly used form or spelling, or in some cases at the abbreviated form.

Words printed in italics in a definition are closely associated with the headword under which they appear and are defined in this position.

Headwords of major importance together with associated topics are defined and discussed in lengthy articles.

Cross references to information elsewhere in the dictionary are indicated by an arrow (➤) preceding selected words or terms in the definition text. Additional references to further information are indicated by a double arrow (➤➤).

Simple cross-reference headwords, including abbreviations and alternative names, refer the reader to the appropriate headword.

Tables in the backmatter give general information concerning, for example, graphical symbols, symbols and units of physical quantities, the electromagnetic spectrum, and major inventions and discoveries in electronics.

A

ABC *Abbrev. for* automatic brightness control. ➤television receiver.

aberration A defect in the image produced by an optical or electronic lens system.

abrupt junction A ➤p-n junction in which the impurity concentration changes abruptly from acceptors to donors (➤semiconductor). In practice such a junction may be approximately realized when one side of the junction is much more highly doped than the other, i.e. a p^+-n or n^+-p junction. This is a *one-sided* abrupt junction.

absolute electrometer ➤attracted-disc electrometer.

absolute temperature ➤thermodynamic temperature.

absolute zero The temperature at which the energy of random motion of the particles in a system at thermal equilibrium is zero. It is the lowest temperature theoretically possibly and is the zero of the ➤thermodynamic temperature scale.

absorption 1. Attenuation of a radiowave due to dissipation of its energy, as by the production of heat. **2.** Attenuation of a beam of light by a crystal due to localized vibrational modes in the crystal resulting from the presence of impurity atoms. This gives rise to characteristic sharp troughs in the transmission or reflection spectra and can be used to analyse the material. Absorption can also occur due to photon-induced electron transitions between different energy bands in a semiconductor and can be used to determine the energy gap.

absorption coefficient For a travelling wave in a lossy medium, the fraction of the power lost per incremental length, given by

$$\alpha = -(dI/I)/dz$$

For a travelling wave in a dielectric waveguide, solutions of Maxwell's equations yield the relationships between the absorption coefficient α and the refractive index of the material with the electrical conductivity and the dielectric constant of the material, i.e. between the optical and electrical characteristics of the material.

absorption loss The magnitude of the ➤absorption of a radiowave, usually expressed in ➤nepers or ➤decibels. ➤unabsorbed field strength.

a.c. (or **AC**) *Abbrev. for* alternating current.

accelerated life test A form of ➤life test of a circuit or device so designed that the duration of the test is appreciably less than the normal expected life of the device. This is achieved by subjecting the item to an excessive applied stress level without altering the basic modes or mechanisms of failure or their relative prevalence. Thermal stress is a commonly applied stress.

accelerating anode ►electron gun.

accelerating electrode Any electrode that accelerates electrons in the electron beam of an ►electron tube. ►►electron gun.

acceleration voltage In general, any voltage that produces acceleration of a beam of charged particles. The term is usually reserved for those devices in which an appreciable acceleration of an electron beam is produced, as in velocity-modulated tubes.

accelerator A machine used to accelerate charged particles or ions in an electric field in order to produce high-energy beams for the study of nuclear structure and reactions. Magnetic fields are used to focus and determine the direction of the beam. Very high energies are achieved using machines in which the beams of particles are subjected to a series of relatively small accelerating voltages. ►linear accelerator; synchrotron.

acceptor *Short for* acceptor impurity. ►semiconductor.

access system In communications, a system that permits users to communicate with a network that services a number of individuals or provides specific services, such as fax, modem, electronic mail, etc., to a number of different users.

access time The time taken to retrieve an item of information from computer ►memory.

a.c.-coupled amplifier ►amplifier.

accumulation mode ►MOS capacitor.

accumulator 1. *Syn. for* secondary cell. ►cell. **2.** A type of ►register.

ACK signal *Short for* acknowledgment signal. ►digital communications.

acoustic coupler A device that allows data or other voice-frequency information to be transmitted across a telephone line. An acoustic coupler is a cradle that supports a standard telephone handset and uses sound transducers to send and receive the audiofrequency signals through the telephone. ►►modem.

acoustic delay line ►delay line.

acoustic feedback Unwanted ►feedback of the sound output of an audiofrequency loudspeaker to a preceding part of a sound-reproduction system via a microphone. The sound waves can be detected and amplified by the electronic circuits in the system; above a critical level oscillations are produced that are heard as an unpleasant and self-sustaining howling noise from the loudspeaker.

acoustic pressure waveform The time variation of acoustic pressure, for example as output from a pressure-sensitive ►microphone.

acoustic wave *Syn.* sound wave. A wave that is transmitted through a solid, liquid, or gaseous material as a result of the mechanical vibrations of the particles forming the material. The normal mode of propagation is longitudinal, i.e. the direction of motion of the particles is parallel to the direction of propagation of the wave, and the wave therefore consists of compressions and rarefactions of the material. The term 'sound wave' is sometimes confined to those waves with a frequency falling within the audible range of the human ear, i.e. from about 20 hertz to 20 kilohertz. Waves of frequency greater than about 20 kHz are ultrasonic waves.

In a crystalline solid an acoustic wave is transmitted as a result of the displacement of the lattice points about their mean position, and the modes of propagation are constrained by the interatomic forces active between the lattice points. The wave is transmitted as an elastic wave through the crystal lattice. The angular frequency, ω, of the wave is related to the wave vector K by the relation

$$m\omega^2 = 2\Sigma_{p>0}C_p(1 - \cos pKa)$$

where m is the mass of an atom, C_p the force constant between planes of atoms separated by p, where p is an integer, and a is the spacing between atomic planes. The range of physically realizable waves that may be transmitted is

$$\pi > Ka > -\pi$$

The limits of this range define the first Brillouin zone for the crystal lattice, and at these limits travelling waves cannot be propagated; standing waves are formed. The energy of the lattice vibrations is quantized. The quantum of energy is the ➤phonon, which is analogous to the photon of energy of an electromagnetic wave. The phonon energy is given by $h\nu$, where ν is the frequency and is equal to $\omega/2\pi$.

A travelling acoustic wave in a solid can be produced by applying mechanical stress to the crystal or as a result of ➤magnetostriction or of the ➤piezoelectric effect. The resulting phonons can interact with mobile charge carriers present in the material. The interaction can be considered as an electric vector, analogous to the electric vector associated with an electromagnetic wave, that extends for about a quarter wavelength distance orthogonal to the direction of propagation of the wave.

acoustic wave device A device used in a signal-processing system in which acoustic waves are transmitted on a miniature substrate in order to perform a wide range of functions. Active and passive signal-processing devices formed on a single semiconductor chip have been produced including delay lines, attenuators, phase shifters, filters, amplifiers, oscillators, mixers, and limiters.

Bulk acoustic waves are acoustic waves propagated through the bulk substrate material. The substrate material consists of a piezoelectric semiconductor, such as cadmium sulphide. The acoustic waves are generated from electrical signals as a result of the ➤piezoelectric effect. The electric field vector of the acoustic wave interacts with the conduction electrons of the semiconductor, which have a drift velocity due to an external applied d.c. electric field. At a sufficient value of the drift velocity the kinetic energy of the drift electrons is converted to radiofrequency energy as a result of interaction with the acoustic field, and amplification of the original signals can result.

Surface acoustic waves (SAW) are propagated along the surface of a substrate. The associated electric field extends for a short distance out of the surface and can interact with the conduction electrons of a separate semiconductor placed just above the surface. The physical separation of the acoustic substrate and the semiconductor allows the materials to be chosen so that the energy dissipation in the system is minimized. The acoustic material is a piezoelectric material that has a high electromechanical coupling coefficient and low acoustic loss. The semiconductor material is one that has high mobility electrons, optimum resistivity, and low d.c. power requirement so that the optimum efficiency is obtained.

acquisition 1. The locating of a signal within a given frequency range. In a ➤phase-lock loop, for example, acquisition is the process by which the loop searches for the reference signal, which is then employed to stabilize the frequency of a different frequency oscillator. **2.** The locating of a signal in space. In ➤radar systems, for example, acquisition is the process of sweeping a receiving antenna in the azimuth and elevation directions searching for a signal from a target.

active Denoting any device, component, or circuit that introduces ➤gain or has a directional function. In practice any item except pure resistance, capacitance, inductance, or a combination of these three is active. ➤➤passive.

active antenna *Syns.* primary radiator; driven antenna. ➤directional antenna.

active area The area of a metal contact through which current flows, or the surface area through which light passes, in a ➤semiconductor device.

active component 1. ➤active. **2.** ➤active current. **3.** ➤active volt-amperes. **4.** ➤active voltage.

active current *Syns.* active component, in-phase component of the current. The component of the alternating current that is in ➤phase with the alternating voltage, alternating current and voltage being regarded as vector quantities. ➤➤reactive current.

active filter ➤filter.

active interval *Syn.* trace interval. ➤sawtooth waveform.

active load A load that is made using an active device, usually a ➤transistor. The active load is used for its high dynamic impedance that is presented to the gain device.

active network ➤network; active.

active region The region of a ➤semiconductor device where electrical activity such as gain or energy conversion takes place, for example the base region of a ➤bipolar junction transistor.

active voltage *Syns.* active component, in-phase component of the voltage. The component of the alternating voltage that is in ➤phase with the alternating current, alternating current and voltage being regarded as vector quantities. ➤➤reactive voltage.

active volt-amperes *Syns.* active component, in-phase component of the volt-amperes. The product of the voltage and the ➤active current or of the current and the ➤active voltage. It is equal to the real power in watts. ➤➤reactive volt-amperes.

activity The ratio of the peak value of the oscillations in a ➤piezoelectric crystal to the peak value of the exciting voltage.

actuacting transfer function ➤feedback control loop.

actuator A device that is used to convert an electrical signal into the appropriate mechanical energy. It is a special case of a ➤transducer.

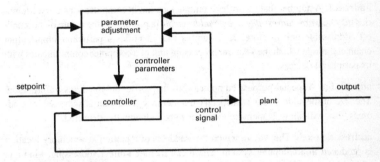

Adaptive control system

adaptive control system A ➤control system in which the controller has adjustable ➤parameters and a mechanism for adjusting the parameters. The controller becomes nonlinear because of the parameter adjustment mechanism. An adaptive control system can be modelled as a system having two loops (see diagram). One loop is a normal ➤feedback with the process and the controller. The other loop is the parameter adjustment loop, which is often slower than the normal feedback loop.

adaptive equalizer ➤equalization.

ADC *Abbrev. for* analogue-to-digital converter.

ADCCP *Abbrev. for* advance data communication control procedure. In communications, a type of ➤protocol developed by the American National Standards Institute. It works at the bit level within a message: particular bits in the message are used to indicate which bits are the message, which bits are the sender and recipient identifiers, and which bits are concerned with error handling.

Adcock direction finder *Syn.* Adcock antenna. A radio direction finder consisting of a number of spaced vertical ➤antennas. The errors due to the horizontally polarized components of the received waves are effectively eliminated as such components have only a minimal effect on the observed bearings.

adder A circuit in a ➤computer that performs mathematical addition. A *full adder* contains several identical sections each of which add the corresponding ➤bits of the two numbers to be added together with a carry digit from the preceding section and produce an output corresponding to the sum of the bits and a carry digit for the next section.

A *half-adder* is a circuit that adds two bits only and produces two outputs; the outputs must be suitably combined in another half-adder in order to produce the correct outputs for all possible combinations of inputs.

If two numbers each consisting of x bits are to be added, a full adder circuit requires $2x$ inputs to x identical sections and $(x + 1)$ outputs in order to perform the addition.

additive synthesis ➤synthesis.

address 1. A number that identifies a unique location in computer ➤memory. Memories may be *word-addressable* or *byte-addressable* depending on the nature of the smallest addressable unit of store. **2.** A number that identifies a particular input/output channel through which the ➤central processing unit of a computer communicates with its peripheral devices.

address bus A special-purpose computer ➤bus that carries only ➤address information. The size of the address bus will specify the memory space that is addressable: n address lines will allow 2^n memory locations to be individually identified.

addressing mode The way in which the ➤address of a particular ➤memory location is produced in a computer system. These can include ➤direct addressing, ➤indirect addressing, ➤relative addressing, and ➤indexed addressing modes. Addressing modes are specified for individual processors; the specification is part of the computer architect's task.

admittance Symbol: Y; unit: siemens. The reciprocal of ➤impedance. It is a complex quantity given by

$$Y = G + jB$$

where G is the ➤conductance, B the ➤susceptance, and $j = \sqrt{-1}$. Since impedance, Z, is given by

$$Z = R + jX,$$

where R and X are the resistance and reactance, respectively, then

$$Y = 1/Z = 1/(R + jX)$$

$$= (R - jX)/(R^2 + X^2)$$

admittance gap A gap in the wall of a ➤cavity resonator that allows it to be excited by a source of radiofrequency energy, such as a velocity-modulated electron beam, or that allows it to affect such a source.

ADSR *Abbrev. for* attack decay sustain release.

aerial ➤antenna.

AES *Abbrev. for* Auger electron spectroscopy.

a.f. (or **AF**) *Abbrev. for* audiofrequency.

a.f.c. (or **AFC**) *Abbrev. for* automatic frequency control.

afterglow ➤persistence.

aftertouch A ➤MIDI control available on some electronic musical instrument keyboards where, having depressed a key to play a note, the application of an additional firm pressure controls an effect such as vibrato (fundamental-frequency modulation) or tremolo (amplitude modulation) on the sound. Aftertouch is either *monophonic*, where the effect is applied to all notes currently depressed, or *polyphonic*, where it is applied individually only to each note that is depressed more firmly.

a.g.c. (or **AGC**) *Abbrev. for* automatic gain control.

ageing *Syn. for* burn-in. ➤failure rate.

AI *Abbrev. for* artificial intelligence.

air bridge A method of forming interconnections or ➤crossovers in MMICs (➤monolithic microwave integrated circuits). The bridge consists of a layer of plated metal with no material other than air between it and the slice. Advantages include a low parasitic capacitance, immunity to edge profile problems, and the ability to carry substantial current. The major steps in the formation of an air bridge are shown in the diagram. A layer of ➤resist is deposited on the slice and processed to produce the required pattern of interconnections. This is then covered by a very thin layer of metal, usually using ➤sputtering techniques (Fig. *a*). A second layer of resist is added, and processed to leave the thin metal layer exposed in those areas where the plating is to take place (Fig. *b*). The slice is then plated (Fig. *c*). The presence of the thin metal layer

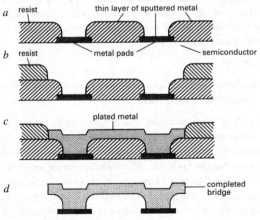

Steps in the formation of an **air bridge**

ensures that the plating current is carried to all parts of the slice. Finally the layers of photoresist and thin metal are removed to leave the required interconnections (Fig. *d*).

air capacitor A ►capacitor that uses air as the main dielectric.

air gap A gap in a magnetic circuit that increases the inductance and saturation point. An air gap is a necessity where moving parts are involved.

airport surveillance radar (ASR) ►precision approach radar.

Alcomax *Trademark* A material used for permanent magnets because of its exceptionally high coercivity. It consists of an alloy of iron, nickel, aluminium, cobalt, and copper.

algorithm A planned set of instructions or steps in a computer ►program that is designed to solve a particular problem.

aliasing A distorting effect that occurs when an analogue signal f(*t*) is sampled digitally at a sampling frequency less than twice the signal frequency (►Nyquist rate): a signal f(*t*′) is retrieved from the sampled information that differs from the original input signal. The retrieved signal – the *alias signal* – has a frequency that corresponds to the harmonics of the high-frequency components of f(*t*). *Anti-aliasing* is any action taken to reduce or remove aliasing effects. ►►anti-aliasing filter.

allowed band ►energy bands.

alloyed device A semiconductor device, such as a transistor or diode, that contains one or more ►alloyed junctions.

alloyed junction A semiconductor junction formed by bonding metal contacts onto a wafer of semiconductor material and then heating to produce an alloy: it is a method commonly used for germanium diodes and transistors and in early silicon devices, the devices being termed *alloyed transistors* or *diodes*. It has been generally superceded by the ►planar process for all but special-purpose items, although it is a useful technique for gallium arsenide devices.

all-pass network ►network.

Alnico *Trademark* A material used for high-energy permanent magnets. It is an alloy of nickel, iron, aluminium, cobalt, and copper.

ALOHA system A satellite communications system conceived by the University of Hawaii to interconnect computers using a random multiple-access protocol. The random-access concept has four basic modes. Any user can transmit at any time, encoding the message with an error-detection cde. After transmission the sender waits for an acknowledgment signal (ACK). If two messages are transmitted at the same time, or overlap in time, data received is incorrect and a collision is said to have occurred. When a collision is detected a negative acknowledgment (NACK) signal is generated by the receivers. When a NACK is received by a sender a retransmission is required.

The retransmission is delayed by a random delay to reduce the probability that the re-transmitted signals will collide. This system is sometimes referred to as *pure ALOHA*.

The pure ALOHA system performance can be improved if the transmitters are synchronized, as in the case of *S-ALOHA* (slotted-ALOHA). In S-ALOHA the retransmission is delayed by a random integer number of slot times, where a slot time is slightly longer than the ALOHA message packet length. A further improvement can be obtained if time slots can be reserved for use by a transmitter. This is the basis of *R-ALOHA* (reservation ALOHA).

alpha current factor *Syn. for* common-base forward-current transfer ratio.

alpha cut-off frequency Symbol: f_α. The frequency at which the common-base forward-current transfer ratio, α, of a ►bipolar junction transistor has fallen to $1/\sqrt{2}$ (i.e. 0.707) of the low-frequency value.

alphanumeric Any letter of the English alphabet or any of the decimal digits 0 to 9.

alternating current (a.c. or AC) An electric current whose direction in the circuit is periodically reversed with a ►frequency, f, independent of the circuit constants. In the simplest form the instantaneous current varies with time:

$I = I_0 \sin 2\pi f t,$

where I_0 is the ►peak value of the current. AC is measured either by its peak value, its ►root-mean-square value, or more rarely by the *current average* – the algebraic average of the current during one positive half cycle.

alternating-current generator A ►generator for producing alternating currents or voltages. ►►synchronous alternating-current generator.

alternating-current motor A ►motor that requires alternating current for its operation.

alternating-current resistance ►effective resistance.

alternating-current transmission 1. A method of transmitting mains electrical power at constant voltage around the country on the national grid. Alternating-current transmission allows the use of transformers for step-up and step-down of the voltage. **2.** A method of transmission used in television in which the direct-current component of the luminance signal (►colour television) is not transmitted. A ►direct-current restorer must be used in this form of transmission. ►►direct-current transmission.

alternative routing In a communication system, the use of a second channel or path that can be used if the primary channel becomes unavailable for any reason. For example, in a telephone connection between two subscribers there are usually a number of different routes that can be used to make the connection. The route selected depends on the level of usage of the system as a whole and whether any physical damage exists on any of the routes.

alternator ►synchronous alternating-current generator.

ALU *Abbrev. for* arithmetic/logic unit. ➤central processing unit.

alumina Aluminium oxide, symbol: Al_2O_3. In solid-state electronics it is used as a dielectric, as in thin-film capacitors, or as the gate dielectric in MOSFETs. In valve electronics it was used as an insulator due to its excellent electrical and thermal resistance.

aluminium Symbol: Al. A metal, atomic number 13, that is extensively used in electronic equipment and devices for contacts and interconnections on integrated circuits. It is a good conductor and is ductile, malleable, resistant to corrosion, lightweight, easily evaporated onto surfaces, and abundant.

aluminium antimonide Symbol: AlSb. A semiconductor with useful properties up to operating temperatures of 500 °C.

a.m. (or **AM**) *Abbrev. for* amplitude modulation.

AMI PCM *Abbrev. for* alternate mark inversion pulse code modulation. ➤pulse modulation.

ammeter An indicating instrument for measuring current. Common types are ➤moving-coil and ➤moving-iron instruments and ➤thermocouple and ➤hot-wire instruments. Most ammeters are shunted ➤galvanometers. ➤digital ammeter.

amorphous silicon An essentially noncrystalline form of silicon. Amorphous silicon has no long-range order of the silicon atoms in the solid, unlike a crystal. Very short-range order, up to about 30 nanometres, can be obtained. The material does not have an ➤energy band structure as such, but can be described internally as having ranges of electron energies of relatively high ➤mobility separated by a mobility gap. Thus doped amorphous silicon can be produced in which the majority charge carriers can be defined, and hence electronic devices can be made in amorphous silicon. The carrier mobilities are generally very low, so the speed of response of the devices is limited. It is deposited as a thin film using low-temperature or plasma-enhanced ➤chemical vapour deposition techniques, and can be deposited very uniformly over very large areas using these techniques.

Amorphous silicon is used for making ➤thin-film transistors, which are employed as driver devices in large-area ➤flat-panel display technology; it can also be used for making ➤solar cells over large areas, for high electrical output though at low efficiency.

amp *Short for* ampere.

ampere Symbol: A. The ➤SI unit of electric ➤current defined as the constant current that, if maintained in two straight parallel conductors of infinite length and negligible cross section and placed one metre apart in a vacuum, would produce between these conductors a force equal to 2×10^{-7} newton per metre of length.

ampere balance ➤Kelvin balance.

ampere-hour Symbol: Ah. A unit for the quantity of electricity obtained by integrating current flow in amperes over the time in hours for its flow; it is used as a measure of battery capacity.

Ampere–Laplace law ➤Ampere's law.

ampere per metre Symbol: A/m. The ➤SI unit of ➤magnetic field strength. It is the magnetic field strength in the interior of an elongated uniformly wound solenoid that is excited with a linear current density in its winding of one ampere per metre of axial distance.

Ampere's circuital theorem ➤Ampere's law.

Ampere's law 1. *Syn.* Ampere–Laplace law. The force between two parallel current-carrying conductors in free space is given by

$$dF = \mu_0 I_1 ds_1 I_2 ds_2 \sin\theta / 4\pi r^2$$

where I_1 and I_2 are the currents, ds_1 and ds_2 the incremental lengths, r the distance between the incremental lengths and θ the angle (see diagram); μ_0 is the ➤permeability

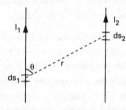

Ampere's law

of free space. Ampere's law thus relates electrical and mechanical phenomena. ➤➤Coulomb's law. **2.** *Syn.* Ampere's circuital theorem. The work done in traversing a closed circuit that encloses a current I is given by

$$\oint \boldsymbol{B}.d\boldsymbol{s} = \mu_0 I$$

where μ_0 is the ➤permeability of free space, $\boldsymbol{B}$ is the magnetic flux density, and $d\boldsymbol{s}$ an incremental length.

The total current flowing is given by the integral of the current density flowing in the area bounded by the loop. In a medium of ➤magnetization $\boldsymbol{M}$, the total current density, $\boldsymbol{j}_T$, is given by the sum of the real current density, $\boldsymbol{j}$, and the equivalent magnetization current density, $\boldsymbol{j}_M$, where

$$\boldsymbol{j}_M = \text{curl } \boldsymbol{M}$$

Since

$$\oint \boldsymbol{B}.d\boldsymbol{s} = \int \text{curl } \boldsymbol{B}.d\boldsymbol{A}$$

where $d\boldsymbol{A}$ is an increment of area, then

$$\int \text{curl } \boldsymbol{B}.d\boldsymbol{A} = \mu_0 \int \boldsymbol{j}_T.d\boldsymbol{A}$$

and thus

$$\text{curl}(\boldsymbol{B} - \mu_0 \boldsymbol{M}) = \mu_0 \boldsymbol{j}$$

Since **H,** the magnetic field strength, is defined by

$$\mu_0 H = B - \mu_0 M$$

Ampere's law, in differential form, may be written as

$$\text{curl } H = j$$

ampere-turn Symbol: At. A unit of ►magnetomotive force equal to the product, *NI,* of the total number of turns, *N,* in a coil through which a current, *I,* is flowing.

amplification The reproduction of an electrical signal, usually at an increased intensity, by an electronic device.

amplification factor Symbol: μ. In an active electronic device used to deliver a constant current to a load, the ratio of the incremental change in output voltage required to maintain the constancy of the current to a corresponding incremental change in the input voltage. μ is equal to $-A_v$, where A_v is the voltage ►gain of the device.

amplifier A device or circuit that produces an electrical output that is a function of the corresponding input parameter and increases the magnitude of the input signal by means of energy drawn from an external source: an amplifier produces ►gain. The type of amplifier depends on the particular input and output qualities. A *current amplifier* provides current gain; this type of amplifier has a low input impedance and a high output impedance. A *voltage amplifier* provides voltage gain; this type has a high input impedance and a low output impedance. A *transimpedance amplifier* provides an output voltage for an input current, with gain measured in ohms; this type has a low input impedance and a low output impedance. A *transadmittance amplifier* provides an output current for an input voltage, with gain measured in siemens; this type has a high input impedance and a high output impedance.

A *power amplifier* is an amplifier that produces simultaneous voltage and current gain. The method of operation of a power amplifier is described by the *class* of the amplifier; for example, a ►class A amplifier has the active amplifying devices switched on for the whole of the cycle of the alternating input signal, whereas a ►class B amplifier will switch on one of a pair of active amplifying devices only during positive parts of the signal, and switch on the other device only during the negative parts of the cycle (►class AB, C, D, E, F, G, H, S amplifier).

A *linear amplifier* produces an output signal that is a linear function of the input signal, otherwise the amplifier is *nonlinear*. In practice all linear amplifiers will have nonlinearities that produce some distortion of the output.

A *multistage amplifier* consists of several amplifying circuits or stages coupled together to form an amplifier. A *direct-coupled* or *DC-coupled amplifier* couples the various stages together directly with no other components between the individual circuits, requiring careful matching of the voltages and currents between the stages. Such an amplifier will amplify steady voltages and currents. An *AC-coupled amplifier* uses other components between the stages, relieving this matching constraint; capacitors are often used to provide a DC-block between the stages. AC-coupled amplifiers have zero gain at DC.

A *limiting amplifier* has a very high gain for a small range of input signal, so that the output is substantially a constant amplitude. This amplifier is very nonlinear, but is useful for removing unwanted amplitude modulation from signals.

amplifier stage A relatively small-gain amplifier that is coupled to other similar sub-circuits in cascade, i.e. the output of one stage forms the input of the next, in order to provide a larger overall gain or to optimize impedance matching.

amplitude 1. Strictly, the ➤peak value of an alternating current or wave in the positive or negative direction. The difference between extreme values in a complete cycle is the *peak-to-peak amplitude.* **2.** The value of an alternating current or wave in the positive or negative direction at a particular moment. **3.** ➤wave.

amplitude compandoring In audio communication systems, a means of increasing the performance of a communications channel by compressing the audio signal before modulating it onto a carrier for transmission and expanding it after demodulation in the receiver. The technique is commonly used in land-mobile communication systems. ➤data compression.

amplitude distortion ➤distortion.

amplitude equalizer A network used to modify the ➤magnitude response of a circuit or system such that the combined ➤amplitude response is flat to within a specified value over a specified frequency range. In this way amplitude variations can be reduced or equalized.

amplitude fading ➤fading.

amplitude modulation (a.m. or AM) A type of ➤modulation in which the amplitude of the ➤carrier wave is varied above and below its unmodulated value by an amount proportional to the amplitude of the signal wave and at the frequency of the modulating signal (Fig. *a*).

If the modulating signal is sinusoidal then the instantaneous amplitude *e,* of the amplitude-modulated wave may be given as

$$e = (A + B \sin pt) \sin \omega t$$

where A is the unmodulated carrier amplitude and B the peak amplitude variation of the composite wave, p is equal to $2\pi \times$ modulating signal frequency and ω to $2\pi \times$ carrier frequency. If the *modulation factor* or *index*, m, is defined by $m = B/A$ then the modulated wave may be given as

$$e = (1 + m \sin pt)A \sin \omega t$$

The modulation factor is sometimes quoted as a percentage of the carrier signal amplitude and is then termed the *percentage modulation.*

If the peak amplitude of the composite wave equals the peak amplitude of the unmodulated carrier, the instantaneous amplitude of the composite wave will equal zero

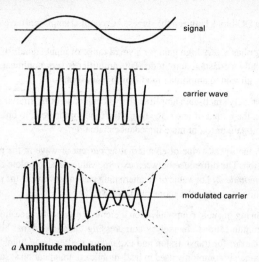

a **Amplitude modulation**

at its minimum. The ratio of *B* to *A* is then unity and the percentage modulation is 100%. If the amplitude of the modulating signal, *B*, is further increased, a condition known as *overmodulation* results and a 'gap' in the signal occurs (Fig. *b*). This gap produces a type of distortion called *sideband splatter* and results in the transmission of frequencies that are outside the normal allocated frequency range for the transmission. This type of distortion can cause severe interference to other adjacent stations.

A variation on conventional amplitude modulation results from the recognition that the amplitude-modulated signal consists of a carrier and two *sidebands*, one above the carrier and the other below. The actual information is contained in the sidebands and, if the carrier and one sideband is eliminated before transmission, the resulting form of transmission is called ►single-sideband or SSB. The carrier alone can also be suppressed, resulting in ►suppressed carrier modulation.

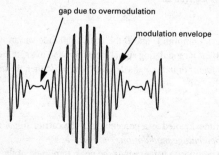

b Overmodulation in **amplitude modulation**

If the modulating signal is not sinusoidal but is made up of discrete levels, the resulting modulation is known as *amplitude shift keying* (ASK). Any number of discrete amplitude signal levels can be used. If only two levels are used, one of which is zero, the result is *binary amplitude shift keying*; this is also known as *on-off keying*. An alternative form of binary amplitude modulation is *two-tone modulation*, in which the carrier is always transmitted. Instead of the carrier simply being turned on or off, the carrier is amplitude modulated by two different sinusoidal signals representing either a one, or a mark, and a zero, or a space. The two frequencies are commonly separated by 170 hertz. Two-tone modulation is commonly used in telegraphy systems.

A composite modulation called *quadrature amplitude modulation* (QAM) results if the carrier wave is simultaneously amplitude- and phase-modulated (►phase modulation). If the modulating signal is divided into a number of discrete levels, the resulting modulation is called *quadrature phase shift keying* (QPSK). A further variation on this type of modulation results where the phase alignment is varied and an offset is introduced; the result of this is *offset quadrature phase shift keying* (OQPSK). The advantage of QPSK and OQPSK is that they reduce out-of-band interference. Out-of-band performance can be improved further by reducing the change in phase at each transition. Such schemes are called *minimum shift keying* (MSK).

Amplitude modulation may be achieved using a radiofrequency class C amplifier or oscillator. The carrier wave is produced in the amplifier or oscillator. The modulating signal is superimposed on it by varying either anode voltage (*anode modulation*) or the grid bias (*grid modulation*) in proportion to the modulating signal if a valve is being used in the transmitter. Alternatively the modulating signal is superimposed on the carrier by varying the base or collector voltages in proportion to the modulating signal if a bipolar transistor is being used in the transmitter. In either case the output of the circuit is then an amplitude-modulated radiofrequency wave.

►frequency modulation.

amplitude response The variation with frequency of the ratio of the magnitude of the output from a circuit or system to that of the input.

amplitude shift keying (ASK) ►amplitude modulation.

AM receiver A ►radio receiver that detects amplitude-modulated signals.

analogue circuit A circuit in which the output varies continuously as a given linear or nonlinear function of the input.

analogue computer ►computer.

analogue delay line ►delay line.

analogue gate ►gate.

analogue signal A ►signal that varies continuously in amplitude and time.

analogue switch Effectively, a solid-state ►relay made using ►transistors; either ►bipolar junction transistors or ►MOSFETs can be used. In the 'switch open' condition the transistor is in the cut-off condition, or below threshold for MOSFETS; in the 'switch closed' condition the transistor is switched fully on – in saturation – and can pass large current with only a small voltage drop. Typical resistance in the on condition is less than 100 ohms.

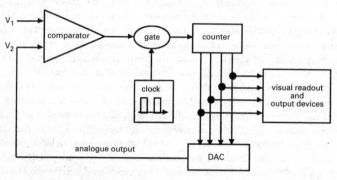

4-bit **analogue-to-digital converter**

analogue-to-digital converter (ADC) A device/circuit/IC that converts a continuous ►analogue signal into a discrete ►binary signal. There are many forms of ADC, including continuous balance converters, voltage-to-frequency converters, and dual ramp or integrating converters. A simple continuous balance converter is shown in the diagram. The counter is initially set to zero so that signal V_2 from the ►digital-to-analogue converter (DAC) section of the circuit is zero. When the unknown input V_1 is greater than V_2, the comparator provides an output voltage that opens the gate and allows pulses to be applied to the counter. So long as the gate remains open, clock pulses are fed to the counter and V_2 continues to increase. When V_2 equals V_1, the comparator output falls to zero and closes the gate. This 'freezes' the number stored in the counter, which can then be displayed on a digital readout device. In practical forms of the device, the counter is one that can count 'up' and 'down' to allow changes in V_1 to be followed.

analogue transmission The transmission of a signal in continuously variable form as opposed to the two discrete levels used in ►digital communications.

anaphoresis ►electrophoresis.

AND circuit (or **gate**) ►logic circuit.

Anderson bridge A ►bridge, modified from the ►Maxwell bridge, in which an inductance, L, is compared with a capacitance, C. Using a ►null-point detector instrument I (see diagram), resistances R_4 and X are adjusted until, when the bridge is balanced,

$$R_2R_3 = R_1R_4; \qquad L = C[R_2R_3 + (R_3 + R_4)X]$$

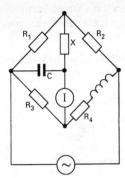

Anderson bridge

anechoic chamber A conducting chamber used for ➤electromagnetic compatibility testing. An antenna is placed in the chamber along with the product under test, which is then subjected to electromagnetic fields in order to determine the ➤radiated susceptibility of the product. All of the walls, the floor, and the ceiling are covered with material designed to absorb reflections that may interfere with the direct energy radiated from the antenna to the product. The action of the chamber is to screen the high fields inside from electrical equipment external to the chamber.

A *semianechoic chamber* is similar to the anechoic chamber but the conducting floor is not covered with the absorbing material. Here the chamber is used in a similar mode to the ➤open-area test site to determine the ➤radiated emissions of the product. The semianechoic chamber has the advantage that there are no ambient signals present but it may limit the volume available for the test.

angle modulation ➤modulation.

angle of flow The portion of the cycle of an alternating voltage, expressed as an angle, during which current flows.

angstrom Symbol: Å. A unit of length equal to 10^{-10} metre.

angular frequency Symbol: ω. The frequency of a periodic phenomenon expressed in radians per second. It is equal to the frequency in hertz times 2π.

anion An ion that carries a negative charge and, in electrolysis, moves towards the anode, i.e. travels against the direction of conventional current. ➤➤cation.

anisotropic 1. Denoting crystalline material whose properties, including conductivity, permittivity, and permeability, vary with direction relative to the crystal axes as a result of the crystal structure. **2.** More generally, denoting any material or process, such as ➤etching, whose properties are nonuniform with direction.

anisotype heterojunction ➤heterojunction.

anode The positive ➤electrode of an electrolytic cell, discharge tube, valve, or solid-state rectifier; the electrode by which electrons leave (and conventional current enters) a system. ➤cathode.

anode current The current flowing from the ➤anode to the ➤cathode of a device such as a solid-state rectifier.

anode dark space ➤gas-discharge tube (diagram).

anode glow ➤gas-discharge tube (diagram).

anode stopper ➤parasitic oscillations.

ANSI The American National Standards Institute. ➤standardization.

antenna *Syn.* aerial. The part of a radio system that radiates energy into space (*transmitting antenna*) or receives energy from space (*receiving antenna*). An antenna together with its ➤feeders and all its supports is known as an *antenna system*. There is a great variety of specially designed antennas, most of which are described by their shape, e.g. umbrella, clover leaf, H. L. T, cigar, and corner antennas. The most important types of antennas are ➤dipoles and ➤directional antennas. ➤Yagi antenna.

antenna array An arrangement of radiating or receiving elements so spaced and connected to produce directional effects. Very great directivity and consequently large ➤antenna gain can be produced by suitable design. An array of elements along a horizontal line is referred to either as a *broadside array* or *endfire array* depending on whether the directivity is in the horizontal plane at right angles to or along the line of the array, respectively. Arrays are commonly designed to have both horizontal and vertical directivity. The horizontal directivity is determined by the horizontal arrangement of antenna elements while the vertical directivity is dependent on the number of elements arranged in tiers (or stacks) one above the other.

The performance of an antenna array is indicated by the ➤antenna pattern of the system. The direction of maximum transmission or reception is given by the major lobe of the antenna pattern (➤steerable antenna).

antenna current The root-mean-square value of the current measured at a specified point in an antenna, usually either at the feedpoint or at the current maximum.

antenna efficiency *Syn.* radiation efficiency. The ratio of the power radiated by an antenna, at a specified frequency, to the total power supplied to it.

antenna factor The ratio of an incident electric field at the surface of an antenna to the received voltage at the antenna terminals. This enables easy conversion of a voltage measured from the antenna terminals to the field present at the antenna, particularly for ➤electromagnetic compatibility testing.

antenna feedpoint impedance The ➤impedance of an antenna at the point at which it is fed. The real part of this impedance is the antenna feedpoint resistance and the imaginary part is the antenna feedpoint reactance.

antenna gain 1. (in transmission) The ratio of the power that must be supplied to a reference antenna compared to the power supplied to the antenna under consideration in order that they produce exactly similar field strengths at the same distance and in the same specified direction (usually the direction of maximum radiation). **2.** (in reception) The ratio of the signal power produced at the receiver input by the given antenna to that produced by a reference antenna under similar receiving conditions and transmitting power.

In both cases the reference antenna must be specified.

antenna pattern *Syn.* radiation pattern. A diagram that represents the distribution of radiation in space from any source, such as a transmitting ➤antenna, or conversely shows the effectiveness of reception of a receiving antenna. A practical antenna neither radiates nor receives equally in all directions. Its antenna pattern is a plot of relative ➤gain as a function of direction, and may be made in terms of either its voltage response or power response; it may be produced in either polar or Cartesian coordinates. The antenna pattern for transmission is identical with that for reception.

A typical antenna pattern consists of one or more *lobes* that represent regions of enhanced response of an antenna in a particular plane. The *main* or *major lobe* is the lobe that contains the region of maximum radiation intensity or maximum sensitivity of detection of an antenna. It usually points forwards along the direction of propagation of the radiation. Other lobes are called *side lobes*. A *back lobe* is a lobe that points in the reverse direction to the direction of propagation. A typical antenna pattern is a polar figure of eight (see diagram). Point A (r, θ) on the diagram indicates the sensitivity of the antenna represented by r at an angle θ to the direction of propagation of the radiation.

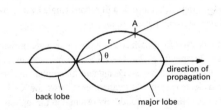

Figure of eight **antenna pattern**

antenna polarization The polarization of an antenna in a given direction is defined in one of two ways. If the antenna is being excited and is radiating, its polarization is defined to be the ➤polarization of the radiated wave. If the antenna is receiving energy, its polarization is said to be the polarization of an incident wave from the given direction that results in maximum available power at the antenna terminals. If a direction is not stated, the polarization is taken to be the polarization in the direction of maximum gain.

antenna radiation resistance The power radiated by an antenna divided by the mean square value of the current at a given specified reference point on the antenna, usually the feedpoint or a current antinode. This resistance takes into account the energy consumed by the antenna system as a result of radiation.

antenna resistance The total power supplied to an antenna divided by the mean square value of the current at a given specified reference point on the antenna, usually the feedpoint or a current antinode. This resistance takes into account the energy consumed by the antenna system as a result of radiation and other losses.

antenna system ➤antenna.

antenna temperature The effective ➤noise temperature of an antenna, which is a consequence of the ➤antenna radiation resistance.

anti-aliasing ➤aliasing.

anti-aliasing filter A low-pass ➤filter used to ensure that the maximum frequency of an analogue signal input to or output from a digital ➤sampling system does not exceed half the sampling frequency being used, or the ➤Nyquist frequency, in order to avoid ➤aliasing. The ➤cut-off frequency of an anti-aliasing filter is usually set to be just less than the Nyquist frequency.

anticapacitance switch A ➤switch that is designed to present the minimum possible series capacitance to a circuit when in the open position.

anticathode *Syn. for* target electrode. ➤X-ray tube.

anticoincidence circuit A circuit with two or more input terminals that produces an output signal when a signal is received by only one input; there is no output when a signal is received by each input either simultaneously or within a specified time interval, Δt. An electronic counter incorporating such a circuit will record single events and is termed an *anticoincidence counter.* ➤➤coincidence circuit.

antiferromagnetism An effect observed in certain solids whose magnetic properties change at a certain characteristic temperature, T_N, known as the *Néel temperature.* Antiferromagnetism occurs in materials that have a permanent molecular ➤magnetic moment associated with unpaired electron spins. At temperatures above the Néel temperature thermal agitation causes the spins to be randomly orientated throughout the material, which becomes paramagnetic: it obeys the Curie–Weiss law approximately but is characterized by a negative Weiss constant, θ, i.e.

$$\chi = C/(T + \theta),$$

where χ is the ➤susceptibility and C a constant (➤paramagnetism).

At temperatures below the Néel temperature interatomic exchange forces (➤ferromagnetism) cause the spins to tend to line up in an antiparallel array. The susceptibility of the material depends on the crystalline structure, the temperature, and, in a single crystal, the direction of application of the external magnetic flux. In a single crystal of antiferromagnetic material, the antiparallel arrangement of spins can be considered as two interlocking sublattices, each spontaneously and equally magnetized in opposite directions (Fig. *a*). The sublattices may be represented by magnetic vectors

M_A and M_B, each of which produces a flux experienced by the other. Thus the flux experienced by sublattice A may be represented as

$$B_A = -\lambda M_B$$

where λ is a constant.

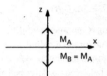

The equivalent magnetization vectors

a Antiparallel arrangement of magnetic moments in simple cubic lattice

If an external magnetic flux density, B, is applied in a direction parallel to the direction of the spins, at temperatures near the Néel temperature the efficiency of the interaction is reduced by thermal effects: the spins are not all completely antiparallel so the material has a small positive magnetic susceptibility, $\chi_\parallel$, falling from a maximum at the Néel temperature as the temperature is reduced. At temperatures near absolute zero the antiparallel array becomes saturated due to the greater efficiency of the interaction and the susceptibility falls to zero (Fig. *b*).

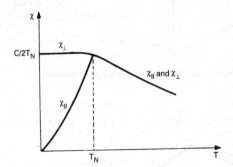

b Variation of susceptibility with temperature in a single antiferromagnetic crystal

If the magnetic flux is applied in a direction perpendicular to the spins, each magnetic vector turns towards B (Fig. *c*). The angle of rotation, α, is determined by the components of the magnetization vectors M_x and M_z and the components of flux B_x and B_z

and is independent of the temperature. The overall magnetization of the specimen is $2M_x$, where M_x is the component of each sublattice in the direction of $\boldsymbol{B}$; it can be shown that

$$2M_x = B/\lambda$$

and that

$$\lambda = 2\mu_0 T_N/C$$

where μ_0 is the ➤permeability of free space and C the Curie constant from the Curie-Weiss law for the material. The susceptibility $\chi_\perp$ is thus equal to $C/2T_N$, i.e. it is not dependent on temperature below the Néel temperature (Fig. *b*).

The behaviour as described above assumes two equal sublattices. Some crystal structures, e.g. face-centred cubic, have more than two sublattices and the stable saturation array therefore may not be completely antiparallel; this leads to more complicated behaviour.

In a polycrystalline sample the susceptibility is a compromise between the two extreme conditions described above, with a small dependence on magnetic field strength at low temperatures, and antiferromagnetic materials are characterized by a maximum susceptibility at the Néel temperature (Fig. *d*).

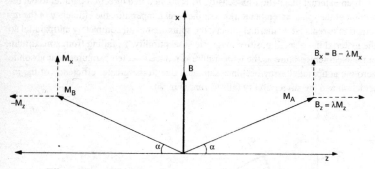

c Effect of external flux, $\boldsymbol{B}$, applied perpendicular to magnetic moments

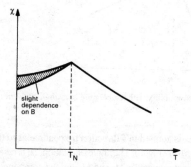

d Variation of susceptibility with temperature of polycrystalline specimen

antihunting circuit A circuit designed to prevent oscillation in a ➤feedback control loop, thus stabilizing it.

anti-interference antenna system A receiving ➤antenna system designed so that only the antenna itself abstracts energy from the incident radiation and no energy is collected by any feeders or supports of the system. The effects of interference can be greatly reduced by locating the antenna remotely from sources of man-made interference, such as electric motors.

antijam margin The difference in signal level between the ➤jamming signal strength and the receiver signal power in satellite communication systems or radar systems.

antijamming Reduction in the effects of ➤jamming.

antinode ➤node.

antiphase Two periodic quantities of the same frequency are said to be in antiphase if the ➤phase difference between them is one half-period, i.e. $180°$.

antipodal signal A digital signal $s_1(t)$ that is the negative of another signal $s_2(t)$ such that $s_1(t) = -s_2(t)$. Antipodal signals are commonly used in ➤bipolar signalling.

antiresonance A trough in a ➤frequency response indicating a frequency where a system responds with a minimum amplitude to a periodic driving force.

APCVD *Abbrev. for* atmospheric pressure chemical vapour deposition. ➤chemical vapour deposition

aperiodic circuit A circuit that contains both inductance and capacitance but is not capable of ➤resonance since the total energy loss in the circuit exceeds the critical value above which resonance does not occur, i.e. the damping of the circuit exceeds critical damping (➤damped).

aperiodic damping *Syn. for* overdamping. ➤damped.

aperiodic signal *Syn.* nonperiodic signal. A signal that does not repeat itself at regular intervals as does a ➤periodic signal. An aperiodic signal is random in nature and exhibits a continuous frequency spectrum.

aperiodic waveform A waveform (➤wave) where the variation of a quantity does not repeat regularly with time. ➤➤periodic waveform.

aperture antenna *Syn.* slot antenna. An ➤antenna formed by a hole in a conducting plane that is excited by a ➤feed connected at points around the edge. The points may be fed via a conducting tube as in a ➤horn antenna. Aperture antennas have an advantage over other antennas in that they may have a very low profile. They are thus very useful in aerospace applications where they can be mounted flush with the skin of an aircraft. In these cases they can easily be covered with a ➤dielectric material for protection.

aperture distortion ➤distortion.

aperture grille ➤colour picture tube.

apparent power ➤power.

Appleton layer *Syn. for* F-layer. ➤ionosphere.

applications software ➤software.

arbitrary unit 1. A unit, such as the ➤kilogram, that is defined in terms of a prototype. **2.** A unit used to indicate a relative quantity where a specific unit is impractical, 'counts' or 'number of instances' being examples.

arc A luminous electric discharge in a gas with a characteristically high current density and low potential gradient. An arc occurs (is struck) when the gas along its path becomes a ➤plasma. The arc is drawn from a localized spot on the cathode, resulting in heating of the cathode. Thermionic emission occurs with a resultant drop in potential across the tube. Arc-discharge tubes can therefore carry very large currents at voltage drops of tens of volts only. ➤➤gas-discharge tube.

arc discharge ➤gas-discharge tube.

architecture ➤computer architecture.

arcing contacts Auxiliary contacts in any type of ➤circuit-breaker switch that operate in conjunction with the main contacts, closing before and opening after them, so as to protect them from arc damage.

arcing horn *Syn.* protective horn. A horn-shaped conductor fitted to an insulator in order to prevent damage to the latter in the event of a power ➤arc arising from fault conditions. It is also an element of a ➤horn gap.

arc lamp A lamp that utilizes the brilliant light accompanying an ➤arc as a source of illumination. The colour of the light produced will depend on the gas inside the discharge tube.

arcover ➤flashover.

Argand diagram ➤complex plane.

argon Symbol: Ar. An inert gas, atomic number 18, that is extensively used as the gas in gas-filled tubes and as a sputtering agent in etching and depth-profiling processes.

armature 1. *Syn.* rotor. The rotating part of an electric ➤generator or ➤motor. **2.** Any moving part in electrical equipment that closes a magnetic circuit or that has a voltage induced in it by a magnetic field. An example is the moving contact in an electromagnetic ➤relay.

armature relay ➤relay.

ARQ *Abbrev. for* automatic repeat request. ➤digital communications.

Arrhenius equation A physical model that expresses the variation, $R(T)$, of some device parameter or material property as a function of operating temperature, T:

$$R(T) = A \exp(-A_E/kT)$$

where k is the Boltzmann constant. The constant A_E is a measure of the average behaviour of a population of items and is the 'empirical' activation energy of the physical process. A is a constant and is the intercept of the plot of logarithm of the parameter versus reciprocal temperature.

arsenic Symbol: As. An element of group V of the periodic table. It is a donor impurity in silicon and is a component of III–V ►compound semiconductors.

artificial antenna *US syn.* dummy antenna. A device that simulates all the electrical characteristics of an actual antenna except that the energy supplied to it is not radiated. (It is usually dissipated as heat in a resistor.) It allows adjustments to be made to a ►transmitter or ►receiver before connecting up to the actual antenna.

artificial intelligence (AI) The study and development of computing applications for tasks that would be described as requiring intelligence if they were done by people. Many of these applications involve systems that are capable of learning, adaptation, or self-correction.

artificial line An electrical network consisting of resistance, inductance, and capacitance that simulates the characteristics of a ►transmission line at any particular frequency.

artificial satellite ►satellite.

artwork The required patterns for integrated-circuit manufacture produced in a form suitable for reduction to masks. The patterns for each layer are produced using ►computer-aided design layout tools. ►lithography.

ASCII *Abbrev. for* American standard code for information interchange. A 7-bit code that provides different binary values for the 128 (2^7) characters and control codes. For example, the character 'A' is represented by $100\ 0001_2$ in ASCII, and 'a' is represented by $110\ 0001_2$.

ASIC *Abbrev. for* application-specific integrated circuit. An integrated circuit designed for a specific application, rather than a generalized mass-produced circuit.

ASK *Abbrev. for* amplitude shift keying. ►amplitude modulation.

aspect ratio 1. The ratio of the width of a television picture to the height. Most countries, including the UK and US, have adopted an aspect ratio of 4:3. ►►HDTV. **2.** The ratio of the width to the length of the channel in ►field-effect transistors.

ASR *Abbrev. for* airport surveillance radar. ►precision approach radar.

assembler ►assembly language.

assembly language The principle machine-oriented language. As each different type of computer has a different architecture (e.g. ►registers, data paths, ►flags), they also have different assembly languages. Assembly language is a symbolic language in which mnemonic codes are used to represent the ➤machine-code instructions available in the computer's ➤instruction set. Thus there is a one-to-one correspondence between the mnemonic instruction codes and the actual machine codes. Each assembly language has its own particular *assembler* program that converts the assembly language into a binary code (machine code) that the processor understands.

associative memory *Syn.* content-addressable memory. A computer storage system that locates data by searching its memory cells for a specific bit pattern. The user supplies the bit pattern by giving two values: an *argument* and a *mask*; the argument holds the value being sought, and the mask specifies which argument bits to use when comparing the argument to the value in each cell.

astable multivibrator ➤multivibrator.

astatic galvanometer *Syn.* Broca galvanometer. A very sensitive type of ➤galvanometer that is a ➤moving-magnet instrument. It has two magnetic needles, NS and N'S', made as nearly equal as possible and suspended on the same axis in opposition. When the magnets are deflected from their stable point the restoring couples on them due to the earth's magnetic field act in opposite directions with a very small resultant couple. The coils carrying the current to be measured are wound round the magnets in opposite senses (see diagram) so that the couples, due to magnetic fields produced by the coils, add. Very small currents cause a noticeable deflection of the magnets that is detected by means of a light spot reflected along a scale from a small mirror attached to the suspending fibre.

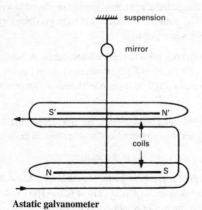

Astatic galvanometer

Aston dark space ➤gas-discharge tube (diagram).

asymmetric Denoting a periodic quantity that has a direct component.

asymmetric mode In two-wire mains plus earth transmission systems, asymmetrical mode refers to signals that appear simultaneously, in phase on the two mains lines, and return via the earth path. ➤symmetric mode.

asymmetric transducer ➤transducer.

asymmetric two-port network ➤two-port network.

asymptotic Approaching a given value or condition. A straight line that approaches a two-dimensional curve but never actually touches it is said to be asymptotic to the curve.

asynchronous Denoting a form of timing control in a circuit, device, or system in which, in a sequence of operations, a specific operation begins on receipt of a signal indicating that the preceding operation has been completed. ➤➤synchronous.

asynchronous logic ➤synchronous logic.

asynchronous transmission In ➤digital communications, transmission of packets of data where each packet is synchronized by a ➤header at the start of the packet and a tail at the end of the packet. ➤➤synchronous transmission.

atomic number Symbol: Z. The number of protons in the nucleus of an atom. The position of an element in the ➤periodic table, and hence its chemical properties, is determined by atomic number.

ATR switch *Short for* anti-transmit-receive switch. ➤transmit-receive switch.

attack decay sustain release (ADSR) A series of parameters that control the waveform envelope of a sound produced by an electronic musical instrument. When a note starts, for example by depressing a key, its ➤amplitude rises to a level and over a time set by the *attack* parameter, then it falls over a time set by the *decay* parameter to the *sustain* level (see diagram). This level is maintained until the note stops, for example by releasing the key, when the amplitude falls to zero over a time set by the *release* parameter.

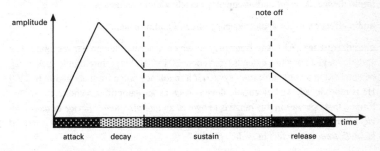

Attack decay sustain release

attenuation The reduction in magnitude of any electrical parameter of a signal, particularly electromagnetic radiation, on passing along any transmission path. The amount of attenuation is given by the ratio of the value of the parameter at the output to the corresponding value at the input under specified conditions. Attenuation results from the resistance present in the transmission path. It can be deliberately introduced into a transmission channel in order to reduce the magnitude of unwanted components of the parameter under consideration (►attenuator). Attenuation can also result from unwanted ►dissipation in the transmission path. ►►absorption.

attenuation band ►filter.

attenuation constant *Syn.* attenuation coefficient. Symbol: α. The rate of exponential decrease in amplitude of voltage, current, or field-component in the direction of propagation of a plane progressive wave, at a given frequency. If I_2 and I_1 are the currents at two points a distance d apart (I_1 being nearer the source of the wave), then

$$I_2 = I_1 \exp(-\alpha d)$$

α is usually expressed in ►nepers or ►decibels.

attenuation equalizer A ►network that throughout a specified frequency band provides compensation for attenuation ►distortion.

attenuator A ►network or ►transducer designed to produce distortionless attenuation of an electrical signal. It may be variable or fixed. (The latter is also called a *pad*.) Attenuators are usually calibrated in ►decibels.

atto- Symbol: a. A prefix to a unit, denoting a submultiple of 10^{-18} of that unit.

attracted-disc electrometer *Syn.* absolute electrometer. An electrometer that measures potential difference in terms of fundamental mechanical quantities. The potential difference to be measured is applied across two parallel metal discs and the force of attraction between them is measured.

audioconference ►teleconference.

audio device *Short for* audiofrequency device. ►audiofrequency.

audio effect *Short for* audiofrequency effect. ►audiofrequency.

audiofrequency (AF) Any frequency to which a normal human ear responds. The audible range in practice extends from about 20 to 20 000 Hz. Intelligible speech can be obtained in a communication system if a frequency range from about 300 to 3400 Hz is reproduced. Any electronic device, such as an ►amplifier, ►choke, or ►transformer, that operates in this range is known as an *audiofrequency device* or *audio device*; similarly any effect, such as ►distortion, that involves audiofrequencies is termed an *audiofrequency effect* or *audio effect*.

audiometer An instrument used to measure both hearing loss due to deafness and the masking produced by noise. Many forms exist, the most common being a system involving the production of a sound of known frequency and intensity in a telephone earpiece. Frequency and intensity are both variable, and the instrument may be calibrated to read hearing loss directly in operation.

audio signal An electrical signal having the frequency of an ►acoustic wave to which the human ear responds. ►►television.

Auger electron spectroscopy (AES) A method of spectroscopy that detects the electrons produced by the ►Auger process. The Auger electrons have a mean free path of only 1–3 nanometres, and their energy is typical of the material producing them. The Auger process dominates in low atomic number materials. Auger electron spectroscopy therefore is useful in the detection of low atomic number atoms in the surface of a semiconductor. Excess electrons are produced in the material by exciting the atoms with an electron beam, and the resulting Auger electrons are detected. The energy spectrum of the detected electrons can be correlated to the atomic species in the material. A *scanning Auger microprobe* (SAM) uses a focused electron beam to excite the material, which is then scanned across the surface. Lateral resolution using a SAM probe is limited by the diameter of the electron beam used, and can be as good as 50 nanometres. The composition of a material as a function of depth below the surface can be determined by using an ion beam to sputter material from the surface (►sputter etching) and a SAM to continuously detect the Auger electron spectra produced.

AES is a flexible analytical tool. It is used to detect physical defects due to unwanted particles in a material, to detect contaminants causing high contact resistance, to examine bonds between an IC and external leads, to study the composition of thin films, and to provide information about the chemical state of materials (for example, elemental and oxidized silicon produce different spectra).

Auger process ►recombination processes.

autocatalytic plating ►electroless plating.

autocorrelation ►correlation.

automatic brightness control ►television receiver.

automatic contrast control ►television receiver.

automatic control Any system or device that carries out operations automatically in response to the output of other electronic devices, such as sensing devices, computers, or discriminator circuits. ►Feedback of the output signal is frequently employed to provide the controlling signal.

An *automatic controller* measures any variable quantity or condition and produces an output designed to correct any deviation from the desired value. The *threshold signal* is the minimum input signal to which the control system responds by producing a corrective signal. A system that is actuated by electrical signals is known as an *electric controller*. If the electrical signals are used to excite electromagnets that determine all the basic functions of the device, it is termed a *magnetic controller*. In electrical switching, a desired performance is maintained by automatically opening and closing ►switches in a given sequence.

automatic direction finding ►direction finding.

automatic frequency control (a.f.c. or AFC) A device to maintain automatically the frequency of any source of alternating voltage within specified limits. The device is 'error-operated' and usually consists of two parts: one part is a frequency ►discrimi-

nator that compares the actual and desired frequencies and produces a d.c. output voltage proportional to the difference between them, with sign determined by the direction of drift; the other part is a ➤reactor that forms part of the ➤oscillator tuned circuit
and is controlled by the discriminator output in such a way as to correct the frequency
error.

automatic gain control (a.g.c. or AGC) *Syn.* automatic volume control. A device for
holding, for example, the output volume of a radio receiver or the recording level on
a tape recorder substantially constant despite variations in the input signal. The term
also covers the process involved. A variable gain element in the receiver is controlled
by a voltage derived from the input signal. Variations in the size of the input signal cause
compensatory changes in gain in the receiver. *Biased automatic gain control* is a
process that comes into operation only for signals above a predetermined level.

   In radio receivers with automatic gain control, the AGC action can result in maximum gain when no carrier is being detected. This can produce a significant amount of
noise that is heard by the listener. It can also result in receivers that automatically scan
for the next radio transmission. To avoid the annoying noise when no carrier is being
detected, the output of the receiver is switched off or *muted*. ➤squelch circuit.

automatic grid bias ➤grid bias.

automatic noise limiter A circuit in a radio receiver designed to limit the effect of impulse ➤noise.

automatic tracking A method of holding a radar beam locked on target while the target range is being determined.

automatic tuning control A type of ➤automatic frequency control in a radio receiver
that adjusts the tuning to the correct setting for a given received signal when the
manually operated adjustment has been set only approximately for that signal. It holds
the tuning at the correct setting despite any small drift in the components of the receiver
and any small drift in the frequency of the received signal.

automatic volume compressor ➤volume compressor.

automatic volume control ➤automatic gain control.

automatic volume expander ➤volume compressor.

autotransductor A ➤transductor in which the main current and control current are carried in the same windings.

autotransformer A ➤transformer that has a single winding, tapped at intervals, rather
than two or more independent windings. Part of the winding is thus common to both
primary and secondary circuits (see diagram). The voltage V_2 across a tapped section
is related to the total applied voltage V_1 by

$$V_2/V_1 \propto n_2/n_1$$

where n_2 is the number of turns in the tapped section and n_1 the total number in the winding.

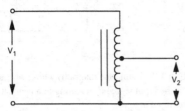

Single-phase **autotransformer**

availability In digital communication systems, a measure of the number of communication channels available to a particular transmitter. For example, it could be the number of telephone network connections that could be used to route an incoming call.

available power ➤maximum power theorem.

avalanche *Syn.* Townsend avalanche. A cumulative ionization process in which a single particle or photon produces several charge carriers, each of which in turn gains sufficient energy from an accelerating field to produce more charge carriers, and so on. A large number of charged particles is thus produced from the initial event. The phenomenon is utilized in, for example, the avalanche ➤photodiode and ➤IMPATT diodes.

avalanche breakdown A type of ➤breakdown that occurs in a ➤semiconductor. It is caused by the cumulative multiplication of free charge carriers under the action of an applied electric field. Some of the carriers gain enough energy to liberate new electron-hole pairs by ➤impact ionization, which in turn can generate further pairs, i.e. an ➤avalanche takes place.

The avalanche process occurs when a *critical field* is reached across the semiconductor. The critical field is the voltage gradient that just causes avalanche in a given specimen of semiconductor and is typical for the material used. For a given semiconductor the value of the applied voltage required to produce the critical field – the breakdown voltage – is a function of the ➤doping concentration and the thickness of the specimen.

Avalanche breakdown is the breakdown that occurs in a reverse-biased ➤p-n junction. For a given junction the breakdown voltage is a function of the doping concentrations on each side of the junction and on the width of the ➤depletion layers associated with the junction. A more highly doped junction has a lower value of breakdown voltage than a junction with lighter doping. The avalanche process is initiated by carriers contributing to the small reverse saturation current.

avalanche photodiode ➤photodiode.

a.v.c. *Abbrev. for* automatic volume control. ➤automatic gain control.

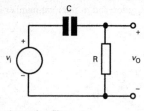 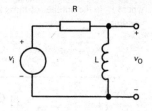

Averaging filters

averaging filter A type of simple RC or LR circuit (see diagram) whose ►time constant is much greater than the period of the input signal v_i, resulting in a near-DC output voltage v_o that is the average of the input signal.

avionics Electronics applied to airborne systems including aviation and space applications.

Ayrton shunt ►universal shunt.

B

Babinet's principle A principle originally stated for optics but applicable to other radiating situations. It states that when the field behind a screen with an opening is added to the field of a complementary structure, the sum is equal to the field when there is no screen.

back electromotive force (back e.m.f.) An e.m.f. that opposes the normal flow of current in a circuit.

backgating *Syn.* backside gating. A phenomenon that arises in closely packed monolithic ➤integrated circuits where a negative bias on an ➤ohmic contact can affect nearby ➤FET devices. The substrate can be biased and acts as a ➤gate on the back of the FET, affecting the source-drain current.

background counts Counts registered by a radiation ➤counter in the absence of the radiation source to be measured. These counts may arise from other radiation sources, naturally occurring background radiation, contamination of the counter itself, or spurious signals in the electronic circuitry of the counter or from a combination of these.

back heating ➤magnetron.

backing store ➤memory.

backlash The incomplete rectification of an alternating current in a thermionic valve due to the presence of positive ions in the residual gas in the valve.

back lobe ➤antenna pattern.

back porch ➤television.

backward diode ➤tunnel diode.

backward-wave oscillator ➤travelling-wave tube.

balance controls 1. Variable components used in electrical instruments and bridges in order to obtain conditions of electrical equilibrium in the circuit. **2.** Variable components used to equalize the outputs of two or more similar circuits, such as those in a stereophonic sound reproduction system.

balanced amplifier *Syn. for* push-pull amplifier. ➤push-pull operation.

balanced line A ➤transmission line consisting of two conductors whose voltages at any point on the line are equal but opposite in polarity with respect to earth potential.

balanced modulator ➤ring modulator.

balanced two-port network ➤two-port network.

balanced-wire circuit A circuit having two sides that are symmetrical with respect to earth and other conductors and are electrically alike.

balance method ➤null method.

ballast lamp ➤ballast resistor.

ballast resistor A resistor manufactured from material having a large positive ➤temperature coefficient of resistance. It is constructed so as to have a substantially constant current over a range of voltages, and may be used as a current ➤regulator; it is connected in series with a circuit to absorb small changes in the applied voltage and stabilize the current in the circuit. One consisting of a resistor seated in an evacuated or gas-filled envelope of glass or metal is called a *ballast lamp*.

ballistic galvanometer A ➤galvanometer that measures the quantity of electricity Q, flowing during the passage of a transient current I, where

$$Q = \int_0^\infty I \mathrm{d}t$$

The value of Q is deduced from the deflection, θ, of the moving part of the instrument. For a moving-coil instrument, $Q \propto \theta$.

balun *Syn.* balancing transformer. Acronym from *bal*anced *un*balanced. A device that is used to couple a balanced impedance, such as an antenna, to an unbalanced transmission line, such as coaxial cable. A balun is required in order to prevent asymmetrical loading of the balanced impedance and the induction of currents on the exterior of the unbalanced transmission line.

banana jack-and-plug A single-conductor jack-and-plug system in which the plug, with a sprung metal tip, somewhat resembles a banana in shape.

band 1. A specific range of frequencies used in communications for a definite purpose, such as the longwave band in radio; certain frequencies within the band are assigned to different transmitting stations and the receiver may be tuned to any desired frequency within the band. ➤frequency band. **2.** A closely spaced group of atomic energy levels. ➤energy bands. **3.** A closely spaced group of molecular energy levels that appear as fluted bands separated by dark spaces in the spectra of compounds. Under higher resolution the bands resolve into fine spectral lines.

band edge *Syn. for* cut-off frequency.

band gap *Syn. for* forbidden band. ➤energy bands.

band-limited channel In communication systems, a transmission channel that has a limited frequency bandwidth and hence limited information capacity. An example is the ➤voice-grade telephone channel.

band-pass filter ➤filter.

band-reject filter *Syn. for* band-stop filter. ➤filter.

bandspread A technique used in communications receivers that allows the receiver to select a transmission of a narrow-frequency bandwidth in close proximity to other transmissions: the bandwidth is spread over a wider physical or electrical range so that tuning is made easier. An example is the use of a gearing system to desensitize the rotation of the tuning adjustment in order to allow finer frequency selection. A similar effect can be obtained by a coarse and fine electrical adjustment where electronic tuning is employed.

band-stop filter *Syn.* band-reject filter. ➤filter.

band switch ➤turret tuner.

band-to-band recombination ➤recombination processes.

bandwidth 1. The band of frequencies occupied by a transmitted modulated signal (➤modulation) and lying to each side of the ➤carrier-wave frequency. The bandwidth of a transmission channel is a measure of the information-carrying capacity of the channel. **2.** The amount of deviation of frequency that an ➤antenna array is capable of handling without a mismatch. **3.** The band of frequencies over which the power amplification in an amplifier falls within a specified fraction (usually one half) of the maximum value. **4.** ➤receiver.

bank 1. A number of devices of the same kind, connected so as to act together. **2.** An assembly of fixed contacts that is used in automatic switching in telephony to form a rigid unit in a selector or similar device with which wipers engage.

Barkhausen effect An effect observed in ferromagnetic materials whereby the magnetization of the specimen proceeds as a series of finite jumps when the magnetizing flux is increased steadily. The effect supports the domain theory of ➤ferromagnetism: the spin magnetic moments present in the material can only have certain allowed orientations; the minute jumps correspond to the spins changing from one allowed orientation to the next. If all possible directions were allowed the magnetization would proceed smoothly.

The effect can be demonstrated by winding the specimen with two coils. When the current in the primary coil is increased steadily to produce a smoothly increasing magnetic flux density, the fluctuations in the magnetization can be shown by connecting the secondary coil to a sensitive cathode-ray oscillograph.

barrage reception A method of reception used in telecommunications in which the receiving antenna consists of an array of several ➤directional antennas of different orientations. The received signal is input from selected antennas in the array that are chosen so as to minimize interference from a particular direction.

barrel distortion ➤distortion.

barrel shifter A ➤logic circuit that has data and control inputs and data outputs. The output equals the input, rotated by a number of bit positions specified by the control inputs. The circuit can shift its input by any amount from 0 to $2^n - 1$ bits, where n is the number of input/output bits. It is usually necessary to specify whether the shifting operation is to the right or to the left.

barrier height *Abbrev. for* Schottky barrier height. ➤Schottky diode, Schottky effect.

Bartlett window *Syn.* triangular window. ➤windowing.

base 1. *Short for* base region. The region in a ➤bipolar junction transistor between the ➤emitter and ➤collector into which minority carriers are injected. The electrode attached to the base is the *base electrode*. ➤➤ semiconductor. **2.** *Short for* base electrode.

base address An ➤address specifying the start of a collection of data units – a block – in computer ➤memory. Thus the base address of an array is the address of its first (or zero) element. Base addresses are used when ➤relative addressing modes are used.

baseband ➤carrier wave.

base electrode ➤base.

base level (of a pulse) ➤pulse.

base limiter *Syn.* inverse limiter. ➤limiter.

base region ➤base.

base stopper ➤parasitic oscillations.

base units ➤SI units.

battery A source of direct current or voltage that consists of two or more electrolytic ➤cells connected together and used as a single unit.
   A *floating battery* is formed from secondary cells and is connected simultaneously to a discharging circuit and a charging circuit. The current in the charging circuit is adjusted so as to balance the loss of charge from the battery to the discharging circuit driven by it. A constant level of charge is therefore maintained in the battery. A floating battery is often used to provide a constant e.m.f. in the discharge circuit, despite fluctuations in the electrical mains supply.
   A *dry battery* is a relatively small portable battery made up from dry cells that can be discharged once. *Rechargeable batteries* can be discharged and recharged many times.

baud A unit of telegraph signalling speed equal to one unit element per second. Thus if the duration of the unit element is $1/n$ seconds then the speed of transmission of successive signals is n bauds. In the case of data transmission, 1 baud is generally equal to 1 bit per second (bps).

Baudot code ➤digital codes.

bayonet fitting 1. A type of pin-and-socket fitting in which the base of a lamp or tube has two pins, diametrically opposite each other, that can be inserted into the socket and rotated in slots in the socket so that the device is held in position. **2.** A type of plug-and-socket ➤BNC connector where the pins are on the socket.

BBD *Abbrev. for* bucket-brigade device.

BCD *Abbrev. for* binary-coded decimal.

BCH codes *Abbrev. for* Bose–Chaudhuri–Hocquenghem codes. A family of error-correcting codes. ➤digital codes.

BCS theory ➤superconductivity.

beacon A signal station or the signal transmitted by such a station, which acts as a reference point. A beacon that transmits an identifiable signal is a *code beacon*. A *homing beacon* is one that guides an object, such as an aircraft, to a target, such as an airport. At the airport, signals from the *localizer beacon* associated with the instrument landing system are picked up enabling the aircraft to be guided to land. A beacon that employs radar signals is a *radar beacon* and one employing radiofrequency waves is a *radio beacon*. The receiver used for detecting the signals from a beacon is a *beacon receiver*.

bead thermistor ➤thermistor.

beam A narrow stream of essentially unidirectional electromagnetic radiation (as in a radiowave) or charged particles (as in an electron beam).

beam angle ➤cathode-ray tube.

beam bending An unwanted effect that occurs in television ➤camera tubes. The electron beam used to scan the target area can be deflected from its intended position by the electrostatic charges stored on the target. This can result in misalignment of the picture image in the receiver with respect to the original optical image.

beam coupling The production in a circuit of an alternating current between two electrodes upon passage of an intensity-modulated electron beam. The *beam-coupling coefficient* is the ratio of the alternating current produced to the beam current.

beam lead A connecting lead on a silicon chip that is formed chemically and cantilevered across a void or space on the chip. The lead may be cantilevered either from the chip to the interconnection pattern or vice versa.

beam switching ➤lobe switching.

beamwidth A parameter used to specify the width of the maximum beam of an ➤antenna. It is usually the *half-power beamwidth,* which is the angle between the two directions in which the radiation intensity is one-half the maximum value of the main beam, where the direction of the maximum and the half-power points are in the same plane. If it is used otherwise, such as the angle between the 10-decibel points, the specific points on the ➤antenna pattern should be described.

beat frequency ➤beats.

beat-frequency oscillator ➤beats.

beating ➤beats.

beats The periodic signal produced by interference when two signals of slightly different frequencies are combined. The amplitude is equal to the sum of the original amplitudes; the frequency (the *beat frequency*) is equal to the difference between the original frequencies. The production of beats is termed *beating* and is achieved, for example, by using a *beat-frequency oscillator*. This device incorporates two radio-frequency oscillators, one producing a fixed frequency wave and the other a variable frequency; the output is produced by beating together the two frequencies.

Beating is also the basis on which a musical instrument is tuned by reducing the beats with a reference to 0 Hz by adjusting, for example, the string or pipe length.

bel Symbol: B. ►decibel.

beryllium oxide *Syn.* beryllia. Symbol: BeO. An insulating material that has a high thermal conductivity (about half that of copper) and is used in heat sinks.

Bessel filter A type of electrical ►filter based on a mathematical power series known as the Bessel polynomial. The filter is ►maximally flat in the pass-band and it also has ►linear phase response and consequently a flat ►group delay response. The ►amplitude response approximates a Gaussian response.

beta circuit ►feedback.

beta current gain factor *Syn.* common-emitter forward-current transfer ratio; forward-current gain. Symbol: β. The short-circuit current-amplification factor in a bipolar transistor with ►common-emitter connection:

$$\beta = (\partial I_C / \partial I_B) \qquad V_{CE} \text{ constant}$$

where I_C is the collector current and I_B is the base current; the collector voltage, V_{CE}, is constant. β is always greater than unity and practical values up to 500 are used.

beta cut-off frequency The frequency at which the ►beta current gain factor has fallen to $1/\sqrt{2}$ of its low frequency value. The beta cut-off frequency is considerably lower than the ►alpha cut-off frequency.

BFSK *Abbrev. for* binary frequency shift keying. ►frequency modulation.

bias *Short for* bias voltage. A voltage applied to an electronic device to ensure that it operates on a particular portion of its ►characteristic curve.

biased automatic gain control *Syn.* delayed automatic gain control. ►automatic gain control.

bias voltage ►bias.

BiCMOS *Syn.* merged CMOS/bipolar. Integrated circuits that contain both ►bipolar junction transistors and ►complementary MOS transistors (CMOS). The combination of both types of device on the same chip has a wide variety of functions and allows the advantages of both processes to be exploited. Bipolar circuits are inherently faster than CMOS circuits and have much better analogue performance. Bipolar transistors are preferred for ►operational amplifiers, ►comparators, ►multipliers, and high-speed logic circuits, such as ►emitter-coupled logic (ECL). CMOS circuits are preferred where low power dissipation and high ►packing densities are required, as with memory counters,

registers, and random logic. Merging the two types of circuit allows combinations of
different circuits to be formed on the same chip. It also allows the production of circuits
where the characteristics of both types of device are needed, for example mixed ana-
logue-digital circuits or logic circuits where part of the circuit demands the high speed

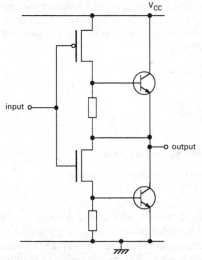

BiCMOS output buffer

of bipolar logic, such as fast clock circuits or input/output buffers. A merged logic output buffer is shown in the diagram. BiCMOS circuits provide improved system performance and a reduction in the number of components, hence reduced chip sizes and
lower costs.

biconical antenna An antenna that has characteristics of a ➤broadband antenna and
is formed by placing two conducting cones of infinite extent together, point to point,
with a source placed between the points. In practice the cones are truncated and are
formed of a number of cylindrical rods, typically six, to approximate the ideal structure. The truncation, size of the centre gap, and number of rods define the upper and
lower frequencies of useful performance.

The *bow-tie antenna* is a simplified form of this where each cone is formed of a
flat triangular sheet of metal or even a tube forming the boundary of this triangle.
The bow-tie antenna exhibits lower broadband characteristics than the full biconical.
➤➤Bilog antenna.

bidirectional network *Syn. for* bilateral network. ➤network.

bidirectional transducer *Syn. for* bilateral transducer. ➤transducer.

bidirectional transistor A ➤transistor that has substantially the same electrical characteristics when operated with the emitter and collector interchanged.

BiFET ➤integrated circuit.

bifilar suspension A form of construction of an instrument in which the movable part is suspended by two threads, wires, or strips, arranged so that the restoring force is produced mainly by gravity.

bifilar winding A method of winding consisting of two contiguous insulated conductors connected so that they carry the same current in opposite directions. This results in a negligible magnetic field being produced. The technique is commonly used to wind noninductive resistors (see diagram).

Bifilar winding

bilateral network *Syn.* bidirectional network. ➤network.

bilateral transducer *Syn.* bidirectional transducer. ➤transducer.

bilevel resist ➤multilevel resist.

Bilog antenna *Trademark* A combination ➤broadband antenna in which a ➤log-periodic antenna is used for the higher part of its frequency range and a ➤bow-tie antenna is joined to the back of the transmission line to provide a lower limit to the frequency range than with just a log-periodic of comparable size. The addition of the bow-tie element necessitates a ➤balun behind the bow-tie in order to ensure that the structure does not become unbalanced in the lower part of its operating frequency range. The Bilog antenna is commonly used in ➤electromagnetic compatibility testing due to its compact size and wide frequency range.

bimetallic strip A device consisting of two metals with different coefficients of expansion riveted together; an increase in temperature causes the strip to bend. One end of the strip is held rigidly and the movement of the other end may be used to open and close contacts, particularly in thermostats, or to move the pointer of a pointer-type thermometer. A thermal instrument that utilizes the deformation of a bimetallic element, heated directly or indirectly by an electric current, is a *bimetallic instrument*.

binary ASK ➤amplitude modulation.

binary code A rule for transforming a message from one symbolic form into another in which each element is represented by one or other of two distinct states, values, or numbers.

binary-coded decimal (BCD) A number representation in which 4 ➤bits encode a single decimal digit and the 4-bit unit is the smallest allowed piece of information. For example,

$$4_{10} = 0100_{BCD}, \ 10_{10} = 0001 \ 0000_{BCD}, \ 25_{10} = 0010 \ 0110_{BCD}$$

binary FSK ►frequency modulation.

binary logic circuit ►logic circuit.

binary notation A method of numerical representation with two as the base and thus having only two digits (0 and 1). These symbols may be easily represented electronically by two voltage levels in a circuit (►logic circuit) and binary notation is therefore used in ►computers. The radix point in a binary system is the *binary point* and is analogous to the decimal point in decimal notation. For example,

$$101.0101_2 = 2^2 + 2^0 + 2^{-2} + 2^{-4} = 5.3125_{10}$$

binary point ►binary notation.

binary PSK ►phase modulation.

binary scaler ►scaler.

binary signal A digital signal whose voltage at any particular time will be at one of two discrete levels.

binding energy 1. Symbol: E_B. The total energy released when protons and neutrons bind together to form a nucleus. **2.** The energy required to release an extranuclear electron from a nucleus. The energy required to strip all the extranuclear electrons from a nucleus is the *total electron binding energy.*►ionization potential.

BIOS *Abbrev. for* basic input/output system. That part of the input/output system on a PC or similar processor that directly controls the hardware interface devices; it is usually found in ►ROM. BIOS gets the computer started before the operating system is loaded from the hard drive. The BIOS in a computer system is designed and placed into ROM by the computer manufacturer and is not alterable by the user of the computer.

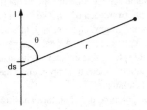

Biot–Savart law

Biot–Savart law The magnetic flux density, *B,* at a point distance r from a current-carrying conductor in free space when a current I flows in an element of length ds is given by

$$dB = \mu_0 I(ds \times r)/4\pi r^3 = \mu_0 I ds \sin\theta/4\pi r^2$$

where μ_0 is the permeability of free space (see diagram). Integration of the Biot–Savart law gives Ampere's circuital theorem (►Ampere's law).

biphase PCM *Syn.* Manchester code. ►pulse modulation.

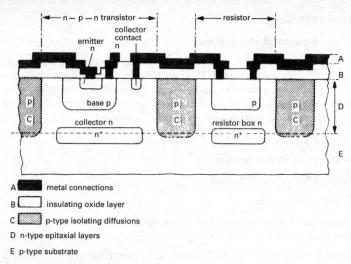

A ▬ metal connections

B ▭ insulating oxide layer

C ▨ p-type isolating diffusions

D n-type epitaxial layers

E p-type substrate

Cross section of typical **bipolar integrated circuit**

bipolar integrated circuit A type of monolithic ►integrated circuit based on ►bipolar junction transistors. A section of a typical circuit is shown in the diagram. The substrate (E) is formed from a wafer of ►semiconductor (p-type is shown) and has buried n⁺ (highly doped) regions selectively diffused into it. These regions serve to reduce the collector series resistance of the completed transistors. The n-type epitaxial layer (D) is then grown on the substrate, and this in turn has an insulating oxide layer (B) grown on it. ►Photolithography is used to etch the oxide layer in the desired positions and isolating diffusions (C), of the same semiconductor type as the substrate, are made to isolate the individual components from each other. The individual components are formed by oxide growth, photolithography, etching, and diffusion of the appropriate type of impurity, in turn, into the epitaxial layer. A final passivating oxide layer is grown. This has windows etched into it to enable contacts to be made to the semiconductor; the desired interconnection pattern is made by evaporating a metal layer (A), usually aluminium, on to the oxide and etching.

The substrate is held at the most negative potential possible, thus causing the collector-substrate junctions to be reverse biased and preventing current flow across them. This ensures isolation of the individual components. The n-type region surrounding the resistor – the resistor box – is held at the most positive potential possible, thus preventing current flow across the resistor-resistor box junction by ensuring that it too is reverse biased.

bipolar junction transistor (BJT) One of the two major classes of ►transistor. It is a ►semiconductor device that consists of two ►p-n junctions back-to-back in close proximity to each other, with one of the regions common to both junctions. This forms either a *p-n-p* or *n-p-n transistor*. The three regions in the transistor are called *emitter*, *base*, and *collector*, as shown in Fig. *a*.

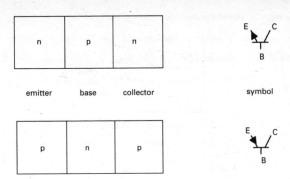

a Schematic of a BJT showing principal regions, and circuit symbol

In normal or *forward active* operation of a BJT, the base-emitter p-n junction is forward biased and the base-collector junction is reverse biased. ➤Majority-carrier current flows across the forward-biased emitter-base junction. The emitter is much more heavily doped than the base region, so that most of the total current flow across the base-emitter junction consists of majority carriers from the emitter injected into the base. These injected carriers become minority carriers in the base region, and will tend to recombine. The ➤recombination is minimized by making the base region very narrow, so that the injected carriers can diffuse across the base to the reverse-biased base-collector junction, where they are swept across the junction into the collector, to appear in the outside circuit as the collector current. The magnitude of the collector current, I_C, depends on the number of majority carriers injected into the base from the emitter, and thus current is controlled by the base-emitter p-n junction voltage. The output (collector) current is therefore controlled by the input (base-emitter) voltage V_{BE}: the output circuit of the transistor can be modelled as a voltage-controlled current source (➤dependent sources). The input circuit looks like a p-n junction diode.

In principle, the transistor can be operated in the *reverse active* mode by reversing the connections. But in fact the transistor is not completely symmetrical in practice: the emitter is very heavily doped to maximize emitter ➤injection, and the collector is relatively lightly doped so that it can accommodate large voltage swings across its reverse-biased junction. While the electrical characteristics are similar in appearance, the forward characteristics show much greater gain, as expected.

If both junctions are reverse biased, the transistor behaves like an open switch, with only the p-n junction reverse leakage currents flowing. If both junctions are forward biased, there is injection of carriers into the base region from both sides, and a low resistance is presented to current flow in either direction: the transistor behaves like a closed switch, and the base stores the injected charge.

The electrical characteristics of a bipolar junction transistor are illustrated in Fig. *b*, for the ➤common emitter connection, showing the regions of operation.

The bipolar junction transistor can be used to provide linear voltage and current amplification: small variations of the base-emitter voltage and hence the base current at the input terminal result in large variations of the output collector current. Since the

b BJT electrical characteristics (common emitter)

transistor output has the appearance of a current source, the collector can drive a load resistance and develop an output voltage across this resistance (within the limits of the supply voltage). The transistor can also be used as a switch in digital logic and power switching applications, switching from a high-impedance 'off' state in cut-off, to a low-impedance 'on' state in saturation. In practice, full saturation conditions of base-collector forward biased are generally avoided, to limit the ►carrier storage in the base and reduce the switching time.

Bipolar junction transistors find application in analogue and digital circuits and ►integrated circuits, at all frequencies from audio to radio frequency; ►heterojunction bipolar transistors are used for higher frequencies such as microwave applications.

bipolar logic circuit ►logic circuit.

bipolar signalling In digital communications, the use of two signals – one the negative of the other – to represent a binary one and binary zero. ►►antipodal signal.

bipolar transistor *Short for* bipolar junction transistor.

biquad filter *Short for* biquadratic filter. A general term for a second-order filter (►order) based on a ►transfer function that is the ratio of two quadratic equations – hence the term biquadratic. The filter transfer function $H(s)$, from which low-pass, high-pass, band-pass, and band-stop filters can be realized, is given by:

$$H(s) = (a_2 s^2 + a_1 s + a_0) / (s^2 + b_1 s + b_0)$$

where s is the complex frequency variable (►Laplace transform).

bistable 1. Having two stable states. **2.** *Short for* bistable multivibrator. A circuit having two stable states. ►flip-flop.

bit Acronym from *bi*nary dig*it*. One of the digits 0 or 1 used in ►binary notation. It is the basic unit of information in a computer or data processing system.

bit line ►RAM.

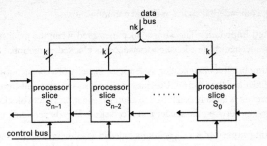

General structure of an nk-bit bit-slice microprocessor

bit-slice processor A special form of computer processor. The fundamental bit-slice device is a microprocessor slice, which is the execution unit for a simple CPU with a small data word size, typically 4 bits. The key property of a k-bit processor slice is the fact that n copies of the slice can be interconnected in a simple and regular manner to form an nk-bit processor that performs the same functions as a single slice on nk-bit rather than k-bit operands. The processor slices are interconnected in the form of a cascade circuit, or one-dimensional array (see diagram).

The term 'bit slicing' arises from the similarity between the processor array and a large processor that has been sliced into n identical subprocessors, each operating on a k-bit portion or slice of the data being processed by the array. As indicated in the diagram, the same control lines are connected to all slices so that, in general, they all perform the same operations at the same time; these operations are independent of n, the number of slices in the processor array. The k-bit data buses that transmit operands to or from the slices are simply merged to form kn-bit data buses for the array.

BJT *Abbrev. for* bipolar junction transistor.

black box Any self-contained unit or circuit in an electronic device that may be treated as a single package. The action of such a package may be approached mathematically in terms of its input and output characteristics, irrespective of its internal elements.

black level ➤television.

Blackman window ➤windowing.

black-out A temporary loss of sensitivity of any electronic device following the passage of an intense transient signal.

black-out point ➤cut-off.

blanking The rendering of a device or channel ineffective or inoperative for a desired time. For example, blanking eliminates the return trace from the screen of a cathode-ray tube. There are usually two blanking components to eliminate the horizontal and vertical components of the return trace.

blanking level ➤television.

Bloch walls ➤ferromagnetism.

block code ►digital codes. ►►digital communications.

blocked impedance The ►impedance measured when the mechanical motion of an electromechanical or acoustic ►transducer is blocked. ►►motional impedance.

blocking In telecommunication systems, the inability of a connection to be made between the sender and the receiver. Blocking, or congestion, can result from a busy line where the receiver is already connected to another sender. In system blocking, all the connection routes between the sender and receiver are already servicing other connections.

blocking capacitor A ►capacitor included in a circuit in order to limit the flow of direct current and low-frequency alternating current while allowing the passage of higher-frequency alternating current. The capacitance is chosen so that there is a relatively small ►reactance at the lowest frequency at which the circuit is designed to operate.

blocking oscillator A type of oscillator in which blocking occurs after completion of (usually) one cycle of oscillation and lasts for a predetermined time. The whole process is then repeated. Fundamentally it is a special type of ►squegging oscillator and has applications as a ►pulse generator and a ►timebase generator.

BNC connector A type of bayonet locking connector used extensively on modern instruments and equipment. It connects with a quarter-turn twist and completes both the shield (earth) circuit and the inner conductor (signal) circuit simultaneously. It is used for signals up to frequencies in the region of 4000 megahertz and is available in both panel-mounting and cable-terminating varieties.

board *Short for* printed circuit board.

Bode diagram A diagram in which gain or phase shift in a feedback-control system is plotted against frequency. It is so called because of *Bode's theorem,* which shows the interdependence of phase angle and rate of change of gain of a network at a desired frequency.

Bode equalizer An ►attenuation equalizer designed so that one simple control varies the amount of equalization in the same proportion for all frequencies within the range.

Bode's theorem ►Bode diagram.

body capacitance A ►capacitance caused by the proximity of a human body to a circuit.

bolometer A small resistive element capable of absorbing electromagnetic power. The resulting temperature rise is used to measure the power absorbed.

Boltzmann constant Symbol: k. A fundamental constant having the value 1.380 658 $\times 10^{-23}$ joule per kelvin.

In a system in which n_1 particles have an energy E_1 and n_2 particles have an energy E_2 then

$$n_1 = n_2 \exp[-(E_1 - E_2)/kT]$$

where T is the thermodynamic temperature. This is the *Boltzmann distribution law.*

Boltzmann distribution law ►Boltzmann constant.

bonded Denoting metal parts in a circuit that are connected together electrically so that they are at a common voltage.

bonded silvered mica capacitor ►mica capacitor.

bonding pads (or **bond pads**) Metal pads arranged on a semiconductor chip (usually around the edge) to which wires may be bonded so that electrical connection can be made to the component(s) or circuit(s) on the chip. Bonding is usually effected by thermocompression or ultrasonic bonding. ►►tape automated bonding; wire bonding.

Boolean algebra An algebra introduced by George Boole in 1854 originally to provide a symbolic method for analysing human logic. Almost a century later it was also found to provide a means for analysing logical machines. An algebra is a collection of sets together with a collection of operations over those sets. A Boolean algebra may be defined as a set K of *Boolean values* or constants, along with a set P of three operators AND, OR, and NOT. The set K contains 2^n elements, where n is a nonzero integer, and includes two special elements denoted 0 and 1. Thus in the simplest two-valued Boolean algebra, $K = \{0,1\}$. This is the basis for all digital logic design, from the simplest logic functions to complex microprocessors and computers.

Boolean values ►Boolean algebra.

booster 1. A generator or transformer inserted in a circuit in order to increase (*positive booster*) or decrease (*negative booster*) the magnitude or to change the phase of the voltage acting in the circuit. **2.** A repeater station that amplifies and retransmits a broadcast signal received from a main station, with or without a change of frequency.

bootstrapping A technique used in a variety of applications in which a capacitor – the *bootstrap capacitor* – is used to provide 100% positive ►feedback for alternating currents across an amplifier stage of unity gain or less. Bootstrapping is used for control of the output signals by using the positive feedback to control the conditions in the input circuit in a desired manner.

  Bootstrapping is commonly used in circuits that generate a linear timebase, particularly in a sawtooth generator. A simple sawtooth generator consists of a capacitor that is charged by means of an input load resistor and discharged by a periodic step voltage. As the capacitor is charged, the voltage increases exponentially and as the voltage increases, that across the input load (and hence the charging current) drops correspondingly. The output is approximately linear provided that only a small portion of the charging characteristic is used. The linearity may be improved by using a bootstrap circuit to maintain a constant charging current. A typical circuit is shown in Fig. *a*. The output is taken from an ►emitter follower, capacitively coupled via the bootstrap capacitor C_1 to the input load resistor R. As the output voltage rises, the voltage at the node between R and R_1 also rises; the voltage across R and hence the charging current is therefore maintained substantially constant.

  Bootstrapping may also be used in ►MOS logic circuits in order to optimize the voltage swing between the high and low logic levels. A typical bootstrapped circuit is

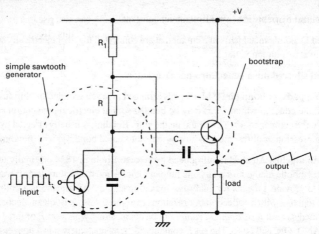

a **Bootstrapping** in a sawtooth generator

shown in Fig. *b*. The output voltage V_o at point X is high when a low logic level is applied to the gate of the transistor T_s and is determined by the value of the gate and threshold voltages of the transistor T_1. In the absence of the bootstrap capacitor, C, and the load transistor, T_L, the gate voltage V_{G1} of T_1 is equal to V_{DD} and

$$V_o = V_{DD} - V_T$$

where V_T is the threshold voltage. If a load transistor T_L is present then

$$V_o = V_{DD} - 2V_T$$

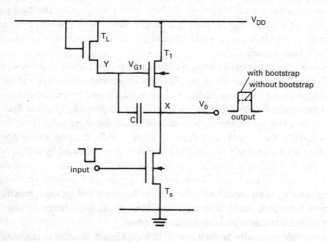

b **Bootstrapping** in MOS circuits

If V_{G1} can be increased to a value greater than V_{DD}, then V_o can also rise to a maximum possible value equal to V_{DD}. Bootstrapping is used to achieve this effect. A bootstrap capacitor, C, is connected between points X and Y (Fig. b). As V_o rises from the low logic level, V_{G1} also rises because of the positive feedback provided by C. This causes T_L to switch off and point B is thus isolated from the supply bus. V_{G1} can therefore rise to a value greater than V_{DD} and as V_{G1} increases, V_o also increases until it reaches V_{DD}. The total voltage swing at X is therefore optimized.

Bootstrapping is also used in high input impedance amplifiers, such as an emitter follower or a field-effect transistor stage, in order to maximize the a.c. input impedance. The a.c. input signal to the base or gate electrode can tend to flow through the bias resistors used to provide the d.c. bias for the base or gate unless bootstrapping is provided to prevent this effect.

borocarbon resistor ➤boron resistor.

boron resistor *Syn.* borocarbon resistor. A resistive ➤film resistor that has a small percentage of boron introduced into the carbon film to add stability.

boundary conditions Conditions that specify how the tangential and normal components of the field in one material are related to the components of the field across a boundary in another material.

bow-tie antenna ➤biconical antenna.

BPSK *Abbrev. for* binary phase shift keying. ➤phase modulation.

branch ➤network.

branch instruction An instruction that causes a break in the sequential execution of instructions in a computer program by altering the content of the ➤program counter, either conditionally or unconditionally. A *conditional branch* is a break in sequential execution that depends on the outcome of a branch test (such as whether the value in a register is zero); an example is

IF x > 10 THEN branch ELSE no branch

branch point *Syn. for* node. ➤network.

breadboard model A rough assembly of discrete components, often attached temporarily to a board, for testing the feasibility of a circuit, system, or design principle.

break 1. An accidental interruption of a broadcast programme. **2.** To open a circuit suddenly by means of a switch or other device. **3.** The minimum gap between the contacts needed for the circuit of such a device to be open.

breakdown 1. A sudden catastrophic change in the properties of a device rendering it unfit for its purpose. **2.** A sudden disruptive electrical discharge in an insulator or between the electrodes of an electron tube. **3.** A sudden change from a high dynamic resistance to a much lower value in a semiconductor device.
➤➤avalanche, second, thermal, and Zener breakdown.

breakdown voltage The voltage under specified conditions at which breakdown occurs.

break frequency *Syns.* band edge; corner frequency; cut-off frequency; critical frequency. The point at which the angular frequency of a signal is equal to the reciprocal of the circuit ➤time constant.

bremsstrahlung Electromagnetic radiation produced when an electron is suddenly decelerated by a nuclear field. ➤X-rays.

Brewster angle The angle at which an electromagnetic wave incident upon a material does not have part of itself reflected. It is dependent on the material type and the ➤polarization of the incident wave: the Brewster angle has different values depending on whether the electric field is perpendicular or parallel to the plane of incidence. In the case of a nonmagnetic material, the Brewster angle does not exist for perpendicular polarization: if an unpolarized beam of light is incident upon a surface at the Brewster angle, the parallel polarization is then completely transmitted and only perpendicularly polarized light is reflected. The Brewster angle is then also called the *polarizing angle*.

bridge An assembly of at least four circuit elements, such as resistors, capacitors, etc., together with a current source and a null point detecting device (see diagram). Each of the circuit elements is arranged in one *arm* of the bridge. When the bridge is balanced, i.e. zero response is obtained from the null detector, there is a calculable relationship between the values of elements in the arms given by

$$(Z_1/Z_2) = (Z_3/Z_4)$$

An unknown element may therefore be very precisely measured by comparison with known standards. The current source may produce either direct or alternating current. Bridge networks form the basis of many measuring instruments.

For resistance measurements ➤Wheatstone bridge, Kelvin double bridge, Carey–Foster bridge. For capacitance measurements ➤Wien bridge, de Sauty bridge,

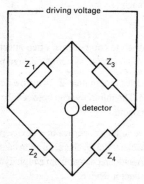

Bridge circuit

Schering bridge. For inductance measurements ►Anderson bridge, Hay bridge, Maxwell bridge, Owen bridge. For mutual inductance measurements ►Campbell bridge, Felici balance, Hartshorn bridge.

►►resonance bridge; Wagner earth connection.

bridged-H network ►two-port network.

bridged-T network ►two-port network.

Bridgeman method ►horizontal Bridgeman.

bridge network ►network.

bridge rectifier A ►full-wave rectifier circuit in the form of a bridge, with a rectifier in each arm (see diagram).

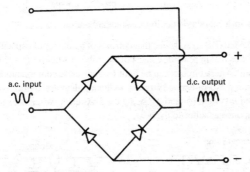

Bridge rectifier circuit

brightness control ►television receiver.

British Standards Institute (BSI) ►standardization.

broadband ►wideband.

broadband antenna Any antenna possessing good radiation characteristics that extend over a reasonably large range of frequencies in comparison to a generic ►dipole antenna. ►►biconical antenna; bow-tie antenna; log-periodic antenna; Bilog antenna.

broadband dipole An antenna composed of two conducting elements forming essentially a dipole configuration but possessing broadband characteristics. A simple form of broadband dipole is a cylindrical dipole in which the radius of the elements is made thicker to give a crude broadband performance. More effective variants include the ►biconical and ►bow-tie antennas.

broadcasting Radio or television transmission to the public. Specific ►frequency bands are available for public broadcasts and are assigned in accordance with international agreements.

broadside array ➤antenna array.

Broca galvanometer ➤astatic galvanometer.

brush A conductor, often made from specially prepared carbon, that provides electrical contact with a moving conductor, usually between the rotating and stationary parts of electrical machines. The contact resistance between the brush and the moving surface is the *brush contact resistance*. In electrical machines this resistance usually decreases with increasing current density resulting in a roughly constant voltage drop at the contact surface.

brush contact resistance ➤brush.

brush discharge A luminous electrical discharge that appears as a number of branching threads surrounding a conductor. Brush discharge occurs when the electric field around the conductor is not sufficiently large to produce a ➤spark.

BSI *Abbrev. for* British Standards Institute. ➤standardization.

bucket-brigade device (BBD) A device that consists of a number of capacitors linked by a series of switches that in practice consist of ➤bipolar or ➤field-effect (MOS) transistors (see diagram). These circuits can be built from discrete components but are invariably manufactured as ➤integrated circuits. ➤Clock pulses are applied to close the switches, a two-phase system (ϕ_1 and ϕ_2) being used. As each switch is closed charge is transferred from one capacitor to the next.

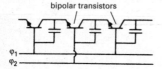
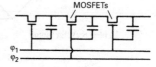

Bucket-brigade devices

Bucket-brigade devices are frequently used as ➤delay lines in both digital and analogue systems since the amount of charge stored may vary continuously from zero to the limit set by the magnitude of the capacitance and the operating voltage. The capacitors are provided in practice by using the collector-base capacitance or drain-gate capacitance of the transistors used in the circuit.

buffer An isolating circuit interposed between two circuits to minimize reaction from the output to the input. Usually it has a high input impedance and low output impedance. It may be used to handle a large ➤fan-out or to convert input and output voltage levels.

buffer memory A memory in a computer printer, disk drive, etc., in which data can be stored temporarily during a transfer.

bug A fault in a computer ➤program or in computer equipment. To *debug* the program or system is to find and correct any faults.

built-in field *Syn.* diffusion potential. ➤p-n junction.

bulk acoustic wave ➤acoustic wave device.

bulk-channel FET ➤field-effect transistor.

buncher ➤velocity modulation; klystron.

bunching ➤velocity modulation.

buried layer A layer of high-conductivity ➤semiconductor material diffused into the substrate layer during the manufacture of ➤bipolar integrated circuits and ➤bipolar transistors. The buried layer is located below the ➤collector and serves to reduce collector resistance.

burn-in *Syn.* ageing. ➤failure rate.

burst signal *Syn. for* colour burst. ➤colour television.

bus 1. *Syns.* busbar; busline. A conductor having low impedance or high current-carrying capacity to which two or more circuits can be separately connected or which can connect several like points in a system, as with an *earth bus*. Buses frequently feed power to various points. The name was first applied to metal bars of rectangular cross section supported on porcelain insulators. **2.** A set of wires associated with a computer along which signals travel in parallel. Usually these signals are grouped by function, such as data signals on a ➤data bus or address signals on an ➤address bus.

bushing An insulator used to form a passage for a conductor through a partition.

bus network A line structure in data networks where a number of users or devices are all connected together by a single transmission line or combinations of lines. The data bus can be a single pair of conductors down which serial data is transmitted or multiple conductors down which parallel data is transmitted.

Butterworth filter A type of filter with a flat passband response. ➤filter.

button mica capacitor ➤mica capacitor.

BW *Abbrev. for* bandwidth.

by-pass capacitor A shunt capacitor that is used to provide a comparatively low impedance path for alternating current. The magnitude of the capacitance determines the frequency that it passes. Such a capacitor is used to prevent a.c. signals reaching a particular point in a circuit, or to separate out a desired a.c. component.

byte A fixed number of ➤bits taken together and treated as a unit in a computer. A byte is a subdivision of a ➤word, and is almost always 8 bits.

byte-addressable ➤address.

C

cable An assembly of conductors that has some degree of flexibility; the conductors are insulated from each other and enclosed in a common binding or sheath. The most common types of cable are ➤paired cable and ➤coaxial cable although other configurations may be used. A *composite cable* is one in which the gauge of the conductors and/or the type of construction is not the same throughout its cross section.

cable track locator A device used to locate buried telephone cables. An audiofrequency oscillator, known as a bleeper, is used to send a signal along the cable in question and the bleeps are detected using a portable receiver and headset. The position of the buried cable is along the track corresponding to the maximum detected signal intensity.

cache memory A small-capacity high-speed buffer that holds copies of the most frequently accessed values in computer ➤memory. It is placed between a memory system and its user in order to reduce the effective access time of the memory system. Its performance is based on two assumptions: firstly, small memory is faster than larger memory; secondly, spatial and temporal locality apply to data accesses – if an item of data has just been accessed it will be accessed again soon and data close to it will be more likely to be accessed than data further away.

CAD *Abbrev. for* computer-aided design.

cadmium cell ➤Weston standard cell.

calibration The determination of the relationship between the indicated value on a measuring instrument and the true value of the being measured. This may result in the adjustment – or calibration – of the measuring instrument to a set value against a reference input. The true value is the value that would be obtained if all sources of error were eliminated.

CAM *Abbrev. for* content-addressable memory. ➤associative memory.

camera tube The device contained in a ➤television camera that acts as an optical–electrical ➤transducer and converts the *optical image* of the scene to be transmitted into electrical *video signals*. Most camera tubes are ➤electron tubes: the two basic types of tube are the ➤image orthicon and the ➤vidicon tubes from which many other tubes have been developed.

A wide variety of camera tubes is available. The main differences between them are in the target material and the velocity of the electrons used to produce the video signal. *Photoemissive camera tubes* have a target coated with photoemissive material (➤photoemission). *Photoconductive camera tubes* are coated with material that exhibits ➤photoconductivity. *Low-electron-velocity tubes* are most often used but *high-electron-velocity tubes,* such as the iconoscope, are also produced.

The performance of camera tubes depends very greatly on the scanning system employed. Beam alignment is achieved by using small coils to ensure that the electron beam emerging from the electron gun is central. Deflection of the beam is provided by deflection coils controlling the horizontal and vertical directions. These coils are supplied with a ➤sawtooth voltage causing a linear scan with very rapid return to the start of the scanning position. Focusing coils are also provided to ensure a small cross section when the beam reaches the target; an extra electrode ensures that the beam is essentially perpendicular to the target surface. In low-electron-velocity tubes this electrode decelerates the beam so that it is essentially stationary at the target.

A ➤solid-state camera has also been developed in which the transduction element is an array of ➤charge-coupled devices. This type of camera is very much smaller and lighter than those containing electron tubes and is typically the size of an ordinary hand-held photographic camera.

➤➤image dissector.

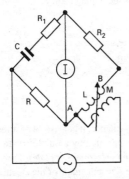

Campbell bridge

Campbell bridge An a.c. ➤bridge that is used to measure a mutual inductance, M, by comparison with a standard capacitance, C. The resistances R_1 and R_2 are varied until null deflection is obtained on the indicating instrument I (see diagram). At balance

$$L/M = (R + R_1)/R$$

$$M/C = RR_2$$

where L is the self-inductance of the coil AB.

capacitance Symbol: C; unit: farad. The property of an isolated ➤conductor or a set of conductors and ➤insulators whereby it stores electric charge. A charge of Q coulombs will increase the voltage of an isolated conductor by V volts. The capacitance is defined as the ratio Q/V and is determined by the size and shape of the conductor. It is constant for a given isolated conductor.

If the isolated conductor is placed near a second conductor or a semiconductor but is separated from it by air or some other insulator, the system forms a ►capacitor. An electric field is produced across the system and the potential difference between the conductors is determined by this field. The capacitance, C, is defined as the ratio of the charge on either conductor to the potential difference between them. ►►mutual capacitance; impedance; series; parallel.

capacitance integrator ►integrator.

capacitance strain gauge *Syn. for* variable capacitance gauge. ►strain gauge.

capacitance-voltage curves (C-V curves) Curves demonstrating the relationship between the junction capacitance and applied voltage of a ►p-n junction diode or a ►metal-semiconductor junction diode (a Schottky diode). These curves are used to analyse the material structure of the diode; in particular the potential barrier height of the junction can be found, and also the variation with depth of the majority ►carrier concentration within the ►semiconductor, known as the doping profile.

capacitive coupling ►coupling.

capacitive feedback ►feedback.

capacitive load ►leading load.

capacitive reactance ►reactance.

capacitive tuning ►tuned circuit.

capacitor A component that has an appreciable ►capacitance. It consists of an arrangement of at least two conductors or semiconductors separated by a ►dielectric (an insulator). The conductors or semiconductors are known as *electrodes* or *plates*. The value of the capacitance of a given device depends on the size and shape of the electrodes, the separation between them, and the relative ►permittivity of the dielectric. Most types of capacitor have a value, which may be variable, determined by the geometry of the device. Some capacitors however have a value that is also a function of the voltage across the device or of the operating frequency. The dielectric material may be solid, liquid, or gaseous. ►►ceramic, chip, electrolytic, metallized film, metallized paper, mica, MOS, and plastic-film capacitors; varactor.

capacitor microphone *Syn.* electrostatic microphone. A type of microphone in which a diaphragm forms one plate of a capacitor. Variations of sound pressure cause movements of the diaphragm and these alter the capacitance. Corresponding changes in potential difference across the capacitor are thus produced.

capacity 1. ►channel capacity; memory capacity. **2.** *Obsolete syn. for* capacitance.

capture ratio ►cochannel rejection.

carbon Symbol: C. A nonmetal, atomic number 6, that exists in two allotropic crystalline forms: diamond and graphite. In diamond form the ►resistivity (5×10^{14} ohm cm) falls within the range of insulators. In graphite form it is a poor conductor, resistivity about 1.4×10^{-3} ohm cm, and in granular form exhibits a variation of resistance with pressure.

In graphite form it is used to form resistors, in microphones, and as a filament of electric lights.

carbon-composition resistor *Syn. for* carbon resistor. ►resistor.

carbon-film resistor ►film resistor.

carbon microphone A ►microphone that utilizes the variation of contact resistance of granular ►carbon with applied pressure. There are several types of carbon microphone, including the telephone transmitter, in which sound-pressure variations are transmitted to the carbon granules through a diaphragm; the corresponding changes of resistance are detected by fluctuations in a current passed through the granules.

carbon resistor *Syn.* carbon-composition resistor. ►resistor.

cardioid microphone *Syn.* unidirectional microphone. A ►microphone with a direc-

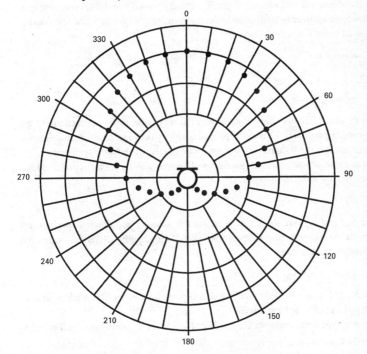

Idealized directional response of a **cardioid microphone** (scale in degrees)

tional response pattern described by $(1 + \cos\phi)$, where ϕ is the angle of incidence of the sound. Such a pattern can be considered to be the combination of an ➤omnidirectional microphone response (1 all round) and a ➤figure-eight microphone response ($\cos\phi$). The overall response is approximately heart-shaped (see diagram), hence the term cardioid. Practically, such a response is difficult to achieve for frequencies above which the wavelength becomes comparable with the physical dimensions of the microphone itself.

Carey–Foster bridge A modification of the ➤Wheatstone bridge that measures the difference in resistance between two nearly equal resistances. The resistances are placed in the ratio arms of the bridge and the balance point found on a slide-wire. The resistances are then switched and a new balance point found. The resistance difference is proportional to the distance between the balance points.

carrier 1. *Short for* charge carrier. A mobile ➤electron or ➤hole that transports charge through a metal or a ➤semiconductor and is responsible for its conductivity. **2.** *Short for* carrier wave.

carrier concentration *Syn.* carrier density. The number of charge ➤carriers in a ➤semiconductor per unit volume, in practice usually quoted as numbers per cubic centimetre. In an intrinsic (i-type) semiconductor the number of holes, p, and of electrons, n, is equal to the *intrinsic density*, n_i:

$$n = p = n_i$$

where

$$n_i^2 = N_c N_v \exp(-E_g/kT)$$

where N_c and N_v are the effective densities of energy states in the conduction and valence bands respectively, E_g is the difference in energy between the conduction and valence bands, k is the Boltzmann constant, and T the thermodynamic temperature.

In an extrinsic semiconductor the electrical neutrality of the sample is preserved and in a sample that contains impurities,

$$N_A^- + n = N_D^+ + p$$

where N_A^- and N_D^+ are the numbers of ionized acceptor and donor impurities, respectively. At relatively high temperatures most of the impurity atoms are ionized and it is possible to state that

$$n + N_A = p + N_D$$

where N_A and N_D are the total numbers of acceptors and donors. The product $np = n_i^2$ is independent of added impurities.

In an n-type semiconductor the number of electrons, n_n, at thermal equilibrium is given by

$$n_n = |N_D - N_A|$$

provided $N_D > N_A \gg n_i$

The number of holes, p_n, in an n-type semiconductor is given by

$p_n = n_i^2/n_n$

In a p-type semiconductor the situation is reversed:

$p_p = |N_A - N_D|$

provided $N_A > N_D \gg n_i$

The number of electrons is given by

$n_p = n_i^2/p_p$

The carrier concentrations of i-, n-, and p-type semiconductors are shown schematically in the diagram, where E_A and E_D are the locations of the ►energy levels of the acceptor and donor impurities respectively.

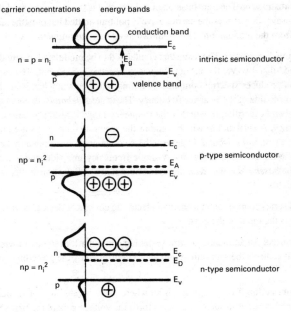

Carrier concentrations in semiconductors

carrier density *Syn. for* carrier concentration.

carrier frequency The frequency of a ►carrier wave. The carrier frequency is usually the average frequency, known as the *centre frequency*, of the transmitted signal.

carrier mobility ►drift mobility.

carrier storage *Syn.* charge storage. An effect occurring in a ➤p-n junction under forward bias. In a semiconductor with a relatively long bulk lifetime, excess minority carriers injected across the junction remain near the junction as a net concentration of charge. When ➤reverse bias is applied to the junction the effect of the carriers stored near the junction is to produce a reverse current across the junction substantially greater than the normal reverse saturation current; this current flows until all the stored charges have been removed either by recombination or by crossing back across the junction under the influence of the reverse bias. The time interval between application of the reverse bias and cessation of the reverse current surge is the *storage time* of the junction.

When several diodes are used together in a rectifier circuit that also contains inductive elements, the carrier storage can result in undesirable transients in the circuit and limits the frequency at which such circuits can be operated. The storage effect is however utilized in the ➤step-recovery diode.

carrier suppression ➤carrier transmission.

carrier transmission The transmission of a signal that is the result of ➤modulation of a ➤carrier wave. Sometimes the carrier wave is not transmitted but only the sidebands resulting from the modulation. This is known as *carrier suppression*.

carrier wave *Syn.* carrier. The wave that is intended to be modulated in ➤modulation, or, in a modulated wave, the carrier-frequency spectral component. The process of modulation produces spectral components falling into frequency bands at either the upper or lower side of the ➤carrier frequency. These are *sidebands,* denoted *upper* or *lower sideband* according to whether the frequency range is above or below the carrier frequency. A sideband in which some of the spectral components are greatly attenuated is a *vestigial sideband.* In general these components correspond to the highest frequency in the modulating signals. A single frequency in a sideband is a *side frequency.* The *baseband* is the frequency band occupied by all the transmitted modulating signals.

cascade A series connection of a chain of electronic circuits or devices so that the output of one is the input of the next.

cascade control An automatic control system in computer technology in which each control unit controls the succeeding unit and is controlled by the preceding one in the chain.

cascode connection A two-stage amplifier where the two transistors or vacuum tubes are connected in series in such a way as to force the voltage gain of the first active device to be unity. This connection minimizes the input capacitance of the amplifier by eliminating the ➤Miller effect, which would otherwise multiply the influence of the internal feedback capacitance of the first device by its voltage gain. Wide-bandwidth amplification and high-speed switching can therefore be achieved.

Typical configurations include a ➤common-emitter amplifier loaded by a ➤common-base amplifier, using bipolar transistors, and a ➤common-cathode amplifier loaded by a ►common-gate amplifier, using vacuum tubes. In each case, the current gain of the second active device is unity, but the overall power gain of the pair of devices is comparable to that of a single device due to current gain in the first stage and voltage gain in the second stage; this performance is maintained up to the transition frequency of the transistors.

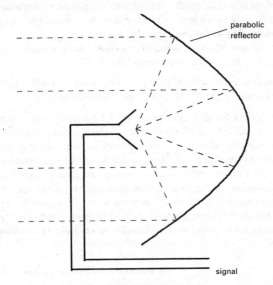

Cassegrain feed to a microwave dish antenna

Cassegrain feed A configuration that may be employed in microwave antennas using a parabolic 'dish' to achieve the beam focus and gain. It involves the use of a secondary reflector and an axial ►feed; the principle is illustrated in the diagram, which shows how the length of the feed to the antenna can be substantially reduced by the Cassegrain feed.

catastrophic failure ►failure.

catcher ►klystron.

cathode The negative ►electrode of an electrolytic cell, discharge tube, valve, or solid-state rectifier. The electrode by which electrons enter (and conventional current leaves) a system. ►►anode.

cathode dark space *Syn. for* Crookes dark space. ►gas-discharge tube.

cathode follower ►emitter follower.

cathode glow ►gas-discharge tube (diagram).

cathode-ray direction finding ►direction finding.

cathode-ray oscilloscope (CRO) An instrument in which a variety of electrical signals are presented on the screen of a ►cathode-ray tube for examination. The signal under examination is used to deflect the electron beam of the CRT in one direction (usually the vertical) while another known signal is used in the other direction. The composite signal is shown on the screen. Visualization of the input signal is achieved using the output from a sweep generator, usually called a timebase generator (►timebase), and selecting the appropriate sweep speed. The sweep may be generated as a ►sawtooth waveform or initiated by an external trigger pulse.

More complicated cathode-ray oscilloscopes often include a delayed trigger, access to the X-deflection plates, and beam-intensity modulation facilities.

cathode rays Streams of ►electrons emitted from the cathode in an evacuated tube containing a cathode and anode. They were first observed in ►gas-discharge tubes.

cathode-ray tube (CRT) A funnel-shaped evacuated ►electron tube that converts electrical signals into a visible form. All CRTs have an ►electron gun to produce an electron beam, a grid that varies the electron beam intensity and hence the brightness, and a luminescent screen to produce the display (see diagrams). The electron beam is moved across the screen either by deflection plates or magnets. The *deflection sensitivity* of the tube is the distance moved by the spot on the screen per unit change in the deflecting field. High-frequency applications, as in the ►cathode-ray oscilloscope, usually employ electrostatic deflection; electromagnetic deflection is used when high-velocity electron beams are required, as in ►television or ►radar receivers, which need a bright display.

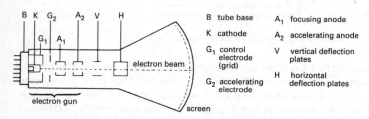

a **cathode-ray tube**: electrostatic focusing and deflection

Focusing of the beam may be done electrostatically or electromagnetically or by a combination of methods. A greater degree of focusing is required when the electron beam is deflected towards the edges of the screen. The point at which the electron beam comes to a focus is the *crossover area* and the solid angle of the cone of electrons emerging from this area is the *beam angle*. For convenience the deflection and focusing coils are often mounted around the narrow neck of the tube as a single unit, termed a *scanning yoke*. Such an arrangement reduces the overall physical dimensions of the assembly and is particularly important when the tube contains more than one electron beam, as in the *double-beam CRT* or some forms of colour picture tube, and therefore requires more than one set of coils.

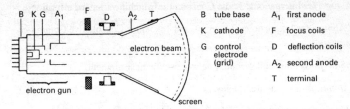

b CRT: electromagnetic focusing and deflection

cathodoluminescence The emission of light when substances are bombarded by cathode rays (electrons). The frequency of light emitted is characteristic of the bombarded substance.

cation An ion that carries a positive charge and, in electrolysis, moves towards the cathode, i.e. travels in the direction of conventional current. ➤anion.

cat's whisker *Obsolete. See* point contact.

Cauer filter *Syn. for* elliptic filter.

cavity resonator *Syns.* rhumbatron; resonant cavity. The space within a closed or substantially closed conductor that will maintain an oscillating electromagnetic field when suitably excited externally. The resonant frequencies are determined by the size and shape of the cavity. The whole device has marked resonant effects and replaces tuned resonant circuits for high-frequency applications for which the latter are impracticable.

C band A band of microwave frequencies ranging from 4.00–8.00 gigahertz (IEEE designation). ➤frequency band.

CCCS *Abbrev. for* current-controlled current source. ➤dependent sources.

CCD *Abbrev. for* charge-coupled device.

CCD filter A ➤transversal filter realized using ➤charge-coupled devices (CCDs).

CCD imaging ➤charge-coupled device.

CCD memory A ➤solid-state memory that consists of one or more ➤shift registers formed from ➤charge-coupled devices and that is used to store digital information. CCD memory is inherently slower, cheaper, and more compact than ➤RAM and is suitable for applications that are either serial in nature or that do not require the fast operating speeds of RAM.

High operating speeds can be achieved using a number of short CCD shift registers that are clocked in parallel, each of which is connected to its own sense/refresh circuits. A particular register may be selected using suitable decoding circuits.

CCI First-generation current conveyor. ➤current conveyor.

CCII Second-generation current conveyor. ➤current conveyor.

CCIR *Abbrev. for* International Radio Consultative Committee. ➤standardization.

CCITT *Abbrev. for* International Telegraph and Telephone Consultative Committee. ➤standardization.

CCS *Abbrev. for* centum call second. ➤network traffic measurement.

CCTV *Abbrev. for* closed-circuit television.

CCVS *Abbrev. for* current-controlled voltage source. ➤dependent sources.

CD *Abbrev. for* compact disc. ➤compact disc system.

CDMA *Abbrev. for* code-division multiple access. ➤digital communications.

CD-ROM *Abbrev. for* compact disc, read-only memory. A ➤compact disc that is used for the storage of digital information in a manner similar to ➤ROM in that it can be written to only once but the information is retained permanently. The information is read from disc by means of a *CD-ROM drive* in a computer system (an audio CD player cannot handle CD-ROMs). The stored information may represent text, sound, or still or video images; a combination of these forms can be used in an integrated way as a *multimedia* system, accessed by a suitably equipped personal computer. CD-ROMs are also used for the distribution of data, images, and software.

Ceefax *Trademark. See* teletext.

cell 1. *Syns.* electrolytic cell; voltaic cell. A device that produces electricity by chemical means, consisting of a pair of plate electrodes in an electrolyte. In a *primary cell* the chemical action is not normally reversible, the current being produced as a result of the dissolution of one of the plates. A *secondary cell* has reversible chemical action and is charged by passing a current through it. The rate and direction of the chemical action is determined by the value of the external voltage. The *volt efficiency* of a secondary cell is the ratio of the voltage developed by it during the discharge to the average voltage supplied to it during the recharging cycle.

The *cell internal resistance* is the resistance offered to the passage of current inside the cell. If the open-circuit e.m.f. is E, then the potential difference, U, across the cell when current flows is given by

$$U = ER/(R + r),$$

where R is the external resistance and r the internal resistance.

A portable cell that has the electrolyte in the form of a nonspillable jelly or paste is known as a *dry cell*. Most cells contain a liquid electrolyte and are sometimes termed *wet cells*. Dry cells are used in the batteries for torches, portable radios, etc. **2.** Any device that can generate a direct electromotive force from a nonelectrical source of energy, particularly from light energy. Examples include ➤solar cells and ➤photovoltaic cells.

cell constant The area of the electrodes in an electrolytic ►cell divided by the distance between them. This equals the resistance in ohms of the cell when filled with a liquid of unit resistivity.

cell internal resistance ►cell.

cellular communications A ►telecommunication system in which a number, often a large number, of transmitters are spread over a wide region to enable a mobile radio system to operate anywhere within that region. The radio system will first decide which transmitter provides the greatest signal at a particular point; it will then transmit to and receive from that transmitter until the signal becomes weak, when it will attempt to transfer to the transmitter that now provides the greatest signal. The radio system may be a *mobile phone* or a *mobile multimedia* system comprising personal computer and transmitter; where the system is designed for stand-alone personal use it is sometimes referred to as a *personal communications device*. The transmitters may be at fixed land-based sites, connected to a main telecommunications network, such as the telephone system, by means of land lines or microwave link; alternatively they may be satellite-based, in which case the signal is then relayed to another earth-based station.

cellular mobile radio ►cellular communications.

cellular phone ►cellular communications.

centi- Symbol: c. A prefix to a unit, denoting a submultiple of 10^{-2} of that unit: one centimetre equals 10^{-2} metre.

central limit theorem A statistical theorem which states that, in general, the probability distribution of the sum of a number of statistically independent random variables approximates to the ►Gaussian distribution as the number of variables becomes very large, irrespective of the distributions of the individual random variables. The central limit theorem is commonly used in the understanding and analysis of ►noise in communication systems.

central processing unit (CPU) The portion of a ►computer that controls the operation of the entire computer system and executes the arithmetic and logic functions contained in a particular ►program. A CPU usually consists of two units: the *control unit* organizes the data and program storage in the ►memory and transfers data and other information between the various parts of the system; the *ALU* (arithmetic/logic unit) executes the arithmetic and logic operations, such as addition, multiplication, and comparison.

centre clipper A circuit that sets boundary values on a signal in the centre of the input range, and gives an output proportional to the input for inputs that are outside this centre range.

centre frequency 1. In a communications system, the average frequency of the transmitted signal when information is modulated onto a carrier for transmission. The carrier frequency is usually the centre frequency. **2.** In analogue electronics, the midpoint of the frequency range over which a circuit works when the circuit has a response that is frequency dependent. **3.** In spectrum analysis, the frequency that corresponds to the middle of the frequency span.

centre tap A connection made to the electrical centre of an electronic device such as a resistor or transformer.

centum call second (CCS) ➤network traffic measurement.

cepstrum The spectrum of the logarithm of a frequency spectrum; the term arises from a letter reversal in the word spectrum. A cepstrum is often used in speech analysis for ➤fundamental frequency and ➤formant analysis. The X-axis has the dimension of time and is referred to as *quefrency* (from letter reversal in 'frequency'), and regularly spaced peaks in the cepstrum are known as *rahmonics* (from letter reversal in 'harmonics'). Low-frequency components, such as the fundamental frequency in speech, have a high quefrency, and high-frequency components, such as formant peaks, have a low quefrency, so they can be separated by means of a *lifter* (from letter reversal in 'filter').

ceramic capacitor A capacitor utilizing a ceramic as the dielectric material. The behaviour of the capacitor is determined by the electrical properties of the ceramic used; these vary widely but most have high permittivity allowing the capacitors to be smaller than most other types.

ceramic filter A highly selective bandpass ➤filter using a ➤piezoelectric crystal, often lead zirconate titanate, PZT, as the ➤resonator element. The electrical equivalent circuit of the ceramic filter is an LC circuit. Ceramic filters are available for common ➤intermediate frequencies. They are less expensive than comparable crystal filters, but offer lower ➤Q factor or ➤selectivity.

ceramic pick-up ➤pick-up.

cermet Acronym from *cer*amic and *met*al. A mixture of an insulating material and a highly conducting material. Ceramics such as oxides and glasses are used as the insulator and metals such as chromium and silver are the conductors. Cermets are used for the production of thick films and thin films, particularly for use in ➤film resistors.

CGA *Abbrev. for* colour graphics adapter. A colour video display standard introduced by IBM for its first PC, with a resolution of 320×200 ➤pixels and four colours.

CGS system A system of ➤units, now obsolete, in which the fundamental units of length, mass, and time were the centimetre, gram, and second. In order to include and completely define electric and magnetic quantities a fourth fundamental quantity is required. As a result two mutually exclusive systems of CGS units arose: the *CGS electromagnetic units* (emu) and the *CGS electrostatic units* (esu).

In the emu system the fourth quantity was the ➤permeability of free space, μ_0, which was chosen to be unity. In the esu system the fourth quantity was the ➤permittivity of free space, ε_0, also chosen to be unity. The choice of the four fundamental quantities completely defined each system but since, by Maxwell's equations, it can be shown that

$$\mu_0\varepsilon_0 = 1/c^2$$

where c is the speed of light, the systems were mutually exclusive. All mechanical quantities however, such as force and work, had the same units in either system. To differentiate between the two systems the prefix *ab-* was used to denote units in the emu system and the prefix *stat-* for units in the esu system. The ratio of ab- unit to equivalent stat- unit for the primary CGS units, such as current, was equal to c or $1/c$, where c was measured in cm/s. The ratio for secondary units was some power of c.

CGS units were superseded by the ►MKS system of units, from which the system of ►SI units was developed.

chaff ►confusion reflector.

channel 1. A specified frequency band or a particular path used in communications for the reception or transmission of electrical signals. **2.** A route along which information may travel or be stored in a data-processing system or a computer. **3.** (in a ►field-effect transistor) The region connecting ►source and ►drain, the conductivity of which is modulated by the ►gate voltage. **4.** (in a p-n-p ►bipolar junction transistor) The spurious extension of n-type ►base across the surface of the ►collector to the edge of the chip. This results in excessive leakage currents and may be overcome using a ►channel stopper. The inevitable presence of positive charges at the interface of the high-resistivity p-type collector with the passivating oxide layer causes the formation of the n-type channel on the surface of the collector.

channel capacity The amount of information that can be transmitted over a communication channel. Channel capacity cannot be exceeded without error but in theory can always be attained with an arbitrarily small probability of error. ►Shannon–Hartley theorem; digital communications.

channel code ►compact disc system.

channel stopper 1. (in a p-n-p ►bipolar junction transistor) A means for limiting ►channel formation by surrounding the n-type ►base entirely with a ring of highly doped low resistivity p-type material. **2.** (in a ►MOS integrated circuit) A region of highly doped material of the same type as the lightly doped substrate. This increases the field ►threshold voltage and inhibits the formation of spurious ►field-effect transistors caused when interconnections pass between adjacent ►drain regions.

characteristic A relationship between two values that characterizes the behaviour of a device, circuit, or apparatus. The relations are most commonly produced for ►transistors and ►two-port networks. They are usually plotted in the form of families of graphs (*characteristic curves*) relating the currents obtained to the voltages applied for a range of operating conditions.

The electrode characteristic shows the relationship between current and voltage at an electrode of the device, for example drain current versus drain voltage in a ►field-effect transistor. The *transfer characteristic* shows the relationship between the current (or voltage) at one electrode and the voltage (or current) at a different electrode, for example drain current versus gate voltage in a FET. The *static characteristic* shows, for example, collector current versus base voltage in a ➤bipolar junction transistor, all other applied voltages being held constant, i.e. under static conditions. The *dynamic characteristic* relates the current from one electrode and the voltage on another, under dynamic conditions. ➤transfer parameters.

characteristic curve ➤characteristic.

characteristic impedance ➤iterative impedance; transmission line.

charge Symbol: Q; unit: coulomb. A property possessed by some elementary particles causing them to exert forces of attraction or repulsion on each other. The types of force exerted by charged particles are differentiated by the terms *negative* and *positive*, the natural unit of negative charge being possessed by the ➤electron. The ➤proton has an equal amount of positive charge. A charged body contains an excess or lack of electrons with respect to its proton content.

Two electric charges, when close to each other, exert a force on each other. This force is given by ➤Coulomb's law and can be explained in terms of the ➤electric field surrounding the charges. If a charge, Q, is placed in an electric field E it experiences a force QE.

charge carrier ➤carrier; semiconductor.

charge-coupled device (CCD) An array of closely spaced ➤MOS capacitors. Information is represented as a packet of charge stored in a capacitor, rather than as a voltage or current. In operation, this charge can be transferred from capacitor to capacitor in a controlled manner by application of a suitable sequence of voltage pulses to the capacitors: the CCD behaves like an analogue ➤shift register. Typically a three-phase clock sequence is used (see diagram): every third electrode is clocked synchronously to maintain a physical separation between the charge packets, minimizing corruption of the data packets. Some charge is lost during each transfer due to recombination at the surface, diffusion out of the capacitor, etc., and so the data must be restored periodically along the array.

CCDs can be used for signal processing applications, such as ➤transversal filters, and in imaging applications, where the MOS capacitors are charged by exposure to light or other electromagnetic radiation and the image data is then read out in a serial manner.

charge density 1. *Syn.* volume charge density. Symbol: ρ; unit: coulomb per metre cubed. The electric charge per unit volume of a medium or body. **2.** *Syn.* surface charge density. Symbol: σ; unit: coulomb per square metre. The electric charge per unit area at the surface of a medium or body.

charge storage ➤carrier storage.

charge-storage diode ➤step-recovery diode.

charge-transfer device (CTD) A semiconductor device in which discrete packets of charge are transferred from one location to the next. Such devices can be used for the short-term storage of charge in a particular location provided that the storage time is short compared with the recombination time in the material. Several different types of charge-transfer device exist, the main classifications being ►charge-coupled devices and ►bucket-brigade devices. Applications of charge-transfer devices include short-term memory systems, ►shift registers, and imaging systems. Information is usually only available for serial access.

chart recorder ►graphic instrument.

Chebyshev filter A type of filter with an ►equiripple passband response. ►filter.

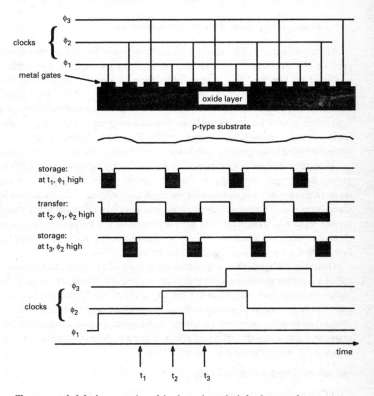

Charge-coupled device: operation of the three-phase clock for data transfer

chemical vapour deposition (CVD) A technique for depositing thin films of metal, semiconductor, or insulator, employed in the manufacture of semiconductor devices and integrated circuits using ►planar process technology. The films are produced from gas-phase source materials that react on the surface of the sample to produce the film. A simple reactor is shown in the diagram: the heating can be provided directly, by radiant heaters, or by induction heating using a radiofrequency signal.

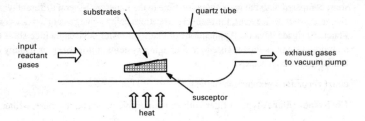

Reactor tube for **chemical vapour deposition** process

The *deposition rate* of the film is governed by the temperature of the sample and the pressure of the reactant gases inside the reactor. These parameters can also affect the quality of the resulting film, in terms of its chemical purity and number of defects present in the crystalline film. Metallic films are usually polycrystalline. Insulating films such as silicon dioxide or silicon nitride, used for isolation between metal layers or as capacitor dielectrics, are amorphous layers. Semiconductor layers can be deposited using CVD techniques: amorphous or polycrystalline films can be deposited, the crystalline order controlled by temperature with lower temperatures producing less-ordered films.

➤Epitaxial crystals can be grown on semiconductor surfaces using CVD techniques, the reactants and conditions chosen to enable the reaction at the semiconductor surface to produce a layer following the crystalline ordering of the underlying layer; this process is also known as *vapour phase epitaxy* (*VPE*). Often the reactants are inorganic materials, though recently the use of organic compounds containing metal atoms has been employed for VPE of ➤compound semiconductors – this process being called *organo-metallic vapour phase epitaxy* (*OMVPE*).

Generally these reactions are carried out at reduced pressure, and the process is called *low pressure chemical vapour deposition* (*LPCVD*). Faster deposition occurs at higher pressures, and some processes can take place at atmospheric pressure: *atmospheric pressure chemical vapour deposition* (*APCVD*). Reaction rates can also be speeded up by using energetic reactant gases: the use of a ➤glow discharge or plasma at low pressure to provide highly energetic and reactive species is employed in *plasma-enhanced chemical vapour deposition* (*PECVD*).

chemiluminescence ➤luminescence.

Child's law ➤thermionic valve.

chip *Syn.* die. A small piece of a single crystal of ➤semiconductor material containing either a single component or device or an ➤integrated circuit. Chips are commonly sliced from a much larger ➤wafer of substrate, in which many of the components or circuits have been produced in a regular array. Chips are not normally ready for use until suitably packaged.

chip capacitor A capacitor component consisting of a small ceramic ➤dielectric with the two contacting electrodes placed directly onto the dielectric, at each end of the component. They are of small size: standard sizes are denoted by 0603, 0402, etc., which refers to the length and width dimensions, respectively, in millimetres. Chip capacitors have no connecting leads and are soldered directly onto the ➤printed circuit board using ➤surface mount technology. These components therefore have much lower ➤parasitic inductance and capacitance than similar-valued leaded components, and are used primarily for high-frequency or high-precision circuits.

chip-enable A control input to an integrated circuit that ➤enables the chip's operation; it is often labelled CS. Chip-enables are used in all computer systems when more than one device (memory or input/output) is connected to the same wire, such as an ➤address or ➤data bus.

chip inductor An inductor component consisting of a small piece of ➤ferrite material as the inductor, with the two contacting electrodes placed directly onto the ferrite, at each end of the component. Like chip capacitors, chip inductors have no connecting leads and are soldered directly onto the ➤printed circuit board using ➤surface mount technology. These components therefore have much lower ➤parasitic inductance and capacitance than similar-valued leaded components, and are used primarily for high-frequency or high-precision circuits.

chip resistor A resistor component consisting of a small piece of metal film resistor material with the two contacting electrodes placed directly onto the resistor, at each end of the component. Like chip capacitors, chip resistors have no connecting leads and are soldered directly onto the ➤printed circuit board using ➤surface mount technology. These components therefore have much lower ➤parasitic inductance and capacitance than similar-valued leaded components, and are used primarily for high-frequency or high-precision circuits.

choke 1. *Syn.* choking coil. An ➤inductor used to present a relatively high impedance to alternating current. Audiofrequency and ➤radiofrequency chokes are used in audio- and radiofrequency circuits, respectively. *Smoothing chokes* are used to reduce the amount of fluctuation in the outputs of rectifying circuits. Smoothing chokes whose impedance varies with the current passing through them are called *swinging chokes*. **2.** A groove cut to a depth of a quarter wavelength in the metal surface of a ➤waveguide to prevent the escape of microwave energy.

choke coupling ➤coupling.

choking coil ➤choke.

chopped impulse voltage ➤impulse voltage.

chopper amplifier An ➤amplifier that amplifies direct-current signals by first converting them into alternating current and then using normal a.c. amplifying techniques. The conversion is achieved using a system of ➤relays or a suitable ➤vibrator.

chrominance signal ➤colour television.

chrominance subcarrier ➤colour television.

chronotron An electronic device that measures the time interval between events. A pulse is initiated by each event and the time interval is determined by the position of the pulses along a transmission line.

cipher An algorithm used for encryption of information or, in its inverse form, for decryption. It is one of the earliest forms of encryption. Used by Julius Caesar during the Gallic wars, Caesar cipher employed a simple alphabetic shift in which the transmitted letter was the desired letter shifted a number of places in the alphabet. For example, to transmit 'many' using a two-letter shift results in the transmission of 'ocpa' – each letter is shifted two places along the alphabet with wrap around at z. To decrypt the code all that is required is the number of shifts.

CIRC *Abbrev. for* cross-interleave Reed–Solomon code. ➤digital codes.

circuit The combination of a number of electrical devices and conductors that, when connected together to form a conducting path, fulfil a desired function such as amplification, filtering, or oscillation. A circuit may consist of discrete components or may be an ➤integrated circuit. Some circuits, such as ➤charge-coupled devices, can only be produced in integrated form.

A circuit is *closed* when it forms a continuous path for current. When the path is broken, as by a switch, the circuit is *open*. Any constituent part of the circuit other than the interconnections is a *circuit element*.

circuit-breaker A device such as a contactor, switch, or tripping device that is used to ➤make or ➤break a circuit under normal or fault conditions. An unwanted arc is often produced as the circuit-breaker operates and this can be minimized using a *magnetic blow-out* device. This device is fitted to the circuit-breaker and produces a magnetic field in the neighbourhood of the arc, thus causing the pathlength of the arc to increase and thereby extinguishing it rapidly. When used for fault conditions an automatic break and manual make system is commonly used.

circuit diagram A diagram that represents the function and interconnections of a circuit. Each circuit element is represented by its appropriate graphical symbol. ➤Table 1 in the back matter.

circuit element ➤circuit.

circuit parameters ➤parameters.

circular polarization ➤polarization.

circular scanning ➤scanning.

CISC *Abbrev. for* complex instruction set computer. In essence, a description for any computer that lacks the properties of being a ➤RISC (reduced instruction set computer). In general, a CISC is a computer with the following features: complex ➤addressing modes, instructions with many sizes, instructions that have memory-to-memory or memory-to-register operations, and a ➤microprogrammed control unit.

CISPR *Abbrev. for* International Special Committee on Radio Interference.

clamping circuit A circuit used to shift an a.c. waveform up or down by adding a d.c. voltage to the positive or negative peak value of the a.c. signal. ➤direct-current restorer.

clamp-type mica capacitor ➤mica capacitor.

Clark cell A mercury-zinc standard cell with e.m.f. defined as 1.4345 volts at 15 °C. It has been superseded by the ➤Weston standard cell.

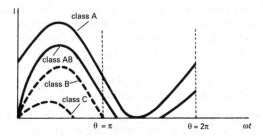

Angle of flow of output current for different amplifier classes

class A amplifier A linear ➤amplifier in which output current flows over the whole of the input current cycle, i.e. the ➤angle of flow equals 2π. These amplifiers have low distortion but low efficiency. Distortion can occur with large-signal operation due to the device transfer characteristics becoming nonlinear (see diagram below).

class AB amplifier A linear ➤amplifier in which the output current flows for more than half but less than the whole of the input cycle, i.e. the ➤angle of flow is between π and 2π. At low input-signal levels class AB amplifiers tend to operate as ➤class A and at high input-signal levels as ➤class B amplifiers.

class B amplifier A linear ➤amplifier operated so that the output current is cut off at zero input signal, i.e. the ➤angle of flow equals π, and a half-wave rectified output is

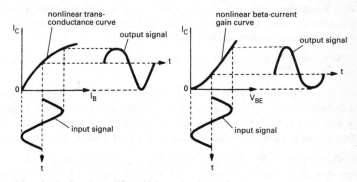

Distortion in **class A amplifier** with large-signal operation

produced. Two transistors are required in order to duplicate the input waveform successfully, each one conducting for half of the input cycle (►push-pull operation). Class B amplifiers are highly efficient but suffer from crossover ►distortion.

class C amplifier A nonlinear ►amplifier in which the output current flows for less than half the input cycle, i.e. the ►angle of flow is less than π. Although more efficient than other types of amplifier, class C amplifiers introduce more ►distortion.

class D amplifier An ►amplifier operating by means of pulse-width modulation (►pulse modulation). The input signal produces a square wave modulated with respect to its ►mark space ratio. ►Push-pull switches are then operated by the modulated square wave so that one switch operates with a high input level and the other with a low input level. The resultant output current is proportional to the mark space ratio and hence to the input current. In theory class D amplifiers are very highly efficient but they require an impractically high speed of switching to avoid distortion.

class E amplifier A switching ►amplifier used for radiofrequency amplification; it employs only one ►transistor, with an output ►resonator to control the frequency of operation and harmonic content of the output. In this arrangement the output capacitance of the transistor can be incorporated into the load resonator network. Theoretical efficiency of 100% can be obtained with an ideal switching behaviour.

class F amplifier An amplifier that is similar to ►class C where the amplifying ►transistor acts as a ►current source, but it operates in a switched manner. An output ►resonator is used to control the frequency of operation and harmonic content of the output.

class G amplifier A ►class B amplifier that uses a low-voltage power supply for small signals, switching dynamically to a higher-voltage supply as larger signals are present.

class H amplifier An amplifier that is similar to ►class G, except that the voltage supply is boosted dynamically rather than switched to a higher-voltage supply.

class S amplifier A device that uses a ►class A amplifier with a limited current output capability, with a ►class B amplifier connected so as to make the load appear to be within the current drive capability of the class A stage.

clear *Syn.* reset. To restore a memory or storage device to a standard state, usually the zero state.

clipping A form of amplitude ►distortion that results in the flattening of the output waveform. Clipping can be caused by improper setting of the ►bias voltage, especially in transistor circuits. When clipping occurs at the output of an amplifier and is due to the input-signal amplitude being too large, the amplifier is said to be *overdriven*.

clock An electronic device that generates periodic signals that are used to synchronize operations in a ►computer or to monitor and measure properties of the circuits involved. The master frequency generated by a clock is the *clock frequency*. The regular pulses applied to the elements of a ►logic circuit to effect logical operations are called *clock pulses*.

The use of clock pulses in order to drive any particular electronic circuit, device, or apparatus is known as *clocking* and the driven circuit, etc., is described as *clocked* or *synchronous*.

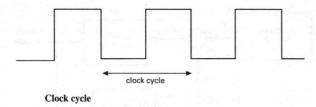

clock cycle

Clock cycle

clock cycle The time taken for the clock signal in a computer to return to its original state and start a new sequence (see diagram).

clocked circuit ►clock.

clocked flip-flop ►flip-flop.

clock frequency ►clock.

clocking ►clock.

clock pulse ►clock.

clock skew ►skew.

closed circuit ►circuit.

closed-circuit television (CCTV) A ►television system, other than broadcast television, that forms a closed circuit between television camera and receiver. Closed-circuit television has many industrial and educational applications, for example in security systems.

closed-loop control system Any ►control system in which the output has an effect upon the input quantity. To improve the performance of a system with respect to maintaining the output quantity as close to the desired quantity as possible, a person is replaced by a mechanical, electrical, or other form of comparison unit. The comparison between the reference input and the feedback signals results in an actuating signal that is the difference between these two quantities (see diagram). The actuating signal maintains the output at the desired value.

closed-loop frequency response A plot of the ►frequency response of a control system that has a ►closed-loop configuration.

closed-loop gain The gain obtained from an amplifier (often an ►operational amplifier) by creating a feedback loop around the amplifier by the use of an external network.

By contrast, the gain of the amplifier without this feedback network is termed the *open-loop gain.* ➤gain-bandwidth product.

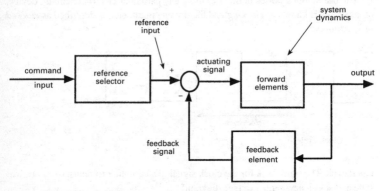

Closed-loop control system

closed-loop voltage gain The ratio of the output voltage to the input voltage of an ➤operational amplifier with a feedback loop created by an external network.

clutter Unwanted signals that are caused by noise, unwanted images, or echos and appear on the display of a ➤radar system.

CLV *Abbrev. for* constant linear velocity. ➤compact disc system.

CMOS ➤MOSFET; complementary transistors.

CMOS logic circuit ➤MOS logic circuit.

CMR *Abbrev. for* common-mode rejection.

CMRR *Abbrev. for* common-mode rejection ratio.

coarse scanning ➤radar; scanning.

coax *Short for* coaxial cable.

coaxial cable A cable formed from two or more coaxial cylindrical conductors insulated from each other. The outermost conductor is often earthed. Coaxial cables are frequently used to transmit high-frequency signals, as in television or radio, since they produce almost no external fields and are not significantly affected by them. A terminal to which a coaxial cable may be connected is a *coaxial terminal.*

coaxial pair ➤pair.

coaxial resonator ➤waveguide.

coaxial terminal ➤coaxial cable.

cochannel rejection *Syn.* capture ratio. A measure of how well a radio receiver can reject unwanted signals that are on the desired frequency channel. The rejection may typically be only a few decibels.

Cockcroft–Walton generator (or **accelerator**) A simple linear ►accelerator producing a potential difference of about 800 kV (d.c.). It consists of cascaded rectifier circuits and capacitors, to which a lower (a.c.) voltage is applied.

code ►digital codes.

code beacon ►beacon.

codec *Short for* coder-decoder. A circuit that is used in ►digital communications systems – in telephone and video systems – to convert an incoming analogue signal into an encoded representation as a digital stream (the coder), and converts an incoming stream of digital signals into an analogue signal (the decoder). At the transmitting end, the analogue signal is ►sampled at a sufficiently high rate for good reproduction. The sampled signal is converted by an ►analogue-to-digital converter and the resulting data stream is transmitted. At the receiving codec, a ►digital-to-analogue converter recreates a close approximation to the original analogue signal.

code distance ►digital codes.

coder ►pulse modulation.

code rate ►digital codes.

coding gain ►digital codes.

coding weight ►digital codes.

coercive force ►magnetic hysteresis.

coherent detection *Syn.* coherent demodulation. ►modulation.

coherent oscillator ►coherent radiation.

coherent radiation Radiation in which the waves are in ►phase both spatially and temporally. A *coherent oscillator* is one that produces very pure well-defined oscillations, as in a ►laser.

coil A conductor or conductors wound in a series of turns. Coils are used to form inductors or the windings of transformers and motors.

coil antenna ►loop antenna.

coil loading ►transmission line.

coincidence circuit A circuit with two or more input terminals that produces an output signal only when an input signal is received by each input either simultaneously or within a specified time interval, Δt. An electronic ►counter incorporating such a circuit records events occurring together and is termed a *coincidence counter*. Such a device, if fed by the outputs of two radiation detectors, may be used to determine the direction of radiation or to detect cosmic-ray showers. ➤anticoincidence circuit.

cold cathode A cathode from which ➤field emission of electrons occurs at ambient temperatures due to the application of a sufficiently high voltage gradient at its surface. An electron tube containing a cold cathode is a *cold-cathode tube.*

cold emission ➤field emission.

collector 1. *Short for* collector region. The region in a ➤bipolar junction transistor into which ➤carriers flow from the ➤base through the collector junction. The electrode attached to this region is the *collector electrode.* ➤➤semiconductor. **2.** *Short for* collector electrode.

collector-current multiplication factor The ratio of enhanced current flow in the collector of a transistor as a result of minority carriers entering from the base with sufficient energy to create extra electron-hole pairs, to the current carried by minority carriers at the collector voltage. This ratio is usually unity but under high field conditions increases rapidly as the ➤avalanche breakdown voltage is reached.

collector efficiency ➤common-base forward-current transfer ratio.

collector electrode ➤collector.

collector region ➤collector.

colour burst *Syn.* burst signal. ➤colour television.

colour cell ➤colour picture tube.

colour code A method of marking electronic parts, such as resistors, with information for the user. The value, tolerance, voltage rating, and any special characteristic of the component may be indicated using coloured bands or dots painted on it. ➤Table 2 in the back matter.

colour coder ➤colour television.

colour decoder ➤colour television.

colour flicker ➤flicker.

colour fringing The presence of unwanted coloured fringes round the edges of objects viewed on the screen of a ➤colour picture tube. Fringing is particularly noticeable when a monochrome signal is received. It is minimized using a colour killer (➤colour television).

colour killer ➤colour television.

colour picture tube A type of ➤cathode-ray tube designed to produce the coloured image in ➤colour television. The coloured image is produced by varying the intensity of excitation of three different ➤phosphors that produce the three primary colours red, green, and blue and reproduce the original colours of the image by an additive colour process.

The three-gun colour picture tube consists of a configuration of three ►electron guns, the *red gun, blue gun,* and *green gun,* that are tilted slightly so that the electron beams intersect just in front of the screen. Each electron beam has an individual electron lens system of focusing and is directed towards one of the three sets of colour phosphors. There are several different types of colour picture tube, the main differences being in the configuration of electron guns and arrangement of the phosphors on the screen.

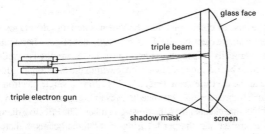

a Colourtron

One main type is the *dot matrix tube,* an example of which is the *colourtron.* It has a triangular arrangement of electron guns and has the phosphors arranged as triangular sets of coloured dots (Fig. *a*). A metal *shadow mask* is placed directly behind the screen, in the plane of intersection of the electron beams, to ensure that each beam hits the correct phosphor (Fig. *b*). The mask acts as a physical barrier to the beams as they progress from one location to the next and minimizes the generation of spurious colours by excitation of the wrong phosphor.

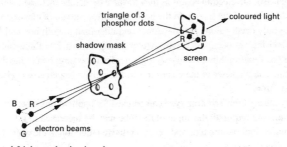

b Light production in colourtron

The other main type of three-gun colour picture tube is the *slot matrix tube,* which has the electron guns arranged in a horizontal line. The phosphors are arranged as vertical stripes on the screen (Fig. *c*) and the shadow mask is replaced by an *aperture grille*

of vertical wires. This type has advantages in focusing the beams but has a smaller field
of view than the triangular arrangement of electron guns.

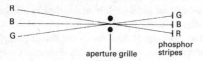

c Slot matrix tube: horizontal arrangement of electron guns

The *Trinitron* is a type of colour picture tube that has certain advantages over
three-gun tubes. It has a single electron gun with three cathodes aligned horizontally,
an aperture grille, and vertically striped phosphors. The cathodes are tilted towards the
centre so that the electron beams intersect twice, once within the electron lens focusing system and once at the aperture grille (Fig. *d*). This allows a single electron lens
system to be used for all three beams, thus needing fewer components. The system
is therefore much lighter and cheaper than the three-gun tubes. The effective diameter
of the electron lens is greater and sharper focusing of the three beams is therefore
possible.

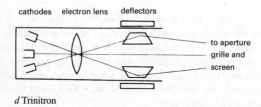

d Trinitron

Misconvergence of the electron beams as they traverse the screen increases with the
distance from the centre of the screen. In the horizontal arrangement of electron guns
misconvergence only occurs along the line direction rather than in both line and field
directions (►television) as occurs in the triangular configuration. The three-cathode
arrangement of the Trinitron however allows a greater lens aperture than in the three-
gun arrangement. The diameter of the electron tube for a given screen size is also reduced in the Trinitron.

Cathode-ray tubes of the Trinitron type can be used for applications where multiple electron beams are required; the angle of the tube may be increased to give a relatively wide-angle colour picture tube and hence a relative reduction in the overall size
of colour television receivers.

The colour quality and definition of the picture on the screen of a picture tube depends greatly on the dynamic convergence of the beams and the size of the *colour cell*.
Scanning of the three electron beams across the screen is effected by a system of deflection coils to which ►sawtooth waveforms are applied in synchronism with the transmitter, the flyback signal being blanked. Extra *convergence coils* are frequently used

to ensure the correct convergence of the beams at the shadow mask or aperture grille. A system of *dynamic focusing* is also used in which the voltage applied to the convergence coils is varied automatically according to the relative position of the spots on the screen; this minimizes misconvergence.

The size of the colour cell is the smallest area on the screen that includes a complete set of the three primary colours. A smaller colour cell is available with horizontally aligned electron guns or cathodes than is possible in the triangular configuration.

colour saturation control ➤television receiver.

colour separation overlay (CSO) A technique used in colour television for superimposing part of one scene on another. When a particular colour, such as blue, occurs in one scene viewed by a camera, the output of another camera filming a different scene is automatically switched in to replace the areas of the chosen colour in the original picture. All other colours are transmitted normally from the first camera. The technique is widely used for achieving special effects.

colour television A television system that produces a coloured image on the screen of a ➤colour picture tube. An additive colour reproduction process is used on the screen whereby three primary colours – red, green, and blue – are combined by eye to produce a wide variety of colours. The apparent colour of the image depends on the relative intensities of the three primary colours and a properly adjusted colour ➤television receiver approximates the original colours of the transmitted scene.

Three separate video signals are produced by a colour ➤television camera. These signals are used to produce a composite signal that is broadcast and is received by a colour receiver. The receiver extracts the original video information from the composite signal and modulates the intensities of the three electron beams of the colour picture tube in order to excite the appropriate red, green, or blue phosphors on the screen.

The composite signal transmitted in colour television needs to be compatible with black and white (monochrome) receivers. It is therefore composed of two parts: the *luminance signal* and the *chrominance signal*. The luminance signal contains brightness information. It is obtained by combining the outputs of the three colour channels and is used for amplitude modulation of the main picture carrier frequency. This produces the black and white image. The colour information is contained in the chrominance signal, which is transmitted using a subcarrier wave at a frequency chosen to cause the least interference on a monochrome set. The chrominance signal is obtained by combining, in a *colour coder* circuit, fixed specified fractions of the separate video signals into sum and difference signals. Two ➤quadrature components of the chrominance signal are produced and used for amplitude modulation of the *chrominance subcarriers*. The subcarriers are suppressed at transmission. The original information is extracted from the chrominance signal in the *colour decoder* in the receiver. The *frequency overlap* is the range of transmitted frequencies that are common to both the luminance and chrominance channels.

The *composite colour signal* contains the luminance and chrominance signals; it also contains synchronizing pulses for line and field scans as well as a *colour burst* signal. The colour burst establishes a phase and amplitude reference signal that is used to demodulate the chrominance signal. In colour receivers the chrominance circuits are disabled by the *colour killer* when a black and white signal is being received. This ensures that only luminance information reaches the tube and prevents colour fringing on the image.

➤PAL; SECAM; NTSC; television; colour picture tube; television camera; camera tube.

colourtron ➤colour picture tube.

Colpitt's oscillator ➤oscillator.

coma ➤distortion.

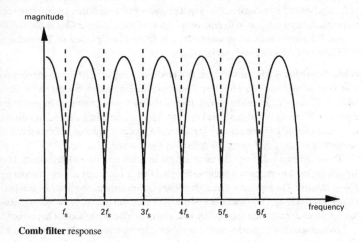

Comb filter response

comb filter A type of ➤notch filter in which the nulls occur periodically across the frequency band, much like an ordinary comb has periodically spaced teeth (see diagram).

combinational logic A ➤logic circuit whose output at a specified time is a function only of the inputs at that time and does not depend on the history of inputs to the circuit. Hence it is a circuit without any ➤memory components or any ➤feedback loops. ➤sequential circuit.

combination-tone distortion *Syn. for* intermodulation distortion. ➤distortion.

common-anode connection ➤common-collector connection.

common-base connection *Syn.* grounded-base connection. A method of operating a ➤bipolar junction transistor in which the ➤base is common to both the input and output circuits and is usually earthed (see diagram). The ➤emitter is used as the input ter-

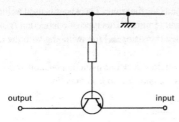

Common-base connection

minal and the ➤collector as the output terminal. This type of connection is commonly used as a voltage amplifier stage.

The equivalent connection for a ➤field-effect transistor is *common-gate connection*.

common-base current gain ➤common-base forward-current transfer ratio.

common-base forward-current transfer ratio *Syns.* collector efficiency; alpha current factor; common-base current gain. Symbol: α. The ratio of the collector current I_c to the emitter current I_e of a bipolar junction transistor operated in ➤common-base connection.

common branch *Syn. for* mutual branch. ➤network.

common-cathode connection ➤common-emitter connection.

common-collector connection *Syn.* grounded-collector connection. A method of operating a ➤bipolar junction transistor in which the ➤collector is common to both the input and output circuits and is usually earthed (see diagram). The ➤base is used as the input terminal and the ➤emitter as the output terminal. This type of connection is used for the ➤emitter follower.

The equivalent connection for a ➤field-effect transistor is *common-drain connection* and for a ➤thermionic valve is *common-anode connection*.

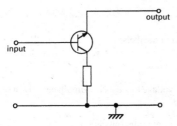

Common-collector connection

common-drain connection ➤common-collector connection.

common-emitter connection *Syn.* grounded-emitter connection. A method of operating a ➤bipolar junction transistor in which the ➤emitter is common to both the input and output circuits and is usually earthed (see diagram). The ►base is used as the input terminal and the ►collector as the output terminal. This type of connection is used for power amplification with a nonsaturated transistor and for switching with the transistor in saturation.

The equivalent connection for a ►field-effect transistor is *common-source connection* and for a ►thermionic valve is *common-cathode connection.*

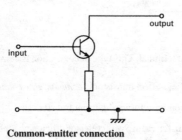

Common-emitter connection

common-emitter forward-current transfer ratio ►beta current gain factor.

common-gate connection ►common-base connection.

common-impedance coupling ►coupling.

common-mode chokes ►electromagnetic compatibility.

common-mode currents ►electromagnetic compatibility.

common-mode rejection (CMR) The ability of a ►differential amplifier to reject, suppress, or zero-out ►common-mode signals.

common-mode rejection ratio (CMRR) A figure of merit of a ►differential amplifier, defined as the ratio of the magnitude of the differential mode gain A_d to the magnitude of the common-mode gain A_{cm}:

$$CMMR = |A_d|/|A_{cm}|$$

The value of the CMRR is often given in decibels:

$$CMRR = 20 \log_{10}(|A_d|/|A_{cm}|)$$

common-mode signal Any d.c. or a.c. voltage that appears simultaneously in two other separate signals.

common-source connection ►common-emitter connection.

communication ►telecommunication; digital communications.

communications satellite An artificial unmanned satellite in earth orbit that provides high-capacity communication links between widely separated locations on earth. International telephone services and the relaying of television programmes are achieved by transmitting microwave signals, suitably modulated, from a ground station to an orbiting satellite and back to another ground location. Transmission of digital data over satellite links is also now of major importance.

The first satellites simply reflected or scattered the microwave beam back to another ground station. In present-day systems the signal is amplified and its frequency changed by a ➤transponder before it is retransmitted to earth.

Transmissions by communications satellites must lie within the ➤radio window, and various frequency bands are allocated for this purpose. One widely used frequency band is located at about 6 GHz for upward or *uplink* transmission and about 4 GHz for downward or *downlink* transmission; it is 500 MHz wide and is divided into repeater channels of various bandwidths. A band at 14 GHz (uplink) and 11 or 12 GHz (downlink) is also much in use, mainly for fixed ground stations. Another band located at approximately 1.5 GMz, bandwidth 80 MHz, is used with small mobile ground stations.

There are now hundreds of communications satellites: these are usually either ➤geostationary earth orbit satellites (GEOS), which move in a very high (36 km) equatorial orbit and appear stationary with respect to points on the earth's surface, or ➤low earth orbit satellites (LEOS). There is growing use of ➤digital communications.

➤Solar cells form the primary power supply for satellites with a back-up of batteries for use during the brief periods of solar eclipses. The operational lifetime of a modern satellite should be at least five years. Redundancy of all the essential subsystems is required although it is not necessary for all the components when there is a large number present, as with the transponders.

A geostationary satellite can be maintained in a stable attitude by spinning it about an axis parallel to the earth's axis. The high-gain antennas may be mounted on a platform that rotates about the spin axis but in the opposite direction. The antennas then appear stationary with respect to the earth, at their desired orientation. Parabolic reflectors allow spot-beam transmission to regions of limited size, such as W Europe, which have high communication traffic densities. Transmission and reception over a hemisphere is effected using conical horn antennas.

The ground stations must be situated some distance from terrestrial microwave relay systems to avoid radio interference. The antennas are often large ➤parabolic reflectors, or dishes; those used in the Intelsat system, such as at Goonhilly in Cornwall, have apertures of 25 to 30 metres. The antennas should be steerable to compensate for perturbations of the orbits caused by gravitational effects of moon and sun. Television programmes can also be relayed to homes equipped with small nonsteerable dish antennas or to cable TV systems, while businesses can use VSATs (very small aperture antennas) for digital data communications.

Present-day systems provide simultaneous *multiple access* to one satellite from a large number of ground stations within one coverage zone. This is achieved, for example, by time-division multiple access (TDMA) or frequency-division multiple access (FDMA) (➤digital communications); in the former case, a ground station does not share the transponder power with other stations and can operate at close to saturation where the transponder is most efficient.

commutation The transfer of current from one path in a circuit to a different one. The transfer is often done in a periodic and automatic manner.

commutation switch A ➤switch that automatically switches from open to closed and vice versa repetitively. It is most often used for ➤pulse modulation.

commutator ➤motor.

compact disc system (CD system) A high-quality sound-reproduction system that uses light to detect audio signals encoded by *digital recording* on a 120 mm metal *compact disc* (CD). A CD system differs from other sound-reproduction systems in that there is no physical contact between the pick-up and the recording, thus minimizing wear; the information layer is buried below the surface of the disc, which minimizes errors in the sound reproduction due to dust or other marks on the surface. The original analogue audio information is sampled and quantized using an analogue-to-digital converter (➤sampling; quantization), and the coded digital data is then recorded. The audio signals are encoded in the form of a spiral track of minute pits that are impressed into one surface of the disc at the time of manufacture; the narrow track spirals outwards from the centre of the CD. For playback, the disc is loaded into a *CD player* and secured to a spindle on which it can be rotated. The system operates by using *CLV*, i.e. a constant linear velocity of track relative to pick-up so the rotational speed is a function of the radius of the track, and varies as the pick-up moves across the disc.

The essential component of the pick-up is a small low-power ➤semiconductor laser continuously emitting coherent light, which is focused as a small spot onto the reflecting surface of the disc (Fig. *a*). Reflected light from the disc is modulated by the

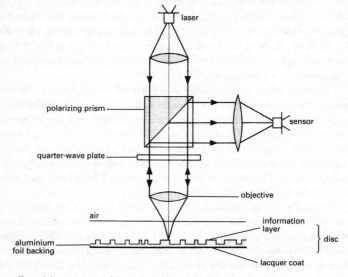

a Essential components of the optical pick-up of a **compact disc system**

code impressed on the track, and is then detected by a ➤phototransistor to produce an electric signal corresponding to the recorded information. These signals are then converted back into audio signals. In order to maintain high-quality sound output very sophisticated error control systems are required to ensure that focusing and tracking integrity are maintained, and that the disc is rotated at the correct speed.

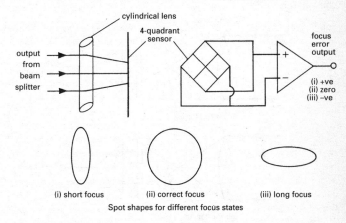

(i) short focus   (ii) correct focus   (iii) long focus

Spot shapes for different focus states

b Cylindrical lens method

The resolution of the pick-up depends critically on the spot size: disc warp and irregularities in the thickness will cause an out of focus condition with resulting loss of sound quality and ➤crosstalk from adjacent tracks. A *focus servo system* is used to move the lens along the optical axis to keep the spot in focus. The *focus error signal* is generated using one of two main methods. The cylindrical lens method (Fig. *b*) has a cylindrical lens placed between the beam splitter and the photodetector. The image reaching the sensor will be circular only when the focus is correct, otherwise it becomes an ellipse whose aspect ratio changes as a function of the state of focus. The sensor is split into four quadrants, connected as shown. The focus error signal is generated from the difference between the outputs, and the data signal is the sum of all four outputs. The knife edge and dual prism methods are the second means of generating the focus error signal. They also require split sensors, mounted beyond the focal point. In the knife edge method (Fig. *c*) a knife edge is positioned at the point of correct focus and the outputs of the two sensors compared to produce the focus error signal. The dual prism method is essentially similar but replaces the knife edge with a dual prism and three sensors.

Accurate tracking of the beam is also required, and a *track-following servo system* is used to keep the spot centralized on the track. Tracking errors arise from various sources: the track separation is smaller than the accuracy to which either the player spindle or the central hole in the disc can be manufactured; a warped disc will be tilted relative to the beam at the surface and the apparent position of the track relative to the pick-up will constantly change as the disc rotates; external forces outside the CD player can induce vibrations that tend to disturb the tracking.

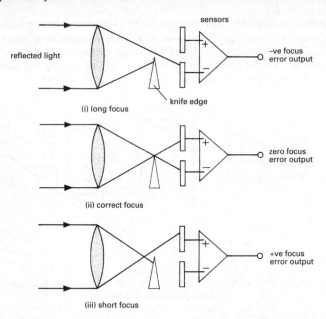

c Knife edge method

During the recording process, the audio signal is sampled at a rate of 44.1 kHz and a form of ►pulse code modulation is used to convert the samples into a coded form which modulates a high-frequency clock pulse. The clock operates at 4.3218 MHz. The majority of compact disc systems use *eight to fourteen modulation* (*EFM*) in which any combination of eight real data bits is uniquely described by a pattern of 14 *channel bits*. A further three *packing bits* are interposed between each pattern to separate them. The digital modulation code produced is known as the *channel code*. The transitions between ones and zeros of the channel code are used to produce bump edges in the surface of a master disc. The bumps are translated into pits when the CD impressions are manufactured. The edges are detected by the optical system in the CD player to produce corresponding transitions in the replay signal. The replay signal must then be accurately decoded to produce the audio output.

The first step in the process is to compare the detected signal with a reference voltage. This process is termed *slicing*. This recreates the binary channel code. A phase-locked loop running at the clock frequency counts the number of clock pulses between transitions and recreates the patterns of 14 channel bits. These are decoded back to data bytes using a ROM or array of gates. The data bytes are then fed through a digital-to-analogue converter to recreate the original audio signals. The packing bits are used to determine the start position of each 14 bit pattern, and a regular synchronizing pattern is also added to lock the readout circuits to the symbol boundaries.

The actual layout of the coded data on the disc is much more complex than a simple sequential layout in order to allow for errors in the readout data to be corrected, and to reduce noise due to contamination and surface scratches. The audio data is coded into data blocks or frames of 33 patterns, each following a synchronization pattern. The audio samples use 24 of the 33 bytes, and 8 of the bytes are redundancy bytes forming the basis of the error correction system; the first byte of each data block is used as a subcode to produce a running time display. The sampling rate of 44.1 kHz, producing 16-bit words in left and right channels, results in 176.4 Kbytes of audio data per second.

The error correction system has to deal with errors due to scratches, which can affect many bits of code (*burst errors*), and random errors caused by imperfect pressing of a bump edge. This latter can result in the conversion of one pattern into another and therefore up to 8 bits of data can be in error. Extra bits are added to the coded information (*redundancy*) and can be used to correct damaged data bits when playing back. A *code word* consists of the total of data and redundant bits. The value of the redundant bits is calculated from the data itself according to a code known as the *Reed–Solomon code*. The resulting code words are then interleaved within each data frame to reduce the effect of burst errors; a large error causes slight damage to many code words rather than severe damage to one. This system is known as *cross-interleave Reed–Solomon code* (CIRC). The Reed–Solomon codes used to correct errors are a very powerful tool. Error correction is necessary because the effect of digital errors results in a sound rather like vehicle ignition interference in radio reception and is unacceptable to a listener. The development of compact disc systems has always taken into account that the finished product should not require any particular handling problems. Where data corruption is so bad that the error correction cannot cope, the system contains muting circuits that operate to reduce the gain of the CD player.

compandor ▸volume compressor; amplitude compandoring.

comparator A circuit, such as a ▸differential amplifier, that compares two inputs and produces an output that is a function of the result of the comparison.

compensated semiconductor A semiconductor with properties similar to an intrinsic ▸semiconductor, produced using ▸doping compensation.

compensating leads An extra pair of leads that are used in conjunction with a sensitive measuring instrument connected to a ▸bridge circuit. The leads are identical to those that connect the measuring device, such as a ▸resistance thermometer, to the bridge and run alongside them. They are shorted at the instrument end and connected in series with the balancing element of the bridge circuit. Any resistance in the working leads is thus balanced by the compensating leads and resistance changes that occur in the working leads due to ambient conditions are also produced in the compensating leads.

compensation *Syn.* stabilization. ▸compensator.

compensation network A network that is added into a ▸control system in order to provide cascade and/or feedback compensation to stabilize a system (see diagram). ▸▸compensator.

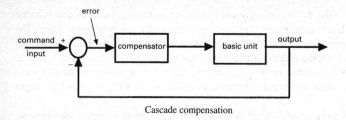

Cascade compensation

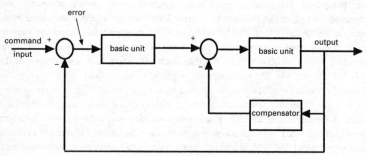

Feedback compensation

Compensation networks

compensator Additional equipment introduced into a system to reshape its ►characteristics in order to improve system performance. The process is called *compensation* or *stabilization*. When the system is compensated it is stable, has a satisfactory ►transient response, and has a large enough ►gain to ensure that the steady-state error does not exceed the specified maximum.

compiler ►programming language.

complementary transistors A pair of transistors of opposite type used together. A pair of n-p-n and p-n-p ►bipolar transistors can be used to achieve ►push-pull operation.

Complementary ►MOSFETs – PMOS and NMOS – can be combined in *complementary MOS (CMOS)* technology. CMOS transistors are used for low-dissipation logic circuits. Such devices are used in conjunction with other similar stages all having essentially capacitive input impedance and therefore zero d.c. current flow. In the basic inverter (see diagram) the p-channel device conducts when the input is low and the n-channel device conducts when the input is high, thus giving an output that is inverted.

complex frequency ►Laplace transform. ►►s-domain circuit analysis.

complex impedance ►impedance.

complex operator ►s-domain circuit analysis.

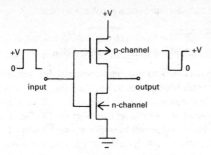

Complementary transistors: CMOS inverter circuit

complex plane A plane described by two orthogonal axes, one axis representing a scale of real numbers and the other axis representing a scale of imaginary numbers. A complex number $a + jb$ (where $j = \sqrt{-1}$) can be geometrically interpreted as a point on the complex plane. A diagram representing the complex plane is often called an *Argand diagram* (see diagram).

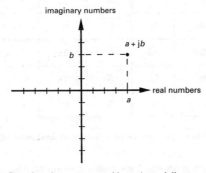

Complex plane represented in an Argand diagram

component A discrete packaged electronic element, such as a resistor, that performs one electrical function. Primarily active electronic elements are usually termed ►devices and passive ones are ►component parts.

component part 1. A discrete electronic element, usually passive. **2.** An independent physical body that contains the realization of an electronic property or properties and cannot be practicably reduced or subdivided without destroying its function.

composite cable ►cable.

composite colour signal ►colour television.

composite conductor A conductor formed from strands of different metals, such as copper and steel.

composition resistor Usually a carbon-composition resistor. ➤resistor.

compound magnet ➤permanent magnet.

compound modulation *Syn. for* multiple modulation. ➤modulation.

compound semiconductor A crystalline ➤semiconductor material that is made from a chemical compound using elements from groups III and V or groups IIB and VI of the ➤periodic table, such that the compounds are ➤isoelectronic with the elemental semiconductors, for example silicon and germanium from group IV. The diagram shows a section of the periodic table with elements used in compound semiconductors; elemental semiconductors are italicized, and elements used in common compound semiconductors are highlighted in bold type. The resulting compounds have similar semiconducting properties to silicon, though some important practical differences arise that enable the use of these materials in the manufacture of special electronic devices and integrated circuits.

IIB	III	IV	V	VI
	B	C	N	O
	Al	*Si*	P	S
Zn	Ga	*Ge*	As	Se
Cd	In	Sn	Sb	Te
Hg	Tl	Pb	Bi	Po

Section of the periodic table showing elements used in **compound semiconductors**

Compound semiconductors are generally characterized by a ➤direct energy gap, which permits radiative transitions between ➤conduction and ➤valence bands to occur, resulting in the emission of ➤electromagnetic radiation. The wavelength of the radiation is related to the ➤energy band gap, and compound semiconductors can be alloyed to produce the appropriate band gaps for emission of visible light, or infrared radiation for optical communications. These materials are therefore used for ➤light-emitting diodes (LEDs) and ➤semiconductor lasers.

Common *III-V compound semiconductors* include ➤gallium arsenide, GaAs, which emits infrared light at about 890 nm, and when alloyed with other III-V semiconductors is used for semiconductor lasers emitting in the range 1000–1550 nm. It can be alloyed with gallium phosphide, GaP, and gallium nitride, GaN, to make green, yellow, and red LEDs. *II-VI compound semiconductors* such as zinc selenide, ZnSe, are used as the basis of blue LEDs. Cadmium mercury telluride, CdHgTe or CMT, has a response in the infrared and is used for thermal imaging devices.

Gallium arsenide and indium phosphide, InP, are also used in microwave devices and ➤integrated circuits, including the ➤transferred electron device, ➤high electron mobility transistor, and ➤heterojunction bipolar transistor.

compressed video Video-frequency signals that are transmitted with considerably lower data rate than would be expected for the frequency. In the transmission of TV signals, for example, standard coding of broadcast-quality signals requires a data rate of up to 90 megabits per second. Compressed versions of the same signal require data rates of around 64 kilobits per second.

compression ➤data compression.

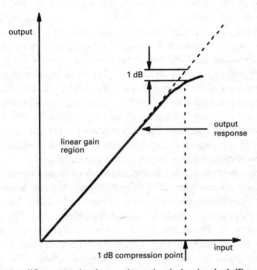

Amplifier output signal versus input signal, showing the 1 dB **compression point**

compression point In an ➤amplifier, the ➤gain for small signals is generally not maintained for large input signals but is reduced; this is called *gain compression*. This nonlinearity is quantified as the input signal level for which the output signal is one decibel (dB) below the linear gain value; this is the 1 dB compression point (see diagram).

compressor *Short for* volume compressor.

computer Any automatic device for the processing of information received in a prescribed and acceptable form according to a set of instructions. The instructions and information are stored in ➤memory. The most widely used and most versatile of these devices is the *digital computer,* which can manipulate large amounts of information at high speed. Its input must be discrete rather than continuous and may consist of combinations of numbers, characters, and special symbols. The instructions – the

➤program – are written in an appropriate ➤programming language. The information is represented internally in binary form.

The development of microelectronics has allowed the corresponding development of a wide range of computers varying in size and complexity according to the required applications. Modern computers range from the *microcomputer* that contains typically a few million ➤logic circuits and a few million ➤words of memory to very large *mainframe computers* typically containing millions of logic circuits and several million words of memory. Such devices are capable of performing millions of operations per second and can serve many users at the same time. Constant improvements in ➤packing densities and subsequent miniaturization of circuits, coupled with improvements in speed of operation of the logic circuits is resulting in ever more powerful microcomputers and dramatic reductions in the physical size of mainframe computers.

Most computer systems consist of three basic elements: the ➤central processing unit (CPU), the main ➤memory, and peripheral devices involved with ➤input/output and permanent storage of information. The CPU controls the operation of the system and performs arithmetic and logic operations on the data. The main memory stores the program and the data in units of ➤bytes or words, each of which has a unique ➤address, so that they may be retrieved quickly by the CPU. A ➤cache memory is employed by high-performance systems: it interacts directly with the CPU and transfers information at extremely high speed. The information currently in active use is held in the cache. A complete computer system consists of the ➤hardware – the electronic and other devices – and complementary ➤software – the set of programs and data.

The *analogue computer* is a device that accepts data as a continuously varying quantity rather than as a set of discrete items required by the digital computer. It is used in scientific experiments, simulation processes, and in the control of industrial processes where a constantly varying quantity can be monitored. A problem is solved by physical analogy, usually electrical. The magnitudes of the variables in an equation are represented by voltages fed to circuit elements connected in such a way that the input voltages interact according to the same equation as the original variables. The output voltage is then proportional to the numerical solution of the problem. It can solve or analyse many types of differential equations.

computer-aided design (CAD) The application of computers to the design and analysis of a product. A CAD program can convert a designer's rough sketches into a more finished form, using input devices such as mouse and light pen.

computer architecture The design of computers and the study of their organization, including instruction-set architecture and hardware-system architecture.

concatenated codes ➤digital codes.

concentration ➤concentrator.

concentration cell A type of ➤cell in which the electrodes are made from the same metal and are immersed in two different concentrations of one salt of the metal. The e.m.f. produced by the metal dissolving in the weaker solution and being deposited by the stronger is dependent upon the substances used and their concentrations.

concentrator In digital communication systems, a circuit or system that, through a process of coding and/or multiplexing, enables more information to be transmitted through a communication channel. The use of such coding and/or multiplexing is known as *concentration*.

condenser *Obsolete syn. for* capacitor.

conditional branch ►branch instruction.

conductance Symbol: G; unit: siemens. The real part of the ►admittance, Y, which is given by

$$Y = G + jB$$

where B is the ►susceptance and $j = \sqrt{-1}$. In d.c. circuits the conductance is the reciprocal of the resistance. The d.c. conductance of a sample of material of ►active area A and length l, and that has a ►conductivity σ, is given by

$$G = \sigma l/A$$

conducted emissions ►electromagnetic compatibility.

conducted interference ►electromagnetic compatibility.

conducted susceptibility ►electromagnetic compatibility.

conduction The transmission of electric (or heat or acoustic) energy through a substance that does not itself move. In electrical conductors, such as metals, it entails the migration of ►electrons; in gases and solids it results from the migration of ►ions. ►hole conduction.

conduction band The band of energies of electrons in a solid in which the electrons can move freely under the influence of an electric field. In metals the conduction band is the highest occupied band. In ►semiconductors it is a vacant band of higher energy than the ►valence band. ►energy bands.

conduction current ►current.

conduction electrons Electrons in the ►conduction band of a solid. ►energy bands.

conductivity Symbol: σ; unit: siemens per metre. A property that describes how easily electric current can flow in a material. It is the product of the charge density and the particle mobility μ:

$$\sigma = ne\mu$$

where the charge density (ne) is the product of the number density of electrons or holes and the charge on each carrier. In a semiconductor both electrons and holes contribute to the overall conductivity. Conductivity is also the ratio of current density J to electric field E in a material:

$$\sigma = J/E$$

which is an expression of ➤Ohm's law. Conductivity is the reciprocal of ➤resistivity. ➤➤conductance.

conductor A material that offers a low resistance to the passage of electric current: when a potential difference is applied across it a relatively large ➤current flows.

cone loudspeaker ➤loudspeaker.

confusion reflector A device used to produce false signals with ➤radar. Strips of paper or metal foil, known as *chaff*, may be used: long strips of metal foil are called *rope* or *window*.

conical scanning ➤scanning.

conjugate branches ➤network.

conjugate impedances Two ➤impedances given by

$$Z_1 = R + jX$$

$$Z_2 = R - jX$$

in which the resistance components, R, are equal and the reactance components, X, are equal in magnitude but opposite in sign.

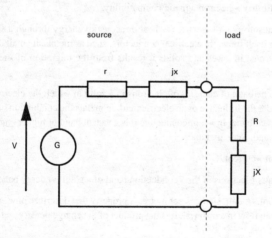

Conjugate matching

conjugate matching A technique used to ensure that the maximum power from a generator of electrical energy is dissipated in the ➤load impedance. The diagram shows a generator with a complex source impedance $(r + jx)$ feeding a load impedance $(R + jX)$. Maximum power is dissipated in the load when

$$R = r \quad \text{and} \quad R + jX = r - jx$$

such that the reactances cancel. Under these conditions the load impedance is said to be the ➤conjugate impedance of the complex source impedance and the circuit is said to be *conjugate matched*.

constantan An alloy of copper (50–60%) and nickel that has a very low ➤temperature coefficient of resistance and a comparatively high resistance. It is used with copper, silver, etc., to form ➤thermocouples and also for precision wire-wound ➤resistors.

constant-current source A circuit that ideally has an infinitely high output ➤impedance so that the output current is independent of voltage. In practice sufficiently high output impedances are only achieved for a limited range of output voltages.

constant failure-rate period ➤failure rate.

constant-R network ➤two-port network.

constant-voltage source A source of voltage that produces a substantially constant value of voltage independently of the current supplied by it. An ideal voltage source has an internal impedance of zero.

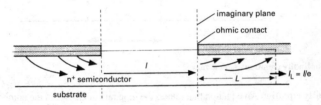

Contact resistance and transfer length

contact The bringing together of two conductors so that current may flow. The resistance at the interface between the conductors is the *contact resistance*. In the case of an ➤ohmic contact between a metal and a semiconductor, the *specific contact resistance* is the contact resistance of a unit area for current flow perpendicular to the contact. In the case of a planar configuration where the final direction of current flow is parallel to the plane of the metal (see diagram), the contact resistance is defined to be the resistance between the metal and an imaginary plane at the edge of and perpendicular to the metal, and the *transfer length*, L, is the distance from the edge of the metal at which the current in the semiconductor has fallen to $1/e$ of its original value.

   If the conductors are made from two different materials a difference of potential will arise when they are placed in contact. This *contact potential* results from a difference in the ➤work functions of the two materials and is usually of the order of a few tenths of a volt. If the contact is made between two semiconductors of different polarities or between a metal and a semiconductor, a built-in field will be produced with an associated contact resistance to current flow. ➤➤p-n junction; metal-semiconductor contact.

contact lithography ➤photolithography.

contact noise ➤noise.

contactor A type of switch used for the automatic making and breaking of a circuit and designed for frequent use.

contact potential ➤contact.

contact resistance ➤contact.

content-addressable memory (CAM) *Syn. for* associative memory.

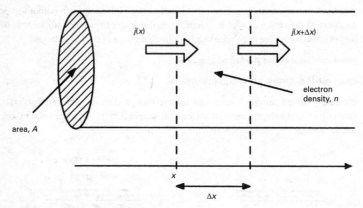

Continuity equation: particle flow

continuity equation An equation that relates the time rate of change in the number of particles in a given volume to the inflow and outflow of particles. In electronics, this usually refers to the continuity of electrons or holes in a ➤semiconductor in which current is flowing. When considering the flow of electrons in a piece of semiconductor, the electron continuity can be determined by considering the flow of electrons through an elemental slice of the semiconductor (see diagram). The rate of change of the number of electrons in the elemental volume of width Δx is equal to the number of electrons flowing in minus the number flowing out, plus the number generated per unit time minus the number that recombine (with holes) per unit time:

$$dN/dt = (\text{inflow} - \text{outflow}) + (\text{generation rate} - \text{recombination rate})$$

where N is the number of electrons. Considering the diagram, this can be written as

$$(dn/dt)A.\Delta x = j(x).A - j(x + \Delta x).A + G.A.\Delta x - R.A.\Delta x$$

where n, the number density of electrons, is multiplied by the elemental volume ($A.\Delta x$) to obtain the actual number; $j(x)$ is the flux of electrons across the plane at x, and G and R are generation and recombination rates of electrons per unit volume, respectively.

If the width of the element, Δx, is small enough, then using the definition of differentiation:

$$dy/dx = [y(x + \Delta x) - y(x)]/\Delta x,$$

the flow terms in j can be replaced, yielding a continuity equation:

$$dn/dt = dj/dx + G - R$$

Normally in electronics, the flux term is expressed as a ➤current density, J, which can be measured. Thermal generation and recombination are processes that are continually occurring in a semiconductor, and are balanced in equilibrium. It is more usual to consider the recombination of excess charge carriers in a semiconductor, or how the carriers in the semiconductor progress from nonequilibrium conditions back to thermal equilibrium. Thus, cancelling the thermal components from G and R, and substituting for current density,

$$dn/dt = (1/e)(dJ/dx) - R_{\text{excess}}$$

which for electrons can be written

$$dn/dt = \mu E(dn/dx) + D(d^2n/dx^2) - (n - n_0)/\tau$$

The first term is derived from ➤drift current, where μ is the electron ➤drift mobility and E is the ➤electric field; the second term is derived from ➤diffusion current, where D is the electron diffusivity; n_0 is the equilibrium electron density and $1/\tau$ is the recombination rate for electrons. A similar expression can be written for holes.

continuous duty ➤duty.

continuous loading ➤transmission line.

continuous wave A single-frequency oscillating signal used in the testing of electronic circuits and products or as the carrier in a communications system.

continuous-wave radar ➤radar.

contrast control ➤television receiver.

control The mechanisms by which a ➤control system exhibits authority on a physical system.

control bus A special-purpose computer ➤bus that carries control information, such as ➤chip enable, write enable, and output strobe, rather than data or addresses.

control electrode An electrode to which a signal is applied in order to produce changes in the currents of one or more of the other electrodes. In a bipolar transistor with ➤common-emitter connection the base electrode is the control electrode; the gate electrode is the control electrode of a ➤field-effect transistor. In a ➤cathode-ray tube it is the ➤modulator electrode. In a ➤thermionic valve it is the *control grid*.

control grid ➤control electrode; thermionic valve.

controlled-carrier modulation ➤floating-carrier modulation.

controlled sources ➤dependent sources.

control system A system that can be classified as either a *process control system* or a *servomechanism*. In a process control system the controlled variable, or output, must be held as close as possible to a usually constant desired value, or input, despite any disturbances. With a servomechanism the input varies and the output must be made to follow it as closely as possible.

control unit ➤central processing unit.

convergence In a multibeam electron tube, such as a ➤colour picture tube, the intersection of the beams at a specified point. Convergence may be achieved electrostatically using a *convergence electrode* or electromagnetically using a *convergence magnet*. When scanning of the beams across the screen of the tube is carried out, the surface generated by the point of intersection of two or more of the electron beams is the *convergence surface*.

convergence coils ➤colour picture tube.

conversion gain ratio ➤frequency changer.

conversion transducer ➤frequency changer.

conversion voltage gain ➤frequency changer.

converter 1. A device for converting alternating current to direct current or vice versa. **2.** A device that changes the frequency of a signal; a ➤frequency changer. **3.** A device, such as an impedance converter, that has different electrical properties at its input and output and may be used to couple dissimilar circuits. **4.** A device, such as a compiler, that changes an information code. **5.** A ➤transducer that converts energy of one type, such as sound waves or electromagnetic radiation, into electrical energy.

convolution A mathematical method of analysing the response of a linear system to any input function. If an input $x(t)$ to a linear system is split up into a succession of rectangular pulses of width $\Delta\tau$ so that the area of a typical pulse at $t = \tau$ is $x(\tau)\Delta\tau$ (see diagram) then this pulse will give rise to a response

$$[x(\tau)\Delta\tau] \, h(t - \tau)$$

The total response $y(t)$ can be expressed as an integral, known as the *convolution integral*, given by

$$y(t) = \int_{-\infty}^{+\infty} x(\tau) \, h(t - \tau) \, d\tau$$

The physical interpretation of the integral can be expressed as follows: the value of the output at a given time t is the integrated effect of the values of the input at all previous times t. This combining of two functions of the same variable is known as convolution and is expressed by the special symbol $\otimes$. Therefore the general expression which defines $\otimes$, known as the *convolution operator*, is given by

$$a(t) \otimes b(t) = \int_{-\infty}^{+\infty} a(\tau) \, b(t - \tau) \, d\tau$$

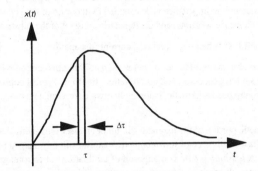

Convolution

convolution integral ➤convolution.

Cooper pair ➤superconductivity.

coplanar process A technique used during the manufacture of LSI ➤MOS integrated circuits and some forms of LSI ➤bipolar integrated circuits, such as I²L. Regions of relatively thick silicon dioxide are used in order to isolate device areas. The coplanar process was developed in order to minimize the vertical projection of the oxide layer. A layer of silicon nitride is deposited on the surface of the silicon wafer and is etched to expose the regions of the surface where thick oxide is required. As oxidation takes place, the effective silicon surface moves downwards and is replaced by a thicker layer of silicon dioxide, so that approximately one third of the oxide is below the original

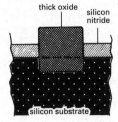

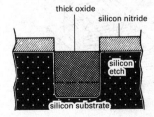

a **Coplanar process** with nitride etch   *b* Nitride plus silicon etch exposed surface level (Fig. *a*). Oxidation may be preceded by etching of the exposed silicon surface so that the final oxide surface is at the same level as the original substrate (Fig. *b*). The silicon nitride is then removed from the rest of the surface and the integrated circuits are formed using normal ►planar-process technology.

coplanar waveguide (CPW) A ►transmission line medium comprising three conductors on the surface of a dielectric substrate: two ground planes and a central signal line, separated by gaps. The characteristic impedance is determined by the ratio of the gap width to the thickness of the dielectric substrate, and the dielectric constant of the substrate.

copper loss *Syn. for* I^2R loss. ►heating effect of a current; dissipation.

coprocessor A support chip manufactured to operate in parallel with the processor of a computer, usually a ►microprocessor, and add functionality to it. Typical coprocessor functions include high-speed arithmetic, mapping hardware for virtual memory, and high-speed graphics.

core 1. The ferromagnetic portion of the magnetic circuit of an electromagnetic device. A simple *ferrite core* is a solid piece of ferromagnetic material suitably shaped into a cylinder, toroid, etc. A *laminated core* is composed of laminations of ferromagnetic material insulated from each other; ►eddy currents are thus reduced. A *wound core* is one constructed from strips of ferromagnetic material wound spirally in layers. **2.** *Syn.* core store. An obsolete type of nonvolatile computer ►memory that consisted of an array of rings of ►ferrite material strung on a grid of wires. The individual rings – *ferrite cores* – were of the order of a millimetre in diameter. Information was stored in the array by causing the direction of magnetization of a core to be either clockwise or anticlockwise, corresponding to the binary digits one or zero. There was ►random access to the memory locations. Information was input and output by electronic means using the wire grid to read or write.

core loss *Syn.* iron loss. The total energy dissipation in the ferromagnetic ►core of an inductor or transformer. The energy loss is mainly due to ►eddy currents and hysteresis loss (►magnetic hysteresis) in the core.

core-type transformer A transformer in which most of the ►core is enclosed by the

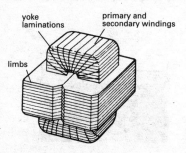

Single-phase **core-type transformer**

windings. The core is made from laminations; usually the yoke is built up from a stack of laminations and the windings are formed around this (see diagram). Once the windings have been formed extra laminations are added to form limbs around each winding and thus complete the core. ➤shell-type transformer.

corner frequency *Syn. for* cut-off frequency.

corner reflector antenna A special form of reflector (➤directional antenna) in which two reflecting plates are joined so as to form a corner. The feed element, usually a ➤dipole, is placed within the region of the corner depending on the application. The included angle is often 90°, in which case the corner reflector will return the received signal in exactly the same direction as it received it; this makes it useful as, for example, a passive target in radio or communication applications.

correlation The process of comparison of signals to establish how alike they are, or the extent of correspondence so found. When two signals are compared with each other, the process is known as *cross correlation*, and if a signal is compared with itself it is referred to as *autocorrelation*. The *correlation function* is the result of multiplying the signals together for different values of time delay between them. A peak in the correlation function indicates a delay at which the signals have a degree of periodicity, and the delay associated with such a peak gives the time period.

correlation function ➤correlation.

cosine potentiometer ➤potentiometer.

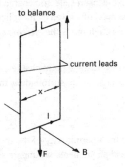

Cotton balance

Cotton balance *Syn.* electromagnetic balance. An absolute means of measuring ➤magnetic flux density, *B*, in air; it is capable of extremely high accuracy. This method can only be used for fairly strong fields that are uniform over a reasonable volume.

A long rectangular coil is suspended from an analytical balance with the lower end of the coil suspended in the magnetic flux that is to be measured (see diagram). The field is directed horizontally with the lower edge of the coil directed perpendicularly to the lines of *B*. The long sides of the coil experience no vertical force due to *B*, since they are vertical, and act as leads for the lower edge. Provided that these leads are sufficiently long the upper edge of the coil experiences a negligible force due to $\boldsymbol{B}$. Thus the net vertical force, $\boldsymbol{F}$, is just the force on the lower edge of the coil and is given by

$$\boldsymbol{F} = I\int\boldsymbol{B}\mathrm{d}x$$

where I is the current through the coil.

The value of I is measured using a standard resistance and potentiometer and the force $\boldsymbol{F}$ by the change in the balance reading when the current is reversed in direction. The value $\int\boldsymbol{B}\mathrm{d}x$ measured is the integrated flux density along the lower edge; the value at any point can be found if the flux-density distribution is known. If the flux is uniform along the length of the lower edge the value of $\boldsymbol{B}$ is

$$\int\boldsymbol{B}\mathrm{d}x/x$$

where x is the length of the lower edge.

Magnetic flux densities of about 0.5 tesla have been measured using this method with an accuracy of a few parts in 100 000.

coulomb Symbol: C. The ►SI unit of electric ►charge, defined as the charge transported through any cross section of a conductor in one second by a constant current of one ampere. Charge, Q, can then be given as

$$Q = \int I\mathrm{d}t$$

where I is the current.

coulombmeter *Syn.* coulometer. An instrument that measures electric charge by the amount of material deposited electrochemically: the mass of a given element liberated from a solution of its ions during electrolysis by one coulomb is the *electrochemical equivalent*.

Coulomb's law The mutual force, F, between two electrostatic point charges, q_1 and q_2, that results from the interaction of the electrostatic fields surrounding them is given by

$$F = q_1q_2/4\pi\varepsilon r^2$$

where r is the distance between the charges and ε the ►permittivity of the medium. Coulomb's law thus relates electrical and mechanical phenomena. ►►Ampere's law.

coulometer ►coulombmeter.

counter 1. A device that detects and counts individual events, such as charged particles. The term is applied to the detector and to the instrument itself. A single event is converted into a pulse, and these pulses are then electronically counted. **2.** *Syn.* digital counter. Any electronic circuit that counts electronic pulses.

In both cases, the average rate of occurrence of events counted is the *count rate*. The *counter lag time* is the delay between the primary event and the occurrence of the count. The *resolving time* is the minimum time between the occurrence of successive primary events that can be successfully counted.

counter/frequency meter An instrument that can be used as a counter or frequency meter by counting the number of events or periods occurring in a given time. It contains a frequency standard, usually a ➤piezoelectric oscillator. The time between events may also be counted by comparing the number of standard pulses occurring in the same time as a given number of cycles of the frequency standard.

counter lag time ➤counter.

count rate ➤counter.

coupling The interaction between two circuits so that energy is transferred from one to the other. In *common-impedance coupling* there is an impedance common to both circuits (Figs. *a*, *b*).

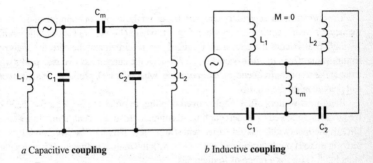

a Capacitive **coupling**   *b* Inductive **coupling**

The impedance may be a capacitance (*capacitive coupling*), a capacitance and a resistance (*resistance-capacitance coupling*), an inductance (*inductive coupling*), or a resistance (➤direct coupling). The impedance may be a part of each circuit or connected between the circuits. In *mutual-inductance coupling* the circuits are coupled by the mutual inductance, M, between the coils L_1 and L_2 (Fig. *c*). The coils used are often those of a transformer. The use of two separate coils between amplifier stages rather than a transformer is termed *choke coupling*. *Mixed coupling* is a combination of mutual-inductance coupling and common-impedance coupling.

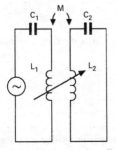

c Mutual-inductance **coupling**

The *coupling coefficient, K,* is defined as

$$K = X_m/\sqrt{(X_1 X_2)}$$

where X_m is the reactance common to both circuits and X_1 and X_2 are the total reactances, of the same type as X_m, of the two circuits. For Fig. *a*:

$$K = L_m/\sqrt{[(L_1 + L_m)(L_2 + L_m)]}$$

For Fig. *b*:

$$K = -C_m/\sqrt{[(C_1 + C_m)(C_2 + C_m)]}$$

For Fig. *c*:

$$K = M/\sqrt{(L_1 L_2)}$$

The current in the secondary circuit depends on the degree of coupling and the frequency. *Critical coupling* occurs when $KQ = 1$, where Q is the ➤Q factor of the circuit. A single peak occurs at the resonant frequency of the circuit and the current has its optimum value. *Overcoupling* occurs when $K > 1/Q$; the current has two side peaks with a dip at the resonant frequency. *Undercoupling,* when $K < 1/Q$, produces a smaller central peak than the optimum.

Band-pass ➤filters often employ overcoupling, in order to pass a narrow band of frequencies, followed by undercoupling, to compensate for the central dip. In a ➤tuned circuit, the bandwidth passed varies with frequency. This may be overcome by employing mixed coupling using a capacitance with the mutual inductance to give a constant bandwidth for a range of frequencies.

Cross coupling is unwanted coupling between communication channels, circuits, or components, particularly those with a common power supply. The removal of unwanted signals, especially those due to cross coupling, is called *decoupling.* It is usually achieved using a series ➤inductance or a shunt ➤capacitor.

coupling coefficient ➤coupling.

coverage area The area around a telecommunication-system transmitter within which the signal from the transmitter is considered to be useable for that system.

CPU *Abbrev. for* central processing unit.

CPW *Abbrev. for* coplanar waveguide.

Cramer's rule. A rule used in the solution of linear algebraic equations. ➤simultaneous equations.

CRC *Abbrev. for* cyclic redundancy check.

crest factor ➤peak factor; pulse.

crest value ➤peak value.

critical bandwidth The effective bandwidth of the human auditory filter at a given centre frequency.

critical coupling ➤coupling.

critical current density The maximum current that can flow in metal conductors which are in contact with a semiconducting substrate without affecting the long-term reliability of the device. At current densities higher than the critical value ➤electromigration can occur.

critical damping ➤damped.

critical field 1. ➤avalanche breakdown. **2.** ➤magnetron.

critical frequency *Syn. for* cut-off frequency.

CRO *Abbrev. for* cathode-ray oscilloscope.

Crookes dark space *Syn.* cathode dark space. ➤gas-discharge tube.

crossbar switch A switch that can simultaneously connect a number of processors to a number of memory banks (or other devices).

cross correlation ➤correlation.

cross coupling ➤coupling.

crossed-field microwave tube *Syn.* M-type microwave tube. ➤microwave tube.

cross modulation ➤modulation.

crossover The point at which two conductors, insulated from each other, cross paths.

crossover area ➤cathode-ray tube.

crossover distortion ➤distortion.

crossover frequency ➤crossover network.

crossover network *Syn.* dividing network. A type of filter circuit that divides the frequency range passed between a number of paths. Frequencies in a specified range pass through one path and those outside that range through other paths. The frequency at which the output passes from one channel to another is the *crossover frequency* and at that frequency the outputs are equal. Such networks are widely used with loudspeakers to separate the bass, mid-range, and treble components.

crosstalk Interference due to cross ➤coupling between adjacent circuits or to ➤intermodulation of two or more carrier channels, producing an unwanted signal in one circuit when a signal is present in the other. It is common in telephone, radio, and many other data systems.

   Crosstalk is classified as *near-end* and *far-end crosstalk* and in speech communication systems as *intelligible* and *unintelligible crosstalk*. Near-end crosstalk is measured at the input or sending terminal. Intelligible crosstalk in a communication system is crosstalk that can be understood by a listener and has a greater interfering effect than unintelligible crosstalk because it diverts the person's attention. Unintelligible crosstalk cannot be understood by the listener and is often classed as miscellaneous noise.

CRT *Abbrev. for* cathode-ray tube.

crystal-controlled oscillator ➤piezoelectric oscillator.

crystal-controlled transmitter A transmitter in which the ➤carrier frequency is produced by a ➤piezoelectric oscillator.

crystal filter A ➤filter that uses ➤piezoelectric crystals to provide its resonant or antiresonant circuits.

crystal growth furnace A furnace that is specially designed to produce a particular temperature profile needed to grow large single crystals from molten material in a controlled manner.

crystal loudspeaker A ➤loudspeaker in which the sound waves are produced by the mechanical vibrations of a ➤piezoelectric crystal.

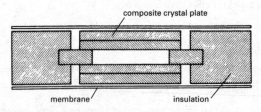

Crystal microphone

crystal microphone A type of microphone containing a ➤piezoelectric crystal in the form of two plates separated by an air gap (see diagram). Sound pressure variations cause displacements of the crystals and corresponding e.m.f.s are produced across them. Greater sensitivity is obtained using a separate diaphragm that is mechanically coupled to the centre of the crystal but this latter construction is more directive and has a poorer frequency response than the former.

crystal oscillator ➤piezoelectric oscillator.

crystal puller ➤liquid-encapsulated Czochralski.

CSMA *Abbrev. for* carrier-sense multiple access. ➤digital communications.

CTD *Abbrev. for* charge-transfer device.

Curie point *Syns.* magnetic transition temperature; ferromagnetic Curie temperature. ➤ferromagnetism.

Curie's law ➤paramagnetism.

Curie–Weiss law ➤paramagnetism.

current Symbol: I; unit: ampere. A flow of electric charge or, quantitatively, the rate of flow of electric charge. A *conduction current* is a current flowing in a conductor due to the movement of electrons or ions through the material, usually under the influence of an applied field. The net current is the algebraic sum of the charges. ➤displacement current.

A *unidirectional current* is one that always flows in the same direction in a circuit. A unidirectional current of more or less constant magnitude is a ➤direct current. ➤alternating current.

current amplifier An ➤amplifier that provides current gain. It has a low input inpedance and a high output impedance. ➤dependent sources.

current average ➤alternating current.

current balance ➤Kelvin balance.

current-controlled current source (CCCS) ➤dependent sources.

current-controlled voltage source (CCVS) ➤dependent sources.

current conveyor A four-terminal ➤active device that can be used to perform analogue signal processing functions by means of steering or controlling current flows. The device can be thought of as a dual of the ➤operational voltage amplifier or op-amp used in voltage-mode circuits. The current conveyor is shown in the diagram. The first-generation current conveyor, CCI, operates in the following manner: a current input to terminal X causes an identical current to flow into the other 'input' terminal Y; these are low-impedance inputs. The input current is also 'conveyed' to the output terminal Z; this is a high-impedance ➤current source output. The second-generation current conveyor, CCII, subsequently developed, is similar except that no current flows into terminal Y. This has proved to be a more versatile device.

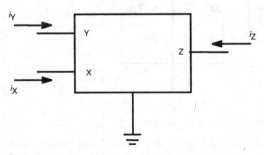

Current conveyor

current density Symbol: j or J; unit: amperes per square metre. A vector quantity that is equal to the ratio of the current to the cross-sectional area of the current-carrying medium. The medium may be either a conductor or a radiation beam. The current density is defined either at a point or as the *mean current density*.

current feedback *Syn.* series feedback. ➤feedback.

current limiter ➤limiter.

current mirror A transistor circuit in which the current in one side of the circuit is forced to be identical to the current in the other side of the circuit: a 'mirroring' action. The simplest current mirror comprises just two transistors, the diagram showing an example using ➤bipolar junction transistors.

Because the bases and emitters of the two transistors are connected together, the base-emitter voltages must be identical. If the transistors are the same, then the collector currents must be the same. The input transistor has the base and collector connected together – a diode-connected transistor – allowing its base-emitter voltage to be determined explicitly. The output of the current mirror is taken as the collector load of the second transistor, providing a high-impedance current source output that is relatively insensitive to loads.

More complex current mirror circuits have been devised that provide even greater insensitivity to loading, an example being the *Wilson current mirror*.

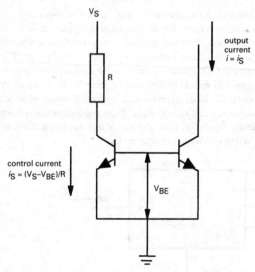

Current mirror using two bipolar transistors

current-mode circuits An approach to analogue or digital ➤signal processing in which the control, steering, or switching of currents is employed to carry out the electronic function or process. This is in contrast to more traditional circuits where the control of voltages is used to process the signal or information; for example, logic levels are traditionally expressed in terms of voltages. These are *voltage-mode circuits*.

current probe A device used to measure ➤common-mode current flowing along a wire or collection of wires. It consists of a ferrite ➤core that clamps around the wire(s) and a coil wrapped around the core. A magnetic field is induced around the core as specified by ➤Ampere's law; the magnetic field induces a voltage around the coil as a result of Faraday's law (➤electromagnetic induction) and this voltage is then measured. From a calibration using a standard current, the measured voltage can be used to calculate the current flowing along the wire(s).

current regulator ➤regulator.

current relay ➤relay.

current saturation ➤saturation current.

current source Ideally, a two-terminal circuit element through which the current is constant and independent of the voltage between its terminals.

current transfer ratio ➤common-base forward-current transfer ratio.

current transformer *Syn.* series transformer. ➤transformer; instrument transformer.

cut-off *Syn.* black-out point. The point at which the current flowing through an electronic device is cut off by the ➤control electrode. In a transistor, the cut-off point is the minimum base current at which the device conducts. In a valve it is the minimum negative grid voltage (the *grid base*) required to stop the current. In a cathode-ray tube the cut-off bias is the bias voltage that just reduces the electron-beam current to zero. In all cases the values are dependent on the conditions at the other electrodes, which must be specified.

cut-off frequency 1. The frequency at which the attenuation of a passive network changes from a small value to a much higher value; this is the theoretical cut-off frequency. The effective cut-off frequency is that frequency where the ➤insertion loss between two specified impedances has risen by a stated amount compared to the value at a reference frequency. An active network has the same cut-off frequency as a passive one with the same inductances and capacitances. **2.** *Syns.* break frequency; band edge; corner frequency; critical frequency. The point at which the angular frequency of a signal is equal to the reciprocal of the circuit ➤time constant. **3.** In a ➤field-effect transistor used for microwave applications, that frequency at which the current gain of the FET becomes zero. **4.** ➤waveguide.

cut-off voltage (of a television ➤camera tube) ➤target voltage.

cut-out A switch, especially a protective device that is operated, usually automatically, under fault conditions, such as when a circuit draws excessive current.

C-V curves ➤capacitance-voltage curves.

CVD *Abbrev. for* chemical vapour deposition.

CW *Abbrev. for* continuous waveform: a signal that is continuous with time.

cycle One complete set of changes in the values of a recurring variable quantity that repeats regularly, such as an alternating current.

cycle time The time required for a computer device, such as a bus or storage system, to complete an operation linked to the physical characteristics of the device. Typically the cycle time will consist of a number of ►clock cycles of the computer.

cyclic redundancy check (CRC) A digital code in which extra bits are added to each fixed-length block of data so that the data may be checked for errors occurring during transmission or some other process. The bits are calculated from the contents of the block both before and after the process. ►digital communications.

cylindrical winding A type of winding in which a coil is helically wound, either as a single layer or in multiple layers. The axial length of the coil is usually several times its diameter. ►disc winding.

Czochralski method ►liquid-encapsulated Czochralski.

D

DAC *Abbrev. for* digital-to-analogue converter.

daisy chaining A way of connecting more than one computer device to a single control line, such as a line carrying ►interrupt signals. Only one device contacts the control line directly; the next device contacts the first device, and so on, with all devices forming a single chain. If a device has a request, it blocks any requests from attached devices but sends its own request; otherwise, it passes the request it receives to the device it contacts, or to the control line if it is the first device in the chain.

damped Denoting the progressive diminution of a free oscillation due to an expenditure of energy. *Damping* is used to mean both the cause of the energy loss – friction, eddy currents, etc., – and the progressive decrease in amplitude. The amount of damping is termed *critical* if the system just fails to ocillate (see diagram). Greater or lesser degrees of damping lead to *overdamping* and *underdamping* respectively. The *damping factor* of underdamped oscillations is the ratio of the amplitude of any one of the damped oscillations to that of the following one. The natural logarithm of the damping factor, the *logarithmic decrement,* is sometimes quoted.

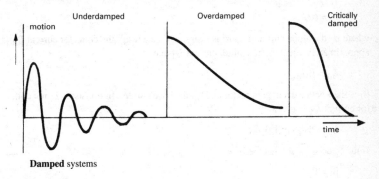

Damped systems

damping ►damped; instrument damping.

damping factor *Syn.* decrement. ►damped.

daraf Symbol: F^{-1}. The reciprocal ►farad, used as a unit to measure the reciprocal of capacitance, i.e. ►elastance.

dark conduction Low-level conduction that occurs in a photosensitive material when it is not illuminated. If such material is used as a photocathode in devices such as ➤photocells, it gives rise to a *dark current*.

dark current ➤dark conduction.

dark resistance ➤photocell.

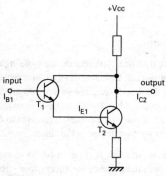

Darlington pair

Darlington pair A compound connection of two ➤bipolar junction transistors that operates as if it were a single transistor with an extremely high forward-current transfer ratio. The input signal is applied to the base of transistor T_1 and the emitter current supplies the input to transistor T_2 (see diagram). In normal transistor operation

$$I_C = \alpha I_E$$

where α (the alpha current factor) is near unity; as a result the collector current I_{C1} is approximately equal to the emitter current I_{E1} and

$$I_{E1} = \beta_1 I_{B1}$$

β_1 is the ➤beta current gain factor and I_{B1} the base current. In transistor T_2 the collector current I_{C2} is given by

$$I_{C2} = \beta_2 I_{B2} = \beta_1 \beta_2 I_{B1}$$

If the transistors are matched, i.e.

$$\beta_1 = \beta_2$$

the overall beta current gain of the combination is given by the square of the value for a single transistor.

d'Arsonval galvanometer A direct-current galvanometer in which the current to be measured passes through a small rectangular coil suspended between the poles of a permanent horseshoe magnet. The magnetic field produced in the coil reacts with the field of the magnet producing a torque and causing the coil to rotate about the vertical axis in the field. D'Arsonval movement is used in many forms of galvanometer since it combines a high degree of sensitivity with low resistance and high damping.

DAT *Abbrev. for* digital audio tape. A sound-reproduction system that uses magnetic tape to store digitally recorded audiofrequency signals. The signals are sampled, pulse code modulated, and stored as 16-bit digital words as in ►compact disc systems, but unlike compact disc the DAT recorders can both record signals and replay prerecorded tapes.

The DAT system uses a miniature rotating tape head to record slanted tracks on a slow-moving tape. The information is packed on the tape with a much higher density than in conventional tape recording systems and demands extremely high accuracy in manufacture. The tape used is similar to that used in ►video recorders. DAT recorders normally use a 48 kHz sampling rate both for recording and replay, but can also replay and sometimes record at the 44.1 kHz sampling rate used by CD systems and can record and sometimes replay at 32 kHz as used by DBS (►direct broadcast by satellite) systems. The recorder detects automatically the appropriate sampling frequency and switches accordingly.

data bus A computer ►bus that is used to transfer data. Each signal line of the bus can transfer one ►bit. The *width* of the data bus (the total number of lines) usually determines the ►word length of the processor, as with 16-bit, 32-bit, 64-bit processors.

data communications equipment (DCE) The equipment that enables connections between a sender and receiver to be established, maintained, and disconnected in a communications system. ►►data terminal equipment.

data compression The process of reducing the amount of data that represents a signal with minimum degradation to the signal when it is reconstructed. This may be of interest in order to reduce transmission bandwidth or storage requirements. One example is speech data compression using linear predictive coding (►linear prediction).

dataflow architecture A computer architecture in which the primary control derives from the availability of the data rather than from the execution of a program. Architects design dataflow computers for executing dataflow graphs; they are thus an alternative to ►von Neumann computers.

data processing The automatic or semiautomatic organization of numerical data on a routine basis. A data-processing system is any system that can receive, transmit, or store data; many systems can also perform mathematical operations upon the data and tabulate or indicate the results.

Analogue and digital ►computers are examples of data-processing systems but the term usually refers to systems that perform a limited number of functions automatically, particularly systems that are used to control the operation of other systems.

data rate The rate at which data can be transferred between devices in a communication or computer system.

data sheets A means by which manufacturers of electronic devices and components convey information about their products to potential customers, such as circuit designers. The data sheet will contain details including physical dimensions, pin connections, absolute device ratings, and basic performance data. Some data sheets also include typical applications and possible circuit configurations.

data terminal equipment (DTE) Equipment that can be used as a data transmitter or data receiver in a communication system. Data terminal equipment is normally connected to ►data communications equipment (DCE).

Day modulation A means of doubling the use of a radio channel by transmitting two ►carrier waves in ►quadrature, each separately modulated with different signals.

dB *Symbol for* decibel.

dBm A ►decibel value with reference to a power level of 1 milliwatt:
$$dBm = 10 \log_{10}(P/10^{-3})$$

DBR laser *Short for* distributed Bragg reflector laser (*syn. for* distributed feedback laser). ►semiconductor laser.

DBS *Abbrev. for* direct broadcast by satellite (or sometimes direct broadcasting satellite).

dBSPL *Abbrev. for* decibel sound pressure level, referenced to the threshold of hearing. ►sound pressure level.

dBu A ►decibel value with reference to a voltage level of 0.775 volts:
$$dBu = 20 \log_{10}(V/0.775)$$

dBV A ►decibel value with reference to a voltage level of 1 volt:
$$dBV = 20 \log_{10}(V/1)$$

dBW A ►decibel value with reference to a power level of 1 watt:
$$dBW = 10 \log_{10}(P/1)$$

d.c. (or **DC**) *Abbrev. for* direct current.

d.c.-coupled amplifier *Syn. for* direct-coupled amplifier. ►amplifier.

DCE *Abbrev. for* data communications equipment.

d.c.-level restorer ►direct-current restorer.

d.c. voltage Informal term for *direct voltage,* i.e. a unidirectional voltage that is substantially constant.

dead Denoting a conductor or circuit that is at earth potential. One not at earth potential is termed *live.*

dead-beat instrument ►instrument damping.

dead time The time interval immediately following a stimulus during which an electrical device does not respond to another stimulus. A correction for the dead time must be applied to the observed count rate in a ►counter to allow for events occurring during this period.

de Broglie waves A set of waves that are associated with a moving particle and represent its behaviour in certain situations, as when a beam of particles undergoes ►diffraction. The wavelength is given by

$$\lambda = h/mv$$

where h is the Planck constant and m and v the mass and velocity of the particle.

debug ►bug.

debunching The spreading of electrons in an electron beam or in a velocity-modulated tube that results from their mutual repulsion. The angle of spread of the electron beam is the *divergence angle*. ►►klystron; travelling-wave tube.

Debye length In a medium such as a ►semiconductor containing fixed charges and mobile charges, the distance required for a significant change in mobile carrier population under equilibrium conditions; the neutral equilibrium values of charge carrier density are either increased or reduced. The Debye length is a result of the screening effect around a fixed charge such as a donor impurity due to electrostatic attraction between it and the mobile carriers, causing them to cluster around the site of the impurity ion and mask its presence. Hence the electric field surrounding the ion declines much more rapidly than it would in the case of the unscreened ion. Fig. *a* illustrates the concept of Debye length for a positive ion. The ion has a charge $+q$ and is surrounded by a cloud of mobile carriers – electrons in this case. The cloud is about a Debye length in radius and contains an integrated charge of $-q$.

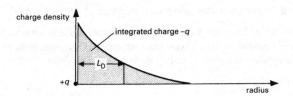

a **Debye length** for an ion, charge $+q$

In the case of a semiconductor where both electrons and holes are present, the general form for the Debye length is given by

$$L_D = \sqrt{[kT\varepsilon/q^2(n_0 + p_0)]}$$

where k is the Boltzmann constant, T the thermodynamic temperature, ε the permittivity of the material, q the value of the fixed charges, and n_0 and p_0 the neutral equilibrium numbers of electrons and holes respectively.

From this, in an ►extrinsic semiconductor where the charge carriers are predominantly due to the presence of impurities an approximate value can be derived as

$$L_{De} \approx \sqrt{[kT\varepsilon/q^2(N_D - N_A)]}$$

for n-type material, where N_D and N_A are the numbers of donor and acceptor impurities respectively.

For ►intrinsic material the value is

$$L_{Di} \approx \sqrt{[kT\varepsilon/2n_i q^2]}$$

where n_i is the intrinsic carrier density for one type of carrier.

The Debye length can be used as a scaling factor to derive a curve that describes the manner in which the carrier density changes from its neutral equilibrium value to near zero, for example at the edge of a ►depletion region, and hence is a measure of the departure in practice from the sharp edge to the depletion region that is often assumed (Fig. b).

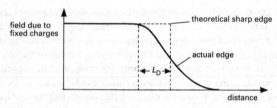

b **Debye length**, depletion region

deca- Symbol: da. A prefix to a unit, denoting a multiple of 10 of that unit.

decade scaler ►scaler.

deci- Symbol: d. A prefix to a unit, denoting a submultiple of 10^{-1} of that unit.

decibel Symbol: dB. A dimensionless unit used to express the ratio of two powers, voltages, currents, or sound intensities. It is ten times the common logarithm of the power ratio. Thus if two values of power, P_1 and P_2, differ by n decibels then

$$n = 10 \log_{10}(P_2/P_1)$$

i.e. $P_2/P_1 = 10^{n/10}$

If P_1 and P_2 are the input and output powers, respectively, of an electric network then if n is positive, i.e. $P_2 > P_1$, there is a gain in power; if n is negative there is a power loss.

The *bel,* symbol: B, is equal to 10 decibels. Since it is inconveniently large the decibel is used in practice. If two power values differ by N bels then

$$N = \log_{10}(P_2/P_1)$$

►►neper.

decimation ►downsampling.

decision threshold In the reception and interpretation of a digital signal, the analogue level above which the signal is taken to be binary one and below which it is taken to be binary zero.

decoder *Syn.* pulse detector. ➤pulse modulation.

decoding ➤digital codes; digital communications.

decoupling ➤coupling.

decrement 1. The operation of subtracting one from the value of a variable, usually in a computer program. It is the opposite operation to ➤increment. **2.** *Syn. for* damping factor. ➤damped.

decryption ➤encryption.

de-emphasis ➤pre-emphasis.

deep level transient spectroscopy (DLTS) A technique used to detect energy levels due to traps in the forbidden band of a semiconductor. The general technique is to use either photons, electrons, or an applied electric field to excite carriers into these levels, and to measure either the change in capacitance across the semiconductor, or transient currents produced as the equilibrium state in the material is reestablished. The technique is very useful in establishing activation energies, concentrations, and capture cross sections of electron traps.

deep ultraviolet exposure ➤photolithography.

definition ➤television.

deflection defocusing ➤cathode-ray tube.

deflection plates ➤electrostatic deflection.

deflection sensitivity ➤cathode-ray tube.

deformation potential Unit: electronvolt. An electric potential that can be caused by ➤acoustic waves or phonons travelling through a crystal, and is a measure of the effectiveness of charge-carrier scattering by the acoustic-mode phonons.

degeneracy Symbol: g. A condition that arises when an atomic or molecular system with a number of possible quantized states (➤quantum theory) has two or more distinct states of the same energy. The number of degenerate states each possessing that energy level is the *statistical weight*. Certain semiconductors, for example, exhibit degeneracy in either the valence or conduction bands, in which holes or electrons of the same energy level have different ➤effective masses.

degenerate semiconductor A semiconductor that has the Fermi level located either in the valence or in the conduction band, giving rise to essentially metallic properties over a wide range of temperature. ➤energy bands.

degeneration *Syn. for* negative feedback. ➤feedback.

degradation failure ➤failure.

degree Celsius *Syn.* degree centigrade. Symbol: °C. ➤kelvin.

deinterleaving ➤digital codes.

delay The time interval between the propagation of a signal and its reception; the time taken for a pulse to traverse any electronic device or circuit. In a switching transistor, for example, it is the time between the application of a pulse to the input and the appearance of a pulse at the output. Excessive pulse rise and fall times can reduce the speed of operation of a switching circuit and lead to undesirable delay. A known delay may be introduced into a system deliberately, by means of suitable ➤delay line.

delay distortion ➤distortion.

delayed automatic gain control *Syn. for* biased automatic gain control. ➤automatic gain control.

delayed-domain mode ➤transferred electron device.

delayed sweep ➤timebase.

delay equalizer A network or filter that compensates for the effects of delay ➤distortion and thus maintains the shape of a transmitted wave.

delay line Any circuit, device, or transmission line that introduces a known delay in the transmission of a signal. Coaxial cable or suitable L-C (inductance-capacitance) networks may be used to provide short delay times but the attenuation is usually too great when longer delay times are required.

  Acoustic delay lines are often employed when a longer delay is needed. The signals are converted to ➤acoustic waves, usually by means of the ➤piezoelectric effect. They are then delayed by circulation through a liquid or solid medium before being reconverted into electrical signals. Fully electronic *analogue delay lines* are now being provided by ➤charge-coupled devices. ➤Shift registers and charge-coupled devices may be used for *digital delay lines*.

Dellinger fade-out ➤fading.

delta circuit (or **Δ circuit**) *Syn.* pi (or π) circuit. A configuration of three impedances arranged as shown in the diagram. ➤star circuit.

delta function ➤Dirac delta function.

delta modulation (DM) *Syn.* slope modulation. ➤pulse modulation.

delta voltage ➤voltage between lines.

demodulation ➤modulation.

demodulator ➤detector.

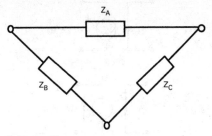

Delta circuit

De Morgan's laws Two mathematical laws of ►Boolean algebra that provide a means of expressing the complement (i.e. negation) of an expression in terms of the complements of individual elements of the expression:

$$\overline{A + B + C + D + ...} = \bar{A} \cdot \bar{B} \cdot \bar{C} \cdot \bar{D} \cdot ...$$
$$\overline{A \cdot B \cdot C \cdot D \cdot ...} = \bar{A} + \bar{B} + \bar{C} + \bar{D} + ...$$

where + and · are the operators AND and OR and a horizontal bar denotes a complement. The laws provide a conversion between NAND and NOR gates (►logic circuit), and also enable a designer to move between positive and negative logic forms.

demultiplexer ►multiplex operation.

denormalization The process of *frequency scaling* and *impedance scaling* applied to a ►normalized low-pass filter. The normalized filter response can be frequency scaled by dividing all the reactive elements of the normalized filter by a *frequency scaling factor*; this factor is given by the ratio of a reference frequency of the desired response to the corresponding reference frequency in the normalized response. Impedance scaling converts the normalized filter circuit element impedances (i.e. *R*, *L*, or *C*) to practical realizable values; the scaling is achieved by the *impedance scaling factor*, which increases the resistance and inductance values and decreases the capacitance values without changing the filter response.

dependent sources Sources of electrical energy whose value is controlled by either a current or voltage located elsewhere in the circuit. An example of a dependent source is the hi-fi amplifier, in which the output voltage is dependent on the input voltage to the amplifier. The symbols shown in the diagram show the four basic types of dependent source.

The *voltage-controlled voltage source* (*VCVS*), sometimes called a ►voltage amplifier, senses an input voltage and produces an output voltage that is related to it by the gain factor *K*.

The *current-controlled current source* (*CCCS*), sometimes called a ►current amplifier, produces an output current that is related to the current flowing in its input by the gain factor *K*.

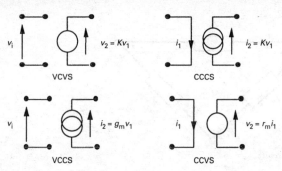

Types of **dependent sources**

The *voltage-controlled current source* (*VCCS*), also known as an *operational transconductance amplifier* (*OTA*) or *transconductance amplifier*, senses an input voltage and produces an output current that is related to it by the constant g_m, which has the dimensions of conductance. The constant g_m is known as the ►transconductance of the amplifier.

The *current-controlled voltage source* (*CCVS*), or *transresistance amplifier*, produces an output voltage that is related to the current flowing in its input by the constant r_m, which has the dimensions of resistance.

depletion layer A ►space charge region in a ►semiconductor due to the fixed ionized dopants, resulting from insufficient local charge carriers. Depletion layers are inevitably formed at the interface between two dissimilar conductivity types of semiconductor, in the absence of an applied voltage (►p-n junction), and at a ►metal-semiconductor contact. The width of the depletion layers increases when reverse bias is applied; the *depletion-layer capacitance* is the capacitance associated with a given depletion layer, which effectively acts as a dielectric when depleted of mobile carriers. Reverse-biased p-n junctions or Schottky diodes can therefore be used as voltage capacitors (►varactor). A depletion layer can also form at the surface of a semiconductor of given conductivity type under the influence of an electric field (►MOS capacitor).

depletion-layer photodiode ►photodiode.

depletion mode A means of operating field-effect transistors in which increasing the magnitude of the gate bias decreases the current. ►field-effect transistor; MOS capacitor. ►►enhancement mode.

deposition The application of a material to a base (such as a substrate) by means of vacuum, electrical, chemical, screening, or vapour techniques. ►►metallization.

derating Reducing the maximum performance ratings of electronic equipment or devices when operated under unusual or extreme conditions. This ensures an adequate safety margin.

derived units ➤SI units.

de Sauty bridge A four-arm ➤bridge used for the direct comparison of capacitances (see diagram). The capacitors are charged or discharged using the key K; if no response is observed from the ballistic galvanometer G, then

$$R_1C_1 = R_2C_2$$

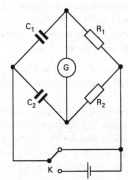

De Sauty bridge

Destriau effect ➤electroluminescence.

destructive read operation ➤read.

detector 1. *Syn.* demodulator. A circuit, apparatus, or circuit element that is used in communications to demodulate the received signal, i.e. to extract the signal from a carrier with minimum distortion. A *linear detector* produces an output proportional to the modulating signal; a ➤square-law detector produces an output proportional to the square of the modulating signal. **2.** Any device used to detect the presence of a physical property or phenomenon, such as radiation.

detune To adjust the frequency of a ➤tuned circuit so that it differs from the frequency of the applied signal.

deviation 1. *Syns.* variation; error. The difference between the observed value of a measurement and the true value. In automatic control systems it is the difference between the instantaneous value and the desired value. **2.** ➤frequency modulation.

deviation distortion ➤distortion.

deviation ratio ➤frequency modulation.

device An electronic part that contains one or more active elements, such as a transistor, diode, or integrated circuit.

DFB laser *Short for* distributed feedback laser. *Syn.* distributed Bragg reflector laser. ➤semiconductor laser.

DFT *Abbrev. for* discrete Fourier transform.

diac *Short for* diode a.c. switch. ➤thyristor.

diagnostic routine ➤program.

dial pulse The pulse of current sent down the line in a telephone circuit to indicate that a number is being dialled. One dial pulse is produced and sent for each unit of each number being dialled; for example, a nine requires 9 dial pulses. Each dial pulse is created by breaking the flow of current in a d.c. circuit within the telephone.

diamagnetism A very weak effect that is common to all substances and is due to the orbital motion of electrons around the nucleus of atoms. Diamagnetism is independent of the temperature of the material.

If a substance is placed in a magnetic flux density *B*, each individual electron experiences a force due to *B* since it is a moving charge. The orbits and velocities of the electrons are changed in order to produce a magnetic flux density that opposes *B* (➤electromagnetic induction). Each orbital electron therefore acquires an induced magnetic moment that is proportional to *B* and is in the opposite sense; hence the sample has a negative magnetic ➤susceptibility.

Diamagnetism causes a reduction of magnetic flux density within a sample; this can be represented schematically (see diagram) by a separation of the lines of magnetic flux as they pass through the material. If a diamagnetic substance is placed in a nonuniform magnetic field it tends to move from the stronger to the weaker region of the field. A bar of diamagnetic material placed in a uniform magnetic field tends to orientate itself with the longer axis at right angles to the field. Purely diamagnetic materials include copper, bismuth, and hydrogen.

Certain materials have a permanent molecular magnetic moment and in such substances the diamagnetism is totally masked by the magnetic properties arising from this permanent moment. ➤paramagnetism; ferromagnetism; antiferromagnetism; ferrimagnetism.

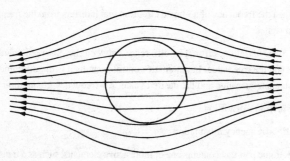

Change in magnetic flux density in a diamagnetic substance

diametrical voltage ➤voltage between lines.

diamond-like carbon A thin-film form of carbon that has a crystal structure of diamond, though it may contain significant crystalline defects and carbon-based impurities. It is produced by ➤chemical vapour deposition. This material has a very wide ➤energy band gap, and behaves like a ➤semiconductor. It is being developed for flat-panel display technologies and high-speed microelectronics.

diaphragm 1. A vibrating membrane used in sound ➤transducers such as microphones and loudspeakers. **2.** A porous partition in an electrolytic cell that separates the electrolytes while allowing ions to pass through.

diaphragm relay ➤relay.

dibit A pair of binary bits transmitted across a channel simultaneously.

dichroic mirror A mirror designed to reflect one designated optical frequency band and transmit all others. It is used in colour TV cameras and for mixers and modulators in optical communication systems.

Dicke's radiometer An instrument that measures microwave noise power precisely by comparing it with the noise from a standard source in a waveguide.

die ➤chip.

dielectric A solid, liquid, or gaseous material that can sustain an electric field and act as an insulator. A *perfect dielectric* is one in which no energy is lost from an electric field applied across the dielectric. An *imperfect dielectric* is one in which ➤electric hysteris losses occur. The displacement, *D,* lags behind the applied field, *E,* resulting in a typical hysteris curve (➤magnetic hysteresis) and in energy losses from the applied field, which appear in the form of heat.

dielectric constant *Syn.* relative permittivity. Symbol: ε_r. A dimensionless property of a material or medium equal to the ratio of the permittivity of the material or medium to the permittivity of free space. ➤permittivity.

dielectric heating A method of heating an insulating material, such as a plastic, in which a high-frequency alternating electric field is applied to the material. The periodic alternation of the electric field causes an alternating ➤dielectric polarization of the specimen, which results in the heating effect. Dielectric heating is usually effected by placing the specimen between specially shaped metal *applicators,* which form the electrodes of a capacitor, and applying the alternating field across them. A single applicator is sometimes used, applied to one surface of the specimen, when local heating only is required. The *heating depth* is the depth below the surface at which the heating effect due to the single applicator is observed. ➤➤induction heating.

dielectric isolation A method of isolating individual regions in an integrated circuit, particularly a ➤bipolar integrated circuit, by surrounding the region with insulating material (dielectric) rather than with isolating ➤diffused junctions. There are many different methods used to achieve dielectric isolation.

dielectric loss angle ►dielectric phase angle.

dielectric phase angle The difference between the phase angle of the alternating (sinusoidal) voltage applied across a dielectric material and the phase angle of the resulting alternating current. The difference between the dielectric phase angle and 90° is the *dielectric loss angle.* The cosine of the dielectric phase angle (or the sine of the dielectric loss angle) is the *dielectric power factor.*

dielectric polarization *Syn.* electric polarization. Symbol: ***P***; unit: coulomb per metre squared. An effect observed in a dielectric in the presence of an applied electric field. The electrons in each particular atom are displaced in the direction opposite to the field, and the nucleus in the direction of the field. (The centre of gravity of the atom remains fixed.) Each atom acquires a dipole moment (►dipole) parallel to and in the same direction as the field. The polarization is defined as the electric dipole moment per unit volume. It is related to the electric field, ***E****,* by the relation

$$P = \chi_e/\varepsilon_0 E$$

where χ_e is the electric ►susceptibility and ε_0 the ►permittivity of free space. χ_e is a tensor and approximately independent of electric field strength in most practical circumstances. In a uniform isotropic medium ***P*** and ***E*** are parallel and χ_e is a scalar constant.

dielectric power factor ►dielectric phase angle.

dielectric resonator (DR) ►dielectric resonator oscillator.

dielectric resonator oscillator (DRO) A microwave transistor ►oscillator that uses a ceramic ►dielectric of high ►Q factor for the resonator element. *Dielectric resonators*

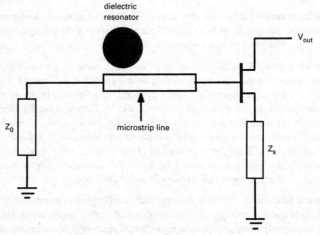

FET common-source **dielectric resonator oscillator**

(DRs) are low-loss, temperature-stable ceramics and resonate in various modes, depending upon the detailed composition and dimensions. The DR is usually a solid cylindrical puck, and in practice this is placed close to the ➤microstrip transmission line to enable magnetic coupling between them; the DR behaves like a high-Q cavity resonator at the oscillation frequency. A schematic FET ➤common-source DRO is shown in the diagram, with the DR resonating the input port.

dielectric strain ➤displacement.

dielectric strength Unit: volts per metre. The maximum electric field that can be sustained by a dielectric before ➤breakdown occurs.

difference transfer function ➤feedback control loop.

differential amplifier An ➤amplifier that has two inputs and produces an output signal that is a function of the difference between the inputs. An ideal differential amplifier produces an output signal of zero when the inputs are identical. In practice a small positive or negative signal may occur. The ➤common-mode rejection ratio is a measure of the ability of a differential amplifier to produce a zero output for like inputs.

differential capacitor A variable capacitor having two sets of fixed plates and one set of moving plates. As the moving plates rotate between the fixed plates, the capacitance to one set of plates is increased while that to the other is decreased.

differential galvanometer A type of ➤galvanometer that gives a deflection that is a function of the difference between two currents. The currents are passed in opposite directions through two identical coils. The difference between the currents determines the magnitude and direction of deflection.

differential-mode currents ➤electromagnetic compatibility.

differential phase modulation ➤phase modulation.

differential relay ➤relay.

differential resistance The resistance of a device or component part measured under small signal conditions.

differential winding Two or more coils or two windings of a single coil arranged so that when carrying a current their magnetomotive forces are in opposition.

differentiator *Syn.* differentiating circuit. A circuit that gives an output proportional to the differential, with respect to time, of the input. ➤integrator.

diffraction A phenomenon occurring when electromagnetic waves or beams of charged particles, such as electrons, encounter either an opaque object or a boundary between two media. The beams are not propagated strictly in straight lines but are bent at the discontinuity. This effect is due to the wave nature of electromagnetic radiation and the ➤de Broglie waves associated with the charged particles. Interference between the diffracted waves produces a *diffraction pattern* of maxima and minima of intensity; the diffraction pattern produced depends on the size and shape of the object causing the diffraction and the wavelength of the incident radiation. It can be used to investigate crystal structures or surface structures.

Electron and *X-ray diffraction* are employed to investigate crystal structures, the diffraction pattern produced being dependent on the spacing of the crystal planes. Electron diffraction is used to assess the structure of a crystalline semiconductor material near the surface. *Low-energy electron diffraction* uses a beam of low-energy electrons incident normal to the surface and the diffraction patterns from the back-scattered electrons are detected. *Reflection high-energy electron diffraction* uses a high-energy electron beam at a very small grazing angle of incidence. In this case the forward-scattered electrons produce the diffraction pattern. Electron diffraction patterns obtained from transmission ►electron microscopy can be used to provide information about crystalline materials throughout their bulk but this latter is a time-consuming method and very difficult to interpret.

X-ray diffraction is employed to detect imperfections in semiconductor crystals by means of *X-ray topography*, using a slit and photographic plate to record a 'map' of the slice on a photographic film. Although lateral resolution is of the order of several micrometres the technique is rapid and nondestructive.

diffused junction A junction between two different conductivity regions within a semiconductor formed by ►diffusion of the appropriate impurity atoms into the material.

diffusion 1. The movement of charge carriers in a semiconductor. ►Fick's law. **2.** The process of introducing selected impurity atoms into designated areas of a ►semiconductor in order to modify the properties of that area. The semiconductor is heated to a predetermined temperature in a gaseous atmosphere of the desired impurity. Impurity atoms that condense on the surface diffuse into the semiconductor material in both the vertical and horizontal directions. The numbers of impurity atoms and distance travelled at any given temperature is well-defined according to ►Fick's law. The interface between two different conductivity regions within a semiconductor is known as a *diffused junction*.

Early diffused devices were formed by performing nonselective diffusions over the entire semiconductor surface; any unwanted regions of the semiconductor were etched away (as in ►mesa transistors) and the junctions were formed below the surface and parallel to it. Modern techniques use the ►planar process of selective diffusion into well-defined areas of semiconductor. The edge of the junction is perpendicular to the surface and the device may be formed along the surface of the material.

Double diffusion is a method of forming diffused junctions in which successive diffusions of different impurity types are made into the same well-defined region of semiconductor. The temperature and diffusion time are adjusted to produce the desired impurity concentration. This technique is used if very precise distances between junctions are required, as in ►DMOS circuits, since the geometry is defined by the diffusion process itself and errors caused by misalignment of successive photographic masks are eliminated.

diffusion constant *Syn.* diffusivity. ►diffusion current; Fick's law.

diffusion current A flow of charge carriers from a region of higher concentration or density to a region of lower concentration, due to the diffusion process described by ➤Fick's law. This flow of charge results in a corresponding electrical current:

$$J_{\text{diffusion}} = eD(\partial n/\partial x)$$

where D is the *diffusion constant* or *diffusivity* of the charge carriers in the material and $\partial n/\partial x$ is the concentration gradient.

diffusion potential *Syn. for* built-in field. ➤p-n junction.

diffusivity *Syn. for* diffusion constant.

digital ammeter An ➤ammeter that displays the values of the current as numbers. The input is usually supplied as an analogue signal and this is repetitively sampled by the ammeter, which displays the instantaneous values. The instrument thus functions as a type of ➤analogue-to-digital converter.

digital audio tape ➤DAT.

digital circuit A circuit that responds to discrete values of input voltage and produces discrete values of output voltage. Usually two voltage levels only are recognized, as in binary ➤logic circuits. ➤➤linear circuit.

digital codes In ➤digital communications, the representation of a signal for transmission down a communications channel. *Coding*, the process of transforming the input message or signal into digital form, follows a defined set of rules; these rules form the codes. Most digital codes result in a signal having two levels – binary one and binary zero. Once coded, the signal is transmitted across the channel, and when received it is *decoded* to restore the original message or signal. The general process of coding as it relates to digital communication is shown in Fig. *a* (➤➤encryption, multiplex operation, modulation).

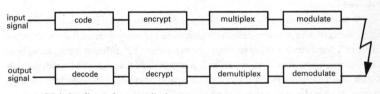

a Digital coding and communications

One of the simplest codes is the *Baudot code* shown in Fig. *b*. This code has 5 bits and so is capable of handling 2^5 or 32 bits of information or 32 different input symbols. It is able to transmit the 26 letters of the alphabet plus all the numbers and common special characters by use of a pair of special codes that switch from one set of characters to the other: a 11111 symbol is transmitted indicating that all the symbols that follow are letters until a 11011 symbol is transmitted, which indicates that all the symbols that follow are special characters.

A		00011
B	?	11001
C	:	01110
D	$	01001
E	3	00001
F	!	01101
G	8	11010
H	#	10100
I	8	00110
J	'	01011
K	(	01111
L	)	10010
M	.	11100
N	,	01100
O	9	11000
P	0	10110
Q	1	10111
R	4	01010
S	bell	00101
T	5	10000
U	7	00111
V	;	11110
W	2	10011
X	/	11101
Y	6	10101
Z	"	10001

Letters: 11111   Figures, punctuation marks, etc.: 11011

b Baudot code

The separation between any two symbols in a code is called the *code distance*; for example, the code distance between symbols 01001 and 01100 is two, indicated by two bits having changed. The *coding weight* is defined as the number of nonzero bits of the symbol.

When a digital signal is to be transmitted down a channel it is frequently divided into *blocks* of *data bits* (also called *information bits* or *message bits*). Each block consists of a fixed number (k) of bits and represents one of 2^k different input characters. Before transmission these data bits have added to them a ►header and a tail; the tail often contains error-detection bits such as ►parity checking bits or parity checking codes. Such codes are known as *block codes*. The ratio of the number of data bits in each block to the total number of bits in the block is called the *code rate*. The use of codes and error-detecting and error-correcting bits gives coded digital transmission a *coding gain* compared with uncoded digital transmission.

Some types of coding refer to how the signal is transmitted down the channel; examples are the *Manchester code*, the name given to biphase pulse code modulation (►pulse modulation), and the *trellis code*, which employs bandwidth expansion to achieve a reduction in transmission errors. Others are codes that intentionally introduce unused symbols to permit error correction as well as error detection. Examples of such *linear block codes* include *Golay codes*, which use one half of the available symbols, and *BCH* (Bose–Chadhuri–Hocquenghem) *codes*, which are a set of codes that allow multiple error correction.

In many situations the sequence of bits to be transmitted will contain a lengthy run of a specific symbol. Rather than code each symbol it is more efficient to describe the run with an efficient substitution code, such codes being called *run-length codes*; for example, a run of spaces (common in text) is substituted by a control character followed by the number of spaces. With *NRZ (nonreturn to zero) codes*, the signal line does not return to zero between a succession of '1' bits.

Enhanced error-correction performance can be achieved by coding an already coded message. The result is a *concatenated code*, an example of which is the *cross-interleave Reed–Solomon code (CIRC)*, a coding system used in ➤compact disc digital audio systems.

If the channel down which the digital signal is to be sent suffers time-dependent distortion such as results from ➤multipath interference, the effect can be reduced by *interleaving* the coded message before transmission and *deinterleaving* upon reception. An example of interleaving is shown in Fig. *c*.

Original coded message comprising 4 symbols, A, B, C, and D:

A   B   C   D

$A_1A_2A_3A_4A_5A_6A_7B_1B_2B_3B_4B_5B_6B_7C_1C_2C_3C_4C_5C_6C_7D_1D_2D_3D_4D_5D_6D_7$

Interleaved coded message of the same 4 symbols:

$A_1B_1C_1D_1A_2B_2C_2D_2A_3B_3C_3D_3A_4B_4C_4D_4A_5B_5C_5D_5A_6B_6C_6D_6A_7B_7C_7D_7$

c Interleaving

In ➤spread-spectrum systems, the input signal is modulated onto a high-speed digital signal called a *PN (pseudonoise) sequence* – a long pseudorandom sequence of symbols – which achieves the wide band-spreading characteristic of spread-spectrum signals. A large number of different PN sequences can be used to transmit simultaneously many different information signals down a single wideband channel; this is achieved by means of CDMA, code division multiple access (➤digital communications). When the PN sequences do not align with each other they are said to be *orthogonal codes*.

digital communications A form of communications over a transmission path or channel in which the information is transmitted in digital form. If an analogue signal is to be transmitted in digital form it must first be coded, i.e. transformed into digital form according to some prescribed rules (➤digital codes). In such cases regeneration of the original signal requires the reverse coding process – decoding. Digital communication has the advantage over analogue transmission of easier regeneration of the information signal. Significantly more noise and distortion can be introduced by the transmission channel before accurate recovery of the original information becomes impossible. The amount of information that can be transmitted over a path or channel is called the *channel capacity*.

Information can be transmitted over a channel in a number of different ways. The information signal can be modulated onto a carrier wave by a process of *digital modulation*. Digital modulation can be achieved using ➤amplitude modulation, in particular amplitude shift keying, ➤pulse code modulation, ➤frequency modulation, in particular frequency shift keying, or ➤phase modulation, in particular phase shift keying. A specific form of digital communication, known as ➤spread spectrum, results when the carrier is itself a digital signal, in particular a repetitive digital code or sequence (➤digital codes). Further variants of spread spectrum include the following: *frequency-hopping spread spectrum*, in which the frequency of the carrier is periodically or randomly changed; *direct-sequence spread spectrum*, in which a carrier is first modulated with the information signal, then the information-modulated signal is again modulated with a wideband spreading spectrum signal; *transmitted-reference spread spectrum*, in which two versions of the spreading carrier are transmitted, one modulated with the information signal and the other unmodulated.

One advantage of digital communications is that, especially in the case of wide-bandwidth channels such as optical fibres, a large number of digital signals can be sent down the same channel; this is usually called *multiple access*. Multiple access can be achieved in a number of different ways. In *time-division multiple access* (TDMA), the information sources that need to be transmitted are divided into segments – packets or blocks – and sent down the channel one segment at a time from each source sequentially, or selected by some alternative scheme. In *frequency-division multiple access* (FDMA), the different information sources are each modulated onto a different frequency carrier. In *code-division multiple access* (CDMA), the digital carrier code sequence is different for each information source. In *carrier-sense multiple access* (CSMA), a common broadcast signal is transmitted down the channel to reduce the risk of packets of information being sent by two different transmitters at the same time; this scheme is also called carrier sense multiple access with collision detection (CSMA/CD) and is commonly used in ➤local area networks.

In telecommunications, where a digital message is being transmitted down a switched network it is normally divided into blocks or packets. Each block or packet has a *header* at the start and a *tail* at the end: the header contains information about the sender, the recipient, the length of the packet and (optionally) the ➤protocol being used; the tail is generated at the sending end of the communication link by determining the ➤parity characteristics of the message or using a ➤cyclic redundancy check (CRC). At the receiving end the data portion of the message is processed in the same way as at the sending end to allow a comparison of the error-checking parity tail. If the two are the same the message is error-free; if not the block or packet has an error.

There are a number of different ways of dealing with the transmission of blocks or packets of data. If the received block or packet is error-free, the receiver can transmit an *ACK* (acknowledgment) signal to the sending end. The transmitting end then sends the next block or packet. This form of transmission is called a *stop-and-wait system*. If the block has an error, the receiver sends a *NAK* (negative acknowledgment) signal

to the transmitter, which then resends the last block. This process is continuously repeated – *ARQ* (automatic repeat request) – until an ACK is received.
➤encryption.

digital computer ➤computer.

digital counter ➤counter.

digital delay line ➤delay line.

digital filter A type of ➤digital signal processor that performs the actions of an electrical ➤filter but on sequences of discrete sampled data as the input signal rather than a continuous signal as in the case of an analogue filter (see diagram). The ADC (➤analogue-to-digital converter) performs the task of converting the continuous signal $x(t)$ into discrete samples. The digital filter processes the digital sequence to perform the required filter response, and the resulting digital output is converted back to a continuous signal $y(t)$ by the action of the DAC (➤digital-to-analogue converter).

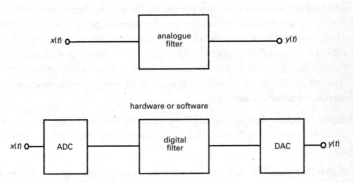

Digital filter compared with analogue filter

The digital filtering process can be achieved in either the ➤frequency domain or the ➤time domain. The frequency domain analysis involves a transformation of the discrete signal into its corresponding spectrum, possibly by the use of a DFT or FFT algorithm (➤discrete Fourier transform). The resulting frequency components are adjusted in accordance with the desired filter characteristic.

In the time domain, which is the most common form of digital filtering, the discrete signal samples are acted upon directly to produce the desired filtering characteristic. There are two methods that can be adopted to obtain the filter characteristic. One method, known as *finite impulse response* (*FIR*), involves the summation of input samples $x(n)$ with the delayed samples of the input $x(n-1)$ scaled by the filter coefficient (which is also known as the *tap weight*); this results in *FIR filters* (also known as *non-recursive filters* or *transversal filters*). Alternatively, a summation of the input samples $x(n)$ with delayed values of the output signal $y(n-1)$ results in filters known as *infinite impulse response* (*IIR*) *filters* (or *recursive filters*).

digital gate ➤gate.

digital inverter ➤inverter.

digital modulation ➤digital communications.

digital recording ➤compact disc system; DAT.

digital signal A signal made up of modes or states that represent a discrete number of levels. In general there are two levels, which can be translated into binary ones and zeros; this is also known as a *binary signal*.

digital signal processing (DSP) The use of a ➤microprocessor specifically designed to process data of the form usually required in ➤signal processing applications. The requirements for processing large amounts of data at the speeds required for speech (sample rate 8–10 kHz), audio (sample rate 50 kHz), telecommunication (sample rate 8 kHz), and video processing systems (sample rate up to 14 MHz) are often outside the performance specification of microprocessors. The DSP microprocessor, or DSP chip, was therefore developed. These chips are characterized by short ➤word length, fast ➤CPUs, and on-chip ➤ADCs and ➤DACs, a ➤Harvard architecture, on-chip ➤ROM and ➤RAM for storing filter coefficients and data, extensive pipelining, dedicated hardware multipliers, but limited program memory. Second-generation DSP architectures have removed on-chip data acquisition equipment, increased program memory space, and accelerated arithmetic operations.

digital switching A method of switching signals by first converting them to a digital signal and then passing them through a communication system such as a digital telephone network.

digital television A television system in which the picture information is encoded into digital form at the transmitter and decoded at the receiver.

digital-to-analogue converter (DAC) A device/circuit/IC that converts a discrete ➤binary signal into a continuous ➤analogue signal. A simple DAC can be built up from a series of weighted resistors. Taking a 4-bit binary word, $B = b_3b_2b_1b_0$, the ➤most significant bit (b_3) is connected to a resistor of value $2R$, bit b_2 to a $4R$ resistor, b_1 to an $8R$ resistor, and so on. The value of the resistor doubles for each less-significant bit. If any bit is at logical 1, a voltage is applied across its resistor and a current inversely proportional to the resistor flows. An ➤operational amplifier circuit can be used to sum the currents and produce a voltage proportional to the total current flowing. The magnitude of the voltage will therefore be directly proportional to the value of the binary number B.

digital voltmeter (DVM) A voltmeter that displays the values of the voltage as numbers. The input is usually supplied as an analogue signal and this is repetitively sampled by the voltmeter, which displays the instantaneous values. The instrument thus functions as a type of analogue-to-digital converter.

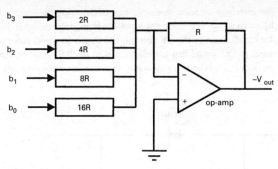

4-bit **digital-to-analogue converter**

digitizer A device that quantizes a continuous signal and represents it in digital form. ➤quantization.

DIL package *Short for* dual in-line package.

diode Any electronic device that has only two electrodes. There are several different types of diode, their voltage characteristics determining their application. Diodes are most commonly used as rectifiers; those used for other purposes have special names.

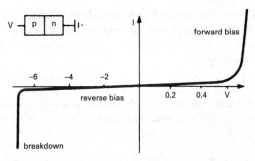

I–V characteristic for semiconductor **diode**

The semiconductor diode consists of a simple ➤p-n junction. Current flows when a forward bias is applied to the diode (➤diode forward voltage) and increases exponentially (see diagram). A straight-line approximation of this forward characteristic allows a resistance value (the *forward slope resistance*) to be calculated from the slope of the straight line. Reverse bias produces only a very small leakage current until ➤breakdown occurs. The now obsolete vacuum or valve diode is a ➤thermionic valve that has an anode and cathode and also passes current only in the forward direction.

➤IMPATT diode; light-emitting diode; photodiode; PIN diode; tunnel diode; varactor; Zener diode.

diode detector *Syn.* envelope detector. A simple way of detecting the presence of a radiofrequency (RF) signal, and extracting any ►amplitude-modulated signal present. The detection is by rectification of the RF voltage, and then a low-pass filter action extracts the baseband or DC signal that is proportional to the RF signal power (see diagram). This type of detector is known as a *square-law detector*. It is inherently wide bandwidth.

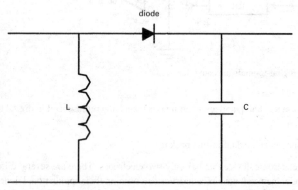

Simple **diode detector** circuit

The detection process can be analysed by considering the nonlinear current-voltage relationship of a ►p-n junction diode or metal-semiconductor (►Schottky) diode. The diode is biased at some voltage V_{DC}; the RF signal that is applied to the detector will be smaller in amplitude:

$$v_{RF} = A \sin(\omega t)$$

where A is the RF signal amplitude, which may be modulated, and ω is the RF frequency. The total voltage across the diode junction is then

$$V = V_{DC} + v_{RF}$$

The diode current can then be found using a Taylor series expansion for the I-V relation, around the bias point. This gives

$$I = I_{DC} + \tfrac{1}{4}\alpha^2(I_{DC} + I_0)A^2 + \alpha(I_{DC} + I_0)A \sin(\omega t) + \\ \tfrac{1}{4}\alpha^2(I_{DC} + I_0)A^2 \cos(2\omega t) + \ldots$$

where $\alpha = e/kT$. The first term on the righthand side is the DC bias current, the second term is the rectified or detected RF signal δI, and the remaining terms are high-frequency RF signals that are filtered out. The detected signal is thus proportional to the square of the RF signal amplitude: hence 'square-law detector'. If the RF signal is amplitude modulated, then the output voltage from the detector will contain this baseband modulation, the RF signals being filtered by the action of the LC low-pass filter.

In addition it can be shown that the detected current is also a measure of the RF power, P_j, dissipated in the junction:

$$\delta I = \tfrac{1}{2}\alpha P_j$$

The above analysis includes a DC bias current and is therefore quite general. In many applications the detector diode is unbiased: $V_{DC} = I_{DC} = 0$. While this does not affect the general result above, it should be noted that the junction resistance R_j is a bias-dependent parameter, and thus the RF power detected will be a function of R_j and hence bias voltage, which can be used to adjust the sensitivity of the detector.

diode drop ►diode forward voltage.

diode forward voltage *Syns.* diode drop; diode voltage. The voltage across the electrodes of a diode when current flows. The current increases exponentially with voltage (►diagram at diode) and therefore the voltage is substantially constant over the range of currents in common use: typically 0.7 volts for silicon p-n diodes. The diode may function as a *voltage reference diode* when it is used to provide a reference voltage, equal to the diode forward voltage, across its terminals.

diode laser *Syn. for* semiconductor laser.

diode transistor logic (DTL) A family of integrated logic circuits in which each input signal comes through a ►diode and the output is taken from the collector of an inverting transistor (►inverter). The basic circuit is a NAND gate (see diagram). If any of the inputs is at the low logic level the corresponding input diode is forward biased and can conduct current away from the diodes D_1 and D_2. The potential at point X is therefore at a value determined by the ►diode forward voltage and represents the low logic level. It is insufficient to forward bias D_2 and no current can flow to the base of the transistor. The transistor is therefore 'off' and the collector voltage is at the high logic level. If all the inputs are high, all the input diodes are reverse biased and

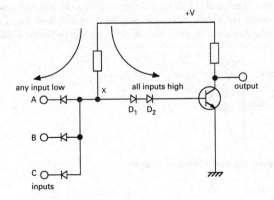

Diode transistor logic NAND circuit

cannot conduct current. The potential at point X is therefore high. Both D_1 and D_2 are forward biased and current flows to the base of the transistor, which turns on and saturates. The collector voltage falls to the low logic level.

The speed of the DTL circuit is slow compared to ➤emitter-coupled logic circuits because the output transistor is operated in the saturated mode: ➤carrier storage at the collector junction causes a delay in the switching time between logic levels. DTL circuits have been largely replaced by ➤transistor-transistor logic circuits.

diode voltage ➤diode forward voltage.

diplexer A device that provides a constant resistive impedance at its input terminal, and directs the output signal to one of two ports depending on the frequency of the signal. It is thus a frequency-selective network with a well-defined resistive input impedance.

dipole 1. *Syns.* electric dipole; doublet. A system of two equal and opposite charges very close together. A uniform electric field produces a torque that aligns the dipole along the field without translation. The product of one of the charges and the distance between them is the *dipole moment* (symbol: p). Dipole moment is related to the electric field strength, E, and the torque, T, by

$$T = p \times E$$

Some molecules have the effective centres of positive and negative charges permanently separated; these are termed *dipole molecules*. Some molecules have an *induced dipole moment* when the presence of a field causes the charge centres to polarize. ➤➤magnetic moment. **2.** *Syn.* dipole antenna. An ➤antenna commonly used for radiofrequencies below 3 gigahertz. It consists of a centre-fed open antenna excited in such a way that the standing wave of current is symmetrical about the mid-point of the antenna. There are several different types: a *half-wave dipole* has a length equal to half the wavelength, λ (see diagram); a *full-wave dipole* has a length of one wavelength; a *folded dipole* consists of two parallel half-wave dipoles separated by a small fraction of the wavelength, connected at their outer ends, and fed at the centre of one of the dipoles; a *multiple folded dipole* consists of more than two parallel half-wave dipoles.

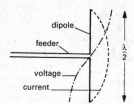

Voltage and current distribution in half-wave **dipole**

dipole antenna ➤dipole.
dipole molecule ➤dipole.
dipole moment *Syn.* electric moment. ➤dipole.

Dirac delta function *Syns.* unit delta function; delta function. A spike or impulse with infinite amplitude for zero time but with unit area. It is designated as $\delta(t)$, which means a unit impulse at time zero, or, more generally as $\delta(t - a)$, which means a unit impulse at time a. Although the delta function is a mathematical conception, it can nonetheless be approximated to a pulse of sufficiently short duration to appear instantaneous in comparison with the time constant of the circuit, and constitutes a powerful analytic tool in circuit theory and analysis.

direct broadcast by satellite (DBS) A method of broadcasting that uses a ►communications satellite in ►geostationary earth orbit as the main transmitter. The signal to be broadcast is transmitted from its point of origin on the earth to the satellite where it is received, amplified, and retransmitted to cover a wide area. It is detected directly by individual receivers using a suitable dish antenna tuned to the DBS signals.

direct conversion receiver A *homodyne* receiver, i.e. a radio receiver in which the local oscillator frequency is the same as the incoming signal carrier frequency, so that mixing results in conversion directly to baseband (►carrier wave). The reception process is thus also known as *zero-IF* (intermediate frequency). ►►heterodyne reception.

direct-coupled amplifier *Syn.* d.c.-coupled amplifier. ►amplifier.

direct coupling *Syn.* resistance coupling. ►Coupling between electronic circuits or devices, such as amplifier stages, that is not frequency dependent; resistive coupling is a form of direct coupling.

direct current (d.c. or DC) A unidirectional current of substantially constant value. ►►current.

direct-current restorer *Syn.* d.c.-level restorer. A device that restores the d.c. component or low-frequency component to a signal that has had its low-frequency components removed by a circuit element with high impedance to direct current. The device may also be used to add d.c. or low frequency to a signal lacking these components.

   Direct-current restoration is used in ►television receivers to reconstruct the original video signal. It is required either to restore the d.c. component to the received signal as in ►alternating-current transmission or to correct for the presence of an unwanted spurious d.c. component.

direct-current transmission A method of transmission used in television in which the direct-current component of the luminance signal (►colour television) is directly represented in the transmitted signal. ►►alternating-current transmission.

direct energy gap ►direct-gap semiconductor.

direct feedback *Syn. for* positive feedback. ►feedback.

direct-gap semiconductor A semiconductor in which the maximum energy of the ►valence band and the minimum energy of the ►conduction band are located at the same value of momentum in ►momentum space, so providing a *direct energy gap*. This permits radiative transitions between the conduction and valence bands to occur, resulting in the emission of ➤electromagnetic radiation. Gallium arsenide and other ➤compound semiconductors are direct-gap semiconductors. ➤indirect-gap semiconductor; energy bands.

directional antenna *Syn.* directive antenna. An ➤antenna that is a more effective transmitter or receiver of energy in some directions than in others. A common method of obtaining such directivity is by using a *passive antenna* and an *active antenna* in conjunction. The active antenna is one connected directly to the transmitter or receiver. The passive antenna is not so connected but influences the directivity by reacting with the active antenna. In transmission there is an induced e.m.f. produced in the passive antenna; in reception the reaction between the antenna elements results from their mutual inductance. A passive antenna placed behind an active antenna is called a *reflector*; one placed in front is a *director*. ➤antenna array; omnidirectional antenna.

directional coupler A ➤waveguide junction with four ➤ports (see diagram). It consists of a primary waveguide 1–2 and a secondary waveguide 3–4. If all ports are terminated in their characteristic impedances (➤transmission line), almost all the power entering port 1 will be transmitted to port 2. There is no power transmitted between ports 1 and 3 or between ports 2 and 4 because there is no coupling between these pairs of ports. However, there may be coupling between ports 2 and 3 and between ports 1 and 4 depending on the structure of the particular coupler.

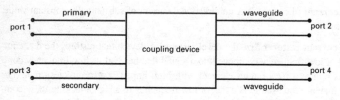

Directional coupler

Usually, directional couplers are designed so that the coupling between ports 1 and 4 and between ports 2 and 3 is as small as possible. In this case the device is often used to measure the power transmitted along a waveguide by connecting ports 1 and 2 in series in the waveguide. In this case a small portion of the power is received at port 4 and, using a calibration curve, the power flowing in the main waveguide can be calculated. The coupling between ports 1 and 4 (and 2 and 3) is required to be small so as not to affect the main waveguide, although typical good values are only 30 to 35 decibels below the coupling between ports 1 and 2.

directional relay *Syn. for* polarized relay. ➤relay.

direction finding *Syn.* radio direction finding. The practice and principle of locating the origin of a radio signal. A discriminating antenna and some form of receiver is required. *Automatic direction finding* carries out the process automatically using either a rotating ►directional antenna or two such antennas at right angles. The rotating antenna is often used in conjunction with a ►cathode-ray tube (CRT) as an indicator to display strength of signal against direction. The direction of maximum strength is the bearing of the radio source. When a CRT is used in this way the term *cathode-ray direction finding* is sometimes applied. *Frame direction finding* employs a loop antenna with a polar figure-of-eight directional response. At the point of zero signal the frame of the antenna points along the direction of propagation.

Direction finding in mountainous areas or in urban areas containing many very tall buildings can be subject to error because of reflections from the mountains or buildings. This is termed the *mountain effect*.

directive antenna *Syn. for* directional antenna.

directivity Symbol: *D*. The value of the ►gain of an ►antenna in the direction of its maximum value. Equivalently it is the ratio of the antenna's maximum radiation intensity over that of an isotropic source. It can then be related to the total radiated power of the antenna so that

$$D = U_{max}/U_0 = 4\pi U_{max}/P_{rad}$$

where U_{max} is the maximum radiation intensity, U_0 is the radiation intensity of an isotropic source, and P_{rad} is the total radiated power.

direct memory access (DMA) A type of computer input/output (I/O) in which a special *DMA controller* (DMAC) transfers data between main memory and the I/O devices. Under normal circumstances, an I/O device requires the ►central processing unit (CPU) to take an active role in the data transfer (Fig. *a*). The transfer rate is then limited by the ►bus read/write cycles, and furthermore involves the CPU in essentially a data movement role rather than a data processing role. The throughput of a bus, in terms of its electrical characteristics, is greatly underutilized and memory components themselves can respond at a far greater speed than that imposed by the bus read/write cycle.

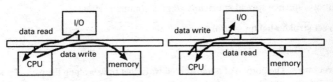

a Data transfer without **direct memory access** controller

DMA attempts to circumvent this bottleneck by allowing data transfer directly between a predetermined peripheral device and memory, without the direct intervention of the CPU during the transfer (Fig. *b*). It requires the use of special circuits – the DMA controller – that can force the CPU to relinquish its role as bus master and can then control the bus to permit a much higher data transfer rate, say 5 Mbytes per second.

The DMAC is directly connected to the I/O device and controls its activity; the DMAC provides all the signals, using the ►control bus, that the memory device requires

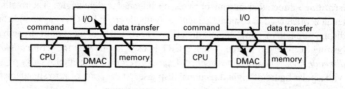

b Data transfer with DMA controller

to carry out memory read/write activities, but at the speed dictated by the DMAC rather than the CPU read/write cycles. When the transfer is from I/O device to memory, the I/O device places data on the ➤data bus under the control of the DMAC; the memory reads from the data bus without modification of its normal control signals. For transfer to I/O device from memory, the memory writes to the data bus without modification of its normal control signals; the I/O device reads data from the data bus under the control of the DMAC. In general, when the DMAC is not commanded to carry out DMA activities, normal programmed data-transfer activity proceeds unimpeded between the two components. The DMAC is connected to the system address, data, and control bus like any other ➤peripheral device, and contains internal registers that the CPU can read from and write to.

There are two common forms of DMA transfer. In *cycle stealing mode*, the DMAC returns control of the bus to the CPU after a single data transfer. In *burst mode*, the DMAC retains control of the bus until it has completed its transfer, without any intervening CPU cycles.

DMA controllers are embedded within many specialized peripheral equipment controllers, such as disk drives.

director *Short for* director antenna. ➤directional antenna.

direct ray The path along which a radiowave travels in the minimum possible time between a transmitting and receiving antenna. ➤➤radio horizon.

direct-sequence spread spectrum ➤digital communications.

direct stroke ➤lightning stroke.

direct voltage ➤d.c. voltage.

direct wave The portion of a transmitted wave that travels along the path of a ➤direct ray. It may suffer from tropospheric refraction. ➤➤ground wave; indirect wave.

disable 1. To place the output(s) of a device in the disconnected or high-impedance state. **2.** An input to a ➤tristate logic device or integrated-circuit chip that disables the device.

discharge 1. To remove or reduce the electric charge on a body such as a capacitor. **2.** The passage of an electric current or charge through a medium, often accompanied by luminous effects. ➤breakdown; gas-discharge tube; arc; spark. **3.** The conversion of chemical energy into electricity in a ➤cell by drawing current from it.

discharge tube ➤gas-discharge tube.

discrete Fourier transform (DFT) A mathematical method of analysing discrete or sampled data signals to determine the frequency spectrum. The DFT is equivalent to the ➤Fourier transform for discrete signals and is given by

$$X(m) = 1/N \sum_{n=0}^{N-1} x(n) \exp(-2j\pi mn/N)$$

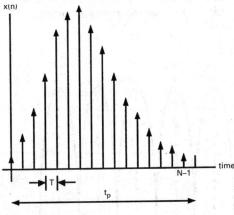

a Discrete signal $x(n)$

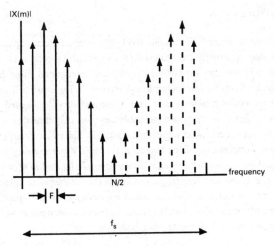

b **Discrete Fourier transform** spectrum $X(m)$

where $x(n)$ is the discrete signal to be analysed, $X(m)$ is the resulting discrete frequency spectrum, N is the number of $x(n)$ samples taken, and exp indicates the exponential function.

A sampled function of time is shown in Fig. *a* and its discrete Fourier transform spectrum in Fig. *b*. The time function is sampled at N points separated by an increment T over an interval $t_p = NT$ to create a discrete function $x(n)$. The resulting spectrum $X(m)$ is periodic with a period $f_s = 1/T$, and contains N components within one period with spacing between components $F = 1/t_p$. If $x(n)$ is a real function, only half or $N/2$ of the spectral components are unique. The integers n and m represent the time and frequency integers that identify the locations in the sequence of the time sample ($t = nT$) and the frequency components ($f = mF$).

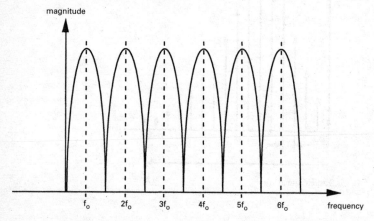

c Finite-length DFT response

A careful investigation of the DFT function shows that many of the multiplication operations are repeated. Algorithms can be devised that exploit this repetition to speed up the transformation by reducing the number of computations required. These high-speed algorithms are known as the *fast Fourier transforms* (*FFT*). The use of FFT analysis does however have a number of sources of error. These include ►*aliasing*, *leakage*, and the *picket-fence effect*. Leakage is an undesirable harmonic distortion that occurs when the time series is not periodic in the sampling interval. The picket-fence effect is due to the frequency response of a finite-length DFT. The DFT can be thought of as a set of narrow band-pass filters whose centre frequencies are located at nf_o (where $n = 0,1,2,3,\ldots N-1$, f_o is the sampling frequency, and N the number of samples), as shown in Fig. *c*. It can be seen that any input signal $x(t)$ coincident with a frequency nf_o will be transformed without distortion. However, the frequency components of $x(t)$ at noninteger multiples of f_o will be transformed with distortion.

discriminator 1. A circuit that converts a frequency-modulated or phase-modulated signal into an amplitude-modulated signal. ►modulation. **2.** A circuit that selects signals with a particular range of amplitude or frequency and rejects all others.

disc thermistor ►thermistor.

disc winding A type of winding for transformers consisting of flat coils in the form of a disc. Disc windings are usually employed for high-voltage applications. ►cylindrical winding.

dish ►parabolic reflector; radio telescope.

disk A storage medium in the form of a circular plate, ►magnetic disks and ►CD-ROMs being the principal examples.

disk drive ►magnetic disk.

diskette *Syn. for* floppy disk. ►magnetic disk.

dislocation An imperfection in the structure of a crystal. When dislocations occur in the crystalline structure of a semiconductor they can have serious deleterious effects. They introduce unwanted energy levels in the forbidden band (traps), they can alter the etching properties of the material, and can seriously change the electrical properties of devices. For example the values of source-drain current and threshold voltage of ►field-effect transistors are strongly dependent on the dislocation density in the semiconducting substrate. The dislocation density in any particular crystal depends on the material used, the purity, and the method of production. Perfect or near perfect small crystals can be produced, but larger crystals are more difficult to produce without dislocations. Large virtually dislocation-free silicon crystals are now being produced, but the dislocation density in large gallium arsenide crystals is significant. Dislocation density can be determined by etching the crystal in a solvent that preferentially etches at dislocations, and then counting the etched pits. Dislocation density is therefore sometimes termed *etch pit density.*

dispersion Any process that separates radiation into components with different frequencies, energies, velocities, or other characteristics. In a transmission system in which the wave velocity is not constant with frequency, the components of a multifrequency signal will travel different distances in a certain time, leading to dispersion of the signal.

displacement *Syns.* electric displacement; dielectric strain; electric flux density. Symbol: D; unit: coulombs per metre squared. A vector quantity defined as

$$D = \varepsilon_0 E + P$$

where E is the electric field strength, P the ►dielectric polarization, and ε_0 the permittivity of free space. In a dielectric medium the total charge within any given closed surface consists of any free charges, ρ_e, in the surface together with the apparent charge density, $-\mathrm{div}\, P$, due to the polarization. The volume charge density becomes $(\rho_e - \mathrm{div}\, P)$ and ►Gauss's theorem relating to a dielectric is written as

$$\text{div } \boldsymbol{E}.d\tau = (1/\varepsilon_0)\!\int\!(\rho_e - \text{div } \boldsymbol{P})d\tau$$

where $d\tau$ is a small volume element. It can therefore be shown that the displacement $\boldsymbol{D}$ is given by

$$\text{div } \boldsymbol{D} = \rho_e$$

In a vacuum $\boldsymbol{P}$ is zero and Gauss's theorem is written as

$$\text{div } \boldsymbol{E} = \rho_e/\varepsilon_0$$

displacement current The rate of change of electric ►displacement in a dielectric with respect to time ($\partial \boldsymbol{D}/\partial t$) when the applied electric field changes. No motion of charge carriers is involved apart from the setting up of electric dipoles (►dielectric polarization). A displacement current gives rise to magnetic effects similar to those of a conduction current and these effects form the basis of Maxwell's electromagnetic theory of light.

display A device for the visual presentation of information, as on the screen of a ►cathode-ray tube, widely used for computer displays, or by using ►scaling alphanumerical displays in cascade, as in a ►digital voltmeter.

dissipation *Syn.* loss. A loss of power due to the tendency of electronic circuits and components to resist the flow of current. In a resistive circuit the power dissipated is equal to I^2R, where I is the current and R the resistance. This is I^2R *loss*. In an inductor or capacitor the *dissipation factor* is the cotangent of the ►phase angle, α, or the tangent of the loss angle, δ. In low-loss components it is almost equal to the power factor, $\cos \alpha$, and can be given approximately by $\sigma/2\pi f\varepsilon$, where σ is the conductivity, ε is the permittivity of the medium, and f the frequency.

Dissipation causes free oscillations to be ►damped and removes the sharpness of cut-off in ►filters. High-frequency industrial heating is made possible because of dissipation of eddy-current energy in conductors (►induction heating) and displacement-polarization energy in dielectrics (►dielectric heating).

A network that is designed to absorb power is a *dissipative network,* as compared to a network that attenuates by impedance reflection. All networks provide some dissipation since entirely loss-free components cannot be made.

dissipation factor ►dissipation.

dissipative network ►dissipation.

distortion The extent to which a system or component fails to reproduce accurately at its output the characteristics of the input. The modification of a waveform by a transmission system or network involves the introduction of features not present in the original input or the suppression or modification of features that are present. Distortion is a significant problem in telecommunication systems. There are several different types of distortion.

Amplitude distortion occurs when the ratio of the root-mean-square value of the output to the r.m.s. value of the input varies with the amplitude of the input, both waveforms being sinusoidal. If harmonics are present in the output waveform only the fundamental frequency is considered.

Crossover distortion occurs in ►push-pull operation when the transistors are not operating in the correct phase with each other.

Delay distortion is a change in the waveform because of the variation of the delay with frequency.

Deviation distortion occurs in frequency-modulated receivers that have an inadequate bandwidth or nonlinear discriminator.

Harmonic distortion is due to harmonics not present in the original waveform.

Intermodulation distortion results from spurious combination-frequency components in the output of a nonlinear transmission system when two or more sinusoidal voltages, applied simultaneously, form the input. Intermodulation distortion of a complex waveform arises from ►intermodulation within the waveform.

Nonlinear distortion is produced in a system when the instantaneous transmission properties depend on the magnitude of the input. Amplitude, harmonic, and intermodulation distortion are all results of nonlinear distortion.

Phase distortion occurs when the phase change introduced is not a linear function of frequency.

Optical distortion of an image is seen in electronic systems, such as ►cathode-ray tubes, television picture tubes, etc., and in facsimile transmission. It is due to errors in the electron-lens focusing systems.

Aperture distortion of an image occurs in a scanning system when the scanning spot has finite dimensions rather than infinitely small dimensions.

Barrel and *pincushion distortion* are seen when the lateral magnification is not constant but depends on image size (see diagram). Barrel distortion occurs when the magnification decreases with object size, pincushion distortion when it increases with object size.

Coma is a plumelike distortion of the spot occurring when the focusing elements of the electron gun are misaligned.

Keystone distortion is due to the length of the horizontal scan line varying with the vertical displacement of the line. It is most pronounced when the electron beam is at an acute angle to the screen and results in a trapezoidal image instead of a rectangular one. It can be removed using suitable transmitter circuits.

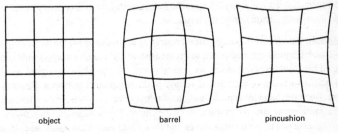

object   barrel   pincushion

Optical **distortion**

Trapezium distortion is a trapezoidal pattern on the screen of a cathode-ray tube instead of a square one and occurs when the deflecting voltage applied to the plates is unbalanced with respect to the anode.

distributed amplifier A multistage ►amplifier, operating at high frequency, in which the inputs and outputs of each stage are taken from tapped ►transmission lines. The diagram is a schematic of a distributed FET amplifier using artificial L-C transmission lines at input and output; the amplifying stages are distributed along the transmission lines. The ►phase velocity along each transmission line must be arranged so that the outputs from each stage add in phase. Distributed amplifiers are very wide bandwidth, providing a reasonable gain up to the transition frequency f_T of each of the transistor stages. In practice, losses and parasitic reactances in the amplifiers and transmission lines limit the upper frequency.

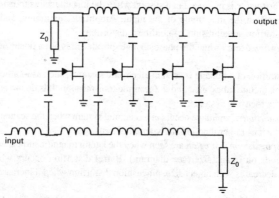

Distributed FET amplifier

distributed Bragg reflector laser (DBR laser) *Syn. for* distributed-feedback laser. ►semiconductor laser.

distributed capacitance 1. The capacitance of an electrical system, such as a transmission line, considered to be distributed along its length. ►distributed circuit. **2.** The capacitance between individual turns on a coil or between adjacent conductors. The distributed capacitance in a coil lowers the inductance of the coil and may be represented by a single capacitor across the terminals.

distributed circuit A type of circuit operating at ►radio or ►microwave frequencies. At such frequencies, traditional discrete resistive and reactive components cannot be used because their dimensions are similar to the wavelength of the signal, resulting in a spatial variation of current and voltage in the component or circuit. Resistance and reactance are therefore realized using transmission-line elements, and the resistance and reactance are *distributed* in space along the circuit. The voltages and currents in the circuit are treated as ►travelling waves that flow along the transmission line. ►transmission line; waveguide; power waves.

distributed feedback laser (DFB laser) *Syn.* distributed Bragg reflector laser. ➤semiconductor laser.

distributed inductance The inductance of an electrical system, such as a transmission line, considered to be distributed along its length. ➤distributed circuit.

distribution control ➤scanning.

dithering The application of a small perturbation or noise to a measurement to reduce the effect of small local nonlinearities.

divergence angle ➤debunching.

diversity gain ➤diversity system.

diversity system A communication system that has two or more paths or channels. The outputs of these are combined to give a single received signal and thus reduce the effects of ➤fading. The *diversity gain* is the gain in reception achieved by using a diversity system.

Frequency diversity employs independent transmission channels on neighbouring frequencies. *Space diversity* employs several receiving antennas spaced several wavelengths from each other. In both cases each antenna supplies its own receiver; the demodulated outputs are then combined. *Polarization diversity* uses antennas that are arranged to receive oppositely polarized waves.

divider A circuit that reduces the number of pulses or cycles by an integer factor.

dividing network ➤crossover network.

D-layer *Syn.* D-region ➤ionosphere.

DLP *Abbrev. for* decode level point. The signal level that results after a digitally coded or modulated signal is converted back to its analogue equivalent.

DM *Abbrev. for* delta modulation. ➤pulse modulation.

DMA *Abbrev. for* direct memory access.

DMOS MOS circuits or transistors that are fabricated using double ➤diffusion. Regions of different conductivity type are formed by successive diffusion of different impurities through the same opening in the oxide layer. DMOS devices are short-channel high-performance devices that were originally developed for microwave applications. They have a very precise channel length that is determined by the double diffusion rather than the inherently less precise method of ➤photolithography.

The speed of operation of a MOS transistor (or ➤MOSFET) is determined by the channel length; for high operating speeds short channel lengths are required. Ordinary MOSFETs formed by a single diffusion (Fig. *a*) suffer from ➤punch-through at short channel lengths since the depletion layer associated with the reverse-biased p-n$^+$ drain junction spreads rapidly through the p-region as the drain voltage is increased. In the DMOS device an n$^-$ substrate is used. The p-regions are formed by diffusion (➤planar process) followed by an n$^+$ source/drain diffusion in which the source diffusion is made into the p-regions. An n⁺-p-n⁻-n⁺ structure is produced (Fig. *b*) in which an n⁻ region, termed the *drift region,* separates the p-region from the n⁺ drain. The drain junction therefore becomes a reverse-biased p-n⁻ junction and the associated depletion layer is almost entirely contained within the n-region.

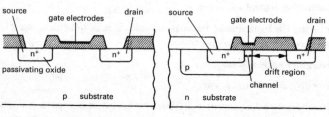

a Cross section of n-channel MOSFET

b Cross section of n-channel DMOS transistor

The ►breakdown voltage of a DMOS device is determined by the width of the drift region up to a theoretical maximum determined by the characteristics of the n⁻ semiconductor (►avalanche breakdown; depletion layer). High-voltage MOSFETs can be formed using relatively wide drift regions. Breakdown voltages up to 300 volts with drift regions of about 25 micrometres have been produced. Devices with relatively short drift regions are produced as integrated circuits for low-power high-speed (up to microwave frequency) applications.

Epitaxial DMOS transistors are formed in an n⁻ epitaxial layer grown on a p-type substrate. Individual transistors on a chip may then be isolated by performing extra p-type isolating diffusions. ➤►VMOS.

domain 1. ►ferromagnetism. **2.** ►Gunn effect. **3.** ►time domain; frequency domain.

dominant mode ►mode.

donor *Short for* donor impurity. ►semiconductor.

dopant ►doping.

doping The addition of a particular type of impurity to a ►semiconductor in order to achieve a desired n-conductivity or p-conductivity: donor impurities are added to form an n-type semiconductor and acceptor impurities to form a p-type. The impurity added is the *dopant.* Doping is carried out by processes such as ►diffusion or.►ion implantation.

doping compensation The addition of a particular type of impurity to a ►semiconductor in order to compensate for the effect of an impurity already present.

doping level The amount of doping necessary to achieve the desired characteristic in a semiconductor. Low doping levels (p, n) give a high-resistivity material; high doping levels (p⁺, n⁺) give a low-resistivity material. ➤►semiconductor.

doping profile *Syn.* impurity profile. The variation with depth of the majority ➤carrier concentration within a semiconductor. ➤Fick's law.

Doppler effect The change in the apparent frequency of a source of electromagnetic radiation (or sound) when there is relative motion between the source and the observer. The observed frequency f' is given by

$$f' = f(c - v_o)/(c - v_s),$$

where c is the speed of light or sound, v_o is the velocity of the observer, and v_s is the velocity of the source.

The effect is utilized in *Doppler navigation,* which is a navigation system (in a moving object) that operates by ground reflection. *Doppler radar* employs the Doppler effect to distinguish between fixed and moving targets: the measurement of the change in the frequency of the reflected wave is used to determine the velocity and direction of the moving target.

Doppler navigation ➤Doppler effect.

Doppler radar ➤Doppler effect; radar.

dot generator A test generator used with a television receiver to adjust the convergence of the ➤picture tube. A pattern of evenly spaced dots or small squares is produced on a dark background and the dynamic focusing (➤colour picture tube) is adjusted until a satisfactory image is formed on the screen.

dot matrix tube ➤colour picture tube.

double amplitude *Syn. for* peak-to-peak amplitude. ➤amplitude.

double-beam cathode-ray tube ➤cathode-ray tube.

double conversion receiver A heterodyne radio receiver in which there are two stages of mixing, and hence two intermediate frequencies (IF) before detection of the information signal. ➤heterodyne reception.

double-current system A telegraph system that reverses the direction of the electric current in order to effect transmission of the signals. ➤➤single-current system.

double diffusion ➤diffusion.

double drift device ➤IMPATT diode.

double-ended ➤single-ended.

double image ➤ghost.

double modulation Multiple ➤modulation involving two carriers only.

double-pole switch A switch that can operate simultaneously in two independent circuits.

double-sideband transmission Transmission of both sidebands generated when a ►carrier wave is amplitude modulated, but not of the carrier itself. Double-sideband suppressed carrier systems are not often used because of the difficulties encountered when reintroducing the carrier at the receiver.

doublet ►dipole.

down conversion Mixing of a (radio) signal with a local oscillator signal to result in a new signal at a lower frequency than the originally received radiofrequency. ►heterodyne reception. ►►up conversion.

downlink The radio-communications path from a moving transmitting device to a fixed receiving site. ►►uplink.

downsampling *Syn.* decimation. The process of decreasing the sampling rate of a sampled signal.

DPNSS *Abbrev. for* digital private network signalling system. The signalling system used on links within or between private telephone exchanges.

DPSK *Abbrev. for* differential phase shift keying. ►phase modulation.

drain The electrode of a ►field-effect transistor through which ►carriers leave the interelectrode space.

DRAM *Abbrev. for* dynamic RAM. ►Solid-state memory that requires refreshing to keep the data active. A memory cell in DRAM generally consists of a capacitor and transistor (see diagram). The capacitor's charge decays, due to leakage, so the system must periodically refresh the charge to maintain the value. When the address line is active, the MOS transistor acts as a closed switch. If the memory is to be read, the voltage on the capacitor is detected on the data line by a sense amplifier. If a write or refresh operation is called for, the data line becomes an input line. When the proper address turns on the MOS transistor in the DRAM cell, the capacitor can be charged or recharged from data-in.

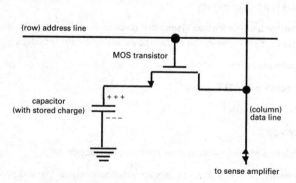

Schematic of one **DRAM** memory cell

D-region *Syn. for* D-layer. ➤ionosphere.

drift 1. The variation with time of any electrical property of a circuit or apparatus. Drift often occurs during warm up or when the device is nearing the end of its useful life. In a voltage regulator or reference standard the variation of output voltage with respect to time is the *drift rate*. **2.** ➤drift current.

drift current The movement of charge ➤carriers in a metal or ➤semiconductor under the influence of an electric field. The resulting average velocity of the charge carriers is known as the *drift velocity*, *v*, which is linearly related to the electric field by the parameter ➤drift mobility, μ:

$$v = \mu E$$

The drift current, *J*, is then the product of the carrier charge density and their average velocity:

$$J = nev = ne\mu E$$

which is an expression of ➤Ohm's law. The parameter $ne\mu$ is the ➤conductivity of the material.

drift mobility *Syn.* carrier mobility. Symbol: μ. A quantity that relates the average drift velocity of charge carriers to the electric field strength. Drift mobility is a material parameter, and electrons and holes generally have different mobilities in a given material. ➤➤drift current.

drift rate ➤drift.

drift region ➤DMOS.

drift space 1. A region in an ➤electron tube that is free of electric or magnetic fields. **2.** ➤klystron.

drift velocity ➤drift current.

driven antenna *Syn. for* active antenna. ➤directional antenna.

driver A circuit or device that provides the input for another circuit or controls the operation of that circuit.

driving impedance ➤motional impedance.

driving-point impedance 1. The ratio at the input terminals of a network of the root-mean square (rms) value of the applied sinusoidal voltage to the rms value of the resulting current between the terminals. **2.** ➤motional impedance.

driving potential ➤photocell.

DRO *Abbrev. for* dielectric resonator oscillator.

droop ➤pulse.

dropper ➤dropping resistor.

dropping resistor *Syn.* dropper. A resistor introduced into a circuit to provide a voltage drop across its terminals and hence reduce the voltage in the circuit.

dry battery ➤battery.

dry cell ➤cell.

dry etching ➤etching.

dry joint A faulty soldered joint that has a high resistance because of a residual oxide film.

DSP *Abbrev. for* digital signal processing.

DTE *Abbrev. for* data terminal equipment.

DTL *Abbrev. for* diode transistor logic.

DTMF *Abbrev. for* dual-tone multifrequency. A method of communications using two different frequency signals, one high and one low, which are combined to indicate a one or zero in the digital data message.

DTW *Abbrev. for* dynamic time warping.

D-type flip-flop ➤flip-flop.

dual in-line package (DIL package) A standard form of package used for integrated circuits. It consists of a ceramic or plastic casing containing a ➤leadframe. The frame is used to form the connections to the bonding pads of the integrated circuit and is connected to output pins arranged in two parallel lines at opposite sides of the package. The number of pins available varies from eight with smaller circuits to about 96 for large circuits. ➤leadless chip carrier; pin grid array; tape automated bonding.

duality Interchangeability of two types of entity in a given system or theory. If two equations that describe the behaviour of two different variables are of the same mathematical form, their solutions will also be identical. The quantities in the two equations that occupy identical positions are known as *dual quantities*. Duality means that a solution for one of these variables can be derived by systematically interchanging symbols with the solution for the other.

dubbing The combining of two sound signals into a composite recording. At least one source of sound will have been prerecorded.

dull emitter ➤thermionic cathode.

dummy antenna ➤artificial antenna.

duplexer A two-channel multiplexer (➤multiplex operation) that uses a ➤transmit-receive (TR) switch so that one antenna may be used for both reception and transmission. The switch protects the receiver from the high power of the transmission. Duplexers are commonly used in ➤radar and ➤personal communications handsets, the TR

switch operating in the time between transmission of the pulse and reception of the return echo.

duplex operation Simultaneous operation of a communications channel in both directions. *Half-duplex operation* occurs when the operation is limited to either direction but not both directions at once. ➤simplex operation.

Dushman equation ➤Richardson–Dushman equation.

dust core A magnetic ➤core that is made from a powdered material such as ➤ferrite.
Such cores have a very low eddy-current loss at high frequencies.

duty A statement of the operating conditions and their durations to which a device is
subjected, including rest and de-energized periods. *Uninterrupted duty* is the operation
of a device without any off-load (➤load) periods. *Continuous duty* is uninterrupted duty
that continues for an indefinite time. *Intermittent duty* has on-load (➤load) periods alternating with off-load periods. When the on-load periods are small in comparison with
off-load periods the intermittent duty is termed *short-time duty*; *periodic duty* occurs
if the load conditions are regularly recurrent. Operation at loads and for durations that
are both subject to wide variation is *varying duty*. The *duty cycle* is a group of variations of load with time. The ratio of the on-load period under specified conditions to
the sum of the on-load and off-load periods is the *duty ratio*.

duty cycle ➤duty.

duty factor (of a pulse train) ➤pulse.

duty ratio ➤duty.

DVM *Abbrev. for* digital voltmeter.

dynamic Denoting any electrical device, circuit, or apparatus in which the electrical
parameters are constantly changing. The term can be applied to those devices that operate with alternating currents and voltages, especially those with a marked frequency
dependence. It is also used to describe components, such as varactors, in which an electrical property varies with the operating conditions, e.g. if the reactance is a function
of the applied voltage. Any device or circuit in which the signals decay over a period
of time unless regenerated, is also termed dynamic. ➤Dynamic memory is an example of this type of device. Any component, circuit, or device that is not dynamic or that
is operated with essentially constant electrical conditions is described as *static*.

dynamic characteristic ➤characteristic.

dynamic focusing ➤colour picture tube.

dynamic impedance *Syn.* dynamic resistance. The impedance at resonance of a parallel ➤resonant circuit. It is purely resistive by definition (➤resonant frequency).

dynamic memory A solid-state memory in which the stored information decays over
a period of time. The decay time can range from milliseconds to seconds depending
on the nature of the device and its physical environment. The memory cells must

undergo refresh operations sufficiently often to maintain the integrity of the stored information. ➤➤DRAM.

dynamic operation ➤MOS integrated circuit.

dynamic RAM ➤DRAM.

dynamic range The range over which an active electronic device can produce a suitable output signal in response to an input signal. It is often quoted as the difference in decibels between the noise level of the system and the level at which the output is saturated (the *overload level*).

dynamic resistance 1. The resistance of any electronic device under normal operating conditions. Many devices exhibit a variation of resistance with frequency. **2.** ➤dynamic impedance.

dynamic sensitivity ➤phototube.

dynamic time warping (DTW) The process of modifying the timescale of a signal representation when looking for a best-fit match with a reference template pattern.

dynamo An electromagnetic ➤generator that produces either alternating or direct current. It consists of a plane coil that is made to rotate in a uniform magnetic field of flux density B. The coil, of area A, rotates with angular velocity ω; an angle θ is subtended

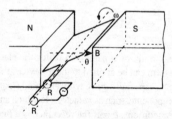

a Slip-ring operation of a **dynamo**

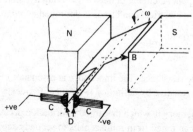

b Carbon-brush operation of a **dynamo**

by the normal to the plane of the coil and **B**. If time $t = 0$ is chosen so that $\theta = \omega t$, then the e.m.f. induced in the coil is given by:

$$V = A\omega B \sin\omega t$$

This is an alternating voltage, of period ω, that is a maximum when the plane of the coil lies in the direction of the magnetic field.

Alternating current will flow if the two ends of the coil are connected to a pair of slip rings, R and R′ (Fig. *a*). The ends of the coil can be connected to a commutator by means of semicircular segments, D, making contact with a pair of carbon brushes, C (Fig. *b*). Then the voltage in each terminal always has the same sign since each segment moves to the next brush as the e.m.f. changes sign. This produces a direct current in a load connected across the brushes. The current is not steady however since the induced voltage is alternating. The fluctuations about the mean value are known as ripple.

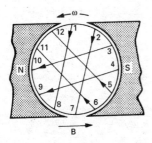

c 12-pole drum armature

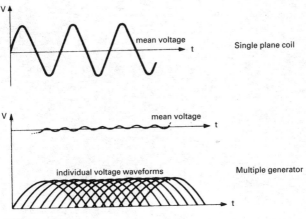

d Voltage waveforms

The ripple may be reduced using a drum armature. This has the coil wound around a drum-shaped armature so as to produce several plane coils symmetrically spaced around the drum (Fig. *c*); each vertical conductor forms a pole. The brushes are placed opposite each pole and are connected in pairs. The effect is to superimpose a number of voltage waveforms all varying slightly in phase (Fig. *d*). The output voltage has a higher mean value and less magnitude of ripple than that of the single coil. The period of the ripple voltage, which corresponds to the time interval at which successive conductors occupy the same position, is also much higher.

dyne The unit of force in the obsolete CGS system of units. One dyne equals 10^{-5} newton.

dynode ➤electron multiplier.

E

early failure period ►failure rate.

Early voltage The intercept point on the voltage axis of the ►common-emitter output characteristics of a ►bipolar junction transistor, when the collector-current curves are extrapolated: the extrapolations converge to a single point. This voltage is a measure of the small-signal output resistance of the transistor, with higher values indicating a high output resistance.

EAROM *Abbrev. for* electrically alterable read-only memory. ►ROM.

earphone A small loudspeaker that is designed to be used very close to the ear. Applications include hearing aids, the receiver in a telephone, a.c. bridge measurements, and use with reproduction systems such as radio. Two earphones used together form a *headset*.

earth *U.S. syn.* ground. **1.** A large conducting body, such as the earth, that is taken to be the arbitrary zero in the scale of electric potential. **2.** A connection, which may be accidental, between a conductor and the earth. An effective earth for electrical equipment is formed by a wire connected to a cold water pipe. Connection may also be made to an *earth electrode,* i.e. a large copper plate buried in moist soil. **3.** The point or portion in an electric circuit or device that is at zero potential with respect to earth. **4.** To connect an electric circuit or device to earth.

earth bus ►bus.

earth capacitance The capacitance between any circuit or equipment and a point at ►earth potential.

earth current 1. A current that flows to earth, especially one that results from a fault in a system. **2.** Any current flowing in the earth. Particular earth currents are associated with ionospheric disturbances. Buried cables sometimes have their lead sheaths corroded by a direct earth current.

earth electrode *Syn.* earth plate. ►earth.

earth fault A fault that occurs when a conductor is accidentally connected to earth or when the resistance to earth of an insulator falls below a specified value.

earth plane ►ground plane.

earth plate *Syn. for* earth electrode. ►earth.

earth potential *Syn.* zero potential. The potential of a large conducting body, such as the earth, taken to be the arbitrary zero in the scale of electric potential.

earth-return circuit A circuit composed of one or more conductors in parallel that connect two points in a ➤telecommunication system and that is completed through earth at the two points.

e-beam lithography *Short for* electron-beam lithography.

e-beam resist ➤electron-beam lithography.

E-bend *Syns.* E-plane; edgewise bend. ➤waveguide.

Ebers–Moll model ➤transistor parameters.

EBIC *Abbrev. for* electron-beam induced current analysis.

E-cell A solid-state timing device consisting essentially of a thimble-shaped electrolytic cell with a central gold electrode, an outer silver-plated electrode, and a paste electrolyte consisting of a suitable silver salt (see diagram). When a current passes, silver is lost from the outer electrode, which is made the anode, and is deposited at the same uniform rate on the central gold cathode. After a given time, determined by the elements of the external circuit, the current reverses and the cathode becomes the anode. When the cathode is 'deplated' the reverse current ceases and the process is once more reversed.

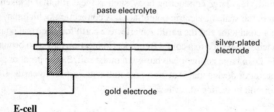

E-cell

An E-cell is a useful and versatile timing device since it can be made very small, is robust, and can be used for a wide range of time intervals simply by altering the external circuit. The desired time interval is usually determined by charging and discharging a capacitor.

ECG *Abbrev. for* electrocardiograph or electrocardiogram.

echo 1. A communications wave that has been reflected or refracted and that has sufficient magnitude and delay to distinguish it from the direct wave. In radio an echo is heard. In television it appears as a ➤ghost on the screen; it may or may not be heard simultaneously. **2.** The portion of a transmitted radar signal reflected back to the receiver.

echo sounding A system based on the same principles as ➤radar but using sound waves instead of radiowaves. ➤sonar.

ECL *Abbrev. for* emitter-coupled logic.

eddy current A current induced when a conductor is subject to a varying magnetic field. Energy is dissipated by eddy currents (*eddy-current loss*), usually appearing as heat, and becomes significant in high-frequency applications. This effect is utilized in ➤induction heating. Eddy currents in a moving conductor react with the magnetic field to produce retardation of the motion and are used to produce damping. They are sometimes called *Foucault currents* (although discovered by Joule).

eddy-current heating ➤induction heating.

eddy-current loss ➤eddy current.

edge connector A track on a ➤printed circuit board that is taken to one edge of the board to form a connector. Each board has several edge connectors, which may be plugged into a suitable socket allowing connections to be made to the circuit on the board.

edge effect Deviation from parallel in the lines of force representing the electric field at the edges of parallel-plate capacitors resulting in a field that is nonuniform at the edges.

edge profile The shape of the semiconductor edges produced after ➤etching away portions of the slice to form ➤mesas, particularly after wet etching. The shape of the edge profile is a function of the crystalline structure of the semiconductor, the crystal orientation, and the particular etchant used.

edge triggering The process of ➤triggering an electronic device, usually digital, with the active edge of a ➤clock pulse. The device, generally a ➤flip-flop, can be either *positive edge triggered* or *negative edge triggered*. The specified edge indicates the initialization of data transfer. In the diagram, with both inputs to the JK flip-flop at logical 1, whenever the clock triggers the device the output on Q will change from logical 1 to logical 0, or vice versa. The polarity of the edge trigger will specify at which edge this change occurs, i.e. positive or negative.

EDS *Abbrev. for* energy dispersive spectroscopy. ➤electron microprobe.

EEG *Abbrev. for* electroencephalograph or electroencephalogram.

EEPROM *Abbrev. for* electrical-erasable programmable ROM. ➤ROM.

effective address The ➤address that the CPU generates to reference an instruction or variable in computer memory using the instruction-provided ➤addressing modes, such as base displacement and indexing, but not using the address-translation mechanisms of ➤virtual memory. The effective address is usually the same as the compiler's ➤logical address but not the same as the computer's physical address.

effective aperture The ratio of the power delivered to an antenna load to the incident ➤power density. The effective aperture is the area that when multiplied by the incident power density gives the power delivered to the load. The effective aperture is not necessarily the same as the physical aperture.

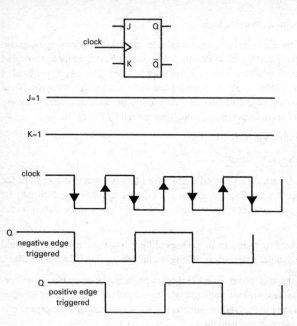

Edge triggering of a clocked JK flip-flop

effective mass Symbol: m^*. The motion of ➤electrons in free space is given by a classical Newtonian relationship,

$$E = p^2/(2m_0)$$

where E, p, and m_0 are the energy, momentum, and mass of the electron. In a crystal, the local potential due to the atoms will modify the dynamic behaviour of the electrons. The motion of electrons in a crystalline solid can be found using ➤quantum mechanics; this results in the ➤energy band structure of the crystal, which is the relationship between electron energy and momentum, E and p, in the crystal. The motion of an electron in the energy bands due to an applied electric field (a force) can be found in a semi-classical manner, as follows. The quantum mechanical part describes the velocity of the electron in the crystal: the ➤group velocity of the electron ➤de Broglie wave is

$$v = \partial E/\partial p$$

The energy dE gained by the electron in travelling a distance $v dt$ under the influence of the force F is

$$dE = Fv dt = F(\partial E/\partial p) dt$$

The acceleration due to the force is

$$dv/dt = (d/dt).(\partial E/\partial p) = (dp/dt).(\partial^2 E/\partial p^2)$$

But since force is the rate of change of momentum, the term $\partial^2 E/\partial p^2$ has the units of (1/mass). The reciprocal of this term is the effective mass of the electron in the energy band of the crystal. The electron is seen to behave like a particle with this effective mass in response to an applied force, such as an electric field. The value of the effective mass is determined by the detailed shape of the energy band structure.

effective radiated power (ERP) The power transmitted by an antenna in a particular direction when the antenna is not omnidirectional, i.e. when it does not radiate equally in all directions. The ERP is the power input to the antenna multiplied by its ➤gain in the direction of transmission.

effective resistance *Syn.* alternating-current resistance. The resistance to alternating current of a conductor or other circuit element. It is measured as the power in watts dissipated as heat divided by the current in amperes squared. It includes the resistance to direct current and resistance due to ➤eddy currents, ➤hysteris, and the ➤skin effect.

effective value ➤root-mean-square value.

EFM *Abbrev. for* eight to fourteen modulation. ➤compact disc system.

EGA *Abbrev. for* enhanced graphics adapter. A colour video display standard with a resolution of 640×350 ➤pixels and 16 colours; it was introduced by IBM for the PC AT.

EGG *Abbrev. for* electroglottogram. ➤electroglottograph.

EHF *Abbrev. for* extremely high frequency. ➤frequency band.

EHT *Abbrev. for* extra high tension. Usually applied to the high-voltage supply for cathode-ray tubes or television picture tubes.

EIA *Abbrev. for* Electronics Industries Association. ➤standardization.

eigenvalue Given an $N \times N$ matrix A, an eigenvalue of A is a number λ with the property $Ax = \lambda x$, where

$$x = (x_1, x_2, \dots x_N)^\mathrm{T}$$

and is called an *eigenvector* of A. Eigenvalues are the possible values for a ➤parameter of an equation for which the solutions will be compatible with the ➤boundary conditions.

eigenvector ➤eigenvalue.

Einstein photoelectric equation ➤photoelectric effect.

Einthoven galvanometer *Syn.* string galvanometer. A ➤galvanometer that has a tightly strung conducting thread between the poles of a strong electromagnet. A current passed through the thread causes it to be deflected at right angles to the magnetic field. The deflection is observed with a high-power microscope. The instrument is highly sensitive and can detect currents of 10^{-11} ampere.

EIRP *Abbrev. for* effective isotropic radiation power. The ►effective radiated power (ERP) of an antenna compared to that of an antenna that radiates equally in all directions.

elastance Symbol: S; unit: farad^{-1}. The reciprocal of ►capacitance.

elastic recoil detection analysis (ERDA) A technique used for surface analysis of a material up to depths of one micrometre. The technique is similar to ►Rutherford back scattering (RBS), but uses a beam of energetic heavy charged particles rather than the alpha particles used in RBS. The ERDA technique is used for detecting and measuring the concentration of light elements in the material, including hydrogen.

elastoresistance A change in the resistance of a material when it is stressed within its elastic limit.

E-layer *Syns.* E-region; Heaviside layer; Kennelly–Heaviside layer. ►ionosphere.

electret A substance that is permanently electrified and has oppositely charged extremities: it is the electrical analogue of a permanent magnet. Electrets have been used in ►electrometers and in ►capacitor microphones.

electrically alterable read-only memory (EAROM) ►ROM.

electric axis The direction in a crystal of maximum ►conductivity. It is the X-axis of a ►piezoelectric crystal.

electric charge ►charge.

electric conduction ►conduction.

electric constant *Syn. for* permittivity of free space. ►permittivity.

electric controller ►automatic control.

electric current ►current.

electric dimensions Of an electronic circuit or structure, quantities that are measured in wavelengths at the frequency of interest. Electric dimensions are useful in determining the properties of a system or the method used to determine these properties. In particular, a system is considered to be electrically small if its dimensions are significantly less than a wavelength, typically a factor 10 to 100 times smaller.

electric dipole ►dipole.

electric dipole moment ►dipole.

electric displacement ►displacement.

electric field The space surrounding an electric charge or varying magnetic field in which another electric charge experiences a perceptible mechanical force. The charge acted upon by the field is assumed to be sufficiently small so that the electrical conditions are not altered by it. ►►electric field strength; Coulomb's law.

electric field strength Symbol: E; unit: volts per metre. The strength of an ➤electric field at a point, measured in terms of the mechanical force per unit charge. The force, F, given by ➤Coulomb's law, is thus related to field strength by:

$$F = eE$$

where e is the electron charge.

The potential difference between two points separated by distance s is given by

$$dV = -E.ds$$

or in more general terms

$$E = -i\partial V/\partial x - j\partial V/\partial y - k\partial V/\partial z$$

where i, j, and k are unit vectors along the x-, y-, and z-axes, respectively.

electric flux Symbol: Ψ; unit: coulomb. The quantity of electricity displaced across a given area in a ➤dielectric. It is given by the scalar product $D.dS$ of the electric ➤displacement, D, and an element dS of area. ➤Gauss's theorem states that

$$\int D.dS = \int \rho_e d\tau$$

where ρ_e is the volume electric charge density in a small volume element $d\tau$. The total flux through an area surrounding a charge q is thus equal to q. This flux is unaltered by the presence of a dielectric medium.

electric flux density ➤displacement.

electric heating The production of heat from electric energy. Methods used include passing a current through a ➤resistance, use of an electric ➤arc, ➤induction heating, and ➤dielectric heating.

electric hysteris loss Electrical loss in a dielectric material due to internal forces in the material produced by a varying electric field. The loss usually appears in the form of heat.

electric image *Syn.* image charge. A concept devised by Kelvin for solving electrostatic problems associated with a point charge in the vicinity of a conducting surface. It can be shown that under certain conditions the charges induced, by the point charge, on the surface of the conductor have electrical effects identical with those that would be produced by an imaginary point charge located at a particular point relative to and below the surface. This imaginary point charge is known as the electric image of the first charge. The force exerted on the first charge, due to the induced surface charges, may be calculated as if it were due to the image and is called the *image force*. In the case of an infinite plane surface the image charge is equal and opposite in sign to the inducing charge and at a distance behind the plane equal to the distance of the actual charge in front of it (see diagram).

electric intensity *Obsolete syn. for* electric field strength.

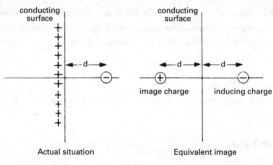

Electric image

electricity The phenomena associated with static or dynamic electric charges, such as electrons.

electric moment *Syn. for* dipole moment. ►dipole.

electric polarization ►dielectric polarization.

electric potential Symbol: V or ϕ; unit: volt. The work required to bring a unit positive charge from infinity to a point in an ►electric field. The potential is one volt when one joule is needed to transfer a charge of one coulomb.

electric screening The surrounding of apparatus with an electrical conductor to prevent interference from unwanted electrical disturbances. The screen is usually a ►Faraday shield.

electric spectrum The colour spectrum produced by an electric ►arc.

electric susceptibility ►susceptibility.

electric transducer ►transducer.

electroacoustic transducer ►transducer.

electrocardiograph (ECG) A sensitive instrument that measures and records the voltage and current waveforms produced by the heart muscles of living animals. The trace produced is an *electrocardiogram*.

electrochemical series If a metal is placed in a molar solution of one of its salts a potential, termed the *electrode potential*, is developed between the metal and the solution. The electrochemical series is the arrangement of chemical elements in order of their electrode potentials (►Table 13, in the back matter). The standard reference is the *hydrogen electrode*, which is given arbitrarily the value zero. This electrode consists of gaseous hydrogen at a pressure of one atmosphere in contact (by means of a platinum electrode) with an acidic solution containing one molar hydrogen ions at 25 °C. Elements that come above hydrogen in the electrochemical series and tend to give up electrons to acquire a positive charge are *electropositive*; those below it, including the halogens, tend to acquire a negative charge and are *electronegative*.

electrode A device that emits, collects, or deflects electric charge carriers. It is usually in the form of a solid plate, a wire, or a grid that controls current into and out of an electrolyte, gas, vacuum, dielectric, or semiconductor. Liquid mercury electrodes are also used. The *electrode current* is the current that flows through a specified electrode, such as a ➤collector, ➤grid, or ➤drain.

electrode current ➤electrode.

electrodeposition ➤Deposition by electrolysis.

electrode potential ➤electrochemical series.

electrodynamic instrument An instrument in which the interaction of magnetic fields produced by currents in a system of movable and fixed coils produces a torque. These instruments operate with both direct and alternating currents.

electrodynamics The study of the mechanical forces acting upon and between currents.

electrodynamometer An electrodynamic instrument that contains two coils, one fixed and one movable, and produces a deflection equal to the square of the current passing through the coils, which are connected in series. It is frequently used as a standard for current or voltage measurements.

For use as a ➤wattmeter, the fixed coil is connected in series with the load and the movable coil shunted across a high noninductive resistance. Transformer coupling to the current circuit may be used to isolate the instrument from the main current. The couple produced is then proportional to the wattage.

electroencephalograph (EEG) A sensitive instrument that measures and records the voltage waveforms produced by the brain. The trace produced is called an *electroencephalogram*.

electroglottogram (EGG) ➤electroglottograph.

electroglottograph *Syn.* electrolaryngograph. A device that monitors the vibration of the human vocal folds in the larynx during speech. As the vocal folds vibrate they come into contact and peel apart regularly once per cycle. Two electrodes are placed externally at the level of the larynx, one either side of the neck. A high-frequency (1–3 MHz) constant-voltage signal is applied to one electrode and the interelectrode current-flow waveform is output; this is known as an *electroglottogram* (EGG). The nature of vocal-fold vibration during normal and pathological speech as well as singing can be studied from the electroglottogram.

electrolaryngograph *Syn. for* electroglottograph.

electroless plating Any plating process performed without the presence of an external applied voltage (➤electroplating). The techniques are particularly appropriate for plating nonconducting or semiconducting surfaces with a noble metal such as gold. The surface to be plated is immersed in an appropriate solution. *Immersion plating* proceeds by the chemical replacement of a low electrode potential metal on the surface by the required higher electrode potential metal (►electrochemical series). The reaction ceases once a thin layer has formed. *Autocatalytic plating* involves the continuous chemical reduction of the more noble metal to form a coating on the base metal. The coating metal may act as a catalyst for the reduction. Both techniques are complex.

electroluminescence *Syn.* Destriau effect. The emission of light by certain phosphorescent substances under the influence of an applied electric field. For example, if a slab of dielectric material within which phosphorescent power is dispersed is sandwiched between transparent plate electrodes, light is produced when voltages of 400 to 500 volts are applied across the plates.

electrolysis The production of chemical change, usually decomposition, when current flows through an electrolyte.

electrolyte A substance that conducts electricity when in solution or when molten because of its dissociation into ►ions. *Strong electrolytes* are compounds, such as mineral acids, that are completely dissociated into ions when in solution. *Weak electrolytes* are compounds that are only partially dissociated in solution. Weak solutions of such compounds are better conductors than strong solutions because of their greater degree of dissociation.

electrolytic capacitor Any ►capacitor in which the dielectric layer is formed by an electrolytic method. The capacitor does not necessarily contain an electrolyte. When a metal electrode, such as an aluminium or tantalum one, is operated as the anode in an electrolytic cell a dielectric layer of the metal oxide is deposited. The capacitor is formed using either a conducting electrolyte as the second electrode or a semiconductor, such as manganese dioxide. The electrolyte used is either in liquid form or in the form of a paste, which saturates a paper or gauze. Tantalum capacitors containing a nonliquid electrolyte are usually in bead form. Electrolytic capacitors have a high capacitance per unit volume but suffer from high leakage currents.

electrolytic cell ►cell.

electrolytic dissociation The reversible separation of certain substances in solution into oppositely charged electrolytic ions. The positively charged ions are ►cations, the negatively charged ones ►anions. Some compounds, such as sodium chloride, are completely dissociated whereas others, such as acetic acid, are only partially dissociated. ►►electrolyte.

electrolytic meter An instrument, such as a ►coulombmeter, that measures electric charge by the amount of material deposited or gas liberated electrolytically. It can be used as an energy meter with a constant voltage supply; it is then calibrated in joules or kilowatt-hours.

electrolytic photocell An electrolytic cell, constructed from certain materials such as selenium electrodes in selenium dioxide solution, that is sensitive to light when a small external d.c. voltage is applied across it. The cell has a linear sensitivity of about one milliampere per lumen of luminous flux.

electrolytic polishing *Syn.* electropolishing. Polishing of a metal by making it the anode of an electrolytic cell and passing current through the cell. Under suitable conditions the protuberances dissolve preferentially leaving a smooth lustrous surface.

electrolytic rectifier A ➤rectifier that has two dissimilar electrodes in an electrolyte. Suitable combinations of electrodes and electrolyte produce a system that conducts much more readily in one direction than the other. A typical combination is the tantalum rectifier, which contains tantalum and lead electrodes and a dilute solution of sulphuric acid as the electrolyte.

electromagnet A device that is magnetized only when an electric current flows in it. It consists of a helical winding around which a magnetic field is produced when current flows through the winding. A ferromagnetic ➤core is almost invariably used: the design of the core is a major factor in determining the magnetic flux density. The winding around which the field is produced is the *field coil;* the current used to produce the field is the *field current.*

electromagnetic balance ➤Cotton balance.

electromagnetic compatibility (EMC) The degree to which an electronic system is able to function compatibly with other electronic systems and not be susceptible to or produce ➤interference. Such interference is commonly termed *electromagnetic interference* (EMI). If the frequency of the interference lies within the radiofrequency part of the electromagnetic spectrum, it is termed *radiofrequency interference* (RFI).

A system may be electromagnetically incompatible with other systems in a number of ways. *Conducted/radiated emissions* are said to occur when the system unintentionally produces signals that are conducted/radiated away from itself. These emissions may cause interference to other systems. The system is said to suffer from *conducted/radiated susceptibility* if signals that are present in its environment are transferred in or onto the system through conduction/radiation and cause the system to fail. In general, *conducted interference* travels along cables, while *radiated interference* travels through the space between source and victim. Conducted interference tends to be more important at lower frequencies (below about 50 MHz), and the radiated interference at higher frequencies.

It is important to note that a system may and probably will receive unintentional signals that do not cause it to fail. In this case the system is electromagnetically compatible with its environment – it is only when it fails to function in its intended manner that it constitutes interference. If a system does not fail under the presence of an unintentional signal, it is said to possess *immunity* from electromagnetic interference.

Generally, the failure of a system takes place when an unintentional signal passes into the system and interacts with an active device in that system. These devices usually interact with each other by way of transmission lines carrying *differential-mode currents*, where the current is carried into the device along one line and returned along the other. Ideally, differential-mode currents are equal in magnitude and oppositely directed along each line of the transmission line. In practice, the transmission line will not be exactly balanced, neither geometrically nor electrically, giving rise to *common-mode currents* that flow with equal magnitude in the same direction along each of the lines.

The presence of common-mode currents along a transmission line can significantly increase the radiated energy from that line, despite the relatively small size of these currents compared to differential-mode currents, because the transmission line is a more efficient radiator of common-mode signals than differential-mode. They thus considerably enhance the radiated interference from a system. Because an efficient radiator is also an efficient receiver of energy, common-mode currents are also easily induced on transmission lines compared to differential-mode currents. The nonideal nature of the transmission lines then ensures that some of the energy in the common-mode currents is converted to differential-mode current and passes into the terminating devices where it may cause failure.

A means of reducing the effects of common-mode radiation or reception is to use *common-mode chokes*, typically made of *ferrite beads*, which are placed around cabling and prevent the flow of common-mode currents while allowing the normal operation of the transmission line.

Another common failure mode is when a system is subject to *electrostatic discharge* (ESD) in which a burst of charge is transferred to the system causing it to fail, typically irrecoverably due to damage of the semiconductor devices. A typical source of ESD is from human contact where the peak voltage during the burst may approach 10 kV. It is particularly important to guard against ESD when assembling or replacing electronic components. In this case a *wrist strap* is often worn; this consists of a conducting lead with one end earthed to the same earth as the equipment being handled and the other end wrapped around the wrist of the person working on the equipment to make a good electrical contact. This ensures that both the person and the equipment are at the same potential and prevents the relative build-up of charge.

A similar type of failure mode can occur due to *electromagnetic pulse* effects from nuclear detonations. Here a pulse of energy is radiated out and may cause large currents to be induced in a system with a high risk of failure and, as in the case of ESD, with a significant probability of irrecoverable damage to the devices in a system.

Most countries have developed EMC standards for electronic products that specify acceptable levels of emissions and immunity. These involve a series of tests on the product to determine the radiated/conducted emissions from the product and, by subjecting it to standard intensities of radiation/current, determine the immunity of the product. In the case of radiated standards, antennas such as the ►log-periodic, ►biconical and ►Bilog may be used to produce/receive the radiation; in the case of conducted standards, typically a *line impedance stabilization network* (LISN) is inserted in series in the a.c. power supply to monitor the currents on the power cable. The LISN provides a reference supply impedance across a range of frequencies.

electromagnetic deflection *Syn.* magnetic deflection. The use of ➤electromagnets to deflect an electron beam. The most common use is in the ➤cathode-ray tube in which two pairs of deflection coils produce vertical and horizontal deflections.

electromagnetic focusing *Syn.* magnetic focusing. ➤focusing.

electromagnetic induction The production of an electromotive force in a conductor when there is a change in magnetic flux through the conductor. The laws of electromagnetism may be expressed as follows.

(i) When a moving conductor cuts the flux of a magnetic field or when a changing magnetic field crosses a conductor an *induced electromotive force* is produced across the conductor.

(ii) *Faraday–Neumann law* (or *Faraday's law*): if a conductor cuts a magnetic flux, Φ, the induced potential difference, V, is proportional to the rate of change $(d\Phi/dt)$ of flux.

(iii) *Lenz's law*: the induced potential difference is in such a direction as to oppose the change that produces it:

$$V = -d\Phi/dt$$

If a current in a circuit varies, the associated magnetic flux also changes in direct proportion causing a back e.m.f. This is *self-inductance*, and the back e.m.f. is given by

$$V = -LdI/dt$$

where I is the current and L is the coefficient of self-inductance, also called *self-inductance*, which is measured in henrys.

The change in flux associated with a varying current can also link with another circuit and produce an e.m.f. in it. This is *mutual inductance*. The induced e.m.f. in a second circuit is given by

$$V_2 = -MdI_1/dt$$

where M is the coefficient of mutual inductance, also called *mutual inductance*, which is measured in henrys. In an ideal mutual inductance between two circuits with self-inductance L_1 and L_2,

$$M^2 = L_1L_2$$

electromagnetic interference (EMI) ➤electromagnetic compatibility.

electromagnetic lens *Syn.* magnetic lens. An ➤electron lens consisting of an arrangement of coils that focuses an electron beam electromagnetically. ➤focusing. ➤➤electrostatic lens.

electromagnetic pulse (EMP) ➤electromagnetic compatibility.

electromagnetic radiation Energy resulting from the acceleration of electric charge and the associated electric and magnetic fields. Transverse sinusoidal electric and magnetic fields are propagated at right angles to each other and to the direction of motion, the instantaneous values of the fields being related to the charge and current densities by ➤Maxwell's equations. These equations define the fields as *electromagnetic waves* propagated through free space at the *speed of light*, symbol c, which is a constant equal to

$$2.997\ 924\ 58 \times 10^8 \text{ metres per second}$$

Just as moving charged particles have associated wavelike features (➤de Broglie waves) so electromagnetic radiation has a wave/particle duality: it may also be considered as a stream of particles (➤photons) that move at the speed of light, c, and have zero rest mass. Although wave motion is sufficient to explain the properties of reflection, refraction, and interference, ➤quantum theory, which is concerned with the particulate nature of electromagnetic waves, must be used to explain phenomena, such as the ➤photoelectric effect, that occur when radiation and matter interact.

The characteristics of the radiation depend on the frequency, ν, of the waves. The frequency, wavelength λ, and speed c are related by

$$c = \nu\lambda$$

The photons have energy E related to the frequency ν by

$$E = h\nu$$

where h is the ➤Planck constant. Energy is exchanged between radiation and matter by absorption and emission in discrete amounts, termed ➤quanta, the energy of each quantum being $h\nu$.

The total range of possible frequencies is defined as the electromagnetic spectrum (➤Table 10, in the back matter). ➤Radiowaves have the lowest frequencies; progressively higher frequencies are associated with infrared radiation, light, ultraviolet radiation, ➤X-rays, through to ➤gamma rays at the highest frequencies.

electromagnetic spectrum The entire range of frequency of ➤electromagnetic radiation. ➤Table 10, in the back matter.

electromagnetic strain gauge *Syn.* variable inductance gauge. ➤strain gauge.

electromagnetic units (emu) ➤CGS system.

electromagnetic wave ➤electromagnetic radiation.

electromagnetism 1. The magnetic phenomena associated with moving electric charges, the electric phenomena associated with varying magnetic fields, and the magnetic and electric phenomena associated with electromagnetic radiation. **2.** The study of these phenomena, relating electricity to magnetism.

electrometer A device, usually based on an ➤operational amplifier, that measures potential difference without drawing much current from the source. These devices can also measure low currents by passing the current through a high resistance. Originally, electrometers were electrostatic devices based on the electroscope.

electromigration Movement of the atoms of a conductor due to the flow of charge carriers: the metal atoms are physically displaced from the solid crystal by collisions with the charge carriers and are moved in the direction of flow of the carriers. This effect occurs at high current densities, and is most noticeable in metals of low atomic mass. For example, the fine interconnecting lines on an integrated circuit have only a small cross-sectional area and can carry a very high current density even at modest currents. Electromigration can cause enough of the aluminium atoms to move so that the conductor is physically broken, resulting in a failure of the component.

electromotive force (e.m.f. or EMF) Symbol: E; unit: volt. The property of a source of electrical energy that causes a current to flow in a circuit. The algebraic sum of the potential differences in a circuit equals the e.m.f., which is measured by the energy liberated when unit electric charge passes completely round the circuit. A battery of e.m.f. E will supply a current I to an external resistance R:

$$E = I(R + r)$$

where r is the internal resistance of the battery. The term 'electromotive force' strictly applies to a source of electrical energy but is sometimes misused as being equivalent to ➤potential difference.

electromyograph A sensitive instrument, such as an ➤electrocardiograph, that measures and records the current waveforms generated by contractions in muscles. The trace produced is an *electromyogram*.

electron A stable elementary particle that has a negative charge, e, of $1.602\ 1773 \times 10^{-19}$ coulomb, mass m of $9.109\ 3897 \times 10^{-31}$ kg, and spin ½. It is the natural unit of electric charge. Electrons are constituents of all atoms, moving around the nucleus in several possible or 'allowed' orbits (➤energy levels). They also exist independently. They are primarily responsible for electrical conduction in most materials (➤energy bands). Electrons moving in one direction under the influence of an electric field constitute an electric current, the direction of conventional current flow being opposite to the direction of motion of the electrons. Electrons were first discovered as ➤cathode rays by Sir J. J. Thomson in 1897.

Electrons are liberated by various effects: in ➤gas-discharge tubes by the ionization of gas molecules; by heating metal filaments (➤thermionic emission); by the action of light, ultraviolet radiation, X-rays, or gamma-rays on matter (➤photoelectric effect); by the application of an intense electric field at the surface of a metal (➤field emission); by bombardment of a surface by high-speed electrons or positive ions (➤secondary emission).

Electrons experience an electric force F_E in the presence of an electric field E; moving electrons experience a transverse force F_H in the presence of a magnetic field. These have values

$$F_E = Ee$$

$$F_H = ev \times B$$

where B is the magnetic flux density vector acting at right angles to v, the velocity vector. These forces are utilized in the ►focusing of electron beams. The electric force also accelerates the electrons.

Electrons interact with matter to produce effects dependent on the velocity of the electrons and the state of matter involved. They may undergo elastic scattering, producing deflection and localized heating due to the energy lost, or inelastic collisions, losing discrete amounts of energy and producing various phenomena: with a gas, excitation and light emission or ionization occurs; with solids and liquids, effects such as the production of X-rays, fluorescence, and secondary electron emission are observed.

An electron beam also has wavelike properties, similar to those of electromagnetic radiation, the wavelength being given by

$$\lambda = h/mv$$

where h is the Planck constant and mv the electron's momentum (►► de Broglie waves).

The antiparticle of the electron is the *positron,* which has a positive charge and a mass equal to that of the electron.

electron affinity 1. Symbol: A or E_a; unit: electronvolt. The energy released when an electron becomes attached to an atom or molecule to form a negative ion. Many atoms or molecules have a positive electron affinity, i.e. the negative ion is more stable than the neutral species. **2.** Symbol: χ; unit: electronvolt. The difference in energy between the bottom of the conduction band, E_c, in a semiconductor and the vacuum level, i.e. the energy of free space outside the semiconductor. ►► work function.

electron beam A beam of ►electrons usually emitted from a single source, such as a ►thermionic cathode. The *electron-beam voltage* at a point in an electron beam is the average, with respect to time, of the voltage between that point and the electron-emitting surface. The *electron-beam d.c. resistance* is the ratio of the electron-beam voltage and the direct-current output of the electron beam. ►► electron gun.

electron-beam d.c. resistance ►electron beam.

electron-beam device Any device that utilizes one or more electron beams as an essential part of its operation. Several electrodes may be present to form, control, and direct the electron beam. The deflected beam is made to strike a fluorescent screen in measuring instruments such as the ►cathode-ray tube. ►► electron microscope; klystron.

electron-beam induced current analysis (EBIC) A technique that is used to detect crystal defects in a semiconductor slice. The technique requires the presence of a ►p-n junction, ►Schottky barrier, or ►MOS capacitor to be successful. An electron beam is scanned across the sample, which leads to the generation of electron-hole pairs. The charges generated are collected by the diode and the resulting current detected. Any defect or inhomogeneity that affects the production or recombination of the electron-hole pairs will also affect the current detected. Use of a ►cathode-ray tube synchronized with the scanning electron beam produces an image of the area scanned. EBIC can be easily added to many scanning ►electron microscopes.

electron-beam lithography (e-beam lithography) A method of ➤lithography that uses energetic electrons instead of light to expose the ➤resist. It is used to form patterns on photomasks for use in ➤photolithography. It can also be used to form patterns directly on the semiconductor – *direct writing*. This latter application is often used to form small features on ➤wafers since greater resolution is possible using an electron beam than with photolithography.

Special resists – *e-beam resists* – are used. These are not usually sensitive to ordinary light and can therefore be handled under normal lighting conditions. Both positive and negative resists are available. Unlike photolithography no mask is used in e-beam lithography. The electron beam is scanned across the slice under computer control in order to expose the resist in the desired pattern. Two main scanning methods are used: *raster scanning*, in which the beam is moved backwards and forwards across the slice in a rectangular pattern and the beam switched on and off as appropriate, and *vector scanning*, in which the beam is moved, under computer control, to trace out the pattern directly. The scanning method is chosen according to the resolution required and the type of pattern to be produced. Different machines are required for the two methods.

Because of the extremely high resolution required in direct writing only a very small portion of the wafer can be exposed at a time. After exposure of one field, another portion of the wafer must be moved into position. Final alignment within the field is performed by the use of *alignment marks* on the wafer which are scanned by the beam and monitored by a detector.

electron-beam voltage ➤electron beam.

electron binding energy ➤ionization potential.

electron capture ➤recombination processes.

electron charge ➤charge; electron.

electron density 1. The number of electrons per unit mass of a material. **2.** The number of electrons per unit volume of a material. ➤carrier concentration. **3.** *Syn.* equivalent electron density. The product of the ion density in an ionized gas and the ratio of the electron mass to the mass of a gas ion.

electron diffraction ➤diffraction.

electronegative ➤electrochemical series.

electron emission The liberation of ➤electrons from the surface of a material. ➤emission.

electron gas The concept that the ➤free electrons in a solid, liquid, or gas may be treated as a gas and compared with a real gas dissolved in the appropriate material. Such an electron gas obeys a totally different energy distribution than a real gas, obeying ➤Fermi–Dirac statistics rather than the Maxwell–Boltzmann statistics obeyed by an ordinary gas. It also has a vanishingly small specific heat.

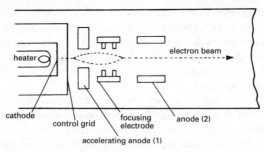

Electron gun

electron gun A device that produces an ➤electron beam and forms an essential part of many instruments, such as ➤cathode-ray tubes, ➤electron microscopes, and ➤linear accelerators, all of which need electron beams for their operation. The source of electrons may be a ➤thermionic cathode, as shown in the diagram, or a ➤field emitter. A series of accelerating and focusing electrodes usually produces a narrow beam of high-velocity electrons.

electron-hole pair A ➤hole created in the valence band of a semiconductor when an electron is thermally excited into the conduction band. Relaxation of a conduction electron into an unoccupied energy level in the valence band results in band-to-band recombination (➤recombination processes). ➤➤semiconductor.

electron-hole recombination Syn. for band-to-band recombination. ➤recombination processes.

electronic device A device that utilizes the properties of electrons (or ions) moving in a vacuum, gas, or semiconductor.

electronic efficiency The ratio of the output power delivered to the load of an ➤oscillator or ➤amplifier at the desired frequency to the average power input to the device.

Electronic Industries Association (EIA) ➤standardization.

electronic mail (e-mail or email) A means of transmitting and receiving messages using a network of computers and the telephone system. A central computer acts as a clearing house for messages: it is accessed by other computers using the telephone lines and can receive and store messages or transmit stored messages on demand. A commonly used medium is the Internet.

electronic memory A ➤memory that has no moving parts and in which the read and write operations are entirely electronic. The earliest forms of electronic memory employed thermionic valves in the memory cells. These were replaced by the much smaller and cheaper ferrite ➤cores, which consumed very much less power. These in turn have been superseded by ➤solid-state memories that are extremely versatile high-speed low-power devices: the growth of microelectronic techniques has made possible the production of semiconductor memories of large memory capacity and extremely small physical size.

electronic news gathering A recording system used in ►television in which scenes outside the television studio are recorded directly on ►videotape rather than on film. A portable ►televison camera and videotape recorder are used, often in conjunction with a mobile transmitter that relays the recording directly to the main control centre.

electronics The study, design, and use of devices that depend on the conduction of electricity through a vacuum, gas, or semiconductor. ►Electrons and ►holes are the most important forms of charge carriers in electronic devices but ►ions also play a part.

electronic switch An electronic device, such as a transistor, that is used as a switch. These devices are usually operated as high-speed switches where a very fast response is required, as for example in a computer.

electronic tuning 1. The use of a signal on a control line of a circuit that changes the frequency of operation of the circuit. Examples of circuits include oscillators and filters where the frequency of the output signal can be changed by varying the voltage on a tuning-control input to the circuit. **2.** The use of an electron beam in one circuit to change the operating frequency of a second circuit. The velocity, intensity, or configuration of the electron beam may be altered to provide the desired frequency change.

electron lens A device that is used for ►focusing an electron beam using either magnetic or electrostatic fields (►electromagnetic lens; electrostatic lens). The two types, which are analogous to optical lenses used with light beams, can be combined in instruments such as electron microscopes. Most electron-beam devices, such as cathode-ray tubes or camera tubes, use an electron lens to provide a sharp narrow beam.

electron microprobe An instrument for analysing the composition of semiconductor material below the surface, using X-rays emitted by the material in response to stimulation by a beam of electrons. The instrumentation is essentially a scanning ►electron microscope (SEM) combined with an X-ray fluorescence detector. The X-ray detector is mounted on a normal SEM and the electron beam of the SEM excites atoms in the material. Both the normal SEM picture and the X-ray spectra are produced simultaneously. The *energy dispersive spectroscope* (EDS) is a sensitive X-ray detector that detects all wavelengths simultaneously. The method is not very sensitive to low atomic number elements, which tend to emit ►Auger electrons rather than X-rays.

   X-ray fluorescence (XRF) is a very similar technique but an X-ray beam is used to produce the X-ray output rather than an electron beam. XRF produces poor lateral resolution because of the lack of focusing of the X-ray beam but it does have very good sensitivity. Once again low atomic number elements are not detected because of the emission of Auger electrons rather than X-rays.

electron microscope An instrument that uses a beam of electrons to investigate a sample in order to achieve a higher magnification and resolution than is possible with an optical microscope.

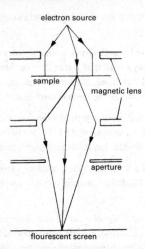

a Transmission **electron microscope**

In the *transmission electron microscope* (TEM) the electron beam is focused by an ➤electromagnetic lens or sometimes an ➤electrostatic lens and has an energy of 50–100 kilovolts (Fig. *a*). A sharply focused image in one plane can only be obtained by using monoenergetic electrons. To avoid energy losses in the beam the sample must be extremely thin (<50 nanometres) so that the scattered electrons that form the image are not changed in energy, and the image therefore appears two-dimensional. A resolution of 0.2–0.5 nanometres is possible.

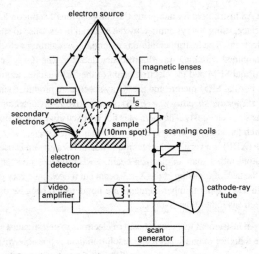

b Scanning **electron microscope**

The *scanning electron microscope* (SEM) has lower resolution and magnification but produces a seemingly three-dimensional image, with great depth of field, from a sample of any convenient size or thickness (Fig. *b*). The sample is scanned by the electron beam, the numbers of resulting secondary electrons being proportional to the geometry and other properties of the sample. These electrons are converted, by means of an electron detector, ►scintillator, and ►photomultiplier, into a highly amplified signal that is used for intensity modulation of the beam of the display cathode-ray tube. The resolution is about 10–20 nm.

Different forms of these two types have been developed. The *scanning-transmission electron microscope* (STEM) combines the high resolution of the transmission instrument with the perspective image of the scanning type. The *scanning-tunnelling microscope* (STM) and the *atomic force microscope* (AFM) have even higher resolution, producing computer-generated contour maps of the sample surface.

Both the TEM and the STEM can be used to produce diffraction patterns from thin samples of crystalline material such as ►semiconductors, rather than the usual image. This can be used to analyse the crystal structure of such materials. ►diffraction.

electron multiplier An electron tube in which current amplification is achieved by means of ►secondary emission of electrons. Primary electrons are released from the cathode by some means, such as the ►photoelectric effect. These are accelerated by a high potential applied to the first anode. The anodes, termed *dynodes,* are made from a good secondary emitter and on impact a greater number of electrons is produced. These are then accelerated in turn by the second and subsequent anodes, to each of which an increasing positive potential is applied (see diagram). A large output pulse is produced at the final anode – the collector – which is operated at a very high potential. ►photomultiplier.

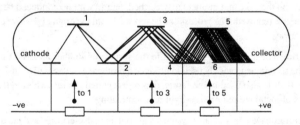

Electron multiplier

electron optics The study of the behaviour of ►electron beams under the influence of magnetic and electrostatic fields in a vacuum or very low pressure gas: the analogue of light beams passing through refractive media. The applied fields form ►electron lenses that are used for ►focusing or defocusing the electron beam.

electron tube An active device in which conduction between two electrodes takes place in an envelope that is sealed or continuously exhausted and contains a gas or a vacuum. Tubes that are evacuated to a high vacuum are termed *hard tubes*; those that contain traces of a gas are *soft tubes*. The concentration of gaseous atoms in soft tubes is sufficient to cause some modification to the characteristics compared with those of hard tubes. Electron tubes frequently contain more than two electrodes. ➤vacuum tube; thermionic valve.

electronvolt Symbol: eV. A unit of energy equal to the energy acquired by an electron when it passes freely through a potential difference of one volt:

$$1 \text{ eV} = 1.602\ 177 \times 10^{-19} \text{ joule}$$

The unit is used principally in atomic, nuclear, and particle physics.

electron voltaic effect A phenomenon similar to the ➤photovoltaic effect in which electrons striking the photocathode of a photocell cause electron emission. At low energies the gain increases rapidly with voltage to a maximum and then decreases.

electro-optical effect ➤Kerr effects.

electro-optical shutter *Syn. for* Kerr cell. ➤Kerr effects.

electro-optics The study of the interactions between the refractive indices of some transparent dielectrics and the electric fields in which they are placed. Changes in the optical properties of dielectrics are produced. ➤Kerr effects.

electrophoresis The movement of colloidal particles in a liquid under the influence of an electric field. Positively charged particles migrate to the cathode (*cataphoresis*) and negatively charged ones to the anode (*anaphoresis*).

electrophorus An early form of simple electrostatic charge generator. A flat dielectric plate ('cake') is positively charged by friction. A metal plate with an insulated handle is placed on the cake and momentarily earthed, taking with it an induced negative charge when it is removed. The process can be repeated until the original positive charge has leaked away.

electroplating The application of a metal to the surface of another material by electrolysis. The technique is widely used to produce a protective or decorative surface on metal objects. It is also used in the fabrication of integrated circuits where relatively thick layers of metal are required. ➤light-assisted plating.

electropneumatic Denoting control systems that contain both electronic and pneumatic elements.

electropolar Having magnetic poles or permanent positive and negative charges, as in a magnet or an electret.

electropolishing ➤electrolytic polishing.

electropositive ➤electrochemical series.

electroscope An electrostatic instrument that detects small potential differences and electric charges. The *gold-leaf electroscope* consists of a pair of gold leaves hanging side by side from an insulated metal support enclosed in a draught-proof case. If a charge is supplied to the support the leaves separate due to their mutual repulsion. A more precise form replaces one of the gold leaves by a rigid metal plate (see diagram). *Lauritzen's electroscope* utilizes metallized quartz fibre as the sensitive element.

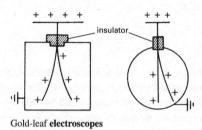

Gold-leaf **electroscopes**

An instrument that is capable of accurate quantitative measurement of potential difference is called an ➤electrometer.

electrostatic adhesion Adhesion between two substances or surfaces due to the presence of opposite charges, which attract each other.

electrostatic deflection The use of electrostatic fields produced between two metal electrodes for deflecting an electron beam. The electrodes used are called *deflection plates* and two pairs of plates, at right angles to each other, are used in electron-beam devices, such as ➤cathode-ray tubes, to provide deflection in two orthogonal directions. ➤➤electromagnetic deflection.

electrostatic discharge (ESD) ➤electromagnetic compatibility.

electrostatic field The electric field associated with charged particles at rest. It is the region, around a distribution of electrostatic charge, in which a stationary charged particle would experience a force. ➤➤Coulomb's law.

electrostatic focusing ➤focusing.

electrostatic generator ➤generator.

electrostatic induction The production of a charge distribution on a conductor when in the vicinity of another charged body under the influence of the associated electric field. If the conductor is in the vicinity of a positively charged body the region nearest to the charged body becomes negatively charged; the more remote regions become positively charged. The reverse effect is seen if the body is negatively charged.

electrostatic lens An ➤electron lens consisting of an arrangement of electrodes that focuses an electron beam electrostatically. ➤focusing. ➤➤electromagnetic lens.

electrostatic loudspeaker ➤loudspeaker.

electrostatic precipitation A method of precipitating solid or liquid particles from a gas. An electrostatic field is applied across the gas between two electrodes, one of which is earthed (usually the positive one). The particles collect on the earthed electrode.

electrostatics The study of electric charges at rest and their associated phenomena.

electrostatic units (esu) ➤CGS system.

electrostatic voltmeter ➤voltmeter.

electrostatic wattmeter ➤wattmeter.

electrostriction A change in the dimensions of a body under the influence of an electric field in a medium of relative ➤permittivity different from its own. Forces of extension or compression may be produced. In a nonhomogeneous field the body will also tend to move: one of higher relative permittivity than its surroundings tends to move into the region of higher field strength and vice versa.

electrothermal instrument ➤thermal instrument.

electrovalent bond ➤ionic bond.

element A substance that consists entirely of atoms of the same atomic number. Over 100 different elements have been identified (➤periodic table). Elements are the basic substances from which compounds are built up by chemical combination: this excludes disruptive processes in which the atomic nuclei comprising the element are disturbed.

elementary particle Any of the particles of matter that cannot be subdivided into smaller particles. Elementary particles are described by a set of quantum numbers describing their intrinsic properties, such as ➤charge and ➤spin. The stable particles are the ➤electron, ➤proton, ➤photon, and neutrino. The ➤neutron is also stable when bound in an atomic nucleus. The electron is the natural unit of electric charge.

Elementary particles can be classified in groups according to the types of *fundamental interaction* to which they are subject. These groups include the hadrons (such as the neutron and proton) and the leptons (including the electron and neutrino). Hadrons are believed to have an internal structure and to consist of *quarks* permanently confined within the hadrons.

elliptical polarization ➤polarization.

elliptic filter *Syn.* Cauer filter. A ➤filter that is characterized by a ➤magnitude response with equal ripples (➤equiripple) in both the pass-band and stop-band, producing a filter with a steeper roll-off or narrower ➤transition band compared to ➤Chebyshev and ➤Butterworth filters. ➤elliptic function.

elliptic function The function that describes the ➤magnitude response $H(\omega)$ of an ➤elliptic filter and is given by:

$$H(\omega) = 1/\sqrt{(1 + \varepsilon R_n^2(\omega))}$$

where ε is a constant that determines the amount of ripple in the pass-band and stop-band, and

$$R_n(\omega) = [\omega(\omega_1{}^2 - \omega^2)(\omega_2{}^2 - \omega^2)\ldots(\omega_k{}^2 - \omega^2)]/[(1 - \omega_1{}^2\omega^2)(1 - \omega_2{}^2\omega^2)\ldots(1 - \omega_k{}^2\omega^2)]$$

e-mail (or **email**) *Short for* electronic mail.

embedded controller A device, such as a laser printer, in which the control processor is an integral part of the device.

EMC *Abbrev. for* electromagnetic compatibility.

e.m.f. (or **EMF**) *Abbrev. for* electromotive force.

EMI *Abbrev. for* electromagnetic interference. ➤electromagnetic compatibility.

emission The liberation of electrons or electromagnetic radiation from the surface of a solid or liquid, usually electrons from a metal. The outer electrons of the atoms in a metal (conduction electrons) move in a random manner among the lattice atoms with no net forces on them in the bulk of the material. Electrons near the surface of the material with directions of motion out of the surface can leave the surface, but then experience a force directing them back to the metal as the metal is left positively charged. The charge on the metal can be considered as an ➤electric image located the same distance inside the metal as the electron is outside it. The force on the electron varies with distance x from the surface (Fig. *a*). As the electron moves out from the surface, work (W) is done to overcome this force (Fig. *b*), where

$$W = \int_0^x F \mathrm{d}x$$

The value W_1 represents a potential barrier that must be overcome by the electron. Electrons can only escape if their energies are greater than W_1. W_1 is related to the ➤work function, Φ, by

$$\Phi = W_1 - E_F$$

where E_F is the ➤Fermi energy.

Normally only a few electrons will have velocities, due to their thermal energy, large enough to escape (Fig. *c*). Emission occurs when sufficient energy is given to the elec-

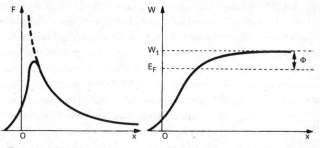

a Force on emitted electron as function of distance from surface

b Work done to overcome force on emitted electron

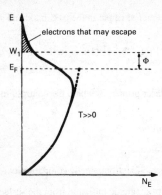

c Number of electrons as a function of electron energy

trons to allow them to escape (as in ►photoemission, ►thermionic, or ►secondary emission) or the potential barrier is distorted by the presence of an intense electric field (as in ►field emission).

emissions *Short for* conducted or radiated emissions. ►electromagnetic compatibility.

emissivity Symbol: ε. When radiation is produced by the thermal excitation of atoms, molecules, etc., the emissivity is defined as the ratio of the power per unit area radiated from the surface to that radiated by a black body at the same temperature.

emitter 1. *Short for* emitter region. The region of a ►bipolar junction transistor from which carriers flow, through the emitter junction, into the ►base. The electrode attached to this region is called the *emitter electrode.* ►semiconductor. **2.** *Short for* emitter electrode.

emitter-coupled logic (ECL) A family of integrated ►logic circuits so called because a pair of transistors coupled by their emitters forms a fundamental part of the circuit. The basic ECL gate has simultaneously the function required and its complement.

A simple OR/NOR circuit is shown in Fig. *a*. Input is via the ►bipolar junction transistors $T_{1a,b,c}$; these are emitter-coupled to transistor T_2 and form a ►long-tailed pair with it. This is an excellent ►differential amplifier. An ►emitter-follower buffer forms the output stage. Transistor T_2 has a fixed bias applied to its base with magnitude halfway between a logical 1 and a logical 0. If a logical 0 is applied to all three input transistors then current flows through T_2 causing a voltage drop across R_2. This in turn produces a logical 0 at the OR output and a logical 1 at the NOR output. If any one of the input transistors $T_{1a,b,c}$ has a logical 1 applied, current flows through that transistor producing a voltage drop across R_1 and the outputs are hence reversed, i.e. a logical 1 occurs at the OR output. Typical values of applied voltages are –1.55 volts (logical 0), –0.75 volts (logical 1), –1.15 volts (fixed bias).

The transistors are operated in nonsaturated mode and the ►delay is exceedingly short (approximately one nanosecond) making ECL circuits inherently the fastest type of logic circuit.

Simpler versions of the original ECL circuits have been designed for VLSI circuits; these have a higher ►packing density and operate with lower voltage swings. Fig. *b*

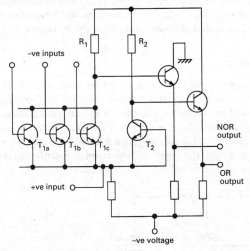

a ECL OR/NOR circuit

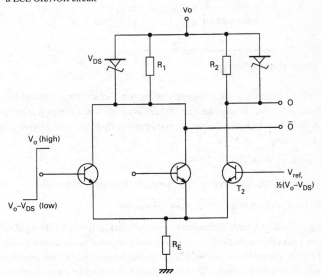

b Low-voltage ECL gate

shows a simple low-voltage ECL gate in which the emitter-follower transistors are replaced by ➤Schottky clamped-load resistors R_1 and R_2. The fixed reference bias applied to transistor T_2 is generated 'on-chip' rather than being supplied externally. The
total difference betwen the 'high' and 'low' logic levels is equal to the forward bias
of the Schottky diode, V_{DS}.

An alternative form of higher packing density ECL circuit uses ECL circuits connected in series (gated); this allows a more complex logic function to be implanted on
a smaller area of chip. This method of *series-gated* circuit design is also widely used
in FET circuitry.

emitter electrode ➤emitter.

emitter follower An amplifier that consists of a ➤bipolar junction transistor with
➤common-collector connection, the output being taken from the ➤emitter (see diagram).
The transistor is suitably biased so that it is nonsaturated and conducting. The emitter
voltage thus has a constant value relative to the base at all times, and the emitter follows the signal applied to the base. The voltage gain of the amplifier is therefore nearly
unity but the current gain is high. The amplifier is often used as a ➤buffer and is characterized by a high input impedance and low output impedance.

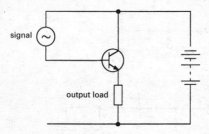

Simple **emitter follower**

The ➤FET analogue is the *source follower* and the ➤thermionic-valve analogue is
the *cathode follower*, although neither of these is as efficient a unity-gain buffer amplifier as the emitter follower, the voltage gain, particularly of the source follower, being
further from unity.

emitter region ➤emitter.

EMP *Abbrev. for* electromagnetic pulse. ➤electromagnetic compatibility.

emu *Abbrev. for* electromagnetic unit. ➤CGS system.

enabling Activating a particular circuit or group of circuits from a larger set of circuits
in order to effect their operation. An *enabling pulse* or *signal* is often used to select the
desired circuit. Examples include selecting a particular input stage from several inputs
in parallel, or selecting a particular semiconductor chip from an array of several similar chips.

encephalograph *Short for* electroencephalograph.

encoder A device or circuit that produces an output in a desired coded form.

encryption The process by which the privacy of a digital communication can be increased by altering the coding of each part of the message – i.e. by *encrypting* the message – such that it has a meaning that can only be converted back to its original form, or *decrypted*, by a recipient with knowledge of the encryption method. ➤cipher; digital communications.

endfire array *Syn.* staggered antenna. ➤antenna array.

energy bands In a crystalline solid, the atoms are close enough together for their ➤energy levels to overlap. As ➤quantum theory requires each electron in a system to have a unique set of ➤quàntum numbers, the individual levels split into bands of energy – known simply as energy bands – to accommodate all the electrons in the solid. Note that the energy bands are not continuous bands of energy, but are ranges of energy containing many closely spaced energy levels.

The formation of the energy bands can be demonstrated mathematically, by solving Schrödinger's equation of ➤quantum mechanics for an electron moving in the periodic potential $V(k)$ that is associated with the atoms in the crystal. The solutions reveal *allowed bands* of electron energies, separated by *forbidden bands* of energy in which there are no travelling wave solutions, indicating that electrons cannot move in the crystal at these energies. The set of energy bands in a particular solid is called the *energy-band structure*.

At a temperature of absolute zero, all electrons occupy the lowest possible energy levels in the energy-band structure, each electron occupying just one energy level (i.e. having a unique set of quantum numbers). The energy bands are then filled up to a certain energy value, called the ➤Fermi energy or Fermi level, E_F. At temperatures above absolute zero, some electrons are able to absorb thermal energy from their surroundings – the crystal; the distribution of electrons in the energy levels is then described by ➤Fermi–Dirac statistics, which yield the probability of finding an electron at a given energy. The way in which the electrons occupy the various bands, and the location of

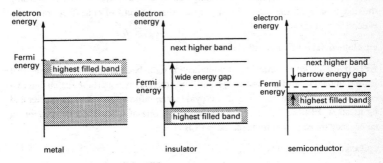

Electron **energy bands** in solids

the Fermi energy, provides the fundamental difference between metals, insulators, and semiconductors (see diagram).

In a metal, the highest energy band containing electrons is only partly filled; this allows the electrons to gain kinetic energy from an applied electric field and move in the solid, giving rise to ►conduction. In an insulator, the highest energy band containing electrons is completely filled; electrons are unable to accept energy from an applied electric field as there are no vacant energy levels within the band for the electrons to move into, so no conduction can take place. A semiconductor is a special class of insulator: while the highest energy band containing electrons is completely filled, the next highest and empty energy band is separated from this band by only a small forbidden energy gap. At absolute zero of temperature a semiconductor is exactly like an insulator, but at higher temperatures a small number of electrons can absorb enough thermal energy to cross the forbidden energy gap to reach the next higher energy band. In this band the electrons are able to accept kinetic energy from an applied electric field, and move to an empty higher energy level, and so conduction can take place. The conductivity is small compared to metals, as only few electrons are able to contribute to the conduction. ►►semiconductor.

energy component 1. (of a current or voltage) ►active current or active voltage. **2.** (of the volt-amperes) ►active volt-amperes.

energy dispersive spectroscopy (EDS) ►electron microprobe.

energy gap Symbol: E_g. A range of forbidden energies, especially that between the highest energy of the valence band and the lowest energy of the conduction band. ►energy bands.

energy levels Discrete values of allowable energy in a quantized system, such as an atom or ►quantum well. Electrons in such systems can only move between the energy levels on absorption or emission of the exact energy difference between the levels. This energy is usually in the form of heat or light (►photons of a specific wavelength), or an applied electric field. ►quantum theory; quantum mechanics. ►►energy bands.

enhancement mode A means of operating field-effect transistors in which increasing the magnitude of the gate bias increases the current. ►field-effect transistor. ►►depletion mode.

envelope detector *Syn. for* diode detector.

EPIRB *Abbrev. for* electronic position-indicating radio beacon. A radio transmitter that is used in the event of an aviation or maritime emergency to alert the search and rescue services to the emergency and to provide some information about the carrier and its location. EPIRBs transmit a coded signal on one of the worldwide recognized emergency transmission frequencies. ►►PLB.

epitaxial growth ►epitaxy.

epitaxial layer ►epitaxy.

epitaxial transistor ►planar process.

epitaxy *Syn.* epitaxial growth. A method of growing a thin layer of material upon a single-crystal substrate, such as silicon, so that the crystal structure of the layer is identical to that of the substrate. The material, which may be the same as the substrate or a different one, is usually deposited from a gaseous mixture (►vapour phase epitaxy). The technique is extensively used in semiconductor technology when a layer (the *epitaxial layer*) of different conductivity to the substrate is required. ►liquid phase epitaxy; molecular beam epitaxy.

E-plane *Syn. for* E-bend. ►waveguide.

EPROM *Abbrev. for* erasable programmable ROM. ►ROM.

equal energy source A source of energy in which the emitted energy is equally distributed throughout the entire frequency range of the source's spectrum.

equalization A means of reducing ►distortion in a system by introducing networks that compensate for the particular type of distortion over the required frequency band. For example, if a loudspeaker system has a poor response to bass frequencies, *bass boost* (amplification of the bass frequencies with respect to the treble frequencies) may be introduced into the system to compensate.

In communication systems, the frequency response of a communication channel is usually not known to sufficient accuracy or is not sufficiently stable to permit a single optimum receiver to be designed. A filter, known as an *equalizer*, is commonly employed to match the receiver to the channel thereby optimizing receiver sensitivity. An *adaptive equalizer* is an equalizer the characteristics of which are dynamically changed to match variations in the channel.

equalizer 1. A network that provides ►equalization. **2.** A low-resistance connection made between two points in an electrical machine to ensure that the points are always at the same potential.

equipotential line A line joining points of equal potential energy or voltage. It is normal to the lines of electric field or force.

equiripple (or **equal ripple**) **1.** A ►magnitude response of a filter circuit that is characterized by equal variations or ripples in the pass band and/or stop band. **2.** *Syn.* equiripple filter. A type of ►filter where equiripple in the filter response is permitted, resulting in a steeper roll-off or narrower ►transition band. ►Chebyshev filters, ►inverse Chebyshev, and ►elliptic filters all possess equiripple responses.

equivalent circuit An arrangement of simple circuit elements that exhibits the same characteristics as a more complicated circuit or device, under specified conditions; it can be used to predict the behaviour of the more complicated system.

equivalent-circuit model ►transistor parameters; modelling.

equivalent network An electrical network that may replace another network without materially affecting the conditions in other parts of the system, but usually at one particular frequency only.

equivalent rectangular bandwidth (ERB) The bandwidth of a filter with a rectangular ➤frequency-response curve that passes equal power and has the same response-curve maximum as the human auditory filter at a given centre frequency. It is otherwise known as the ➤critical bandwidth.

equivalent resistance The value of total ➤resistance that, if placed at a point in a circuit, would dissipate the same power as the total of smaller resistances in the circuit.

erasable PROM ➤ROM.

erase To remove stored information from a location in a storage device such as ➤RAM, ➤magnetic tape, etc.

erasing head ➤magnetic recording.

ERDA *Abbrev. for* elastic recoil detection analysis.

E-region *Syn. for* E-layer. ➤ionosphere.

erlang ➤network traffic measurement.

ERP *Abbrev. for* effective radiated power.

error correction Correction of errors in data transmitted over a communication line or handled by a computer device. *Error-correcting codes* encode data in such a way that a decoder can correct, with high probability of success, any errors produced in the signal by the intervening channel. ➤➤digital codes; digital communications.

error detection Detection of errors in data transmitted over a communication line or handled by a computer device. *Error-detecting codes* encode data in such a way that a decoder can detect, with high probability of success, whether an intervening channel has caused any error in the signal. ➤➤digital codes; digital communications.

Esaki diode ➤tunnel diode.

ESD *Abbrev. for* electrostatic discharge. ➤electromagnetic compatibility.

esu *Abbrev. for* electrostatic unit. ➤CGS system.

etching Chemical erosions of selected portions of a surface in order to produce a desired pattern on the surface. The technique is widely used in microminiature electronics. *Wet etching* utilizes liquids, such as acids or other corrosive chemicals, as the etching agent. The etching process takes place by means of chemical reactions at the surface of the material. This process is limited both by the rate at which the chemical reactions can take place, and by the rate at which the products of the chemical reaction can be removed. These factors are a function of the nature of the liquid used and the temperature at which the etching is carried out. Etching in which the main limiting factor is the reaction rate is known as *reaction rate limited*, *surface limited* or *kinetically lim-*

ited etching. Where the removal rate of the chemical products is the predominant controlling factor the etching is known as *diffusion limited* or *mass transport limited* etching. In certain applications the etching proces can be electrically aided by making the slice to be etched the anode or cathode of an electrolytic cell.

Dry etching uses either chemical or physical reactions between a low-pressure plasma or glow discharge and the surface to be etched. It has several advantages compared to wet etching, but the etching process itself is very complex, and the results can be greatly affected by small variations in the process parameters. Dry etching is capable of patterning smaller geometries than wet etching; lateral etch rates close to zero can be produced under certain conditions, and smooth ►edge profiles can be produced when needed for metal crossovers. Certain semiconductors, particularly gallium arsenide, have no suitable liquid etchants that can produce deep narrow features, and dry etching is particulaly important in these cases.

Plasma etching is any process in which a plasma generates reactive species that then chemically etch material in direct proximity to the plasma. If the chemical reactions are enhanced by the kinetic energy of the ions in the plasma the process is described as *kinetically assisted* chemical reaction. *Reactive ion etching* is similar to plasma etching, but uses only kinetically assisted chemical etching. The applied voltage drop is mainly at the surface of the slice. *Reactive ion beam etching* separates the slices from the plasma by a grid that accelerates the ions created in the plasma towards the slice. The ion energy is higher, and some of the etching is due to physical reactions.

Sputter etching uses energetic ions from the plasma to physically blast (sputter) atoms from the surface. No chemical reactions are involved. *Ion milling* is also a purely mechanical method that uses a roughly collimated beam of energetic ions to erode a surface by bombardment. Ion milling, unlike other dry etching procedures, can be used at angles other than perpendicular to the slice.

etch pit density *Syn. for* dislocation density. ►dislocation.

Ethernet A ►local area network system that allows multiple computers to access a single communications channel to communicate and share resources. The system is capable of a data rate of 10 megabits per second and it uses Manchester coding (►digital codes). The Ethernet system was created jointly by Xerox, Digital Equipment Corporation, and Intel.

Ettinghausen effect The development of a very small transverse temperature gradient in a conductor when carrying current in a magnetic field. The temperature gradient is established in a direction perpendicular to both the magnetic field and the current.

Euler's identity A means of expressing sine and cosine functions in terms of exponentials, and that is stated as

$$e^{j\phi} = \cos\phi + j\sin\phi$$

where $j = \sqrt{-1}$.

eutectic mixture A mixture of elements that retains its composition in both solid and liquid form: there is no segregation of the components. Eutectic mixtures are often used to form electrical contacts, usually ➤ohmic contacts, to semiconductor devices. For example, a gold-germanium eutectic mixture (88:12 w/o Au:Ge) is used as the basis of ohmic contacts to n-type gallium arsenide, where the gold provides the metallic contact and the germanium dopes heavily the surface of the gallium arsenide. The eutectic is evaporated from a solid source and retains its composition on condensation on the semiconductor surface.

E wave *Syn. for* TM wave. ➤mode.

exchange forces ➤ferromagnetism.

excitation 1. The addition of energy to an atom or molecule, transferring it from the ground state to a higher ➤energy level. **2.** The application of a signal to the base or control electrode of a transistor or valve. **3.** The application of voltage to an oscillating crystal. **4.** The application of radiofrequency pulses to a tuned circuit. **5.** The application of current to the winding of an electromagnet in order to produce a magnetic flux. The applied current is called the *exciting current.*

excited state ➤ground state.

exciting current ➤excitation.

exclusive OR gate ➤logic circuit.

expanded sweep ➤timebase.

expanded-sweep generator ➤timebase.

expander 1. A ➤synthesizer that responds only to a ➤MIDI input. It has no keyboard or other performance interface of its own. **2.** A system that ➤upsamples a sampled signal. **3.** *Short for* volume expander. ➤volume compressor. **4.** *Short for* gate expander.

expansion In a telephone or digital communications network, a network or circuit that has more outputs than inputs and thereby expands the incoming channels of the system to a larger number of output channels.

exploring coil *Syn.* search coil. ➤flip-coil.

extrapolated failure rate ➤failure rate.

extrapolated mean life ➤mean life.

extremely high frequency (EHF) ➤frequency band.

extrinsic photoconductivity ➤photoconductivity.

extrinsic semiconductor A semiconductor in which impurities or imperfections determine the charge-carrier concentration. ➤semiconductor. ➤➤intrinsic semiconductor.

eyelet-construction mica capacitor ➤mica capacitor.

F

Fabry–Perot cavity ►laser.

facsimile transmission A method of transmitting any kind of graphic material to produce a pictorial likeness of the object. The system employs *facsimile scanning*: the subject copy is scanned, so providing a successive analysis from which electrical signals are produced. These signals are transmitted to a receiver that converts them into a duplicate image. Early facsimile systems were exclusively analogue but systems now use digital techniques for data encoding and transmission.

Fax services are a form of facsimile transmission that use telephone lines to transmit the data to the receiver, which may be a *fax machine* or a suitably equipped computer. The transmitted signals are produced in digital form and ►pulse code modulation is used. The system is used for high-speed conveyance of documents and communication can be worldwide.

Facsimile telegraph is a telegraph system used for the transmission of pictures. *Phototelegraphy* is a facsimile telegraph system that has special regard to half-tone reproduction.

fader 1. A device that maintains an electrical signal at a constant level while one signal is being *faded out,* i.e. smoothly reduced in amplitude, and another *faded in,* i.e. smoothly increased in amplitude. **2.** A volume slider control on an audio mixing desk.

fading 1. The rise and fall of the received signal at a mobile radio receiver due to the motion of the receiver through a region where the transmitted signal fluctuates rapidly. ►multipath. **2.** Variations in signal strength at a receiver due to variations in the transmission medium. Destructive interference between two waves travelling by two different paths to the receiver is the most common cause of fading; this is termed *interference fading.*

Amplitude fading occurs when all transmitted frequencies are attenuated approximately equally, resulting in a smaller received signal. *Selective fading* occurs when some frequencies are more attenuated than others, resulting in a distorted received signal. *Dellinger fade-out* is a complete loss of the received signal, which may last for minutes or even hours following a burst of hydrogen particles from an eruption associated with a sunspot. This causes formation of a highly absorbent D-layer of the ►ionosphere lower than the regular E- and F-layers.

failure Termination of the ability to perform its required function by a device, component, circuit, or any part or subsystem. The *failure mechanism* is the physical, chemical, metallurgical, or other process resulting in a failure. The predicted or observed result of a failure mechanism on a particular item in relation to the operating condition at the time of failure is the *failure mode*. The predicted or observed result on the function of the item or of related items is the *failure effect*.

Failure is classified according to cause, suddenness, and degree. There are various causes of failure:

misuse failure is attributable to the application of stresses beyond the stated capabilities of the item;

inherent weakness failure is attributable to a weakness inherent in the item itself when subjected to stresses within its stated capabilities;

primary failure is not caused either directly or indirectly by the failure of another item whereas *secondary failure* is;

wear-out failure results from deterioration processes or mechanical wear, the probability of occurrence increasing with time.

Failure can be either sudden or gradual, i.e. unanticipated by prior examination or anticipated. It can also be partial, complete, or intermittent. A failure that is both sudden and complete is termed *catastrophic failure*; one that is both gradual and partial is called *degradation failure.* ➤failure rate; mean time between failures; mean time to failure.

failure rate The number of ➤failures of an item per unit measure of life (cycles, time, etc., as appropriate). For any particular item the failure rate will be based on the results of ➤life tests and one or more of the following will be quoted.

The *observed failure rate* is the ratio of the total number of failures in a single population to the sum of the times during which each item has been subjected to stress conditions. It is associated with particular and stated time intervals and stress conditions. The criteria for what constitutes a failure should be stated.

The *assessed failure rate* is determined as a limiting value of the confidence interval with a stated probability level based on the same data as the observed failure rate of nominally identical items. The following conditions apply: the source of the data should be stated; results may be combined only when all conditions are similar; it should be stated whether one- or two-sided intervals are being used; the upper limit-

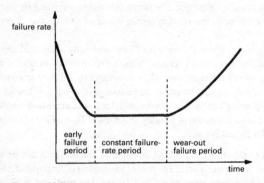

Change of **failure rate** with time

ing value is usually used for failure-rate statements and the assumed underlying distribution should be stated.

The *extrapolated failure rate* is an extension of the observed or assessed failure rate by defined extrapolation or interpolation for durations or stress conditions different from those applying to the conditions of the assessed rate.

The failure rate of an item varies during its lifetime (see diagram). In the *early failure period* the failure rate decreases rapidly. To avoid an early failure in use a manufacturer will operate and test an item in a process, known as *burn-in,* that stabilizes its characteristics. The failure rate increases rapidly in the *wear-out failure period* due to deterioration processes. The *useful life* of an item is the period from a stated time during which, under stated conditions, an item has an acceptable failure rate. For items showing a failure rate pattern as in the diagram, the useful life corresponds to the *constant failure-rate period,* when failures occur at an approximately uniform rate. ➤►mean time between failures; mean time to failure.

fall-time 1. The time required for a ➤logic circuit to change its output from a high level (logical 1) to a low level (logical 0). **2.** *Syn. for* decay time. ➤pulse.

fan-in The maximum number of inputs acceptable by a ➤logic circuit.

fan-out Within a given family of ➤logic circuits, the maximum number of inputs to other circuits that the output of a given circuit can drive.

farad Symbol: F. The ➤SI unit of ➤capacitance. A capacitor has a capacitance of one farad when a charge of one coulomb increases the potential difference between its plates by one volt. The farad is too large for most practical purposes, the submultiples microfarad, nanofarad, and picofarad being more convenient.

Faraday constant Symbol: F. A fundamental constant having the value $9.648\,5309 \times 10^4$ coulombs per mole. It is the quantity of electricity that is equivalent to one mole of electrons and that can deposit or liberate one mole of a univalent ion. It is the product of electron charge and Avogadro constant.

Faraday dark space ➤gas-discharge tube (diagram).

Faraday–Neumann law *Syn.* Faraday's law. ➤electromagnetic induction.

Faraday shield A closed conducting container of arbitrary shape in which the ➤electric field inside is zero whatever external static electric fields are applied to the container. ➤Gauss's theorem.

Faraday's law ➤Faraday–Neumann law.

far-end crosstalk ➤crosstalk.

far-field region A zone extending from infinity to near an ➤antenna within which the radiation from the antenna can be considered to be travelling directly away from the antenna; the electric and magnetic field components are then transverse to the direction from the antenna to that point (➤electromagnetic radiation). For an antenna of size D the far-field region begins at a distance of approximately $2D^2/\lambda$ from the antenna for radiation of wavelength λ. In this region, the ►power density from the antenna falls off as the inverse square of the distance from the antenna while the fields fall off as the inverse of this distance.

The *near-field region* of an antenna is a zone extending from the antenna out to the far-field region. Within the near-field region, the fields generated by the antenna may be oriented in any direction and decrease rapidly, up to the inverse square of the distance from the antenna.

fast Fourier transform (FFT) ►discrete Fourier transform.

fast hopping ►frequency modulation.

fast-recovery diode A diode in which very little ►carrier storage occurs and that may therefore be used to give an ultrahigh speed of operation. Fast-recovery diodes may be formed as p-n junction diodes fabricated from a semiconductor in which the minority-carrier lifetimes are very small. Fast-recovery diodes may also be formed from ►Schottky diodes. The carrier storage in a Schottky diode is negligible as they are majority-carrier devices, and a similar speed of operation is achieved. ►►step-recovery diode.

FATFET A planar field-effect transistor (FET) that has a long gate length compared to gate width and is used to measure ►drift mobility. The device is so called because of its geometry.

fat zero A significant charge packet used in ►charge-coupled devices that corresponds to a zero input sample. Typically a fat zero corresponds to about 10% of the maximum charge packet and is used to minimize distortion of the signal due to surface trapping states in the semiconductor.

fault A defect in any circuit or device that causes an ►error.

fault current A ►current that may flow through a circuit or device as a result of a fault, such as a defect in the insulation. The current may take the form of a ►short circuit, electrical ►surge, current to earth, etc.

fault-tolerant system A system, often an electronic system, that by ►redundancy provides a service complying with the specification in spite of ►faults.

fax 1. A ►facsimile transmission service. **2.** A document sent by fax. **3.** To send or communicate by fax.

FCC *Abbrev. for* Federal Communications Committee. ►standardization.

FDM *Abbrev. for* frequency-division multiplexing.

FDMA *Abbrev. for* frequency-division multiple access. ►digital communications.

FDNR *Abbrev. for* frequency-dependent negative resistor.

feed A system of wires used to carry a signal between an ►antenna and a transmitter/receiver. Often a ►coaxial cable is used, which acts to shield the signal from external interference.

feedback The process of returning a fraction of the output energy of an energy-converting device to the input. The circuit that transmits the feedback signal to the input is the *beta circuit*; the circuit containing the active device, which generates the output signal, is the *mu circuit*.

In the case of an active device, such as a ►transistor, that introduces a gain A in the absence of feedback, *voltage feedback* is employed if a fraction β of the output voltage is returned to the input (Fig. *a*). The effective output voltage is given by

$$v_o = A(v_i + \beta v_o)$$

The overall gain of the combination is then

$$v_o/v_i = A/(1 - \beta A)$$

If β is negative the feedback voltage opposes the input voltage and the process is termed *negative feedback*. The overall gain of the device is reduced but there is a corresponding reduction in the amount of ►noise and ►distortion in the output. If the term $(-\beta A)$ is made large compared with unity the overall gain reduces to $1/\beta$ and is independent of the elements in the mu circuit. An amplifier operated in this manner is very stable and is independent of minor variations in the operating conditions.

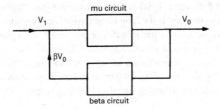

a Voltage **feedback** circuit

If β is positive the feedback voltage reinforces the input voltage and the process is termed *positive feedback*. The overall gain of the device is increased and if the factor $(-\beta A)$ becomes equal to or greater than unity the output voltage becomes independent of any input signal and oscillations occur. The point at which the term $(-\beta A)$ just becomes unity for any given circuit is termed the *singing point*. The overall combination can then be considered to have an effective ►negative resistance (►►oscillator).

Current feedback is a form of feedback in which a fraction of the current output to the load is fed back to the input (Fig. *b*). The effective output current is given by

$$i_o = A(i_i + \beta i_o)$$

and the overall gain is given by

$$i_o/i_i = A/(1 - \beta A)$$

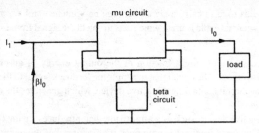

b Current **feedback** circuit

An analysis similar to that of voltage feedback can then be applied.

In a multistage amplifier, feedback may be applied to each individual amplifier stage (*local feedback*) or across the composite device (*multistage feedback*). The phase of the feedback at the input is maintained in the correct relationship to the input by introducing a reactance in the feedback circuit. *Capacitive feedback* uses one or more capacitors and *inductive feedback* uses a self-inductance or a mutual inductance.

Feedback is also used in control systems when a fraction of the controlled parameter is fed back in order to produce any necessary correcting signals (➤feedback control loop).

feedback control loop A method of control, used for many different types of ➤control system, in which a portion of the output derived from a system is fed back to the input circuit in order to control the output signal in a desired manner (see diagram).

The external signal applied to the loop is the *loop input signal*; the controlled signal that is output by the loop is the *loop output signal*. The portion of the signal fed back to the input circuit is the *loop feedback signal*. The feedback signal is mixed with the input signal and produces a *loop actuating signal*, which is used to produce the controlled output. The transmission path between the loop input and the loop output is the *through path*; the path between the loop actuating signal and the loop output is the *forward path*.

The *forward transfer function* is the mathematical relationship between the actuating signal and the controlled loop output. The value of the loop actuating signal is determined by the *loop error*, which is the difference in value between the actual output and the desired value. The *actuating transfer function* is the relationship between the loop input signal and the loop actuating signal. The *difference transfer function* is the relationship between the loop error and the loop input signal. The loop feedback

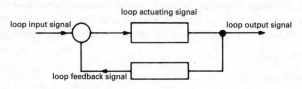

Feedback control loop

signal and the loop input signal are mixed in a suitable manner to maintain the desired transfer functions. In many systems the loop error is used as the loop actuating signal.

feedback oscillator ➤oscillator.

feeder 1. The part of a radio system that conveys the radiofrequency energy from the transmitter to the antenna or from the antenna to the receiver. **2.** An electric line that conveys electrical energy from a generating station to a point of a distributing network without being tapped at any intermediate points. ➤transmission line.

feedforward control A form of ➤control that is used extensively in practice to reduce the effect on the system output of measurable disturbance inputs. These are disturbances that by means of a sensor can be made available as signals, examples being water, oil, or pneumatic supply pressure variations. In many cases feedforward control can give a dramatic reduction of output deviations from the desired value. In the system shown in the diagram, the extra feedforward link G_f acts to reduce the effect of disturbances D on output C, before it changes C.

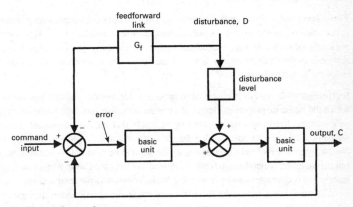

Feedforward control system

feedpoint impedance ➤antenna feedpoint impedance.

feedthrough 1. An electrical contact made between two circuits that are separated by a physical wall in such a way that the through connection does not make electrical contact with the wall. **2.** A contact on a printed circuit board that connects one layer of interconnections with the next layer, passing through the insulating material that separates them. Up to 12 layers of interconnections have been mounted on a single board, although only a double-sided board is commonly used. In an ➤integrated circuit containing multilayer interconnections, feedthroughs can be used to make contact between one layer of interconnections and the next.

Felici balance A type of alternating-current ➤bridge that is used to determine the mutual inductance between the windings of an inductor.

femto- Symbol: f. A prefix to a unit, denoting a submultiple of 10^{-15} of that unit.

Fermi–Dirac statistics A system of quantum statistics that is used to describe the behaviour of solids in terms of a ►free electron model. In this model the most weakly bound electrons of the constituent atoms are considered to behave as a gas subject to certain conditions: the electrons are free to move in any direction through the solid, they do not interact with each other, and are subject to the ►Pauli exclusion principle.

The probability of an ►energy level of energy E being occupied by an electron is given by the *Fermi–Dirac distribution function,* f(E):

$$f(E) = 1/[e^{(E-E_F)/kT} + 1]$$

where E_F is the ►Fermi level, k is the Boltzmann constant, and T the thermodynamic temperature.

The free electron model is used to explain a number of important physical properties of metals.

Fermi energy ►Fermi level.

Fermi level *Syn.* Fermi energy. Symbol: E_F. The maximum electronic energy level that is occupied by an electron in a solid at a temperature of absolute zero. At higher temperatures some electrons are excited into higher energy states. The Fermi level then corresponds to the value of energy at which the Fermi–Dirac distribution function has a value ½. ►energy bands.

ferrimagnetism An effect observed in certain solids, notably garnets and ferrites, in which the magnetic properties change at a certain critical temperature known as the *Néel temperature.* Ferrimagnetism occurs in materials that have a permanent molecular magnetic moment associated with unpaired electron spins. At temperatures above the Néel temperature thermal agitation causes the spins to be randomly orientated throughout the material, which becomes paramagnetic: it obeys the Curie-Weiss law approximately but is characterized by a negative Weiss constant (►paramagnetism).

At temperatures below the Néel temperature ferrimagnetic materials behave in a similar manner to ferromagnetic materials (►ferromagnetism): they show spontaneous magnetization within a domain structure and ►magnetic hysteresis, but the spontaneous magnetization observed is less than that of ferromagnetic materials and does not correspond to full parallel alignment of the individual magnetic moments.

Néel explained this behaviour by suggesting that the interatomic exchange forces are antiferromagnetic (►antiferromagnetism) in nature but that the magnetic moments of sublattices containing the antiparallel spins are unequal, causing a net magnetization.

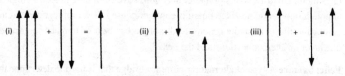

Possible arrangements of magnetic moments on two sublattices

Some possible arrangements are shown in the diagram: (i) unequal numbers of identical moments; (ii) unequal moments; (iii) two identical moments on each sublattice plus one unequal moment on one. The net small spontaneous magnetization causes the material to behave as a weak ferromagnetic material.

Ferrimagnetic materials are technically important because they are usually insulators and hence have low ►eddy-current losses in radiofrequency applications, while exhibiting substantial magnetic moments at room temperature, although less than those of ferromagnetics.

ferrite A low-density ceramic material with composition $Fe_2O_3.XO$, where X is a divalent metal, such as cobalt, nickel, manganese, or zinc. These magnetic materials have very low eddy-current loss and ►cores made from them are used for high-frequency circuits.

ferrite beads ►electromagnetic compatibility.

ferrite core ►core.

ferroelectric crystals Crystals that exhibit electrical properties analogous to certain magnetic properties, such as ferromagnetism. In an alternating electric field very large values of the piezoelectric and dielectric constants are developed, usually in one particular direction within a certain temperature range. These crystals are particularly useful as detectors of vibrations.

ferrofluid A viscous liquid containing a high density of magnetic particles. It can be used to provide a high-vacuum sealant for rotary joints in vacuum systems: the fluid is held in place by permanent magnets, providing a low-friction nonabrading bearing, which enables a rotary motion of a shaft through the walls of the container without causing particulate or oil contamination of the vacuum. This is useful for high-vacuum analysis or microfabrication equipment.

ferromagnetic Curie temperature *Syn. for* Curie point. ►ferromagnetism.

ferromagnetism A phenomenon observed in certain solids in which the magnetic properties change abruptly at a certain characteristic temperature known as the *Curie point*. Below the Curie point the solid exhibits ferromagnetic properties. Above this temperature the thermal energy of the atoms is sufficient to produce magnetic properties typical of ►paramagnetism: the ►susceptibility obeys the Curie-Weiss law approximately, the value of the Weiss constant being close to that of the ►Curie point and a few degrees higher.

The chief ferromagnetic elements are iron, cobalt, and nickel; there are also many ferromagnetic alloys based on these materials. Ferromagnetic materials are characterized by a large positive susceptibility: very large values of magnetization are produced by relatively small magnetic fields, the magnetization varying nonlinearly with field strength (Fig. *a*). Maximum intensity of magnetization (*magnetic saturation*) is achieved at fairly low field strengths and a certain amount of magnetization is retained when the magnetizing field is removed, hence the materials exhibit ►magnetic hysteresis.

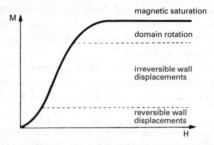

a Typical magnetization curve of a virgin ferromagnetic specimen

Ferromagnetism was first explained by Weiss, who suggested that spontaneous magnetization occurs within ferromagnetic materials due to large interatomic forces acting between neighbouring atoms in the crystal lattice. Below the Curie point these forces can overcome the thermal effects and tend to produce an ordered state. The interatomic forces were discovered by Heisenberg and are known as *exchange forces.* Weiss also postulated that the groups of atoms are organized into tiny bounded regions called *domains.* In each individual domain the magnetic moments of the atoms are aligned in the same direction. The domain is thus magnetically saturated and behaves like a magnet with its own magnetic moment and axis. In an unmagnetized sample the domains are randomly orientated so that the magnetization of the specimen as a whole is zero. The existence of domains has been verified experimentally.

The magnetic moments of the atoms arise from the ►spin of electrons in an unfilled inner shell. In any stable material in the absence of an applied magnetic field, the detailed arrangements of the magnetic moments result from the interaction between the various forces operating within the sample. It is always such as to produce the minimum energy possible. In ferromagnetic materials this minimum energy state occurs when the electron spins of the atoms within a domain are arranged in parallel.

Magnetization of the domains is much harder along certain directions relative to the crystal axes than in others: more energy is required to magnetize the domains lying along these directions. This anisotropy energy is least for small domains. To form the boundaries between the domains however also requires energy because of the exchange forces between neighbouring atoms and this tends to increase domain size. The domain size is determined by a compromise between these competing forces. The boundaries are called

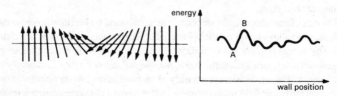

b Representation of spin direction in a Bloch wall with 180° rotation

c Energy as a function of Bloch wall position

Bloch walls and extend over a finite number of atoms, each of whose spins are slightly displaced relative to that of its neighbours (Fig. *b*). The energy state is also affected by the degree of crystalline perfection, the existence of strains and impurities affecting significantly the ferromagnetic behaviour. It can be shown that in any particular material the energy state of the domains is least when the Bloch wall intersects as many dislocations as possible. A typical energy curve as a function of wall position is shown in Fig. *c*. A virgin specimen would have a wall located at the minimum energy position, marked A.

When a magnetic field is applied to a ferromagnetic material the characteristic shape of the magnetization curve (Fig. *a*). is explained by consideration of the domain behaviour. At small values of magnetic field the net effect is to displace the Bloch walls over a few atoms, away from the minimum energy state; thus those domains with spins parallel or nearly parallel to the field grow at the expense of the others (Fig. *d*). If the field is removed the walls tend to move back to the minimum energy state and for small values of applied field the magnetization changes are small and reversible. At larger values of applied field the wall excursions are sufficiently large so that an energy maximum, marked B on Fig. *c*, is passed through and the change becomes irreversible. A single crystal with few dislocations allows much greater reversible wall excursions than a polycrystalline material with many strains and impurities present; it also requires a lower field to produce them. As the magnetic field is increased further a position is reached when further domain growth becomes impossible; further magnetization is only possible by rotation of the magnetic axes of the domains. This is a more difficult process than domain growth because of the crystal anisotropy and above the knee of the magnetization curve the magnetization increases only slowly until saturation is reached.

Ferromagnetic materials are classified as either *hard* or *soft*. Hard materials have a low relative permeability, very high coercive force, and are difficult to magnetize and demagnetize; soft materials have a high relative permeability, low coercive force, and are easily magnetized and demagnetized.

Hard ferromagnetics, such as cobalt steel and various ferromagnetic alloys of nickel, aluminium, and cobalt, retain a high percentage of their magnetization and have a relatively high hysteresis loss (►magnetic hysteresis). They are most suitable for use as permanent magnets, as used in ►loudspeakers. A high degree of dislocation is introduced into their structure during manufacture. Hard materials are frequently heated to high temperatures and then quenched in a suitable liquid to introduce strains.

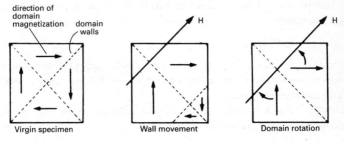

d Representation of magnetization of a ferromagnetic sample

Alternatively they may be produced as compressed powders in which each particle is sufficiently small so as to be a single domain; magnetization can then only proceed by domain rotation since the energy required for wall movement is so great.

Soft ferromagnetics, such as silicon steel and soft iron, retain very little magnetization and have extremely small hysteresis loss. The ease of magnetization and demagnetization makes them very suitable for uses involving changing magnetic flux, as in electromagnets, electric motors, generators, and transformers. They are also useful for ►magnetic screening. Their properties are enhanced by careful manufacture, as by heating and slow annealing, in order to achieve a high degree of crystal purity.

Although the large magnetic moment at room temperatures makes soft ferromagnetic materials extremely useful for magnetic circuits, most ferromagnetics are very good conductors and suffer energy loss from ►eddy currents produced within them. The ideal material for magnetic circuits would be a ferromagnetic insulator. There is also an additional energy loss due to the fact that magnetization does not proceed smoothly but in minute jumps (►Barkhausen effect). This loss is known as *magnetic residual loss* and depends purely on the frequency of the changing flux density, not on its magnitude. ►antiferromagnetism; ferrimagnetism.

FET *Abbrev. for* field-effect transistor.

FFT *Abbrev. for* fast Fourier transform. ►discrete Fourier transform.

fibre-optics system A system that transmits light along the length of an ►optical fibre, a fine filament of glass or clear plastic. A flexible bundle of optical fibres joining an optical modulator to a remotely located optical demodulator can be used as a communications system. The modulator can be any source of visible or infrared radiation, often a ►light-emitting diode (LED) or ►semiconductor laser, which can be modulated by the signal to be transmitted; the signal can be digital or analogue. The demodulator consists of a photosensitive detector that converts the optically modulated signal back into an electrical signal at the receiving end. One of the advantages of a fibre-optics communications system is its very wide information bandwidth.

Fick's law If a concentration gradient of mobile particles exists, there will be a flow of such particles – *diffusion* – from the region of high concentration to the region of low concentration:

$$f = -D(\partial N/\partial x)$$

where f is the particle flux through a plane parallel to the surface, in numbers per square metre per second. D is a constant of proportionality – the *diffusion constant* – and $\partial N/\partial x$ is the concentration gradient.

Since N is also a function of time during the diffusion process, *Fick's second law* can be derived from the first law and is given as:

$$\partial N/\partial t = D(\partial^2 N/\partial x^2)$$

Diffusion also takes place in the y- and z-directions, parallel to the plane of the surface, as soon as a concentration gradient is established below the surface. Diffusion flow will continue until the concentration becomes uniform across the specimen.

This effect is utilized in producing a desired impurity profile in a particular specimen of semiconductor. At normal temperatures the impurity atoms are immobile until heated to a high temperature. A concentration gradient is formed by heating the semiconductor wafer in a gaseous atmosphere of the impurity atoms so that a high concentration exists at the surface. Under such conditions the impurity atoms diffuse into the semiconductor. In practice the diffusion process is stopped by reducing the temperature when the desired profile has been achieved.

field 1. The region in which a physical agency exerts its influence. Typical examples are electric and magnetic fields resulting from the presence of charge or magnetic dipoles. These are *vector fields.* Such a field may be pictorially represented by a set of curves, referred to as ➤lines of force or of flux. The *field density* and direction at a point represent the strength and direction of the field at that point. **2.** (in computing) A set of symbols treated together as a unit of information. **3.** ➤television.

field coil ➤electromagnet.

field current ➤electromagnet.

field density ➤field.

field-effect transistor (FET) One of the two major classes of ➤transistor. It is a three-terminal ➤semiconductor device in which the current flow through one pair of terminals, the *source* and the *drain*, is controlled or modulated by an electric field that penetrates the semiconductor; this field is introduced by the voltage applied at the third terminal, the *gate* (Fig. *a*). The resistance of the channel is controlled by the field, producing a voltage-controlled resistor.

The controlling field is applied to the gate electrode, but must be isolated somehow from the current flow in the channel. The method of isolation yields two basic types of field-effect transistor: the *junction field-effect transistor* (JFET) and the *insulated-gate field-effect transistor* (IGFET). In JFETs the isolation is provided by a reverse-biased ➤p-n junction or a metal-semiconductor ➤Schottky barrier, so the current flow

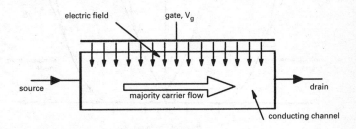

a **Field-effect transistor** operation

across the junction from gate to channel is very small (►MESFET). In IGFETs an insulating layer is placed between the gate electrode and the conducting channel, preventing any current flow between them. The most widely known practical example of the insulated-gate FET is the ►MOSFET (metal-oxide-silicon FET).

The conducting channel in JFETs is in the body of the semiconductor, and these transistors are classed as *bulk-channel FETs*. In MOSFETs the conducting channel lies at the surface of the silicon, at the silicon-oxide interface; these transistors are *surface-channel FETs*.

The FETs are generally described by the type of charge ►carrier that conducts the current in the channel: there are therefore *p-channel* FETs, where holes provide the conduction, and *n-channel* FETs, where electrons conduct. FETs can be further described by the nature of the channel: *depletion-mode* FETs have the conducting channel already present at zero gate voltage, and an appropriate voltage must be applied to close the channel, or turn off the FET; *enhancement-mode* FETs have no conducting path present between source and drain at zero gate voltage, and an appropriate gate voltage must be applied to open the channel. The voltage at which the channel is just becoming conducting is known as the *threshold voltage*, V_T. The presence or absence of the channel at zero volts gate voltage is determined by the details of the ►doping of the channel region of the FET. The general variation of drain current I_D with applied gate voltage V_G for the various types of FET is shown in Fig. *b*.

The output characteristics of an n-channel FET are shown in Fig. *c*. These curves show the *linear* or *triode region* of operation at low values of drain-source voltage V_{DS}, corresponding to the voltage-controlled resistance described above, and the *saturation region* at higher values of V_{DS}. In the saturation region the FET channel is restricted in width, and the current flow cannot increase with further increases in the channel volt-

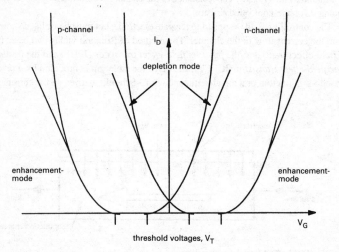

b Drain-current/gate-voltage relationships for FETs

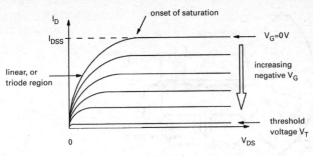

c n-channel FET output (drain-current/drain-source voltage) characteristics

age V_{DS}. The FET behaves like a voltage-controlled current source. This is the region of operation for linear amplification applications. For digital switching applications, such as logic, the FET is switched between the 'off' state, below threshold, to a high-current 'on' state in the triode region.

field emission *Syn.* cold emission. A type of ►emission in which the presence of a large external accelerating electric field reduces the potential barrier at the surface of the emitter (►Schottky effect) and allows electrons to escape from the surface. The potential barrier is shown in the diagram as the work done, W, for an electron to escape.

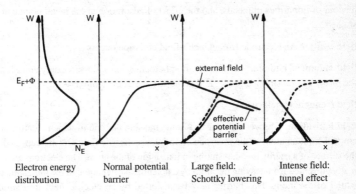

| Electron energy distribution | Normal potential barrier | Large field: Schottky lowering | Intense field: tunnel effect |

The distortion of the potential barrier at sufficiently large values of the field results in an effective narrowing of the barrier and allows the ►tunnel effect to operate: electrons with energies around the Fermi energy E_F (►energy bands) may also be emitted. The current density j has been shown to vary with the electric field E as:

$$j = CE^2 \exp(-D/E)$$

where C and D are approximately constant. The very intense fields required are of the order of 10^{10} volts per metre and are usually only obtained at sharp points on the emitter surface.

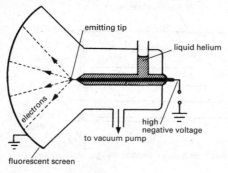

Field-emission microscope

field-emission microscope An instrument for studying the surface of a solid, usually a metal, by causing it to undergo ►field emission. A simple microscope is shown in the diagram. A high voltage is applied in a vacuum between the single crystal solid tip and the curved fluorescent screen. The electrons emitted from the tip form an image on the screen; the different intensities of the image represent areas on the tip that have different work functions. Vibrations of the metal atoms limit the resolution and the tip is usually held at liquid helium or hydrogen temperature. The structure of alloys, the behaviour of impurities in metals, and the effect of adsorption at the metal point can be studied.

field emitter An electrode from which ►field emission occurs.

field-enhanced emission An increase in ►photoemission and ►secondary emission in the presence of a strong electric field at the surface of the emitter.

field frequency ►television.

field ionization A process similar to ►field emission but one in which electrons are emitted from an atom in the presence of an intense electric field and are captured by the surface of a nearby metal, resulting in ionization of the atom. The electrons are normally prevented from leaving the atom by a potential barrier equal to the ionization potential of the atom. This barrier may be distorted by an electric field in a similar manner to that occurring in field emission, allowing electrons to escape.

field-ion microscope An instrument for studying the surface of a solid, usually a metal, by subjecting it to ►field ionization. The form is identical to the ►field-emission microscope but the voltage is applied in the opposite direction. Low-pressure helium is allowed into the microscope and helium ions, formed at the surface of the tip, are accelerated to the fluorescent screen to form the image. Atomic vibrations, which affect the resolution, are minimized by holding the tip at liquid helium or hydrogen temperature; individual atoms of the metal can be resolved. The structure of alloys, the structure and behaviour of surfaces, and adsorption on the metal surface can be studied.

field programmable gate array ►FPGA.

figure-eight microphone A ►microphone with a directional response pattern described by cosφ, where φ is the angle of incidence of the sound (see diagram). Sound picked up from the rear of the microphone is in ►antiphase with that picked up from the front and is thus marked as '–' in the diagram. A ►ribbon microphone generally exhibits a figure-eight response. Practically, such a response is difficult to achieve for frequencies above which the wavelength becomes comparable with the physical dimensions of the microphone itself.

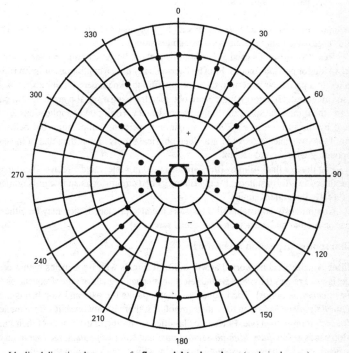

Idealized directional response of a **figure-eight microphone** (scale in degrees)

filament A thread-like body of metal or carbon, particularly the conductor of an incandescent lamp, the cathode of a thermionic valve, or the electrode used to heat an indirectly heated cathode.

film A metal or dielectric coating with a minimal thickness dimension. Thin films have thicknesses in the range one nanometre to one micrometre; thick films range from 10 to 100 micrometres.

film resistor A type of ►resistor that uses a thin layer of resistive material deposited on an insulating core. For low-power applications film resistors are more stable than

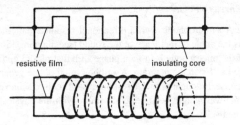

Types of **film resistors**

composition resistors and except for very high precision requirements are smaller and less expensive than accurate wire-wound resistors.

Resistive materials used are crystalline carbon, boron-carbon, and various metallic oxides or precious metals. Film resistors usually have a continuous uniform film applied in a particular pattern to the core, the film thickness determining the resistance (see diagram). Higher resistances are obtained by using a spiral pattern of film on the core, the tighter the spiral the higher the resistance. High-power applications are limited by the film resistor's 200 °C maximum operating temperature but below this limit the resistance achieved for a given physical size is higher than that of the corresponding wire-wound resistor.

Flat thin- or thick-film resistors are also used in ➤integrated circuits. These may be produced in a form suitable for hybrid integrated circuits or as an integral compenent in a fully integrated circuit.

Continuously adjustable film resistors are also produced; they may be either linear or circular and are actuated by a lead screw.

film sputter deposition ➤sputtering.

filter An electrical network that will transmit signals with frequencies within certain designated ranges (*pass bands*) and reject or attenuate signals of other frequencies (*stop* or *attenuation bands*). The frequencies that separate the pass and stop bands are the *cut-off frequencies*, which have the symbols f_c if there is only one cut-off frequency or f_1 and f_2 if more than one. Filters are classified according to the ranges of their pass or stop bands as *low-pass, high-pass, band-pass* and *band-stop filters*; the four main classifications with their corresponding frequency limits are shown in the table.

Type of filter	Pass band(s)	Stop band(s)
low pass	$0 - f_c$	$f_c - \infty$
high pass	$f_c - \infty$	$0 - f_c$
band pass	$f_1 - f_2$	$0 - f_1, f_2 - \infty$
band stop	$0 - f_1, f_2 - \infty$	$f_1 - f_2$

An ideal filter would transmit the pass band without attenuation and completely suppress the stop band, with a sharp cut-off profile. Practical filters however do attenuate the pass band, due to absorption, reflection, or radiation, which results in loss of signal power; neither do they completely suppress the stop bands. A typical curve of output voltage with frequency is shown in Fig. *a* for a simple low-pass filter: V_p is the peak voltage and V_m is the maximum voltage of an ideal filter. The *filter attenuation* is defined as the loss in signal power in decibels or nepers through the filter; the *filter discrimination* is the difference between the minimum value of ►insertion loss in a stop band and the maximum value in a pass band.

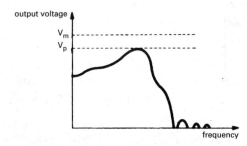

a Low-pass **filter** output

The components of a practical filter may be arranged to give the desired output curve. For example, *Chebyshev* and *Butterworth filters* are band-pass filters with different output characteristics (Fig. *b*). Butterworth filters have a flat response in the pass band whereas Chebyshev filters have some variation of the residual response in the pass band but have a more rapid increase of attenuation giving a sharper cut-off profile.

Filters are *active* or *passive* according to their components. Active filters contain active components, such as operational amplifiers, that introduce some gain into the signal combined with suitable R-C feedback circuits to give them the desired frequency-response characteristic. Most passive filter networks are constructed from impedances arranged in shunt and in parallel (L-C networks). Two basic arrangements are used:

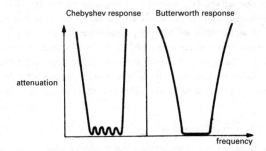

b Band-pass **filters**

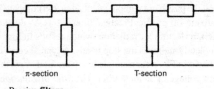

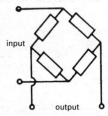

c Passive **filters**

d Lattice **filter**

π-sections and *T-sections* (Fig. *c*). Composite networks are built up from these basic sections and the arrangement is termed a *ladder network* because of the alternation of shunt and parallel sections. Another type of configuration is the *lattice filter* in which the impedance elements are arranged in a bridge network (Fig. *d*).

The bandwidth of a band-pass or band-stop filter is the difference in hertz between two particular frequencies whose geometric mean equals that of the geometric mid-frequency of the pass or stop band. Frequencies exhibiting a particular characteristic, such as the point at which the response is three decibels below the peak value, are usually chosen.

filter attenuation ➤filter.

filter discrimination ➤filter.

filter synthesis ➤network synthesis.

finite-difference time domain (FDTD) A method of computing wave propagation. The region of interest is broken up into a collection of small cubic elements, each of which contains information about the fields and material parameters at that point. Signals are applied to appropriate points within the region and the method proceeds by calculating the effects of these at successive small time steps (hence the term time domain). The method of determining the field variation between spatial and temporal steps is based on mathematical difference techniques applied to finite steps (hence finite difference).

finite impulse response (FIR) ➤digital filter.

finite-state machine (FSM) A design methodology that provides a rigorous way to design sequential digital systems, i.e. digital systems in which one activity or event – an

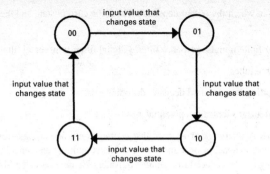

Moore circuit

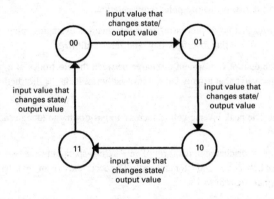

Mealy circuit

Finite-state machines specified by state diagrams

internal *state* – must be completed before the next one begins. An FSM must be able to generate outputs and state transition sequences that are controlled by inputs and the current system state. It can only exist in one of 2^n states, where n is equal to the number of bits in the system state. An FSM is usually specified by means of a *state diagram*, where circles represent the states in a system, and arcs connect reachable states. The arcs are marked with the input condition that causes the change in state, and in some cases the output values for that state transition.

State diagrams are shown for *Moore* and *Mealy circuits*. A Moore circuit is a digital sequential design, using FSMs, in which the output is a function of the state only and is independent of the external inputs. A Mealy circuit is a digital sequential design, again using FSMs, in which the output depends on both the current state and the external inputs.

FIR filter *Short for* finite impulse response filter. ➤digital filter; transversal filter.

firing ➤gas-discharge tube.

firm decision decoding *Syn. for* hard decision decoding.

first ionization potential ➤ionization potential.

first-order network A ➤network or ➤circuit that contains only one energy-storage element (for example, a capacitor or inductor), or contains multiple elements but in such a way that they can be reduced to a single equivalent element. ➤➤second-order network.

fixed-point representation A method of representing numbers in which the decimal or binary point is fixed in a given location and the number contains a constant predetermined number of digits. ➤➤floating-point representation.

flag A 1-bit ➤register that holds processor-status information, examples being carry flag, negative flag, positive flag, overflow flag, and zero-result flag.

flash arc A sudden disruptive transient discharge between the electrodes of a high-voltage ➤thermionic valve that usually but not invariably results in the destruction of the valve.

flashback voltage The peak inverse voltage across a ➤gas-discharge tube needed to produce ionization.

flashover A disruptive discharge in the form of an ➤arc or ➤spark between two electrical conductors or between a conductor and earth. The occurrence of an arc is termed *arcover* and of a spark *sparkover*.

flat-band operation mode ➤MOS capacitor.

flat-panel display *Syn.* flat screen. Video display technology in which the depth of the display equipment is minimal, in comparison with traditional cathode-ray techniques that result in bulky monitors to accommodate the electron guns which irradiate the phosphor front screen. ➤Liquid crystal displays (LCDs) using ➤polysilicon or ➤amorphous silicon technology are most common, offering large screen areas at high resolution. *Gas-plasma screens* are nowadays less popular, being more bulky and expensive than LCDs and offering lower resolution. Each ➤pixel in the gas-plasma display is a small tube containing a gas that is easily ➤ionized. When the pixel is addressed, a ➤glow discharge is created, the colour determined by the gas in the cell.

flat tuning Tuning with a substantially equal response to a range of frequencies.

flatwise bend *Syn. for* H-bend. ➤waveguide.

F-layer, F$_1$-layer, F$_2$-layer *Syn.* Appleton layer. ➤ionosphere.

Fleming's rules Mnemonics in frequent use by practical electricians for the directional relation between current, motion, and magnetic field in a dynamo or electric motor. In each case the thu*M*b represents the direction of *M*otion, the *F*irst finger represents the *F*ield, and the m*I*ddle finger the current, *I*. When these three fingers are held at right angles, the right hand represents the relation in a dynamo (*right-hand rule*) and the left hand represents the relation in an electric motor (*left-hand rule*).

flexible resistor A type of wire-wound ➤resistor in which the wire is wound round a flexible insulating core.

flicker A perceived rapid fluctuation of a displayed image, usually one on a cathode-ray tube such as a television or computer screen. The threshold for perceiving flicker is dependent on the brightness of the observed light and the angle that it subtends to the optic axis. In television flicker prevents complete continuity of the images. In colour television flicker can result from either unwanted variations in the luminance signal – *luminance flicker* – or in the chrominance signal – *colour flicker.*

flicker noise *Syn.* l/f noise. Noise superimposed on ➤shot noise at very low frequencies. Flicker noise voltage is inversely proportional to frequency, but this may change as the frequency falls. The origins of flicker noise include effects resulting from generation-recombination of ➤electron-hole pairs in ➤semiconductors, but l/f noise is observed in many natural systems, including biological systems.

flip-chip A semiconductor ➤chip with thickened and extended ➤bonding pads enabling it to be flipped over and mounted upside down on a suitable substrate, such as a ➤thin-film or ➤thick-film circuit. This results in an improved heat-sinking property. Extension of the bonding pads is achieved by depositing metallic pellets on top of the lead areas in small holes in the overlying oxide layer.

flip-coil The classical means of measuring the value of ➤magnetic flux density, *B,* at a point in air. The flip-coil consists of a number of turns, *N,* wound on a small former of area *A.* If the coil is placed in a magnetic field, with its axis parallel to the direction of *B,* the ➤magnetic flux linking the coil is *NAB.* If the coil is removed very quickly ('flipped') to a point where the flux density is zero, the flux change is just *NAB*; this can be measured using either a ➤ballistic galvanometer or a ➤fluxmeter.

Flip-coils can be constructed with different values of *NA* (turns × area): the appropriate coil can be chosen for the field to be measured, so as to produce a suitable instrument deflection. Accuracy of about one per cent can be achieved using a fluxmeter and about 0.1 per cent with a ballistic galvanometer.

The coil may also be used to investigate the magnetic flux distribution of a magnetic field. It is used in a similar way but the measuring instrument does not need to be calibrated absolutely. The comparative measurements can be plotted to produce a graphical representation of the flux distribution. When used in this way the coil is often termed an *exploring coil*. Exploring coils are commonly used to locate the source of undesired signals in ➤electromagnetic compatibility investigations.

An alternative way of using a flip-coil to measure magnetic flux density is to rotate it rapidly in the field. An alternating voltage, *V,* is produced given by

$$V \propto NAB\omega$$

where ω is the angular velocity of the coil. As this method does not depend on a single throw measurement, less sensitive measuring instruments may be used.

flip-flop A bistable ►multivibrator circuit that usually has two inputs corresponding to the two stable states. It is so called because application of a suitable input pulse causes the device to 'flip' into the corresponding state and remain in that state until a pulse on the other input causes it to 'flop' into the other state.

Flip-flops are widely used in computers as counting and storage elements and several types have been developed. Flip-flops as described above are *unclocked* and are triggered directly by the input pulses. *Clocked flip-flops* have a third input to which a clock pulse is applied. The output state of the device is determined by the state of the inputs at the moment a clock pulse is applied. The basic types of flip-flops are described below.

A *D-type flip-flop* ('D' stands for delay) is a clocked flip-flop with a single input whose output is delayed by one clock pulse: if a logical 1 appears at the input, a logical 1 will appear at the output one clock pulse later.

An *R-S flip-flop* is a flip-flop whose inputs are designated R and S. The outputs corresponding to the various input combinations are shown in the table of Fig. *a*. Logical 1s should not be allowed to appear on the inputs together.

Input		Ouput	
R	S	Q	Q̄
0	0	no change, same as previous state	
1	0	1	0
0	1	0	1
1	1	indeterminate	

a Clocked R–S **flip-flop**

A *J-K flip-flop* is a flip-flop whose inputs are designated J and K (Fig. *b*). These devices are almost invariably clocked and their outputs are the same as the R-S type except when logical 1s appear together at the inputs. In these circumstances the device

J	K	Q
0	0	no change
0	1	0
1	0	1
1	1	toggle

b Clocked J–K **flip-flop**

changes state. The J-K flip-flop together with the D-type flip-flop are the most useful types of flip-flop.

An *R-S-T flip-flop* has three inputs designated R, S, and T. The R and S inputs produce outputs as described above. Application of a pulse to the T input causes the device to change state.

A *T flip-flop* has only one input. Application of a pulse to this input causes the device to change state.

floating Denoting a circuit or device that is not connected to any source of potential.

floating battery ►battery.

floating-carrier modulation *Syn.* controlled-carrier modulation. A type of ►amplitude modulation in which the amplitude of the ►carrier wave does not remain constant but is automatically varied in a manner dependent upon the amplitude of the modulating wave, which is averaged over a short time period. The modulation factor therefore remains substantially constant.

floating-point representation A method of representing numbers by means of a predetermined number of significant digits – the *mantissa* – together with a decimal or binary multiplier – the *exponent*: the number x can be written as

$$x = y \times n^z$$

where y is the mantissa, n is either 10 or 2, and z is the exponent in integer form. ►► fixed-point representation.

floppy disk *Syn.* diskette. ►magnetic disk.

fluorescence ►luminescence.

fluorescent lamp A type of lamp in which light is generated by fluorescence (►luminescence). A common form of fluorescent lamp consists of a ►gas-discharge tube containing a low-pressure gas, such as mercury, with the inner surface of the tube coated with a ►phosphor. When a current passes through the tube the ultraviolet radiation produced strikes the phosphor, which then emits visible radiation. Another type of lamp, the usual sodium vapour or mercury vapour street lamp, does not have a fluorescent coating; electrons in the discharge excite the atoms of vapour and these atoms fluoresce as they decay from the excited states thus produced.

fluorescent screen A type of screen that is used in various electronic devices, such as ►cathode-ray tubes and ►image converters, to convert the electron beam into a visible image. These screens consist of an array of many small-diameter (two to three micrometre) phosphor crystals that emit light when bombarded by high-energy radiation, such as X-rays, or by electrons.

flutter An unwanted type of ►frequency modulation in an audiofrequency system resulting in audible variations of pitch. ►►wow.

flux A measure of the strength of a field of force through a specified area. ►electric flux; magnetic flux.

fluxmeter An instrument that measures changes in magnetic flux. The most usual type is the *Grassot fluxmeter,* which consists essentially of a moving-coil ►galvanometer that is designed so that the restoring couple on the moving coil is negligibly small and electromagnetic damping is large. The galvanometer is used in conjunction with an exploring coil (►flip-coil) of known area. A change in the magnetic flux cutting the exploring coil causes an induced current in the galvanometer coil and hence the latter is deflected. The angle of deflection is directly proportional to the change in magnetic flux through the exploring coil. The instrument is calibrated empirically using a magnetic flux standard.

flyback 1. ►time base. **2.** ►sawtooth waveform.

fly by wire A method of controlling an aeroplane in flight. The flaps, rudder, and other control surfaces of the aeroplane are operated by motors. These motors are controlled by electrical signals that are created as a result of actions by the pilot. This kind of flight control system involves the use of computers to analyse the pilot's intentions and thus work out the right amount of movement of the control surfaces; the computers can override the pilot in situations that would endanger the aeroplane.

flying-spot scanner A device that produces a video signal from an object, such as a film, by scanning the object with a spot of light, which is then focused on a ►photocell to produce corresponding electrical signals. The moving (or 'flying') spot of light is normally produced on the screen of a high-intensity cathode-ray tube used as a light source. Mechanical scanning of the object has also been employed, using a single point source of light, with a suitably perforated rotating disc between it and the object.

flywheel effect The continuation of oscillations in an ►oscillator during the intervals between exciting pulses. It results from electrical inertia, which is analogous to mechanical inertia of a flywheel.

flywheel timebase ►timebase.

f.m. (or **FM**) *Abbrev. for* frequency modulation.

FM receiver A ►radio or ►television receiver that detects frequency-modulated signals (►frequency modulation).

FM synthesis ►synthesis.

focusing The process or a method of making a beam of radiation or particles converge. In an electron-beam device, such as a ►cathode-ray tube, two principal methods of focusing the beam are used.

In *electrostatic focusing* two or more electrodes at different potentials are used to focus the electron beam. The electrostatic fields set up between the electrodes cause the beam to converge; the focusing effect is controlled by varying the potential of one of the electrodes, termed the *focusing electrode.* The electrodes are usually cylindrical, mounted coaxially with the electron tube, and are used in conjunction with deflection plates in instruments such as cathode-ray tubes.

In *electromagnetic focusing* the action of a magnetic field is used to make the electron beam converge. The field is produced by passing direct current through a *focusing coil,* the focusing effect being controlled by varying the current through the coil. The coil is usually one of short axial length that surrounds the tube and is coaxial with it. Electrons of different energies converge at different points along the beam axis. Thus if the electron beam is not monoenergetic, longitudinal spreading of the nominal convergence point results. Electromagnetic focusing is used in conjunction with deflection coils in instruments such as cathode-ray tubes.

focusing coil ➤focusing.

focusing electrode ➤focusing.

foil capacitor A capacitor in which the electrodes are metal foil. The term is most commonly applied to ➤paper capacitors but some polystyrene or polyester ➤film capacitors use foil electrodes and one form of tantalum ➤electrolytic capacitor uses a tantalum foil as one of the electrodes.

folded dipole ➤dipole.

forbidden band ➤energy bands.

forced oscillations Oscillations produced in a circuit that is acted upon by an external driving force, such as oscillations in a resonant circuit coupled to a fixed-frequency oscillator. The resulting oscillations have two components: a transient component, whose frequency is determined by the natural frequency of the circuit and decays rapidly, and a steady component, whose frequency equals that of the external driving force.

If a circuit is acted upon by an external voltage

$$V = V_o \cos\omega t$$

applied at $t = 0$, then the steady-state solution is given by

$$I = (V_o/Z) \cos(\omega t - \phi)$$

where Z is the impedance and is given by

$$Z^2 = R^2 + (\omega L - 1/\omega C)^2$$

and ϕ, the phase angle, is given by

$$\phi = \tan^{-1}[(\omega L - 1/\omega C)/R]$$

The phase differs from that of the applied voltage except when

$$\omega L - 1/\omega C = 0,$$

i.e when

$$\omega = \omega_0 = 1/\sqrt(LC)$$

where ω_0 is the natural frequency of the circuit. This is the resonance condition and the current is then a maximum. ➤resonant frequency; free oscillations.

force factor The ratio of the force required to block the movement of an electro-mechanical ➤transducer to the corresponding current in the electrical system.

formant An acoustic resonance of the human vocal tract, of which there are a number that manifest themselves during speech production when the vocal tract is excited acoustically. The centre frequencies of the lowest three formants in particular change as a result of articulating different speech sounds, such as the vowels in 'bid', 'bed', and 'bad'.

form factor The ratio of the ➤root-mean-square value of an alternating quantity (such as current or voltage) to the half-period mean value, for a half-period beginning at a zero point. A simple sine wave has a form factor equal to $\pi/2\sqrt{2}$ (i.e. 1.111).

forward active operation ➤bipolar junction transistor.

forward bias *Syn.* forward voltage. ➤forward direction.

forward current ➤forward direction.

forward-current gain ➤beta current gain factor.

forward direction The direction in which an electrical or electronic device has the smaller resistance. A voltage applied in a forward direction is the *forward bias;* it produces the larger current, known as the *forward current.* ➤reverse direction.

forward path ➤feedback control loop.

forward slope resistance ➤diode.

forward transfer function ➤feedback control loop.

forward voltage *Syn. for* forward bias. ➤forward direction.

forward-wave amplification ➤travelling-wave tube.

Foucault current ➤eddy current.

Fourier analysis A mathematical method of analysing complex waveshapes or signals into a series of simple harmonic functions, the frequencies of which are integer $(1,2,3,...)$ multiples of the fundamental frequency. An arbitrary periodic phenomenon *u,* of period *T,* may be represented by the *Fourier series* provided that certain conditions – Dirichlet conditions – are satisfied. The Fourier series is then given by

$$u = F(t)$$

where

$$F(t) = \sum_{n=-\infty}^{n=+\infty} a_n \, e^{jn\omega t}$$

where ω is equal to $2\pi/T$, j is the square root of -1, and n is an integer; a_n is the nth coefficient and is given by

$$a_n = (1/T)\int_0^T F(t)\, e^{-jn\omega t} dt$$

The Fourier series may alternatively be written as a series of sines and cosines.

As the period, T, becomes infinitely large so that $1/T$ tends to zero, the Fourier series in its limiting form becomes an integral – the *Fourier integral*. The values of the Fourier series or the Fourier integral are determined by the physical conditions of the phenomenon under consideration.

Fourier analysis is widely used in electronics, where a slightly different representation is commonly used in which the Fourier integral is written as

$$F(t) = \int_{-\infty}^{+\infty} g(\omega)\, e^{j\omega t} d\omega$$

where the function

$$g(\omega) = 1/2\pi \int_{-\infty}^{+\infty} F(t)\, e^{-j\omega t} dt$$

is called the *Fourier transform* of the function $F(t)$. Similarly $F(t)$ is also the Fourier transform of the function $g(\omega)$. ➤discrete Fourier transform.

Fourier integral ➤Fourier analysis.

Fourier series ➤Fourier analysis.

Fourier transform ➤Fourier analysis; discrete Fourier transform.

four-point probe A method of measuring the sheet resistance of slices of semiconductor. Four equally spaced probes are used – the two outer probes supply a small current I and the two inner probes are connected to a high-impedance voltmeter to measure the potential difference V developed. In the ideal case of a homogeneous semi-infinite material the sheet resistance R_s is given by

$$R_s = 2\pi s V/I$$

where s is the spacing between the probes. In practice, the material is not infinitely thick; the conductive layer is usually very thin relative to the probe spacing, and a correction factor of $\pi/\log_e 2$ must be applied. The lateral extent of a slice is not infinite and a second correction factor must be used – the value of this factor is a function of the slice size and positioning of the probe.

four-terminal network ➤two-port network.

four-terminal resistor A standard resistor that has four terminals. Two are used to connect the resistor to the current source and the other two to connect it to a measuring instrument. This arrangement ensures that the potential drop across the resistor is not affected by contact resistances at the terminals.

four-wire circuit A circuit that consists of two pairs of conductors and that forms a simultaneous two-way communication channel between two points of a telecommunication system. One pair of conductors forms the 'go' channel and the other the 'return' channel. In the case of a ➤phantom circuit two pairs or groups of conductors form the circuit. A circuit that operates in the same manner as a four-wire circuit, although not necessarily containing four conductors, is a *four-wire type circuit*. An example is a circuit that contains two wires and that has a different frequency band for each direction of transmission. ➤transmission line; two-wire circuit.

FPGA *Abbrev. for* field programmable gate array. An array of logic cells placed in an infrastructure of interconnections, where each logic cell is a universal function – a functionally complete logic device – that can be ➤programmed to realize a certain function. Interconnections between cells are also programmable, but unlike other programmable logic devices (➤PLDs) these interconnections are of different types and several paths are possible between two given points in the circuit (Fig. *a*); all prediction of the timing is impossible before the final routing of the circuit. Input/output cells are equally programmable, but with fewer possibilities (direction of the information, storage element, electrical level) than logic cells.

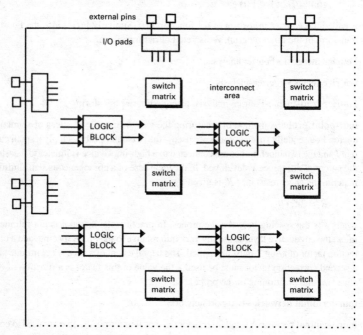

a Corner of a typical **FPGA**

The complexity of the internal logic of an FPGA renders it comparable to a ➤gate array, and its cycle of design is also very similar. Its main advantage is the time of realization: programmable on the spot, an FPGA circuit is in a working state a few minutes after the end of the design, compared to some months for a gate array.

One of the differences between FPGAs and other programmable devices is that one FPGA logic block can be programmed to implement a two- to four-level ➤combina-

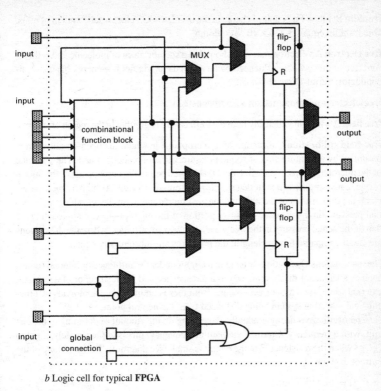

input

input

MUX

flip-flop

R

combinational function block

output

output

flip-flop

R

input   global connection

b Logic cell for typical **FPGA**

tional logic circuit, and the output(s) of that logic block can become the inputs to other logic blocks. An example of the lower-level logic block is shown in Fig. *b*, where there are two flip-flops, nine multiplexers, and a seven-input combinational function unit. The combinational unit can be programmed to realize any two four-input logic functions, or any one five-input logic function (similar to an embedded ➤PLA). The logic block accommodates up to eight inputs and two outputs.

frame 1. The total amount of information presented on a display at a particular time, such as the complete picture in television. **2.** One cycle of a number of pulses that regularly recur in a pulse train used in pulse-train communications.

frame antenna ➤loop antenna.

frame direction finding *Syn.* loop direction finding. ➤direction finding.

frame/field transfer device ➤solid-state camera.

frame frequency *Syn.* picture frequency. ➤television.

franklin Symbol: Fr. A unit of charge in the obsolete CGS electrostatic system of units. One franklin equals 3.336×10^{-10} coulomb.

free electron An electron that is not bound to a specific atom or molecule and is therefore free to move when influenced by an applied electric field. ➤energy bands; semiconductor; Fermi-Dirac statistics.

free-electron paramagnetism ➤paramagnetism.

free field A ➤field in which any boundary effects are negligible in the region of interest.

free-field calibration Calibration of a microphone in which the open-circuit voltage produced by a certain value of sound pressure variation is determined. The presence of the microphone upsets the ➤free field that existed before the introduction of the microphone, causing difficulties in determining the pressure value. At higher frequencies, when the dimensions of the microphone are comparable with the wavelength, the actual pressure on the pressure-sensitive portion of the microphone can approach twice that of the actual pressure in the free wave. At lower frequencies, where the dimensions are small compared to wavelength, the pressures are substantially equal.

free oscillations Oscillations arising in a circuit under the influence of internal forces, such as a capacitor discharging through a resistance and inductance, or of a constant external force, such as a direct voltage. Both these conditions are analogous to a mechanical vibrating system being displaced from the neutral point.

The oscillations decay gradually, depending on the amount of damping in the circuit, with a frequency, f, termed the natural frequency; this is approximately equal to $(LC)^{-2}$ when the resistence R in the circuit is small (see diagram). The amplitude of the current is given by

$$i = A \, e^{-(R/2L)t} \sin\omega t$$

where the angular frequency ω is equal to $2\pi f$, and A is a constant determined by the initial conditions. The maxima of successive oscillations lie on the curve

$$i = A \, e^{-(R/2L)t}$$

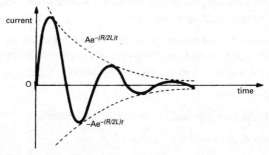

Decay of **free oscillations**

and the amplitudes (i_m, i_{m+1}) of successive maxima of the same sign decrease by a constant ratio. It can be shown that

$$\log_e(i_m/i_{m+1}) = \pi/Q$$

where Q, the ➤Q factor of the circuit, is given by

$$Q^2 = L/R^2C$$

➤forced oscillations.

free space The region used as an absolute standard and characterized by an absence of particles and of gravitational and electromagnetic fields. Free space was formerly referred to as a vacuum. The electric constant and the magnetic constant are the formally defined values of the ➤permittivity of free space and the ➤permeability of free space, respectively. The ➤speed of light in free space is also a defined constant and is the maximum possible value.

F-region *Syn. for* F-layer. ➤ionosphere.

frequency Symbol: ν or f; unit: hertz. The number of complete oscillations or cycles of a periodic quantity occurring in unit time. The frequency is related to the angular frequency ω by the relation $\omega = 2\pi\nu$. The frequency of a periodic quantity, such as an alternating current, is given by the number of times the quantity passes through its zero value in the same sense in unit time.

The frequency of ➤electromagnetic radiation is related to the wavelength, λ, by the equation $\nu = c/\lambda$, where c is the ➤speed of light.

frequency analyser ➤wave analyser.

frequency band A particular range of ➤frequencies that forms part of a larger continuous series of frequencies. The internationally agreed radiofrequency bands are shown in Table *a*. ➤Microwave frequencies range from approximately 0.3 GHz to over 300 GHz. Microwave frequency bands, now defined by IEEE, are given in Table *b*. The entire electromagnetic spectrum is shown in Table 10 in the back matter.

Wavelength	Band	Frequency
1 mm – 1 cm	extremely high frequency; EHF	300 – 30 GHz
1 cm – 10 cm	superhigh frequency; SHF	30 – 3 GHz
10 cm – 1 m	ultrahigh frequency; UHF	3 – 0.3 GHz
1 m – 10 m	very high frequency; VHF	300 – 30 MHz
10 m – 100 m	high frequency; HF	30 – 3 MHz
100 m – 1000 m	medium frequency; MF	3 – 0.3 MHz
1 km – 10 km	low frequency; LF	300 – 30 kHz
10 km – 100 km	very low frequency; VLF	30 – 3 kHz

a **Frequency bands** of radiowaves

Designation	Frequency range in gigahertz
HF	0.003 – 0.030
VHF	0.030 – 0.300
UHF	0.300 – 1.000
L band	1.000 – 2.000
S band	2.000 – 4.000
C band	4.000 – 8.000
X band	8.000 – 12.000
Ku band	12.000 – 18.000
K band	18.000 – 27.000
Ka band	27.000 – 40.000
Millimetre	40.000 – 300.000
Submillimetre	>300.00

b IEEE microwave **frequency bands**

frequency bridge An alternating-current ➤bridge whose balance point is dependent on the frequency at which the measurement is carried out.

frequency changer *Syns.* frequency converter; conversion transducer. **1.** A device that converts alternating current of one frequency to alternating current of another frequency. The *conversion gain ratio* of a frequency changer is defined as the ratio of signal power available at the output to that available at the input; the *conversion voltage gain* is the ratio of the output voltage to the input voltage. **2.** ➤mixer.

frequency compensation Modification of a circuit or device to produce a flat response to a particular range of frequencies.

frequency control ➤automatic frequency control.

frequency converter ➤frequency changer.

frequency-dependent negative resistor (FDNR) A one ➤port device realized using a GIC (➤general impedance converter) whose impedance Z_{in} is given by $1/(s^2D)$, where $s = j\omega$ and D is a constant. A circuit realization is shown in the diagram, where $D = C^2R_2R_4/R_5$. If, for example, $C = 1$ farad and $R_2 = R_5 = 1$ ohm and $R_4 = R$, then Z_{in} becomes equal to $-1/(\omega^2R)$, i.e. a frequency-dependent negative resistor. This device is used extensively to design active ➤filters directly from passive LC ➤prototype filters.

frequency deviation ➤frequency modulation.

frequency discriminator A ➤discriminator that selects input signals of constant amplitude and produces an output voltage proportional to the amount that the input frequency differs from a fixed frequency. Frequency discriminators are commonly used in ➤automatic frequency control systems (when the output is used to correct the frequency) and in ➤frequency-modulation systems to convert the frequency-modulated signals to amplitude-modulated signals. The design of frequency discriminators is such that noise due to amplitude variations in the received signal is almost completely eliminated.

frequency diversity ➤diversity system.

frequency divider A device that produces an output signal whose frequency is an exact integer submultiple of the input frequency.

frequency-division multiple access (FDMA) ➤digital communications.

frequency-division multiplexing (FDM) A form of ➤multiplex operation in which each user of the system is assigned a different frequency band. The transmitted signal contains several ➤carrier waves each of a different frequency and separately modulated with a different input signal. At the receiving end a bank of filters or some other frequency-discriminating circuit is used to extract the individual modulated carriers and, through demodulation, the original signals can be recovered.

frequency domain A term used to refer to a situation in which an electrical signal is represented as a function of frequency. A complex information signal may contain many frequency components. These can be determined using ➤Fourier analysis, which yields the individual single-frequency components of the signal. The components can be observed using a ➤spectrum analyser, which displays signals in the frequency domain.

frequency doubler A ➤frequency multiplier that produces an output signal with a frequency twice that of the input signal.

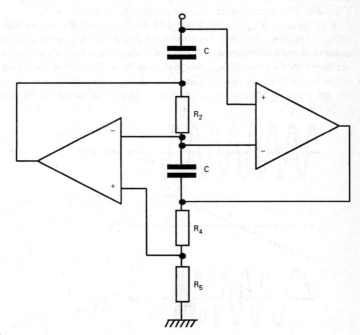

Frequency-dependent negative resistor

frequency-hopping spread spectrum ➤digital communications.

frequency meter An instrument that is used to measure the frequency of an alternating current. The frequency of an electromagnetic wave is commonly measured using a ➤cavity resonator.

frequency-modulated radar ➤radar.

frequency modulation (f.m. or FM) A type of ➤modulation in which the frequency of the ➤carrier wave is varied above and below its unmodulated value by an amount proportional to the amplitude of the signal wave and at a frequency of the modulating signal, the amplitude of the carrier wave remaining constant (see diagram). If the modulating signal is sinusoidal then the instantaneous amplitude, e, of the frequency-modulated wave may be written:

$$e = E_m \sin[2\pi Ft + (\Delta F/f)\sin 2\pi ft]$$

where E_m is the amplitude of the carrier wave, F the unmodulated carrier wave frequency, ΔF the peak variation of the carrier-wave frequency from F due to modulation, and f is the modulating signal frequency. ΔF is called the *frequency swing* and the maximum value (ΔF_{max}) of the frequency swing for which the system has been designed is the *frequency deviation*. The *deviation ratio* is defined as $\Delta F_{max}/f_{max}$; the *modulation index*, β, is given by $\Delta F/f$. If the total ➤bandwidth occupied by the transmission is less than 30 kilohertz, the transmission is said to be *narrowband FM*.

If the modulating signal is not sinusoidal but is made up of discrete levels, the resulting modulation is known as *frequency shift keying* (FSK). Any number of discrete-amplitude signal levels can be used. If only two levels are used, the result is *binary FSK*. In binary FSK the carrier is frequency modulated by two different sinusoidal signals representing either a one or a mark, and a zero or a space (➤mark). The two frequencies are commonly separated by 85 Hz in *narrowband FSK* and by 850 Hz in *wide-*

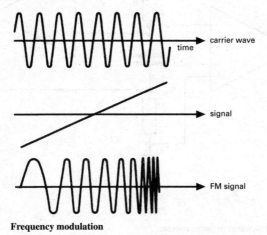

Frequency modulation

band FSK. This *two-tone modulation* is commonly used in telegraphy systems. In the general case where *M* discrete levels of the modulating signal are used, the modulation is called *M-ary FSK*. The rate of switching between the different frequencies affects the system performance; rapid switching is known as *fast hopping*.

Frequency modulation has several advantages over ►amplitude modulation, the most important being the improved ►signal-to-noise ratio. ►►pulse modulation; phase modulation.

frequency multiplexing *Short for* frequency division multiplexing.

frequency multiplier A nonlinear circuit or device that produces an output signal whose frequency is an exact integer multiple of the frequency of the input. Particular cases of frequency multipliers are frequency doublers and triplers.

frequency overlap ►colour television.

frequency pulling ►pulling.

frequency range The range of frequencies at which a circuit or device operates normally. The frequency range at which a particular device is useful depends on the particular operating conditions for the device.

frequency response The complex function that defines how a ►linear system behaves as a function of frequency. It can be separated into an ►amplitude response and a ►phase response.

frequency-response characteristic The variation with frequency of the ►transmission loss or gain of any apparatus, circuit, or device. A series of tests at different frequencies designed to determine the frequency-response characteristic is termed a *frequency run*.

frequency run ►frequency-response characteristic.

frequency scaling ►denormalization.

frequency scaling factor ►denormalization.

frequency selectivity The ability of any circuit or device to differentiate between signals at different frequencies and to select the desired signal.

frequency shift keying (FSK) ►frequency modulation.

frequency spectrum (of electromagnetic waves) ►Table 10, backmatter.

frequency standard, primary A very stable and precise ►oscillator that may be calibrated against national standard frequencies and used as a laboratory standard.

frequency swing ►frequency modulation.

frequency synthesizer ►synthesizer.

frequency transformation The conversion of a ►normalized low-pass filter to a high-pass, band-pass, or band-stop ►filter type using the following substitutions for *s* in the normalized low-pass filter response ►transfer function (where s is the complex operator ►s-domain circuit analysis):

low-pass to high-pass: $s \to \omega_c/s$

low-pass to band-pass: $s \to (s^2 + \omega_c^2)/Bs$

low-pass to band-stop: $s \to Bs/(s^2 + \omega_c^2)$

where ω_c is the cut-off frequency and B is the bandwidth.

frictional electricity *Syn.* triboelectricity. The phenomena associated with electrostatic charge produced by friction between two dissimilar materials, such as glass and silk. *Frictional machines,* such as the Wimshurst machine, were machines designed to produce electricity by friction: they are now obsolete.

Friis transmission equation An equation giving the variation in power between a transmitter and receiver in a ►telecommunication system; this variation is known as *path loss*. In a system with a transmitter of power P_T, ►gain G_T, and a receiver of gain G_R, receiving power P_R, separated by a distance R and operating at a wavelength λ, the Friis transmission equation gives the received power as

$$P_R = P_T G_T G_R (\lambda/4\pi R)^2$$

This specifies the relation in free space; modifications have to be made if the transmission is influenced by other objects, such as the earth's surface.

fringe area A region round a broadcasting transmitter in which satisfactory reception of the broadcast signal is not always obtained.

front porch ►television.

f.s.d. (or **FSD**) *Abbrev. for* full-scale deflection.

FSK *Abbrev. for* frequency shift keying. ►frequency modulation.

FSM *Abbrev. for* finite-state machine.

full adder ►adder.

full-scale deflection (f.s.d.; FSD) The maximum value of the measured quantity for which a measuring instrument is calibrated.

full-wave dipole ►dipole.

full-wave rectifier circuit A ►rectifier circuit that rectifies both the positive and negative half-cycles of the single phase a.c. input and delivers them as unidirectional current to the load (see diagram). ►►half-wave rectifier circuit.

functional packing density ►packing density.

function generator 1. A ►signal generator that produces various specific waveforms for test purposes over a wide range of frequencies. **2.** A unit in an analogue computer that produces an output signal corresponding to the value of a specified function of the independent variable input.

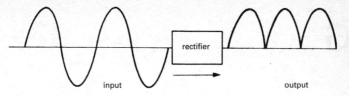

Full-wave rectifier circuit

fundamental *Short for* fundamental frequency.

fundamental frequency 1. The frequency of a sinusoidal component of a periodic quantity that has the same ➤period as the periodic quantity. **2.** The lowest frequency present in a complex vibration. ➤harmonic.

fuse A short length of easily fusible wire that is used to protect electric circuits or devices by melting ('blowing') at a specific current and thus breaking the circuit. The *fuse current rating* is the maximum value of current that the fuse will conduct without melting. The frequency and voltage at which a fuse is designed to operate are specified by the *fuse frequency rating* and *fuse voltage rating*. The *fuse characteristic* is the relation between the current through the fuse and the time taken for the fuse to operate.

fusible-link memory ➤ROM.

fuzzy logic A branch of logic designed to deal with common qualities that do not necessarily fit into mutually exclusive categories such as logic 1 and logic 0. Fuzzy logic provides a framework for capturing indistinctness, as opposed to the uncertainty of randomness. Instead of standard logic values of 1 and 0, fuzzy logic consists of variables that can take on real values μ,

$$0 \le \mu \le 1$$

These new variables cannot be considered logical, except in the limits of 0 and 1; instead μs form part of a *fuzzy variable*, and the other part is an instance of the variable.

A *fuzzy set F* is a collection of ordered pairs:

$$\{x, \mu(x)\}, 0 \le \mu(x) \le 1$$

where $\mu(x)$ is the membership of x in F. The function $\mu(x)$ may be determined arbitrarily or may be computed. For example, the fuzzy set of large real numbers might be defined by

$$\mu(x) = |x|/(|x| + 1)$$

By this measure the number –7 has 0.875 membership in the set of large real numbers.

G

G³ *Abbrev. for* gadolinium gallium garnet. A nonmagnetic form of garnet that can be grown as single crystals and used as a nonmagnetic substrate material for solid-state magnetic circuits.

gain A measure of the ability of an electronic circuit, device, or apparatus to increase the magnitude of a given electrical input parameter. In a power ➤amplifier the gain is the ratio of the power output by the amplifier to the power input to it. For a voltage ➤amplifier that supplies relatively little power to the load, the gain is the ratio of the voltage developed across a specified load impedance to the input voltage. For a ➤directional antenna the gain is the ratio of voltage generated in the direction of maximum sensitivity to that produced by the same signal in an omnidirectional antenna. Gain is often measured in decibels.

gain-bandwidth product (GBP) A means of characterizing an operational amplifier, defined as the product of the low-frequency or d.c. ➤open-loop gain A_L and the ➤cut-off frequency f_c; the diagram shows the open-loop gain response. The gain-bandwidth product is extensively used to compute the bandwidth (BW) of an amplifier when used in a closed-loop configuration:

$$BW \times (\text{➤closed-loop gain}) = GBP$$

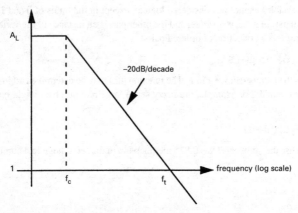

Gain-bandwidth product: open-loop gain response

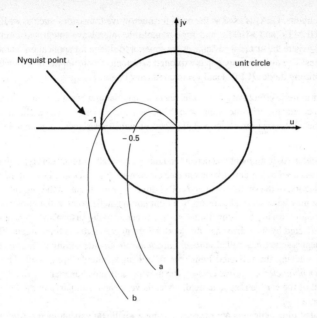

Polar plot showing **gain margin**

gain compression ➤compression point.

gain control *Syn.* volume control. A circuit or device that varies the amplitude of the output signal of an ➤amplifier.

gain margin A measure of the relative stability of a system, defined as the reciprocal of the ➤gain of the system at the frequency at which the ➤phase angle reaches 180°. The gain margin is a measure of the factor by which the system gain would have to be increased for the locus on a polar plot (phase against magnitude) to pass through the −1 point. This can be related to the (−1, 0) point – the *Nyquist point* – on the polar plot (see diagram). In the diagram, the locus **a** has a gain margin of 1/0.5 = 2, the locus **b** has a gain margin of 0.

gain region ➤active region.

galactic noise *Syn.* Jansky noise. ➤radio noise.

gallium arsenide Symbol: GaAs. A ➤compound semiconductor made from equal quantities of group-III material gallium and group-V material arsenic (➤periodic table). This crystalline semiconductor material has a ➤direct energy gap, and a conduction-band structure that permits high electron ➤mobility and ➤saturated velocity; it also exhibits the ➤Gunn effect. In undoped or ➤intrinsic form, GaAs exhibits a high resistivity, which can be used for a semi-insulating substrate in high-frequency transistors and integrated circuits. GaAs is used as the basis for microwave transistors, such as ➤MES-FETs, ➤HEMTs, and ➤HBTs, and for ➤monolithic microwave integrated circuits (MMICs), where the material advantages are superior to silicon for applications at these frequencies. The direct energy gap is exploited in optoelectronic applications, such as ➤light-emitting diodes (LEDs) and ➤semiconductor lasers.

galvanomagnetic effect Any of various phenomena arising when a current is passed through a conductor or semiconductor in the presence of a magnetic field. The effects include the ➤Ettinghausen effect, ➤Hall effect, ➤magnetoresistance, and the ➤Nernst effect.

galvanometer An instrument that detects and measures small currents. One type of galvanometer is a ➤moving-coil instrument (see diagram). This is designed so that the angular deflection of the small suspended coil is directly proportional to the current. The deflection may be observed by means of a light nonmagnetic pointer that moves over a scale or more usually by using a small mirror attached to the suspension. A beam of light is reflected by the mirror and the spot of light moves across a linear scale. This arrangement is sometimes called a *mirror galvanometer*. Greater sensitivity may be obtained by causing the reflected light spot to fall on a suitable ➤photocell. This is known as a *photoelectric galvanometer*. The moving-coil galvanometer is virtually independent of the earth's magnetic field, which is very weak compared to the field of the magnet.

Transient direct currents are measured using a ➤ballistic galvanometer. Alternating currents of frequency less than a kilohertz can be detected and measured with a *vibration galvanometer*. This is usually a moving-coil instrument with light damping, which vibrates about the zero position when the current passes. Currents above a kilohertz can be measured with a ➤thermocouple instrument.

Galvanometers measure small currents. Larger currents are measured using an ➤ammeter. Most ammeters are shunted galvanometers: a resistor is connected in parallel with the galvanometer to reduce its sensitivity. The resistor is known as a *galvanometer shunt*.

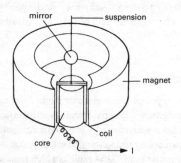

Moving-coil **galvanometer**

galvanometer constant The multiplying factor that must be applied to the scale reading of a galvanometer in order to give the value of the current in amps.

galvanometer shunt ➤galvanometer.

gamma rays (γ-rays) Very high frequency electromagnetic radiation that is emitted spontaneously by certain radioactive elements in the course of a nuclear transition or can be produced in nuclear reactions, as in the annihilation of an elementary particle and its antiparticle. The wavelength of gamma rays emitted by radioactive substances is characteristic of the radioisotope involved and ranges from about 4×10^{-10} to 5×10^{-13} metre. Although gamma rays at one time lay at the extreme low-wavelength end of the electromagnetic spectrum, modern high-voltage generators can now produce X-rays of much shorter wavelength than that of most gamma rays.

Gamma rays undergo no deflection in electric and magnetic fields. Their depth of penetration is controlled by their energy, which depends on the wavelength. The energy of gamma rays is usually measured in ➤electronvolts, and is determined by measuring the maximum energy of photoelectrons that they produce or their diffraction by certain crystal lattices.

ganged circuits Two or more circuits with variable elements that are mechanically or electronically coupled so that they can be operated simultaneously by a single control.

ganging oscillator An oscillator that has a constant output over a wide frequency range and whose frequency may be rapidly adjusted. It is used for testing the accuracy of adjustment of ganged tuned circuits over their tuning range.

gap 1. *Syn. for* forbidden band. ➤energy bands. **2.** The space between the electrodes in any electron tube. A spark gap is a special arrangement of the electrodes so that a ➤spark occurs if the voltage exceeds a predetermined value. A spark gap is often used to divert high-voltage surges and thereby protect a device.

gap length ➤magnetic recording.

gas amplification ➤gas multiplication.

gas breakdown A type of ➤breakdown that occurs in a ➤gas-filled tube when the voltage reaches a given value. Ions in the gas are accelerated by the field and reach high kinetic energies. Little recombination of ions occurs because of the high energies but further ions are produced by collisions between the ions and molecules of the gas; a multiplication effect thus occurs causing rapid breakdown of the gas. The process is analogous to ➤avalance breakdown in a semiconductor.

gas cell ➤gas electrode.

gas-discharge tube A ➤gas-filled tube in which the presence of gaseous molecules contributes significantly to the characteristics of the tube. Normally a gas is a poor electrical conductor but a sufficiently high electric field causes ➤ionization of molecules and atoms in the immediate vicinity of the electrodes; these gaseous ions are attracted to the charged electrodes and a small current – the *preconducting current* – flows.

If a sufficiently high potential difference is applied to the tube ➤gas breakdown occurs and a large current flows across the tube. The potential difference across the tube drops to a relatively small value and the discharge is self-sustaining. The *threshold current* is the value of current at which the discharge becomes self-sustaining; the process of establishing the discharge is termed *firing*. The minimum voltage required for the discharge to continue is the *maintaining voltage*.

Current flows as a result of a multiple collision of ions within the tube: collisions between ions cause excitation and further ionizations, light being produced when excited atoms and ions return to the ground state. Recombination between positive and negative ions also results in a small amount of light emission.

The phenomena observed depend on the pressure of gas in the tube. At pressures near atmospheric pressure a ➤spark passes between the electrodes. If the pressure is reduced, *glow discharge* occurs. At relatively high pressures the mean free path of ions is small; a *positive glow* is observed near the anode and a *negative glow* near the cathode, the centre of the tube being dark. At lower gas pressures the mean free path increases and the positive glow extends across the tube until the whole tube is filled. The colour of the glow discharge is brilliant and characteristic of the gas in the tube. *Glow-discharge tubes* can be used for luminous signs and for lighting purposes.

A further reduction in pressure causes a change in the glow pattern (see diagram). Additional dark regions appear in which electrons, produced by ionization and excitation, have insufficient energy for excitation and the probability of recombination is low. The largest potential drop in the tube is across the *Crookes dark space*. Striations in the positive glow are caused by alternate ionizations and recombinations in the tube.

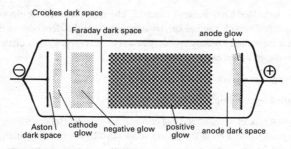

Gas-discharge tube

At very low pressures the Crookes dark space fills the tube and very few collisions occur. The high kinetic energies of the ions cause ➤secondary emission of electrons (originally known as *cathode rays*) from the cathode, and at sufficiently high fields secondary emission of positive ions from the anode. With very high fields ➤X-rays are emitted from the anode as a result of bombardment by high-energy electrons. This phenomenon was utilized in early ➤X-ray tubes.

If the electrodes are placed relatively close together an ➤arc forms across them. The heat generated by the arc causes ➤thermionic emission of electrons from the negative electrode and the current density is therefore very high. The *arc discharges* occur over a very wide range of pressures and the light associated with them is rich in ultraviolet frequencies.

The *Townsend discharge* is a luminous discharge that occurs at lower current densities than the glow discharge; the voltage across the tube is a function of the current density. At low current densities the potential falls uniformly across the tube and the luminous region extends across the tube.

gas electrode An electrode that absorbs or adsorbs a gas so that when in contact with an electrolyte the gas effectively acts as the electrode. A *gas cell* is one that contains a gas electrode.

gas-filled tube An ►electron tube that contains a gas or vapour, such as mercury vapour, in sufficient quantity so that the electrical characteristics of the tube are determined entirely by the gas, once ►ionization has taken place. ►Gas-discharge tubes are one type of gas-filled tube. Gas-filled tubes that operate by virtue of the ionization produced in them when charged particles pass through them are used as detectors of ionizing radiation, examples being the ►ionization chamber and ►Geiger counter.

gaskets Flexible conducting strips used to maintain a conducting path across joints between parts of equipment enclosures. As the parts are secured together the gasket is compressed ensuring that it maintains outward pressure on both parts and good electrical contact.

gas multiplication *Syn.* gas amplification. **1.** The production of additional ions by ions produced in a gas under the influence of a sufficiently strong electric field. **2.** The ratio of the total ionization to the initial ionization as a result of the above process.

gas-plasma screen ►flat-panel display.

gassing The evolution of gas in the form of small bubbles from one or more electrodes during electrolysis. Gassing occurs in an accumulator towards the end of the charging period.

gate 1. *Short for* gate electrode. An electrode or electrodes in a ►field-effect transistor, ►MOS capacitor, ►MOS integrated circuit, ►charge-coupled device, or ►thyristor. **2.** *Digital gate.* A digital circuit that has two or more inputs but only one output. The conditions applied to the inputs determine the voltage level at the output. The output is switched between two or more discrete values. Digital gates are widely used in ►logic circuits, and are then known as *logic gates*. **3.** *Analogue gate.* A ►linear circuit or device that produces an output signal only during a specified interval of the input signal. During this interval the output is a continuous function of the input signal. Analogue gates are widely used in radar and electronic control systems. **4.** An electric signal or trigger that allows a circuit or device to operate. The most common method of gating is to employ a ►clock.

gate array An integrated circuit whose internal structure is an array of ►logic gates with interconnections that are initially unspecified. The logic designer specifies the gate types and interconnections.

gate expander An array of diodes connected to the input stage of a ➤diode-transistor logic gate and used to extend the possible number of inputs to that gate.

gating ➤receiver.

gauss Symbol: Gs or G. The unit of magnetic flux density in the obsolete CGS electromagnetic system of units. One gauss equals 10^{-4} tesla.

Gaussian channel In communication systems, a channel that adds noise of a Gaussian nature to the signal being transmitted, i.e. the noise amplitude over time has a ➤Gaussian distribution.

Gaussian distribution *Syn.* normal distribution. A continuous distribution having a density given by

$$f(x) = (2\pi\sigma 2)^{-\frac{1}{2}} \exp\left(-\frac{1}{2}(x-\mu)^2/\sigma^2\right)$$

where μ is the mean and σ is the standard deviation of the distribution. The distribution is symmetric about the mean. A random variable having this distribution, i.e. having this set of possible values, is said to have a Gaussian or normal distribution.

Gaussian elimination A method used in the solution of linear algebraic equations. ➤simultaneous equations.

Gaussian filter A type of ➤Bessel filter.

Gauss's theorem The electric field strength, *E*, over a closed surface containing electric charges is given by

$$\int E.dS = \Sigma q/\varepsilon_0$$

where d*S* is a small element of area on the surface *S*, ε_0 is the ➤permittivity of free space, and Σq the total charge enclosed within the surface.

If the volume enclosed by the surface contains a distributed charge density, ρ_e, Gauss's theorem becomes

$$\int E.dS = (1/\varepsilon_0)\int \rho_e d\tau$$

where dτ is a small volume element. Since

$$\int E.dS = \int \text{div } E \, d\tau$$

then

$$\text{div } E = \rho_e/\varepsilon_0$$

In a dielectric medium the total charge density includes an apparent charge density due to polarization of the atoms or molecules within the medium and Gauss's theorem becomes

$$\text{div } D = \rho_e$$

where *D* is the electric ➤displacement. This is one of ➤Maxwell's equations.

Two main conclusions can be drawn from Gauss's theorem. Firstly, the electric field inside a hollow conductor containing no charge is zero, the enclosed space being at the same potential as the conductor: electrical apparatus can therefore be shielded from the effects of external electric fields by surrounding it with an earthed conductor. Secondly, any excess static charge on a conductor must reside on the outer surface.

GBP *Abbrev. for* gain-bandwidth product.

Geiger counter *Syn.* Geiger–Müller counter. A gas-filled tube that is used to detect ionizing radiation, especially alpha particles, and to count particles. It contains a low-pressure gas and has a thin wire anode mounted coaxially inside a cylindrical cathode with a potential difference, slightly lower than that required to produce a discharge, maintained across the electrodes. When radiation enters through a thin window the gas along its path becomes ionized. The ions are accelerated by the field and produce an avalanche, which is quickly quenched. The resulting current pulse is amplified and registered by a detector (typically a loudspeaker) or a counting device.

The output is substantially constant for a wide range of voltage; this is the operating region of the tube. The tube is also independent of the energy of the incident radiation in this region.

general impedance converter (GIC) A ►two-port device that is capable of making the input impedance of one of its two ports the product of the impedance terminating its remaining port and the internal impedances of the two-port device. A circuit realization of a GIC using ►operational amplifiers is shown in the diagram overleaf.

generating station ►power station.

generator A machine that converts mechanical energy into electrical energy. An electromagnetic generator (i.e. a ►dynamo) has a coil that is made to move so as to cut lines of magnetic flux. An *electrostatic generator,* such as a ►Van de Graaff generator, has equal and opposite electric charges produced by electrostatic induction or friction; mechanical energy is then used to separate the charges. ►►alternating-current generator.

genetic algorithm An algorithm that emulates biological evolutionary theories to solve optimization problems. In computing terms, a genetic algorithm maps a problem onto a set of (typically binary) strings of characters, each string representing a potential solution. The genetic algorithm then manipulates the most promising strings in its search for improved solutions. A genetic algorithm operates through a simple cycle of stages: creation of a 'population' of strings; evaluation of each string; selection of 'best' strings; genetic manipulation to create the new population of strings.

geometric optics A method of computing the propagation of waves where the waves are treated as having very small wavelengths. A series of rays are allowed to propagate from a source or sources in straight lines. If a ray encounters the surface of an object it is reflected from the surface and its new path followed. From a large set of such calculations the effects of the sources can be derived. Geometric optics is often used as a technique for the construction of realistic images for three-dimensional computer graphics. It becomes less reliable as the wavelength of the wave under consideration becomes larger. ➤geometric theory of diffraction.

geometric theory of diffraction (GTD) A method of computing the propagation of waves based on ➤geometric optics. In this case, as well as accounting for reflection from surfaces, the effects of ➤diffraction from surfaces and around edges is taken into account. This results in reliable models at longer wavelengths compared with methods using geometric optics alone. In particular it allows waves that would diffract into shadowed regions of an object to do so.

GEOS *Abbrev. for* geostationary earth orbit satellite. ➤LEOS.

geostationary earth orbit An orbit in which a satellite will take 24 hours to make one revolution of the earth, so that it appears to be in the same position in the sky at all times: stationary with respect to points on the earth's surface. Geostationary orbits are at a distance of 40 km above the earth's surface, in the plane of the equator. Satellites in such orbits – often referred to as GEOS – are used for broadcast of radio and TV to specific areas of the earth. ➤communications satellite.

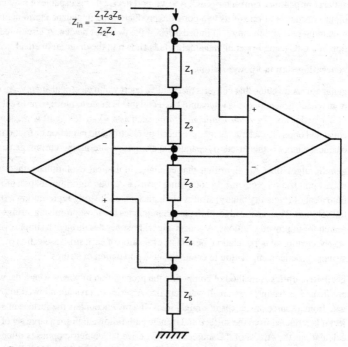

$$Z_{in} = \frac{Z_1 Z_3 Z_5}{Z_2 Z_4}$$

General impedance converter

geosynchronous orbit An orbit around the earth in which a satellite will take 24 hours to complete one revolution. ►geostationary earth orbit.

germanium Symbol: Ge. A semiconducting element, atomic number 32, extensively used for early semiconductor devices. It has now been displaced by silicon as the substrate for most semiconductor devices.

getter A material that has a strong chemical affinity for other materials. Such substances can be used to remove undesirable elements or compounds that are present in small quantities in an environment. Uses include the introduction of barium, often in filament form, into a sealed vacuum system in order to remove residual gases, and the introduction of phosphorus into oxide layers on silicon in order to remove mobile impurities such as sodium. The latter application is particularly important for the stabilization of ►MOSFETs. The process of using a getter is known as *gettering*.

ghost *Syn.* double image. An unwanted double image that appears on the screen of a television or radar receiver. It is caused by ►ground-reflected waves arriving at the receiver slightly later than the ►direct waves.

Gibbs phenomenon The overshoot obtained on a waveform close to a signal discontinuity when its ►spectrum is truncated abruptly (as noted by mathematician Josiah Willard Gibbs).

GIC *Abbrev. for* general impedance converter.

giga- Symbol: G. **1.** A prefix to a unit, denoting a multiple of 10^9 of that unit: one gigahertz equals 10^9 hertz. **2.** A prefix used in computing to denote a multiple of 2^{30} (i.e. 1 073 741 824): one gigabyte equals 2^{30} bytes.

gilbert Symbol: Gb. The unit of magnetomotive force in the obsolete CGS electromagnetic system of units. One gilbert equals 0.7958 amperes (or ampere-turns).

Gilbert cell A ►transistor circuit, devised by Barrie Gilbert, that is based on coupled ►long-tail pair circuits and performs four-quadrant multiplication of two input signals; it is used frequently as a ►mixer circuit, in ►heterodyne radio receivers for example (see diagram overleaf).

glassivation ►passivation.

glitch ►hazard.

global positioning system ►GPS.

glottal airflow The waveform of air flowing through the glottis, the space between the human vocal folds, during speech production. Modelling glottal airflow is an important aspect of electronic speech synthesis.

glow discharge ►gas-discharge tube.

glow-discharge microphone ►microphone.

Golay code ►digital codes.

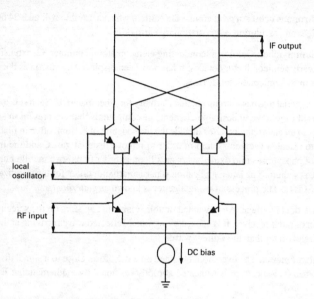

Gilbert cell using bipolar junction transistors

gold-leaf electroscope ►electroscope.

GOS *Abbrev. for* grade of service. A measure of the quality of a communications network, calculated in number of calls lost per 100.

GPIB *Abbrev. for* general-purpose interface bus. ►IEEE-488 standard.

GPS *Abbrev. for* global positioning system. A system of satellites that enables an accurate estimate to be made of the position of a receiving device on or near the earth's surface. Eighteen satellites are used in medium earth orbit at an altitude of approximately 20 000 km with their orbits inclined at 55.6° to the equator such that at least four are always visible from any point on earth. The satellites continuously transmit signals towards the earth: a receiving device monitors the difference in arrival time of the signals and thus calculates its position. The accuracy can be better than a metre horizontally, but is less in the vertical direction.

graded-base transistor A bipolar transistor in which the impurity concentration in the ►base varies smoothly across the base region. The ►doping level is high at the emitter-base junction dropping to a low doping level (therefore high resistivity) at the base-collector junction. This assists the passage of electrons through the base and thus reduces the base recombination current. The high-frequency response of these transistors is good.

graphical symbols Symbols that represent the various types of components and devices used in electronics, telecommunications, and allied subjects. Graphical symbols are shown in Table 1, in the back matter.

graphic equalizer A ➤tone control in which the frequency range is divided into bands. The signal in each band may be adjusted by sliding contacts, the positions of which indicate frequency response.

graphic instrument *Syn.* chart recorder. A measuring instrument that presents the quantity being measured in the form of a graph. The graph is produced either in ink on a suitable paper chart or on the screen of a ➤cathode-ray oscillograph from which a permanent record may be produced by photographing the screen.

graphic panel A master control panel in automatic and remote control systems in which coloured block diagrams are used to show the relation and functioning of the different parts of the control system. It may also have the controls and recording instruments mounted on it in their correct relative positions.

grass *Syn.* picture noise. *Informal.* ➤Noise that occurs in radar receivers due to irregularities in the ➤timebase of the display. These may be due to random fluctuations in the timebase generator or to electrical interference. It results in noise on the picture that resembles grass. ➤➤snow.

Grassot fluxmeter ➤fluxmeter.

gravity cell A type of primary ➤cell that contains two different electrolytes kept apart by their different densities.

Gray code An important nonpositional binary-number code in which there are 2^n codewords, each of n bits, with the code changing only one bit from one number to the next. Gray code is used, for example, to code ➤Karnaugh maps.

grenz rays Soft X-rays produced when electrons are accelerated through 25 kilovolts or less. Grenz rays are produced in many types of electronic equipment, such as colour television sets, but have an extremely low penetrating power.

grid 1. An electrode that has an open structure, such as a mesh or a plate with a hole in it, thus allowing an electron beam to pass through it. ➤thermionic valve. **2.** The nationwide high-voltage transmission line system that interconnects many electricity power stations. It transmits voltages of up to 400 kilovolts. Voltages as high as 735 kV are used in some countries.

grid base (of a thermionic valve) ➤cut-off.

grid bias The potential applied to the grid of a ➤thermionic valve that determines the portion of the characteristic curve at which the valve will operate, or that modifies the cut-off value of the valve. *Automatic grid bias* uses a resistor in either the grid or cathode circuits of the valve to supply the grid bias. The voltage across the resistor is determined by either the grid or cathode current respectively and in turn supplies the potential to the grid.

grid emission The emission of electrons or ions from the grid of a thermionic valve.

grid leak A high resistance connected between the grid and cathode of a thermionic valve that prevents an accumulation of charge on the grid and may also be used to develop the ►grid bias.

grid stopper ►parasitic oscillations.

ground *US syn. for* earth.

ground absorption A loss of energy that occurs during transmission of radiowaves due to absorption in the ground.

ground capacitance *US syn. for* earth capacitance.

ground clutter ►ground return.

ground current *US syn. for* earth current.

grounded-base connection ►common-base connection.

grounded-collector connection ►common-collector connection.

grounded-emitter connection ►common-emitter connection.

ground electrode *US syn. for* earth electrode. ►earth.

ground plane *Syn.* earth plane. A sheet of conducting material that is adjacent to a circuit and is at earth potential. It may be used to provide a low-impedance earth at any point in the circuit. For example, one side of a double-sided printed circuit board may be used as the ground plane. Connections to it from the printed circuit on the other side of the board are easily effected through the board from any desired point in the circuit.

ground potential *US syn. for* earth potential.

ground ray ►ground wave.

ground-reflected wave A radiowave that travels between a transmitting and a receiving antenna situated above the ground and that undergoes at least one reflection from the ground. It can be affected by the properties of the ground and can also undergo refraction in the troposphere. ►►ground wave.

ground reflection Reflection of a transmitted radar wave by the ground before it reaches the target.

ground return Echoes received by a radar receiver due to reflections of the radar wave by the ground or by objects on the ground. Ground return from extraneous sources can lead to *ground clutter* on the radar screen. This type of noise can obscure the target.

ground state The lowest possible ►energy level of a system. If a system has a configuration such that its energy is greater than its ground-state energy, it is said to be in an *excited state*. Many phenomena, including luminescence and semiconduction, depend upon systems being in an excited state.

ground wave *Syn.* ground ray. A radiowave that travels between a transmitting and a receiving antenna situated above the earth. It has two main components: the ►space wave, which includes the ►direct wave and the ►ground-reflected wave, and the ►surface wave. ►indirect wave.

group delay A quantity related to the phase characteristics of a network or circuit and defined as the derivative of the change in phase (ϕ) with respect to frequency, $d\phi/d\omega$.

group operation The operation of all the poles of a multiple switch or circuit-breaker by a single mechanism.

grown junction A ►p-n junction that is formed in a single crystal of semiconductor material while the crystal is being grown from a melt. The amount and type of impurities added to the semiconductor are varied in a controlled manner as the crystal grows. A *grown-diffused junction* is produced by ►diffusion of impurities into the semiconductor after a grown junction has been formed, in order to produce the precise doping profile required.

GTO *Short for* gate-turnoff thyristor. ►thyristor.

guard band A frequency band that is left vacant during broadcasting to minimize mutual interference between two neighbouring frequency bands.

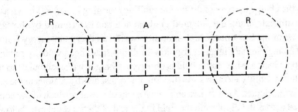

Principle of **guard ring**

guard ring A device used to ensure uniform fields and to define the sensitive volume in absolute electrometers and standard capacitors. It consists of a metal plate surrounding and coplanar with a smaller plate; a narrow air gap separates the two. The diagram shows the lines of force between a plate A surrounded by a guard ring RR, and a parallel earthed plate P. Variations in field at the edges affect only the guard ring RR.

A similar device in the form of an auxiliary electrode is commonly used in semiconductor devices and vacuum tubes.

guard-ring capacitor A standard capacitor that uses a ►guard ring to reduce the edge effect (see diagram overleaf). The *guard-well capacitor* is a special type of guard-ring capacitor used for capacitance below 0.1 picofarad. In this type the guard ring forms a well on which a Pyrex disc is mounted in order to locate the electrode assembly accurately.

Gudden–Pohl effect The transient luminescence observed in a phosphor, previously exposed to ultraviolet radiation, when an electric field is applied to it.

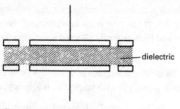

Guard-ring capacitor

guide *Short for* waveguide.

Guillemin effect ➤magnetostriction.

Guillemin line A network that produces pulses with very sharp rise and fall times so that they are almost square.

Gunn diode *Syn. for* transferred electron device.

Gunn effect An effect that occurs when a large d.c. electric field is applied across a short sample of n-type gallium arsenide. At values above the threshold value, typically several thousand volts per cm, coherent microwave oscillations are generated.

The effect results from the ➤degeneracy of the energy levels in the conduction band and is due to the charge carriers of different mobilities forming bunches, known as *domains*. These migrate along the potential gradient at a rate determined by the carrier mobility. Some of the low-energy high-mobility carriers as they gain energy from the applied field change into a higher-energy low-mobility state, i.e. a velocity reduction in response to an increase in electric field is observed. Charges will accumulate to form a domain of low-mobility electrons. As the domain forms, nearby regions in the semiconductor become depleted of carriers and a charge dipole results. Much of the applied field is applied across the dipole and the field in the rest of the material falls below the threshold value. Thus, once one stable domain has formed no others will form while it still exists. The domain drifts towards the anode where it produces a current spike, and the process begins again. The result is that the domains produce microwave current at the output.

gyrator A component that does not obey the ➤reciprocal theorem, i.e it reverses the phase of signals transmitted in one direction but does not affect signals transmitted in the opposite direction. Gyrators are usually used at microwave frequencies, forming part of a waveguide. They may be entirely passive or may contain active elements.

H

halation *Syn.* halo. A glow observed around the spot on the screen of a ➤cathode-ray tube. It is caused by total internal reflections within the thickness of the glass.

half-adder ➤adder.

half-duplex operation ➤duplex operation.

half-IF response ➤IP_2.

half-power beamwidth ➤beamwidth.

half-power point The point on a characteristic curve that corresponds to operation at a power intensity of half the maximum value.

half-wave dipole ➤dipole.

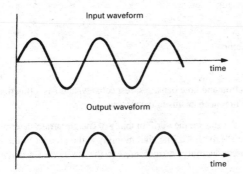

Half-wave rectification

half-wave rectifier circuit A ➤rectifier circuit that delivers unidirectional current to the load only during alternate half-waves of the single-phase alternating-current input (see diagram). Usually the positive half-cycles are rectified.

half-wave voltage doubler A ➤voltage doubler that operates only during the input wave half-cycle.

Hall coefficient ➤Hall effect.

Hall effect If current flows in a bar of conductor or semiconductor that is placed in a magnetic field, where the current flow and magnetic field are at right angles, an electric field is set up across the bar, orthogonal to both current flow and magnetic field directions. This effect arises because the charge carriers in the bar experience a force F due to the magnetic field:

$$F = e.(v \times B)$$

$$= ev_x B_z \text{ in the } y\text{-direction}$$

where v is the drift velocity of carriers of charge e, and B is the magnetic flux density. Charge carriers are thus forced towards one side of the bar, creating an excess ►space charge and hence an electric field E_y in the y-direction, across the bar. In equilibrium the forces due to the transverse electric field and the magnetic field are in balance:

$$eE_y = -ev_x B_z$$

or

$$E_y = -R_H JB$$

where J is the current density along the bar, and R_H is the *Hall coefficient* of the material of the bar and is related to the ►carrier concentration in the bar. The voltage due to this transverse electric field is known as the *Hall voltage*. Measurement of the Hall voltage enables the Hall coefficient to be found, and hence the type and density of the charge carriers in the bar to be determined:

in an n-type semiconductor,

$$R_H = -r/ne$$

in a p-type semiconductor,

$$R_H = +r/pe$$

where n and p are the electron and hole densities, respectively, and r is a function of the detailed charge-scattering mechanisms in the material.

Hall mobility Symbol: μ_H. A value for the ►drift mobility of charge carriers determined from measurements of the ►Hall effect. The Hall mobility is the product of the Hall coefficient, R_H, and the ►conductivity, σ (measured in the absence of a magnetic field):

$$\mu_H = R_H \sigma$$

Hall and drift mobilities are usually very nearly equal. Small differences arise due to the details of the carrier-scattering processes in the material.

Hall probe A convenient means of measuring magnetic flux density, *B,* by utilizing the ►Hall effect. A small sample of a suitable semiconductor, such as indium antimonide or indium arsenide, is used in which the Hall voltage developed across it varies almost linearly with the magnetic flux density. The probe can be used in nonuniform fields.

Hall voltage ►Hall effect.

halo ➤halation.

Hamming code A ➤digital code that helps in the process of error correction. The transmission line of a serial communications system can be susceptible to noise pulses that flip bits to incorrect values. There is a need for altered digital transmission data to be identified and perhaps corrected. Hamming code is one such mechanism. A message of four bits is embedded in a transmission of seven bits. The three extra redundant bits, known as *check bits*, provide enough information that a one-bit error anywhere in the seven bits can be corrected.

In the encoding process, the message bits may be designated $M_3M_2M_1M_0$ and the check bits $C_2C_1C_0$. Three functions have to be identified relating to the Ms and equating to the Cs:

$$C_2 = f_2(M_3, M_2, M_1); \quad C_1 = f_1(M_3, M_2, M_0); \quad C_0 = f_0(M_3, M_1, M_0)$$

There are many Boolean functions f that could be chosen here, and Hamming chose the following: have each C_i be the bit that makes the *parity* – the sum of the bits, either even or odd – of the string $\{C_i, M_x, M_y, M_z\}$ always the same. Then C_i becomes 0 or 1, whichever is needed to make the number of 1s in the set $\{C_i, M_x, M_y, M_z\}$ even (assuming even parity). The table shows the Hamming code for four-bit data.

M_3	M_2	M_1	M_0	C_2	C_1	C_0
0	0	0	0	0	0	0
0	0	0	1	0	1	1
0	0	1	0	1	0	1
0	0	1	1	1	1	0
0	1	0	0	1	1	0
0	1	0	1	1	0	1
0	1	1	0	0	1	1
0	1	1	1	0	0	0
1	0	0	0	1	1	1
1	0	0	1	1	0	0
1	0	1	0	0	1	0
1	0	1	1	0	0	1
1	1	0	0	0	0	1
1	1	0	1	0	1	0
1	1	1	0	1	0	0
1	1	1	1	1	1	1

Hamming code for 4-bit data

Hamming window ➤windowing.

ham radio *Informal* Noncommercial (amateur) radio communication between licensed individuals. The frequencies used are restricted to internationally agreed values in order to prevent interference with commercial broadcast transmission and shipping or aircraft communication.

handset ➤telephony.

handshake In computing, a general method of controlling the flow of information between two communicating processors or devices in which permission to send is requested and granted before the data are sent, and often an acknowledgement of receipt of data is given after the transmission.

Hanning window *Syn.* raised cosine window. ➤windowing.

hard decision decoding *Syn.* firm decision decoding. A form of decoding used in a digital receiving system in which an incoming signal is compared to a fixed voltage level in order that the system can decide whether the incoming signal represents a logical '1' or '0'. The alternative is the use of more sophisticated logic circuitry that will adapt to the signal received and adjust the decoding levels according to the signal strength and noise level.

hard disk ➤magnetic disk.

hardware The physical units, devices, and circuits that make up a computer system. ➤software.

hardware description language (HDL) A computer-based language that allows designers to model their hardware designs in software and simulate them, before building the actual hardware. This approach has been driven by the increasing complexities in hardware. HDL allows designs to be tested, in software, and design faults to be identified and corrected before the final expensive hardware is committed. ➤VHDL.

hardwired Denoting a circuit or circuits that are permanently interconnected to perform a specific function. ➤ROM.

hard X-rays ➤X-rays.

harmonic An oscillation of a periodic quantity, present in a complex vibration, having a frequency that is an integer multiple of the ➤fundamental frequency; the fundamental frequency is also known as the *first harmonic*. An oscillation having a frequency that is an integer submultiple of the fundamental is termed a *subharmonic*. In practice the fundamental need not necessarily be present.

harmonic analyser A device that analyses a periodic function to find its harmonic components, for example in terms of the Fourier series (➤Fourier analysis) corresponding to the function.

harmonic distortion The ➤distortion of a waveform caused by the nonlinear effect of a system or component that creates ➤harmonic frequencies that would not otherwise be present at the output. The extent to which a particular harmonic component distorts a signal is given by the ratio of its amplitude A_n to the amplitude of the fundamental component A_f, expressed as a percentage:

%age nth harmonic distortion = %D_n = $(A_n/A_f) \times 100\%$

The *total harmonic distortion* (THD) is the square root of the sum of the squares of all the individual harmonic distortions:

$$\%THD = \sqrt{(D_2{}^2 + D_3{}^2 + \ldots + D_n{}^2)} \times 100\%$$

where the individual harmonic distortion D_n is specified by the ratio A_n/A_f.

harmonic generator A signal generator that produces ➤harmonics of an input fundamental frequency.

harmonic oscillator ➤oscillator.

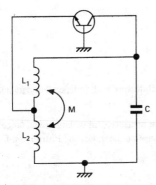

Hartley oscillator

Hartley oscillator A type of ➤oscillator consisting of a transistor with ➤common-emitter connection and a parallel ➤resonant circuit between the emitter and collector (see diagram). Ignoring the resistance of the coil the resonant frequency, ω_0, is given by

$$\omega_0{}^2 \approx C/(L_1 + L_2 + 2M)$$

Hartley oscillators may also be crystal-controlled (➤piezoelectric oscillator, Fig. *d*).

Hartshorn bridge An alternating-current ➤bridge that measures mutual inductance. Direct comparison of a mutual inductance with a standard mutual inductance is virtually impossible because effects, such as the self- and mutual capacitances of the coils, produce voltage components in the secondary circuits in phase with the primary current. This problem is overcome in the Hartshorn bridge by the use of a variable resistor, of resistance r, common to both the primary and secondary circuits of the bridge (see diagram). The standard inductance and the resistance are varied until a balance is achieved with no response registered by the indicating instrument I; assuming that the secondaries are connected in antiphase, then

$$M_1 = M_2$$
$$r = \pm(\rho_1 - \rho_2)$$

where ρ_1 and ρ_2 are the resistances of the mutual inductances M_1 and M_2.

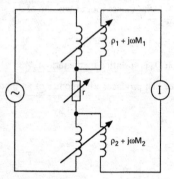

$\rho_1 + j\omega M_1$

r

$\rho_2 + j\omega M_2$

Hartshorn bridge

Harvard architecture A ►von Neumann architecture with independent data and instruction paths to memory (or cache).

Hay bridge A four-arm ►bridge used for the measurement of large inductance (see diagram). At balance, when there is no response on the indicating instrument, I,

$$L_x = R_a R_b C_s / (1 + \omega^2 C_s^2 R_s^2)$$
$$R_x = L_x \omega^2 C_s R_s$$

where ω is the angular frequency.

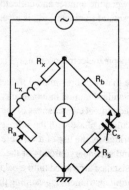

R_x

R_b

L_x

R_a

C_s

R_s

Hay bridge

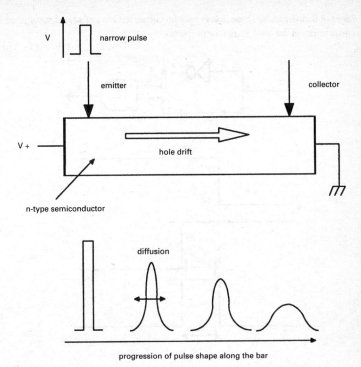

Haynes–Shockley experiment: minority holes in an n-type semiconductor

Haynes–Shockley experiment A classic solid-state experiment that demonstrates the drift and diffusion of charge carriers in a semiconductor. The experimental arrangement is shown in the diagram. A voltage is placed across the length of a bar of n-type semiconductor (for example) to create a *drift field*. ➤Minority carriers, holes in this example, are injected at one end of the bar in a narrow pulse, using either a pulse of light to *photogenerate* the carriers or a short voltage pulse at a rectifying ➤point contact. The injected minority carriers will drift along the bar at the ➤drift velocity associated with the applied electric field. The carriers will also diffuse away from each other, causing the pulse to broaden. The pulse is detected at the downstream end of the bar using a voltage probe – usually another point contact. The broadening of the pulse can be measured and the diffusion of the carriers determined. From these measurements the minority carrier ➤drift mobility and ➤diffusion constant can be calculated.

hazard *Syn.* glitch. An error in the logic state of a ➤logic circuit that causes or could cause the circuit to malfunction. Logic circuits require a finite time to operate and consequently introduce delays into the propagation of information. These delays are generally no more than a few microseconds, and for very high speed logic may be of the order of nanoseconds. These delays may invalidate the laws of ►Boolean algebra and cause errors in the logic state (see diagram).

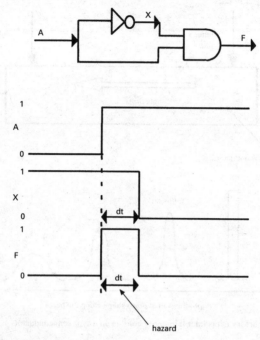

Hazard in a logic circuit

HB *Abbrev. for* horizontal Bridgeman.

H-bend *Syns.* H-plane; flatwise bend. ►waveguide.

HBT *Abbrev. for* heterojunction bipolar transistor.

HDL *Abbrev. for* hardware description language.

HDTV *Abbrev. for* high-definition television. A system in development with increased resolution (between 1000 and 1500 lines) over present ►television systems (625 lines for the ►PAL system), and also with a different aspect ratio, $^5/_3$, compared with $^4/_3$ for present systems.

head A device that records signals on a suitable medium or reads or erases signals already stored on the medium, examples of such media being ►magnetic tape and ►magnetic disk. ►►magnetic recording.

header *Syn.* label. A few bytes at the start of a ►packet of data, transmitted by digital communications, that identify the sender, the receiver, the length of the packet and, in some cases, the ►protocol being used within the packet. ►►digital communications.

head-related transfer function ►HRTF.

headset ►earphone.

hearing aid A complete sound-reproduction system that increases the sound intensity at the ear. Modern hearing aids use a small crystal microphone, battery-powered amplifier, and earpiece. The development of integrated circuits has allowed hearing aids to become very small and light and they may be hidden in a spectacles frame. Good sound quality at sufficient power output for most cases is obtained with volume and tone controls available to the wearer. Special types of hearing aids include those designed to amplify a specific frequency band only and those in which the output from the amplifier is used to produce vibrations in the mastoid bone behind the ear, thus bypassing a defective outer or middle ear.

heat coil A coil that may be used as a switch by the sensing of the temperature of the coil (►heating effect of a current). Such coils are most often used as protective devices that open a circuit when the current through the coil rises above a predetermined value.

heat detector A temperature-sensitive device, such as a thermocouple, that operates an alarm when a predetermined rise in temperature occurs. Heat detectors are used, for example, in fire-alarm and burglar-alarm systems.

heater Any resistor that is used as a heat source when subjected to an electric current. They are used, for example, in indirectly heated cathodes and domestic heating devices. In the latter case the term applies both to the element and the complete unit.

heating depth ►dielectric heating.

heating effect of a current *Syn.* Joule effect. If a current, I, is passed through a circuit of resistance R, electrical energy is dissipated as the current flows from a higher to a lower potential. The rate of dissipation of energy equals I^2R. This is known as I^2R *loss*. The energy appears as heat; if all the electrical energy is converted into a quantity of heat Q, then

$$Q = I^2Rt$$

where t is the time for which the current flows. The temperature does not increase indefinitely since a stable state is reached when the rate of emission from the surface of the resistor just equals the rate of generation of heat.

heat sink A device that is employed to dispose of unwanted heat in a circuit and prevent an excessive rise in temperature. Heat sinks are particularly useful for protecting transistors in power applications.

Heaviside layer (or **Heaviside–Kennelly layer**) *Syn. for* E-layer. ►ionosphere.

hecto- Symbol: h. A prefix to a unit, denoting a multiple of 100 of that unit.

height control A control in a ►television or ►radar receiver that adjusts the overall size of the frame scan.

Heil tube An early type of ►klystron.

helical antenna *Syn.* helix-radio dipole. An ►antenna formed by winding a conducting wire in the shape of a screw thread, producing a helix. Usually the helix is used with a ►ground plane and coaxial transmission line ►feed, the centre conductor of the transmission line being connected to the helix and the outer being connected to the ground plane.

The helical antenna can be used in a variety of modes, the most common two being *normal* (broadside), in which the radiation is linearly polarized, and *axial* (endfire), in which the radiation is circularly polarized (►polarization). The mode is dependent on the circumference and pitch angle of the helix. In axial mode, the antenna is particularly useful for space communications as it will still receive a component of a linearly polarized signal but is insensitive to the actual orientation of the polarization, something difficult to keep constant for satellites, etc. In normal mode the antenna is very common on hand-held mobile communications systems as it performs similarly to a ►monopole antenna with ground plane but can be significantly shorter.

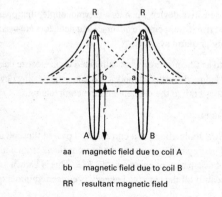

aa	magnetic field due to coil A
bb	magnetic field due to coil B
RR	resultant magnetic field

Helmholtz coils

Helmholtz coils An arrangement of two identical cylindrical coils mounted coaxially a distance r apart. When r just equals the radius of the coils a uniform calculable magnetic field is produced between them when the same current flows through both coils (see diagram). This arrangement was first used in the Helmholtz galvanometer but is now commonly employed when a precise uniform magnetic field is required over a substantial volume.

Helmholtz galvanometer An obsolete form of galvanometer. It is a ►moving-magnet instrument that employs ➤Helmholtz coils instead of a single coil to provide a uniform magnetic field in which the magnet is deflected.

HEMT *Abbrev. for* high electron mobility transistor.

henry Symbol: H. The ➤SI unit of ➤inductance and ➤permeance. The self- or mutual inductance of a closed loop is one henry if a current of one ampere gives rise to a magnetic flux of one weber.

hertz Symbol: Hz. The ➤SI unit of frequency. It is the frequency of a periodic phenomenon that has a period of one second. It replaced, but is equivalent to, the cycle per second (c.p.s.).

Hertzian dipole An infinitesimal current element where the current is assumed to be the same in magnitude and phase at all points along the element length. The concept is often used in the analysis of more complex antennas.

heterodyne receiver ➤heterodyne reception.

heterodyne reception A process whereby a ➤receiver – a *heterodyne receiver* – employs the mixing of the incoming signal with a signal of a different frequency, produced by a *local oscillator* (LO), to create a signal at a frequency that is the difference between the two; this is the *intermediate frequency* (IF). The signal at the IF retains all the modulation of the original signal frequency, and hence this technique can be used in a ➤radio receiver, for example. The heterodyne technique offers advantages over direct detection of the incoming signal including improved ➤selectivity and ➤sensitivity.

A typical heterodyne receiver using two mixing stages is shown in Fig. *a*. The first stage is a filter, called the *preselector*. This is usually a low-pass or band-pass ➤filter to restrict the range of input signal frequencies to the desired range; it also limits the amount of ➤noise received. The next stage is a low-noise ➤amplifier (LNA), used to boost the level of the incoming signal; this improves the sensitivity of the receiver. The (optional) *image filter* is used to remove signals at the ➤image frequency, which is the other frequency that when mixed with the local oscillator will produce a signal at IF. The frequency relationships are illustrated in Fig. *b*, including mixer harmonics and higher-order mixing terms.

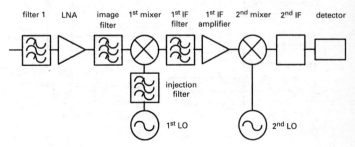

a Typical two-stage heterodyne receiver

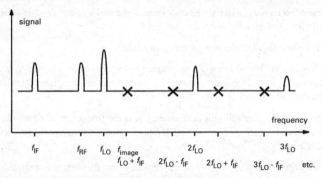

b Frequency relationships in a heterodyne mixing stage

The ►mixer is a nonlinear stage in which the input signal and local oscillator signal are combined to produce the IF. The signal at IF can then be filtered, to define the selected channel bandwidth, and amplified, to improve the sensitivity of the receiver. The combination of IF filtering and IF amplifier is often called the *IF strip*. The overall receiver selectivity is determined by the final-stage IF in a multistage receiver. A receiver may use several stages of intermediate frequencies to improve the performance; different filter types with various characteristics are available in different frequency ranges, and the IFs can be chosen to optimize the filter choice.

heterojunction A junction between dissimilar ►semiconductor materials. A heterojunction would be typically created during the ►epitaxial growth process, resulting in a single semiconductor crystal. The different ►energy band gaps of the two semiconductors in the heterojunction will result in an energy offset between the two ►conduction and ►valence bands. The nature of this offset leads to the classification of two fundamental heterojunction types (see diagram). The heterojunctions can be doped,

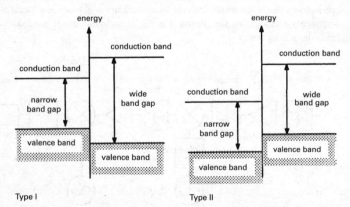

Type I and Type II **heterojunctions**

either the same type – *isotype heterojunction* – or as a ►p-n junction – *anisotype heterojunction.* ►heterojunction bipolar transistor; high electron mobility transistor.

heterojunction bipolar transistor (HBT) A ►bipolar junction transistor that incorporates a wide band gap emitter. The emitter-base junction is a ►heterojunction between semiconductors of different energy band gap. The following are typical material systems:

AlGaAs(emitter)/GaAs(base);
AlInAs/InGaAs;
Si/SiGe

The wider band gap of the emitter reduces significantly the injection of majority carriers from base to emitter, thus maximizing the desired injection of carriers from emitter to base. This eliminates the requirement for a heavily doped emitter to achieve the same result, and consequently allows the base doping to be increased. An increase in base doping is desirable from a device viewpoint as the base resistance can be reduced significantly. This leads to an improvement in the high-frequency performance of the transistor. HBTs are used at radio- and microwave frequencies, in integrated circuit and power applications, and in optoelectronic ICs.

heterojunction FET ►high electron mobility transistor.

heterojunction photodiode ►photodiode.

heterostructure laser ►semiconductor laser.

Heusler alloys Alloys that exhibit ferromagnetic properties although they contain no ferromagnetic material. They contain manganese, aluminium, and zinc or copper.

hex *Short for* hexadecimal.

hexadecimal A number system with 16 elements present, i.e. base 16. These elements are the numbers 0 to 9 and the letters A to F. Therefore $10_{10} = A_{16}$, $11_{10} = B_{16}$, etc., where the subscript indicates the decimal or hexadecimal system. Hexadecimal is often used in computer systems, for example for 16-bit arithmetic.

hexagon voltage *Syn.* mesh voltage. ►voltage between lines.

HF *Abbrev. for* high frequency. ►frequency band.

hidden Markov model (HMM) A statistical model used for automatic pattern matching after a training period when the model is presented with a large number of examples of the patterns to be recognized.

hi-fi Acronym from *high fi*delity. ►reproduction of sound.

high-density packaging The packaging of a given set of electrical components in a very small volume through the application of miniaturization techniques, as by the use of ►integrated circuits or ►surface-mount components.

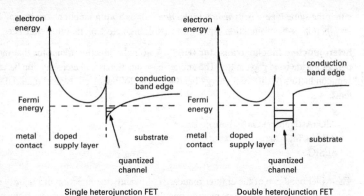

HEMT: schematic of conduction band of heterojunction FETs

high electron mobility transistor (HEMT) *Syn.* modulation-doped field-effect transistor (MODFET). A ➤field-effect transistor in which a ➤heterojunction is used to separate the conducting channel from the doped supply region. The mobility of the electrons in the channel is therefore higher than in conventional FETs because of the absence of carrier scattering by the ionized donors. The following are typical material systems:

AlGaAs (doped supply layer)/GaAs (channel)   single heterojunction FET;
AlGaAs/InGaAs (channel)/GaAs;
AllnAs/InGaAs/InP   double heterojunction FET.

Only n-channel devices are used, since the hole mobility in these III-V ➤compound semiconductors is too low to be useful practically. These structures are illustrated in the diagram.

HEMTs are used for low-noise amplifier and IC applications at microwave and millimetre-wave frequencies. The high-frequency performance is attributed to the improved control of the channel charge modulation by the gate voltage, due to the electron confinement in the narrow ➤quantum well channel.

high electron velocity camera tube ➤camera tube. ➤➤iconoscope.

high fidelity ➤reproduction of sound.

high frequency (HF) ➤frequency band.

high-frequency resistance ➤skin effect.

high-level injection ➤injection.

high-level programming language ➤programming language.

high logic level ➤logic circuit.

high-pass filter ➤filter.

high recombination-rate contact A contact between two semiconductors or between a semiconductor and a metal that has a substantially constant charge-carrier concentration whatever the current density at the contact. This is because the excess minority charge recombination rate is high so that the excess charges generated as the current flows disappear very quickly.

high tension (HT) *Syn. for* high voltage, especially when it refers to the voltage supply to the anode of a ➤thermionic valve, usually in the range 60 to 250 volts.

high-velocity scanning ➤scanning.

high-voltage test A test of the insulation of any electronic device. A voltage in excess of the normal operating voltage is applied across the insulation to ensure that it does not break down.

hi-lo Read diode ➤IMPATT diode.

HMM *Abbrev. for* hidden Markov model.

H-network *Syns.* H-pad; H-section. ➤two-port network.

holding time In telecommunication or digital communication systems, the length of time a circuit or communications channel is occupied by a user to establish a connection and transfer data.

hold time In computing, the length of time that data is required to stay at the input of a device, such as a ➤flip-flop, once the clock has triggered the device. This is required because of the finite time a device requires to input data.

hole An empty energy level in the ➤valence band of a semiconductor due to an electron being lost from the band by thermal excitation (➤electron-hole pair) or being trapped by an acceptor impurity (➤semiconductor). The total current resulting from electrons in a filled valence band is zero:

$$j = nev = e\Sigma_{i=1}^{n} v_i = 0$$

where j is the current density, n and e the electron density and charge, and v the average velocity of electrons in the valence band. If the jth electron is excited to the conduction band or trapped by an impurity then

$$e\Sigma_{i=1, i\neq j}^{n} v_i = -ev_j$$

The net effect of the mobile electrons occupying all but the jth energy level is therefore equivalent to the effect of a single positive electronic charge 'occupying' the vacant level and hence termed a hole. The hole velocity is equal to the velocity of an electron in the same energy level.

Under the influence of an electric field holes can drift through the material by a process of continuous exchange with adjacent electrons (➤hole conduction). They therefore act as mobile positive charge ➤carriers in the valence band and are the majority carriers in p-type semiconductors.

hole capture ➤recombination processes.

hole conduction Conduction of electricity in a semiconductor in which a ➤hole is propagated through the crystal lattice under the influence of an electric field. An adjacent electron moves under the influence of the field and fills the vacancy, leaving a corresponding vacancy behind. The effective movement of the hole due to a process of continuous exchange is equivalent to the movement of a positive charge in the same direction (see diagram). Holes move in the direction of positive field.

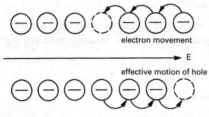

Hole conduction

hole current The current in a ➤semiconductor associated with the movement of ➤holes through the material.

hole density ➤carrier concentration.

hole-electron pair ➤electron-hole pair.

hole injection ➤injection.

hole trap A site in a semiconductor crystal, such as a donor impurity or lattice defect, that can trap ➤holes.

homing beacon *Syn.* locator beacon. ➤beacon.

homodyne receiver *Syn. for* direct conversion receiver. ➤➤heterodyne reception.

homojunction A junction between two regions of opposite polarity types within a semiconductor. ➤p-n junction.

homopolar generator An electromagnetic ➤generator in which the voltage induced in the conductors always has the same sense with respect to the conductors.

honeycomb coil A coil in which the turns are wound in a criss-cross fashion in order to reduce the distributed capacitance.

hook-up *Informal* A connection, usually of a temporary nature, between any electric or electronic circuits.

horizontal blanking ➤blanking.

horizontal Bridgeman (HB) A method of growing ➤compound semiconductor crystals, such as gallium arsenide. The process uses a boat in which the crystal is grown.

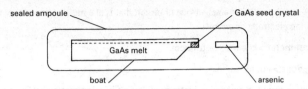

sealed ampoule   GaAs seed crystal

GaAs melt

boat   arsenic

Schematic diagram of **horizontal Bridgeman** growth method

In the case of gallium arsenide, the boat containing either pure gallium or polycrys-talline gallium arsenide is sealed into a long quartz ampoule filled with an inert gas. Pure arsenic is placed in the neck of the ampoule and a seed crystal of gallium arsenide is placed at one end of the boat (see diagram). Heaters are used to cause the arsenic to become gaseous and the material in the boat to melt. A temperature profile is created to cause the melt to be held at the melting point of gallium arsenide and the seed crys-tal region at just below the solidification point. The gaseous arsenic reacts with the gal-lium in the boat. Once the reaction is complete the ampoule and heaters are moved relative to each other so that the temperature front is moved slowly along the length of the boat. Crystal growth occurs from the seed crystal following the temperature front, to produce a large single crystal with a cross-sectional shape matching the shape of the boat. Gallium arsenide produced by this method is sometimes referred to as *boat grown* gallium arsenide.

horizontal hold ➤television receiver.

horizontal polarization 1. Polarization of an electromagnetic wave in which the elec-tric field vector is parallel to the geographic horizon. **2.** The horizontal arrangement of a ➤dipole antenna.

horn antenna A type of ➤antenna formed of a conducting tube of varying cross sec-tion opening out from the feed to the mouth where the radiation is emitted. Various cross sections are used including square, rectangular, and circular with varying characteris-tics. Horns are used as the feed for reflector and lens antennas, as an element in phased arrays, and as a standard for calibration of other high-gain antennas.

horn loudspeaker ➤loudspeaker.

horseshoe magnet A magnet that is in the shape of a horseshoe so that the poles are close together.

hot cathode ➤thermionic cathode.

hot electron A conduction electron in a solid that has an energy such that the electron is not in thermal equilibrium with the crystal lattice, i.e. it is more than a few kT above the ➤Fermi level, where k is the Boltzmann constant and T the thermodynamic tem-perature of the lattice. A *hot hole* is a mobile hole that is similarly not in thermal equi-librium with the lattice.

hot hole ➤hot electron.

hot spot A small portion of an electrode or circuit that is at a higher temperature than the rest of the electrode or circuit.

hot-wire ammeter ➤hot-wire instrument.

hot-wire gauge ➤Pirani gauge.

hot-wire instrument A measuring instrument, basically an ammeter, that utilizes the elongation by heat of a thin wire carrying an electric current. The wire is rigidly clamped so that its elongation can be measured. The elongation is proportional to the temperature rise of the wire, which in turn is proportional to the square of the current. Such instruments can therefore be used for either DC or AC measurements. An extremely large resistance used in series with the instrument allows it to function as a voltmeter.

howl An unpleasant high-pitched audiofrequency tone heard in receivers and caused by unwanted electric or ➤acoustic feedback.

H-pad *Syn. for* H-network. ➤two-port network.

h parameter *Short for* hybrid parameter. ➤network.

H-plane *Syn. for* H-bend. ➤waveguide.

HRTF *Abbrev. for* head-related transfer function. The ➤transfer function imposed on sound arriving at the two ears of a listener due to the physical dimensions and shape of the upper torso and head. The HRTF is unique to an individual listener. The binaural HRTF is the left and right ear frequency response to a sound source at a particular position, and in practice a set of binaural HRTF measurements would usually be made for a sound source placed at a number of positions on a sphere around the head of a listener or a dummy head. Ideally, such measurements should be made in an anechoic room where the walls, floor, and ceiling absorb sound energy fully so that none is reflected back into the space.

H-section *Syn. for* H-network. ➤two-port network.

HT *Abbrev. for* high tension.

HTS *Abbrev. for* high-temperature superconductor. ➤superconductivity.

hum A low-pitched audiofrequency droning noise heard in audiofrequency systems. It usually originates from the mains supply and occurs at frequencies that are ➤harmonics of the mains-supply frequency.

hum modulation Unwanted ➤modulation of a signal by hum arising in an audiofrequency system. It results in ➤noise or ➤distortion of the output.

hunting Fluctuation of a controlled signal about its desired value that results from overcorrection by a control device. The stable desired value is never actually attained.

Huygens' principle A principle that explains ➤diffraction, a phenomena by which waves travelling in straight lines bend around corners; it is named after the Dutch astronomer Christian Huygens. It states that each point on a wavefront can be considered as a source that generates further waves which propagate from the source in all directions. The concept is important in the understanding of radio reception in built-up and mountainous areas.

H wave *Syn. for* TE wave. ➤mode.

hybrid integrated circuit ➤integrated circuit.

hybrid junction A junction between ➤transmission lines, either waveguides or coaxial lines. An ideal junction has no direct coupling between arms 1 and 4 or between arms 2 and 3 (see diagrams). If arm 1 is excited power flows in arm 4 as a result of reflections in arms 2 and 3.

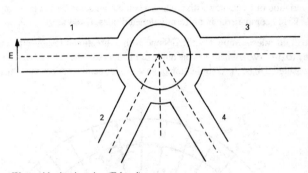

a Waveguide ring junction (E-bend)

The two main types of hybrid junction are the *ring junction,* in which all 4 arms are coplanar (Fig. *a*), and the *T-junction,* in which arm 4 is perpendicular to the plane containing the other three arms (Fig. *b*). If the decoupled arms (1 and 2) are independently matched in impedance and the other two arms (3 and 4) are terminated in their characteristic impedances, Z_0, then all four arms are matched at their inputs.

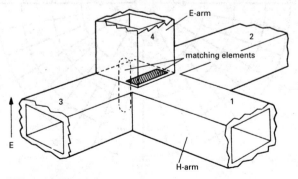

b Waveguide T-junction

hybrid parameters ➤transistor parameters; network.

hybrid-π model ➤transistor parameters.

hybrid ring junction *Syn. for* ring junction. ➤hybrid junction.

hybrid-T junction *Syn. for* T-junction. ➤hybrid junction.

hydrogen electrode ➤electrochemical series.

hydrophone A ➤transducer that produces electrical signals in response to water-borne sound waves.

hyperabrupt varactor A ➤varactor diode in which the doping of the semiconductor is controlled to produce a variation of capacitance, C, with voltage, V, that results in a linear variation of frequency with voltage when the varactor is used in a tuned circuit: $C \propto V^2$. In comparison, in uniformly doped diodes, $C \propto \sqrt{V}$.

hypercardioid microphone A ➤microphone with a directional response pattern described by $(0.5 + \cos\phi)$, where ϕ is the angle of incidence of the sound. Such a pattern can be considered to be the combination of an ➤omnidirectional microphone response

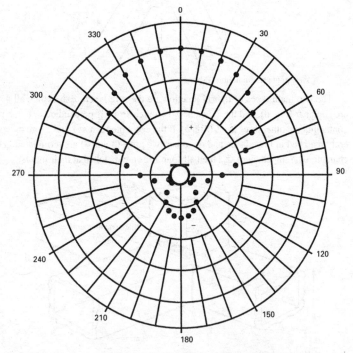

Idealized directional response of a **hypercardioid microphone** (scale in degrees)

attenuated by 6 dB (0.5 all round) and a ►figure-eight microphone response (cosφ). The
overall response is between that of the ►cardioid microphone and ►figure-eight
microphone, and its shape is sometimes described as *cottage loaf* (see diagram).
Sound picked up from the rear of the microphone is in ►antiphase with sound picked
up from the front and is thus marked as '–' in the diagram. Practically, such a response
is difficult to achieve for frequencies above which the wavelength becomes comparable with the physical dimensions of the microphone itself.

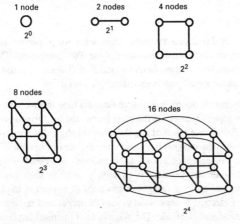

Hypercubes of different degrees

hypercube An interconnection network having the hypercube topology. (A hypercube
in its simplest form is a four-dimensional cube, which may be considered as two
three-dimensional cubes connected at equivalent corners.) The hypercube interconnection topology has 2^N nodes (see diagram), such that each node has a unique binary
address on the range 0 to 2^N-1 and there is a direct connection between every two nodes
whose addresses differ in exactly one bit position; N is the degree of the hypercube.

hysteresis A delay in the change of an observed effect in response to a change in the
mechanism producing the effect. The best-known example of hysteresis is ►magnetic
hysteresis.

hysteresis distortion Distortion introduced by nonlinear hysteresis in one or more elements of a system.

hysteresis factor ►magnetic hysteresis.

hysteresis loop ►magnetic hysteresis.

hysteresis loss 1. ►magnetic hysteresis. **2.** ►electric hysteresis loss.

hysteresis meter An instrument that detects and measures ►magnetic hysteresis.

Hz *Abbrev. for* hertz.

I

IC *Abbrev. for* integrated circuit.

I channel The in-phase signal path in a receiver system, when the incoming signal is converted to an in-phase component and a quadrature-phase (90°) component. This occurs, for example, in the detection of noncoherent digital FSK (frequency shift keyed) signals. The quadrature-phase signal path is called the *Q channel*.

iconoscope A high-electron-velocity photoemissive ➤camera tube in which the target is formed on a thin mica plate. One side of the plate is covered by a *mosaic* of very many small areas of photoemissive material and the other side by a thin metallic *signal electrode* from which the output signal is obtained (see diagram). The optical image produced on the mosaic causes ➤photoemission of electrons from each elemental area. The mica acts as the dielectric for what is effectively an array of elemental capacitors, in which a pattern of positive charges is produced as a result of the photoemission. A high-velocity electron beam is used to scan the target area and discharge each capacitor through the signal electrode. The magnitude of the resulting current is a function of the charge on each target area and hence of the illumination producing the charge pattern.

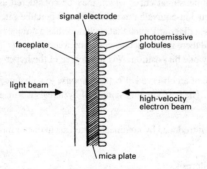

Section of mosaic target of **iconoscope**

The *orthicon* operates in a similar manner to the iconoscope but employs a low-velocity electron scanning beam.

ideal crystal A theoretical crystal that has a perfect and infinite crystal structure and therefore contains no impurities or defects.

ideal diode equation *Syn. for* Shockley equation. ➤p-n junction.

ideal transducer A hypothetical transducer that produces the maximum possible signal power output under specified input and load conditions; thus power loss in the transducer is negligible.

ideal transformer *Syn.* perfect transformer. ➤transformer.

idle component 1. (of current) ➤reactive current. **2.** (of voltage) ➤reactive voltage. **3.** (of volt-amperes) ➤reactive volt-amperes.

IEC *Abbrev. for* International Electrotechnical Commission. ➤standardization.

IEE *Abbrev. for* Institution of Electrical Engineers. ➤standardization.

IEEE *Abbrev. for* Institute of Electrical and Electronics Engineers. ➤standardization.

IEEE-488 standard *Syn.* GPIB (general-purpose interface bus). The standard specification for linking computerized test equipment and instrument controllers together for remote or automated controlling and the transfer of data.

IF *Abbrev. for* intermediate frequency. ➤heterodyne reception; mixer.

iff *Abbrev. for* if and only if. ➤logic circuit.

IF strip ➤heterodyne reception.

IGBT *Abbrev. for* insulated-gate bipolar transistor.

IGFET *Abbrev. for* insulated-gate field-effect transistor. ➤field-effect transistor.

IGMF filter *Short for* infinite-gain multiple-feedback filter. A configuration of filter that is used to construct low-pass, high-pass, and band-pass filters and is similar to the ➤Sallen–Key filter but requires one less component. It has good stability and low output impedance and inverts signals in the pass-band (➤inverse filter).

IIR filter *Short for* infinite impulse response filter. ➤digital filter.

II-VI compound semiconductor ➤compound semiconductor.

III-V compound semiconductor ➤compound semiconductor.

I²L *Abbrev. for* integrated injection logic. A family of ➤bipolar integrated logic circuits that are very compact and provide a very high functional packing density combined with the possibility of high-speed operation. I²L is essentially a development derived from ➤diode transistor logic (DTL). The input diodes and transistors of DTL are combined in I²L into a single multicollector n-p-n transistor, and the current source is provided by a p-n-p transistor rather than a load resistor. The p-n-p transistor is formed in the same area of chip as parts of the multicollector transistor allowing a very compact structure. The multicollector structure provides the isolation between logic gates provided in DTL by the input diodes of the following gates.

A typical structure is shown in Fig. *a* and the equivalent circuit in Fig. *b*. The p-n-p transistor is arranged laterally on the chip; the p-type emitter is common to a large

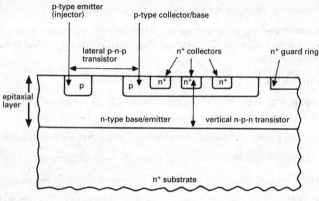

a Cross section of **I²L** gate

number of gates and is termed the *injector*. The n-p-n transistor is a vertical transistor, which is inverted compared to the usual method of fabrication. The n-type epitaxial layer forms the base of the p-n-p transistor and the emitter of the n-p-n transistor and is common to all the gates. An n⁺ ►guard ring surrounds each p-type p-n-p collector/ n-p-n base region containing the n⁺ n-p-n collectors, in order to provide isolation between individual gates. The ►fan-out is determined by the number of multiple collectors.

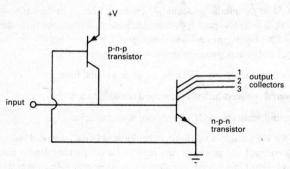

b Equivaqalent circuit of single **I²L** gate

A basic 3-input NOR gate is shown in Fig. *c*. If any of the collectors of gates 1, 2, or 3 is at the low logic level, current from the p-n-p transistor flows to that collector (indicated by the dashed line) and the transistor T_4 is 'off'. The collectors of T_4 are therefore at the high logic level. If all the inputs are high T_4 is 'on' and current from the p-n-p transistor flows through T_4 and the collector voltages are low. The p-n-p transistors are not shown but are indicated by the label 'injection current'.

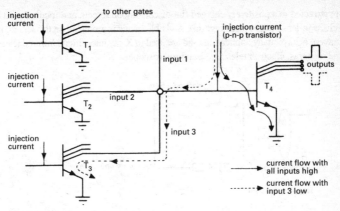

c Basic NOR gate

The difference between the high and low logic levels is determined by the forward voltage of the base-emitter junction of the multicollector transistor and the circuits can operate with a total voltage swing of about 0.7 volts. The power consumed by the circuits is a linear function of the speed at which they are operated and the circuits can be designed to optimize the speed and power at any point within the circuit. The small voltage change renders the I²L circuit susceptible to stray noise pulses or interference when used alone, and on-chip input and output buffer circuits are normally used to convert pulses from ➤transistor-transistor logic to those suitable for I²L and vice versa.

Schottky I²L is a form of I²L in which the collectors are formed as ➤Schottky diodes. Excessive ➤carrier storage at the collector junction is reduced and the speed of operation is therefore increased. The total voltage swing is also reduced due to the characteristically small forward voltage of the Schottky diode. The speed of operation of the Schottky I²L circuits may also be increased using Schottky diodes across the base-collector junction to prevent the output transistors from going too far into the saturated mode, and hence keeping charge-storage times to a minimum.

I²L has found many varied applications. It is fabricated using standard bipolar techniques and other types of circuit, such as LED drivers, operational amplifiers, and oscillators, can be easily produced on the same chip allowing great flexibility.

IM *Abbrev. for* intermodulation.

image *Short for* electric image.

image attenuation constant *Syn.* image attenuation coefficient. ➤image transfer constant.

image charge ➤electric image.

image converter *Syn.* image tube. An ➤electron tube that converts an image outside the visible part of the spectrum, such as an infrared image, into a visible image. The 'nonvisible' image is focused on a photosensitive cathode and produces electrons. These

are attracted to a positively charged fluorescent anode screen, and focused on it by an
►electron-lens system (see diagram). A visible image is produced on the screen. Image
converters have many applications and are used in X-ray intensifiers, infrared telescopes
and cameras, electron telescopes, and microscopes.

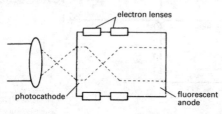

Image converter

image dissector A type of television ►camera tube, little used now, in which electrons
from each portion of a photosensitive plate are focused in turn on a collector electrode
in order to produce a video signal. It differs from the ►iconoscope or ►image orthicon
camera tubes in which the static charge pattern on the photosensitive plate is scanned
by an electron beam.

image filter ►heterodyne reception.

image force ►electric image.

image-force lowering ►Schottky effect.

image frequency An unwanted input frequency that arises from a source other than
that to which a receiver used in ►heterodyne reception is tuned and that causes spurious signals to be output; the unwanted response is generated in the intermediate-
frequency amplifier. The spurious signals are termed *image interference*. The image
frequencies are those frequencies such that the difference between them and the local-
oscillator frequency falls within the bandwidth of the intermediate-frequency amplifier. The characteristics of the particular receiver determine the input conditions that cause
image interference. The image frequencies can be equal to the desired input signal frequency, equal to the intermediate frequency, or to twice the local-oscillator frequency.
    The *image ratio* is the ratio of the image-frequency amplitude to the desired sig-
nal amplitude when identical outputs are produced.

image impedances The values Z_{i1} and Z_{i2} of ►impedance that simultaneously satisfy
the following conditions: if Z_{i1} is connected across one pair of terminals of a ►two-port
network then the impedance between the second pair is Z_{i2}; if Z_{i2} is connected across
the second pair of terminals then the impedance across the first pair is Z_{i1}.

image interference ►image frequency.

image orthicon A low-electron-velocity ►camera tube in which light from a scene is
focused on a ►photocathode consisting of a light-sensitive material deposited on a thin

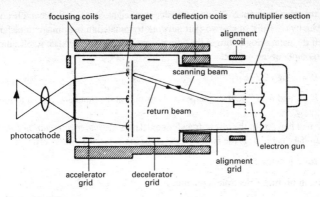

Image orthicon

sheet of glass. Electrons are emitted from the photocathode in proportion to the intensity of the light and are focused on a target consisting of a thin glass disc with a fine mesh on the photocathode side of the disc (see diagram). The impact of electrons from the photocathode causes ➤secondary emission of electrons from the target that is greater than, but proportional to, the original electron density from the photocathode. The secondary electrons are collected by the mesh screen and drained off to a power supply. The target is left with a positive static charge pattern corresponding to the original light image.

The reverse side of the disc is scanned with a low-velocity electron beam produced from an ➤electron gun. Positively charged areas of the target are neutralized by electrons from the beam, which is consequently intensity-modulated by the original picture information. The electron beam is reflected by the target glass and travels back towards the electron gun. It is collected by an electrode that surrounds the electron gun and acts as the final aperture for the scanning beam. This electrode acts as the first dynode of an ➤electron multiplier section from which the video ouput is obtained.

The image orthicon has high sensitivity, the spectral sensitivity is close to that of the human eye, and the speed of response is relatively fast.

image phase constant *Syn.* image phase-change coefficient. ➤image transfer constant.

image potential The potential at a point just outside the surface of a material due to an ➤electric image.

image processing The process of modifying one or more aspects of a representation of an image, often used as a means of enhancing the visual contrast of features of particular interest.

image ratio ➤image frequency.

image recognition The process of automatically recognizing an image, which often involves edge enhancement.

image transfer constant *Syn.* image transfer coefficient. Symbol: θ. A complex quantity, given by (α + jβ), of a ►two-port network terminated in its ►image impedances. It is half the natural logarithm of the complex ratio of the steady-state ►volt-amperes input to output of the network:

$$\theta = \tfrac{1}{2}\log_e(E_1 I_1 / E_2 I_2)$$

where E_1, I_1 and E_2, I_2 are the voltages and currents at the input and output terminals, respectively.

The real part (α) of the image transfer constant is the *image attenuation constant*; the imaginary part (jβ) is the *image phase constant.* ►► propagation coefficient.

image tube ►image converter.

immersion plating ►electroless plating.

immunity ►electromagnetic compatibility.

impact ionization Ionization of an atom or molecule due to the loss of orbital electrons following a high-energy collision. In a semiconductor electron-hole pairs can be generated if the electrons have sufficient energy to enter the conduction band.

IMPATT diode Acronym from *IMP*act ionization *A*valanche *T*ransit *T*ime. A semiconductor diode that acts as a powerful source of microwave power. When a p-n junction is reverse-biased into ►avalanche breakdown, it exhibits negative resistance at microwave frequencies and may be used as a ►negative-resistance oscillator. The differential current is out of phase with the differential voltage due to two effects: following a voltage increment the current builds up with a delay time, t_A, characteristic of the avalanche; the terminal current increment is further delayed by a time t_t (the transit time) during which the carriers are collected by the electrodes. The diode is usually formed so that the current is delayed by half a cycle with respect to the voltage.

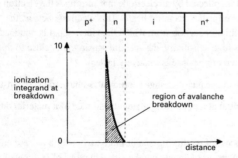

a Typical **IMPATT diode**: Read diode

Although any p-n junction diode will exhibit IMPATT mode operation, typical devices used consist of an avalanching region together with a drift region in which no avalanche occurs. Examples of such diodes are shown in the diagrams. Fig. *a* shows a *Read diode* with structure p⁺-n-i-n⁺. Although both holes and electrons are produced by the avalanche breakdown, only the electrons are given a drift region and collected. This type of device is known as a *single drift device*. Fig. *b* shows a modified Read diode, known as a *hi-lo Read diode*, in which the intrinsic region is replaced by an n-type region. Fig. *c* shows a ►PIN diode in which the avalanche occurs throughout the intrinsic region. An alternative structure where both the holes and electrons are given a drift region is a p^+-p-n-n^+ structure. Avalanche occurs at the centre p-n junction and both holes and electrons are collected. Such a structure is termed a *double drift device*.

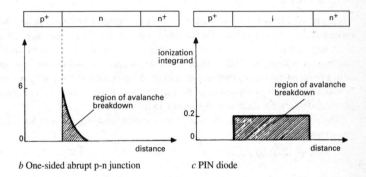

b One-sided abrupt p-n junction   *c* PIN diode

The diode is mounted in a microwave cavity that can have its impedance matched to that of the diode in order to form a resonant system. In an appropriate circuit spontaneous oscillation will occur.

impedance 1. Symbol: Z; unit: ohm. A measure of the response of an electric circuit to an alternating current. The current is opposed by the capacitance and inductance of the circuit in addition to the resistance. The total opposition to current flow is the impedance, which is given by the ratio of the voltage to the current in the circuit.

The alternating current is given by

$$I = I_0 \cos\omega t$$

where I_0 is the peak current and ω is the angular frequency. When ►reactance, due to the capacitance and inductance, is present in a circuit, the voltage will be out of phase with the current and is given by

$$V = V_0 \cos(\omega t + \phi)$$

where ϕ is the ►phase angle. In a circuit containing resistance, R, capacitance, C, and inductance, L, the voltage is given by

$$V_0 \cos(\omega t + \theta) = IR + L(dI/dt) + 1/C \int I dt$$

Solving this equation shows that the current is equal to

$$I = V_0 \cos(\omega t + \phi)/\sqrt{[R^2 + (\omega^2 L^2 - 1/\omega^2 C^2)]}$$

Since the impedance is the ratio of current to voltage then

$$Z = \sqrt{[R^2 + (\omega^2 L^2 - 1/\omega^2 C^2)]}$$

$$= \sqrt{[R^2 + X^2]}$$

where X is the reactance. Z is thus a complex quantity whose magnitude or modulus $|Z|$ is equal to the vector sum of R and X. The *complex impedance* can thus be given by

$$Z = R + jX$$

where j is equal to $\sqrt{-1}$. The real part, the resistance, represents a loss of power due to dissipation. The imaginary part, the reactance, indicates the phase difference between the voltage and current (►j). It is either positive or negative depending on whether the current lags or leads the voltage, respectively. In a circuit containing only resistance, or in a resonant circuit, the current and voltage are in phase and Z is purely resistive. In a circuit containing only reactance the current and voltage are out of phase and Z is purely imaginary, i.e. there is no dissipation in the circuit.

The complex impedance is the ratio of the complex voltage to the complex current, given respectively as

$$V = V_o \exp[j(\omega t + \phi)]$$

$$I = I_o \exp(j\omega t)$$

Hence the complex impedance can be expressed as

$$Z = |Z| \exp(j\phi)$$

$$Z = |Z| (\cos\phi + j \sin\phi)$$

2. ►impedor.

impedance coupling ►coupling.

impedance matching The matching of the ►impedances of the parts of a system to ensure optimum conditions for transfer of power from one part of the system to another.

The maximum power is transferred from an ►amplifier to a load if the load impedance is made the ►conjugate impedance of the amplifier output impedance.

A nonuniform ►filter will transmit a wave undisturbed provided that the ►iterative impedance of each filter section is made equal. If the iterative impedances of the sections are different, loss of power output occurs due to reflections of the wave occurring at the junctions. A small section terminated in its ►image impedances, where one impedance equals the iterative impedance of the filter sections and the other the load impedance, may be used to match the load to the filter.

Reflections of wave power in a ►transmission line are eliminated by making the load impedance equal to the generator output impedance, and the line impedance of the transmission line equal to both the above impedances. Transmission lines of differing line impedances may be joined in a system using a ►quarter-wavelength line.

impedance scaling ➤denormalization.

impedance scaling factor ➤denormalization.

impedor *Syn.* impedance. Any circuit element, such as a resistor, capacitor, or inductance, that is used mainly for its ➤impedance.

imperfect dielectric ➤dielectric.

impulse *Short for* impulse voltage or current. ➤impulse voltage.

impulse current ➤impulse voltage.

impulse flashover voltage ➤impulse voltage.

impulse generator *Syn.* surge generator. An electronic device that produces a single pulse, typically one representative of the ➤surges generated by lightning in a transmission line. A typical impulse generator operates by charging and discharging one or more capacitors.

impulse noise ➤Noise of large amplitude and very short duration that results from a disturbance or series of disturbances.

impulse puncture voltage ➤impulse voltage.

impulse ratio for flashover (or **puncture**) ➤impulse voltage.

impulse response The response of a network, system, or circuit to a voltage spike or impulse (such as a ➤Dirac delta function).

impulse voltage A unidirectional voltage that rises rapidly to a maximum value and then falls to zero more or less rapidly without any appreciable superimposed oscillations. An *impulse current* is a unidirectional current with similar characteristics. Impulses are usually unwanted, being generated by fault conditions in electrical equipment or apparatus or by procedures such as switching on or off. A common source of large impulse voltages is a ➤lightning stroke.

A typical waveshape is shown in Fig. *a*. The maximum voltage, *V*, of the impulse is the peak value. The *wavefront* is the rising portion OA and the *wavetail* is the falling portion ABC. The duration of the wavefront is the time, T_1, required for the voltage to rise from zero to the peak value. The time to half value of the wavetail, T_2, is the time required for the impulse to rise from zero, reach the peak value, and then decay to half the peak value. The waveshape in Fig. *a* is described as a T_1/T_2 impulse. The average rate of increase of voltage with time is the *steepness* of the wavefront. A typical impulse voltage that results from lightning is a 1/50 wave, where T_1 and T_2 are measured in microseconds; standard testing equipment uses a wave with these proportions produced by a surge generator.

An impulse voltage or current frequently causes ➤flashover or puncture in electrical apparatus. If this occurs the impulse voltage collapses rapidly and is described as a *chopped impulse voltage* (Fig. *b*). Flashover or puncture can occur either during the wavefront or during the wavetail. The actual value of the impulse voltage at which

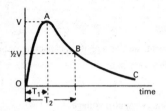

a Waveshape of **impulse voltage**

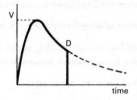

b Waveshape of chopped **impulse voltage**

flashover or puncture occurs during the wavefront is the *impulse flashover voltage* or the *impulse puncture voltage*. If flashover or puncture occurs on the wavetail the peak value of the impulse voltage is quoted. The *time to flashover* or the *time to puncture* is the time interval between the beginning of the impulse voltage or current and the instant at which the wave is chopped (point D in Fig. *b*).

In an insulator used in a.c. power transmission systems, the impulse voltage at which flashover or puncture occurs in the insulator is in general different to the voltage of the a.c. power at which failure occurs. The *impulse ratio for flashover* (or *puncture*) is the ratio of the impulse flashover voltage (or puncture voltage) to the peak value of the alternating voltage at which flashover or puncture occurs. Typical values for commonly used insulators are between 1.5 and 1.2.

impulsive noise *Syn. for* impulse noise.

impurities Foreign atoms in a semiconductor that are either naturally occurring or deliberately introduced. Impurities have a fundamental effect on the type and amount of conductivity of the semiconductor. *Impurity diffusion* is the deliberate diffusion of impurities into selected regions of a semiconductor in order to produce the desired characteristics. The energy levels due to the impurities are *impurity levels*. The presence of impurities affects the mobilities (both ►Hall and ►drift mobilities) of charge carriers due to *impurity scattering* between the carriers and the impurity atoms. ►diffusion; semiconductor.

IM rejection ►IP$_3$.

inactive interval ►sawtooth waveform.

incandescence The emission of visible light from a substance at a high temperature. In the incandescent lamp, for example, when an electric current is passed through a metal or carbon filament, the temperature of the filament is raised sufficiently for incandescence to occur (►heating effect of a current). The term incandescence is also used to describe the emitted radiation itself. ►luminescence.

incident current *Syn.* initial current. ►reflection coefficient.

increment The operation of adding one to the value of a variable, usually in a computer program. It is the opposite operation to ►decrement.

incremental permeability ➤permeability.

index error *Syn.* zero error. An error on the scale of a measuring instrument such that under zero input conditions a reading x is registered by the instrument. In the absence of other errors, a simple correction of $-x$ can be applied to all readings on the instrument.

indicating instrument An instrument that indicates the presence of a variable electrical quantity or measures its value. Examples include the ➤ammeter, ➤galvanometer, and ➤digital voltmeter.

indicator tube An extremely small ➤cathode-ray tube, often with a screen diameter measured in millimetres. It is used to indicate the value of a varying quantity by altering the size or shape of the image on the screen in accordance with the input voltage. A rectangular voltage pulse, termed an *indicator gate,* may be applied to the grid or cathode circuit of the tube in order to sensitize or desensitize it.

indirect-gap semiconductor A semiconductor, such as silicon, in which the maximum energy of the ➤valence band and the minimum energy of the ➤conduction band are not coincident in ➤momentum space. ➤➤direct-gap semiconductor; energy bands.

indirectly heated cathode ➤thermionic cathode.

indirect photoconductivity *Syn.* phonon-assisted photoconductivity. ➤photoconductivity.

indirect ray ➤indirect wave.

indirect wave *Syns.* indirect ray; reflected wave. The portion of a transmitted wave that does not travel directly from the transmitter to the receiver but is reflected by the ➤ionosphere. ➤➤ground wave.

indium phosphide ➤compound semiconductor.

induce To cause an electrical or magnetic effect in one circuit or device by altering the condition of a second circuit or body. ➤electromagnetic induction; electrostatic induction.

induced current A current that flows in a conductor as a result of a changing ➤magnetic flux density, in which the lines of magnetic flux intersect with the conductor. ➤➤electromagnetic induction.

induced dipole moment ➤dipole.

induced electromotive force ➤electromagnetic induction.

induced noise ➤Noise that appears in any circuit, device, or apparatus as a result of ➤electromagnetic induction from other nearby circuits.

inductance 1. Unit: henry. A constant that relates the magnetic flux, ϕ, linking a circuit to the current flowing in the circuit or in a nearby circuit. ➤electromagnetic induction.

The *self-inductance* (symbol: *L*) of a circuit is defined as one henry if the circuit is threaded by a total flux of one weber when a current of one ampere is flowing. The *mutual inductance* (symbol: *M* or L_{12}) between two circuits is defined as one henry when a flux of one weber is present in one circuit due to a current of one ampere flowing in the second circuit. **2.** ➤inductor.

induction *Short for* electromagnetic induction or electrostatic induction.

induction coil A device that uses ➤electromagnetic induction to produce a series of pulses of high potential and approximately unidirectional current. It consists of a primary coil of wire with only a few turns wound on an iron core, surrounded by a coaxial secondary coil of many turns and insulated from it (Fig. *a*). When the current in the primary coil is interrupted suddenly, a large e.m.f. is induced in the secondary. When the primary circuit is remade a much smaller e.m.f. is induced in the secondary in the opposite sense. The relatively high resistance introduced into the primary circuit at break, compared to remake, results in a much smaller ➤time constant in the primary and consequently in the higher e.m.f. The voltage output of the secondary depends on the sharpness of the break. The performance of the coil depends on the type of interrupter used in the circuit. The output from the secondary consists of a series of large pulses corresponding to the breaks in the primary circuit alternating with much smaller inverse pulses at the remake points (Fig. *b*).

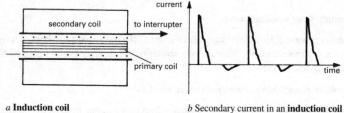

a **Induction coil**   *b* Secondary current in an **induction coil**

induction compass A device that indicates direction. It consists of a small coil that is made to rotate in the earth's magnetic field. The direction indicated by the compass depends on the magnitude of the induced current in the coil.

induction flowmeter A device that measures the rate of flow of a conducting liquid. The liquid is made to flow through a tube, length *L,* placed in a magnetic field of flux density *B* (see diagram). Electrodes are placed on a diameter of the tube and the induced e.m.f., *E,* across this diameter depends on the rate of flow, *v,* of the liquid:

$$E = v \times BL$$

induction heating *Syns.* eddy-current heating; RF heating. Heating caused by induced ➤eddy currents in a conducting material when subjected to a varying magnetic field. The varying magnetic field is commonly produced by an alternating current in a coil, known as the *load coil,* surrounding the load to be heated.

One type of furnace uses induction heating for melting metal: the advantages of this type of heating are that the heat is generated within the metal itself, and the eddy currents set up circulatory movements in the molten metal that have the effect of stirring the melt. ➤dielectric heating.

induction instrument An instrument that depends for its operation on the interaction between a varying magnetic field in a fixed winding and the ➤eddy currents induced by the field in a movable conductor (usually in the form of a disc or cylinder). The resulting torque produces a deflection of the conductor.

induction microphone A type of ➤microphone that has a straight-line conductor as the moving element. The conductor moves in a magnetic field, and the current induced in it by electromagnetic induction depends on the sound pressure causing it to move.

induction motor An alternating-current ➤motor in which the current in the secondary winding, usually the rotor, is produced by ➤electromagnetic induction when alternating current is supplied to the primary winding, usually the stator. Mechanical movement results from the torque produced by the interaction between the rotor current and the magnetic field due to the current in the stator.

inductive Denoting any electric circuit, device, or winding having an appreciable self-inductance (➤electromagnetic induction). In practice it is extremely difficult to produce a total lack of inductance and the term is usually applied to a circuit, etc., in which for a particular application the effect of inductance is not negligible.

inductive coupling ➤coupling.

inductive feedback ➤feedback.

inductive interference A type of interference in a communication system caused by electromagnetic induction from the electric supply system.

inductive load ➤lagging load.

inductive-output device An electronic device in which the output voltage is produced by electromagnetic induction between the current and the output electrode, without the current carriers being collected by the output electrode.

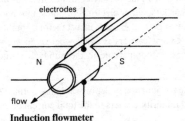

Induction flowmeter

inductive reactance ➤reactance.

inductive tuning ➤tuned circuit.

inductor *Syn.* inductance. A device or circuit element, usually in the form of a coil, that possesses inductance and is used primarily because of that property. ➤➤choke.

inert cell A primary ➤cell that contains the chemicals and other necessary ingredients in solid form, but will not function until water is added to form an electrolyte.

inertia switch A type of ➤switch that operates when an abrupt change in its velocity occurs.

infinite impulse response filter ➤digital filter.

information satellite ➤satellite.

information technology (IT) The technology of the production, storage, and communication of information, using computers and microelectronics.

information theory An analytical technique that determines the optimum amount of information required (i.e. the necessary and sufficient amount) to transmit a message or solve a specific problem in communication, control, or computer systems. The *information content* of a message is the minimum amount of information required to transmit the message with a desired accuracy, in the absence of ➤noise. *Information retrieval* is the means of extracting specific information from stored or transmitted data.

infrared image converter ➤image converter.

infrared radiation The portion of the electromagnetic spectrum of radiation extending from the limit of the red end of the visible spectrum to the microwave region. The infrared region extends from about 730 nanometres to about 300 micrometres in wavelength. Infrared radiation is emitted by hot bodies, and is detected by devices such as ➤bolometers, thermopiles (➤thermocouple), and ➤photocells.

inherent weakness failure ➤failure.

inhibiting input An input applied to a digital ➤gate that prevents any output that might otherwise occur.

initial current *Syn. for* incident current. ➤reflection coefficient.

injection 1. In general, the application of a signal to an electronic circuit or device. **2.** The introduction of excess charge carriers, either electrons or holes, into a ➤semiconductor material so that the total number present exceeds that at thermal equilibrium. In *low-level injection* the number of excess carriers is small whereas in *high-level injection* the number of excess carriers is comparable to the numbers at thermal equilibrium.

injection efficiency The efficiency of a p-n junction under forward bias, defined as the ratio of the current carried by injected minority carriers to the total current across the junction.

in parallel ►parallel.

in phase ►phase.

in-phase component 1. (of a current or voltage) ►active current; active voltage. **2.** (of the volt-amperes) ►active volt-amperes.

input 1. The signal or driving force applied to a circuit, device, computer, machine, or other plant. **2.** The terminals at which this signal is applied. **3.** To apply as an input signal or driving force.

input impedance The ►impedance of a circuit or device presented at its input.

input/output (I/O) The passing of information into or out of the processing unit of a ►computer. Input devices include keyboards, pointing devices such as the mouse, document scanners, magnetic-card readers, and speech-recognition units. The most common output devices are printers and visual displays. ►Magnetic disks and ►magnetic tape are examples of media used for the recording and reading of data. An important function of most I/O equipment is the translation between the processor's signals and the symbols, actions, or sounds understood or generated by people.

in series ►series.

insertion gain ►insertion loss.

insertion loss The loss of power in a load that occurs when a network is inserted between the load and the generator supplying the load. *Insertion gain* occurs when a gain rather than a loss of power results. The loss or gain is usually expressed as the ratio of the power delivered to the load after insertion of the network, to the power delivered to the load before insertion; it is measured in ►nepers or ►decibels. The value usually depends not only on the network parameters but also on the load and generator impedances.

instantaneous automatic gain control A fast-acting automatic gain control in ►radar systems that reduces the ►clutter by responding very rapidly to variations in the mean clutter level.

instantaneous carrying-current The peak value of current that may be carried instantaneously by a switch, circuit-breaker, or similar device at rated voltage under specific conditions.

instantaneous frequency The rate of change of phase of any oscillating electric variable, measured in radians per second divided by 2π. Particular applications are in ►frequency and ►phase modulation.

instantaneous power The value, at the output terminals of a circuit, of the rate at which power is transmitted from that circuit to the next portion of the system.

instantaneous sampling ►sampling.

instantaneous value The value at a particular moment of any quantity, such as current or potential difference, that varies with time.

Institute of Electrical and Electronics Engineers, Inc. (IEEE) ➤standardization.

Institution of Electrical Engineers (IEE) ➤standardization.

instruction fetch The part of a machine cycle during which the ➤central processing unit fetches an instruction from memory and decodes it prior to execution.

instruction register A ➤register in the control unit of a CPU that holds an instruction while the circuitry decodes and executes it. It is not generally part of the computer's instruction-set architecture.

instruction set The repertoire of arithmetic and logical operations of a computer. All programs exist in a computer as sequences of instructions drawn from the instruction set.

instrumentation amplifier A type of amplifier used in measurement and control to amplify the voltage difference between two ➤nodes of a circuit, but without disturbing their individual voltages. The amplifier consists of a number of ➤differential amplifiers connected as shown in the diagram and has an extremely high differential input impedance.

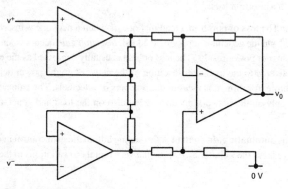

Instrumentation amplifier

instrument damping A process widely used in indicating instruments to absorb energy from the indicator and bring it to rest quickly. A device incorporated in an indicating instrument to provide damping is a *damper*; one that is overdamped so that the needle goes to its true position without wavering gives a *dead-beat instrument*. ➤➤damped.

instrument rating The limits, designated by the manufacturer, between which an instrument can operate without damage. This is not necessarily the same as the ➤full-scale deflection.

instrument sensitivity The response of a measuring instrument expressed as the ratio of the magnitude of the physical response to the magnitude of the quantity measured. It is often stated directly as, for example, divisions per volt.

instrument shunt ➤shunt.

instrument transformer A ➤transformer that is used in conjunction with a measuring instrument. It utilizes the current-transformation (*current transformer*) or voltage-transformation (*voltage transformer*) properties of a transformer. The primary winding forms part of the main circuit and carries the current or voltage to be measured. The secondary winding is connected to the measuring instrument, such as an ammeter or voltmeter. Instrument transformers are used to extend the dynamic range of a.c. instruments and to isolate the instruments from circuits operating at high voltages.

insulate To support or surround a conductor with insulating material so that the current through it is confined to a desired path.

insulated-gate bipolar transistor (IGBT) A power electronic device that has a high-impedance input, like a ➤MOSFET, requiring only a small energy to switch the device. The output resembles that of a ➤bipolar junction transistor (BJT) in the on-state: a high current capability with a low voltage drop resulting in low conduction losses. The structure of the IGBT is a variant of a vertical MOSFET, with an additional ➤semiconductor layer to produce the output bipolar transistor (see diagram).

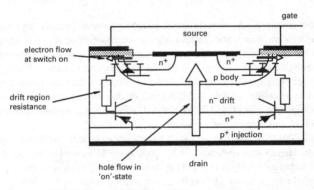

Structure of IGBT showing MOSFET and BJT components

In the 'off-state', the gate-source voltage is below threshold for the input MOSFET, so this device is switched off, and the drain-source voltage is dropped across the reverse-biased ➤p-n junction formed by the body and drift regions. Only a small leakage current flows. In the 'on-state', the gate-source MOSFET is switched on, providing a current path between the n^+ source regions and the drift region. The electron current flowing in this path causes substantial hole injection from the p^+ drain, creating a p-n-p bipolar junction transistor between drain and source. The reverse-biased body-drift region p-n junction is the collector of this transistor. A substantial output current can now flow. Switching times are comparable with power ►BJTs, approximately one microsecond.

insulated-gate field-effect transistor (IGFET) ►field-effect transistor.

insulating barrier A screen of insulating material that is fitted to certain electrical apparatus, such as switch gear or fuses. It prevents the formation of an arc between the apparatus and some other point, which may be an operator or another part of the apparatus, and thus prevents damage.

insulating resistance The resistance between two electrical conductors or systems of conductors that are normally insulated from each other. It is usually of the order of megohms but in the case of cables it is expressed as megohms per kilometre.

insulation Material that is used to insulate an electrical conductor. It is also the process of insulating a conductor.

insulator A material, such as glass or ceramic, that has a very high resistance to electric current so that the current flow through it is usually negligible. An insulator is used to prevent the loss of electric charge or current from a conductor. ►►energy bands.

integer A whole number or, in computing, a binary datum representing a whole number. It is any value from the infinite set

$$\{... - 3, -2, -1, 0, 1, 2, 3, ...\}$$

integrated circuit (IC) A complete circuit, including active and passive electronic devices and their interconnections, that is made on a single substrate. A *hybrid integrated circuit* consists of several separate component parts attached to a ceramic substrate and interconnected either by wire bonds or a suitable ►metallization pattern. The individual parts are unencapsulated and may consist of diffused or thin-film components or one or more monolithic circuits. An IC is much smaller than a circuit made from discrete packaged components and once fabricated an individual component cannot be altered without destroying the entire circuit.

A *monolithic integrated circuit* has all the circuit components made into or on top of a single chip of ►semiconductor. The components are then connected by one or more levels of metallization deposited onto the surface of the IC in an appropriate pattern (►multilevel metallization). Making the circuit directly in the semiconductor enables circuits and structures to be made, and electronic functions to be generated, that would be impossible to realize in discrete form. This is because the reliability of the many individual components required for the circuit is not sufficient to allow the complex circuit formed therefrom to be guaranteed to operate, whereas using the same technology the reliability of the complete IC would be similar to that of each component.

►Silicon is used in the majority of commercial integrated circuits. ►Bipolar integrated circuits are based on ►bipolar junction transistors, and are used for high-speed analogue and digital circuits and for lowest-noise ICs. ►MOSFETs are used in the highest-density ICs, such as ►microprocessors and ►memory ICs, as the small size and low power consumption of the individual transistors permit very complex circuits to be made. Combinations of technologies are used in particular applications; for example, *BiCMOS* technology combines bipolar transistors for output power capability and CMOS transistors for logic operation, while *BiFET* technology combines bipolar circuits with JFETs for low-noise high-speed analogue ICs. ►Gallium arsenide is used for ►monolithic microwave integrated circuits (MMICs) for specialist applications at microwave frequencies.

The complexity of a (digital) monolithic IC is described by the number of components that form the circuit; often this is counted in terms of the number of ►logic gates per IC. *SSI (small-scale integration)* refers to simple circuits of up to about ten components; many GaAs MMICs fall into this category. *MSI (medium-scale integration)* describes circuits of 10–100 components. *LSI (large-scale integration)* refers to circuits of 100 to a few thousand gates; many high-speed silicon and GaAs digital circuits are LSI. ►VLSI (very large scale integration) describes all ICs that are larger than a few thousand gates; this includes most ►microprocessors, ►memory, and ►digital signal processing ICs currently available.

integrated injection logic ►I²L.

integrated Schottky logic (ISL) *Syn. for* Schottky I²L. ►I²L.

integrating array ►solid-state camera.

integrating frequency meter *Syn.* master frequency meter. An instrument that allows a check to be made on a source of alternating voltage by integrating the number of cycles that occur in a specified time interval. Comparison with the number of cycles that should occur in the same time interval indicates whether the prescribed frequency has been maintained.

integrating wattmeter ►watt-hour meter.

integrator A device that performs the mathematical operation of integration so that the output of the device is substantially the integral with respect to time of the input to it.

A *capacitance integrator* utilizes a capacitor, usually in series with a resistor, to perform integration. The voltage across the capacitor C when a direct current I flows into it is given by

$$V = (1/C)\int I dt$$

The capacitor thus integrates the current with respect to time.

intelligent terminal ►terminal.

intelligibility A function of a speech communication system, such as a telephone system, in which it is more important for the received information to be intelligible than for the input to be flawlessly reproduced. The intelligibility is measured by the *syllable articulation score,* which is the percentage of correctly perceived monosyllabic nonsense words uttered in an uncorrelated sequence.

intelligible crosstalk ►crosstalk.

intensity 1. *Short for* magnetic intensity or electric intensity (both obsolete terms). ➤magnetic field strength; electric field strength. **2.** The rate of flow of sound energy through unit area perpendicular to the direction of flow. **3.** The rate of flow of light or other radiant energy emitted or reflected by a surface in a given direction per unit area.

intensity modulation *Syn.* z-modulation. The variation in brilliance of the spot on the screen of a cathode-ray tube in accordance with the magnitude of an input signal.

interaction space A region in an electron tube that roughly corresponds to the inter-electrode space and in which the electrons interact with an alternating magnetic field.

interactive Allowing continuous two-way communication between the user of an ➤online peripheral device, such as a ➤terminal, and a ➤computer. Interactive operation enables a user at a remote location to send and receive information to and from a computer quickly, and to modify the operation of a ➤program during its execution following the production of intermediate results or interrogation. ➤time sharing; real-time operation.

intercarrier system *Syn.* video IF system. A system in a television receiver in which the same intermediate-frequency (IF) stages amplify the sound and video signals (➤mixer).

interconnecting feeder ➤trunk feeder.

interconnection 1. Any method of providing an electrical path between any of the materials (metals, semiconductors, etc.) that combine to form a circuit. **2.** Connections between and external to any functional item that forms a circuit or system of circuits. Functional items include component parts, devices, subassemblies, and assemblies. ➤intraconnection.

interconnector ➤trunk feeder.

interdigitated capacitor ➤monolithic capacitor.

interelectrode capacitance The capacitance between specified electrodes of an electronic device (such as the base and emitter of a bipolar transistor) that may form a small capacitor within the device. The operation of such devices can be significantly affected by the existence of interelectrode capacitances.

interface The electronic circuitry used to connect two or more devices, usually required to compensate for differences in speed, signal levels, and/or codes between the connecting devices. The devices are generally computer components or systems.

interference A disturbance to the signal in any communication system caused by unwanted signals. A common cause of interference in radio reception is the operation of electrical machinery and apparatus, particularly commutating machines and apparatus containing gas-discharge tubes. Television signals frequently suffer serious interference from motor-vehicle ignition systems.

Man-made interference such as that described above can usually be eliminated by fitting special devices (*suppressors*) to the offending apparatus, but interference arising from natural causes, such as changes in the atmosphere, is not easily prevented.

interference fading ➤fading.

interlaced scanning ➤television.

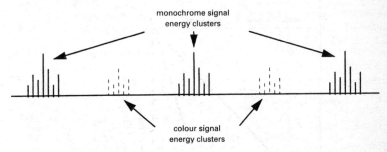

Interleaving of colour television signals

interleaving 1. In ➤colour television signal transmission, the slotting in of colour information into the gaps between the monochrome modulation sidebands of the frequency spectrum (see diagram). **2.** ➤digital codes.

interline transfer device ➤solid-state camera.

interlock A safety device that allows a piece of apparatus to function only when predetermined conditions are fulfilled.

intermediate frequency (IF) ➤heterodyne reception; mixer.

intermittent duty ➤duty.

intermodulation (IM) The mixing of different frequency components of a signal in a nonlinear component or ➤active device in a circuit, producing unwanted frequency components. If two frequencies f_1 and f_2 are applied to a nonlinear element, then a num-

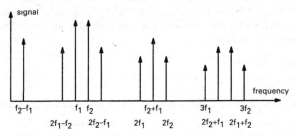

a Spectrum of **intermodulation** products up to 3rd order

ber of new frequency tones will be generated, as shown in Fig. *a*. These new frequency components are called *intermodulation products*, and are ➤distortions of the original signals. The IM products are described in terms of their *order*: for example, 3rd order means all 3-fold combinations of the two mixing frequencies, i.e. $3f_x$ and $2f_x \pm f_y$. An nth-order nonlinearity will produce nth and lower-order products.

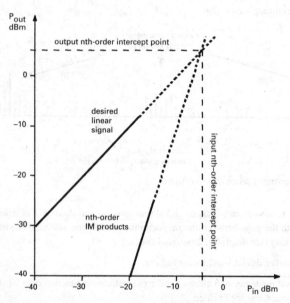

b Determination of the nth-order intercept point

The nth-order *intercept point* represents a fictitious signal amplitude where the extrapolated values of the magnitudes of the desired input signal and the unwanted nth-order distortion components are equal. The nth-order intercept point is written IP_n and measured in terms of power. The definition of intercept point is illustrated in Fig. *b*. The distortion products increase by n dB in amplitude for every 1 dB increase in the linear signal. Both input and output intercept points are often quoted: the output intercept point equals the input intercept point plus the stage gain. Knowing the value of the nth-order intercept point and the input power level P_{in}, the (nth-order) *intermodulation level* can be found:

$$P_{IMn} = n.P_{in} - (n - 1).IP_n$$

➤IP_2; IP_3.

intermodulation distortion *Syn.* combination-tone distortion. ➤intermodulation; distortion;.

intermodulation products ➤intermodulation.

internal photoelectric effect ➤photoconductivity.

internal resistance Symbol: r; unit: ohm. A small resistance possessed by a cell, accumulator, or dynamo. It is given by

$$r = (E - V)/I$$

where E is the e.m.f. generated by the device, V is the potential difference across the terminals, and I is the current. ➤cell.

international ampere Symbol: A_{int}. The former standard of electric current defined as the constant current that, when passed through an aqueous solution of silver nitrate, will deposit silver at the rate of 0.001 118 00 grams per second. It was replaced in 1948 as the standard unit by the ➤ampere:

$$1\ A_{int} = 9.9985\ \text{amperes}$$

International Electrotechnical Commission (IEC) ➤standardization.

international ohm ➤ohm.

International Radio Consultative Committee (CCIR) ➤standardization.

International Standards Organization (ISO) ➤standardization.

international system A former system of units that expressed the values of electrical quantities in terms of the ➤international ampere, international ➤ohm, centimetre, and second.

International Telecommunication Union (ITU) ➤standardization.

International Telegraph and Telephone Consultative Committee (CCITT) ➤standardization.

international units ➤international system.

international volt ➤volt.

Internet The worldwide matrix of connecting computers.

interpolation A method that enables additional values to be obtained between sampled values of a signal.

interrogating signal ➤transponder.

interrupt In computing, an ➤asynchronous execution initiated by an external device, such as a timer or an I/O controller.

interrupter A device, such as an ➤induction coil, that periodically interrupts a continuous current. ➤make-and-break.

interstage coupling ➤Coupling between successive stages of a multistage amplifier employing several amplifying stages in cascade. The type of coupling chosen (direct, resistive, etc.) depends on the particular design of amplifier stage used.

intersymbol interference A type of ►interference occurring in digital communication systems as an effect of ►bandwidth. In such systems the communication channel always has a limited bandwidth. Digital signals transmitted down the channel are therefore distorted by this band-limiting effect. If the channel bandwidth is of a similar magnitude to the signal bandwidth, the distortion will be such that one digital symbol will overlap the next, producing intersymbol interference. The result of intersymbol interference on a digital system will be an increased error rate. As the interference is a bandwidth effect, increasing the signal power will not reduce the level of interference.

intraconnection An electrical connection inseparably associated with circuit elements within a component part. ►interconnection.

intrinsic conductivity The conductivity of a ►semiconductor associated with the semiconductor material itself and not contributed by impurities. At any given temperature, equal numbers of electrons and holes are thermally generated and these give rise to the intrinsic conductivity. The numbers of charge carriers are dependent on temperature and the conductivity, σ, is a function of temperature:

$$\sigma = A \exp(E_g/2kT),$$

where A is a constant that depends on the particular material, E_g is the band-gap energy, k is the Boltzmann constant, and T the thermodynamic temperature.

In an extrinsic semiconductor, the intrinsic conductivity is usually negligible compared to the extrinsic conductivity at normal working temperatures. At sufficiently high temperatures however the numbers of thermally generated carriers become much larger than those contributed by impurities and the semiconductor becomes intrinsic (►intrinsic temperature range). In practice, at room temperature, the intrinsic carrier concentration in silicon is about 10^{10} per cm^3. The minimum impurity carrier concentration obtainable in silicon is about 10^{14} per cm^3, hence the extrinsic carrier concentration is about 10^4 times greater than the intrinsic value.

intrinsic density ►carrier concentration.

intrinsic mobility The mobility of charge carriers in an intrinsic semiconductor.

intrinsic photoconductivity ►photoconductivity.

intrinsic semiconductor *Syn.* i-type semiconductor. A pure semiconductor that has equal concentrations of ►electrons and ►holes under conditions of thermal equilibrium. Absolute purity is unobtainable in practice and nearly pure materials are termed intrinsic. ►semiconductor. ►intrinsic conductivity; intrinsic temperature range; extrinsic semiconductor.

intrinsic temperature range The temperature range in which the electrical properties of a semiconductor are not essentially modified by impurities in the crystal. For a pure sample of ►intrinsic semiconductor, the intrinsic temperature range is the whole range of working temperatures. For an extrinsic semiconductor at most temperatures, the impurities contribute most of the charge carriers; at sufficiently high temperatures however the intrinsic carrier concentration rises to such a level that the extrinsic conductivity becomes negligible compared to the ➤intrinsic conductivity and at that temperature the semiconductor becomes intrinsic.

inverse Chebyshev filter A filter that provides a ➤maximally flat passband and an ➤equiripple stopband. It has similar roll-off or transition band characteristics to a Chebyshev filter (which has an equiripple passband response) but has a better ➤group delay characteristic. ➤➤filter.

inverse feedback *Syn. for* negative feedback. ➤feedback.

inverse filter A filter with a ➤transfer function that is the reciprocal of a given filter.

inverse gain The ➤gain of a ➤bipolar junction transistor when it is connected in reverse, that is, so that the emitter acts as the collector and the collector as the emitter. The inverse gain is usually less than the gain normally obtained, due to the higher doping level of the emitter compared with the collector. This causes the emitter to have a higher ➤injection efficiency into the base than has the collector.

inverse limiter *Syn. for* base limiter. ➤limiter.

inversion The production of a layer of opposite polarity at the surface of a semiconductor under the influence of an electric field, usually an applied one. Spontaneous inversion can occur in the surface of p-type material when it is in contact with an insulating layer due to the presence of positive ions in the insulator. Inversion can only occur when sufficient mobile minority carriers are present in the semiconductor material, otherwise a depletion layer forms. The phenomenon is utilized for formation of the channel of an insulated-gate ➤field-effect transistor. ➤➤MOS capacitor.

inverter 1. A device or circuit that converts direct current into alternating current. **2.** *Syn.* linear inverter. An amplifier that reverses the polarity of a signal, i.e. that introduces a 180° phase shift. **3.** *Syns.* digital inverter; NOT circuit. A ➤logic circuit that inverts the level of the input signal, i.e. that produces a low output for a high input and vice versa.

inverting amplifier A configuration of an ➤operational amplifier (see diagram) where the gain G of the amplifier is given by $-(R_f/R_{in})$. ➤➤noninverting amplifier.

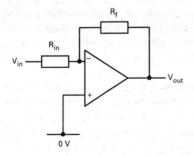

Inverting amplifier

inverting transistor A bipolar or MOS transistor used as an analogue or digital ►inverter. Operation as an analogue inverter results from the 180° phase shift introduced by both types of transistor between the input and output. A digital inverter may be formed with a bipolar transistor by applying the input to the base. A high input causes ►saturation with a resulting drop in the collector (output) voltage. With a low input level the saturation ceases and the output therefore rises. A MOSFET is used as a digital inverter essentially by operating it as a switch that connects the output to a point of low potential difference (usually earth potential). A high voltage level input to the gate causes the conducting channel to form across the transistor and therefore the drain (output) potential falls to the low value of the source. A low input level, which must be less than the threshold voltage of the transistor, causes the drain to be isolated from the source. Since the drain is connected to the power supply by means of a dropping resistor the drain potential then rises to the high level.

I/O *Abbrev. for* input/output.

ion An atom, molecule, or group of atoms or molecules that has an electric charge. Negative ions (►anions) contain more electrons than are necessary for electrical neutrality of the atom or group; positive ions (►cations) contain fewer.

ion-beam analysis A method of analysing the surface of a material, such as a thin film or a semiconductor, using a beam of ions. Various techniques have been developed including elastic recoil detection analysis and Rutherford back scattering.

ion-beam lithography A method of ►lithography similar to ►electron-beam lithography except that the electron beam is replaced by a beam of ions. The ions are heavier than the electrons used in e-beam lithography and therefore suffer far less scattering within the resist and produce very few low-energy secondary electrons. They can also supply more energy to the resist, which results in greater resist sensitivity and reduced writing time. There are difficulties in producing collimated ion beams and with the effect of the ions on the semiconductor substrate below the resist layer. ►Multilevel resists are needed to eliminate the latter effect.

ionic atmosphere The atmosphere, consisting of ions, that surrounds an individual ion in an electrolyte. In the absence of an electric field each anion is surrounded by a symmetrical accumulation of cations and vice versa. When an electric field is applied the anions migrate to the anode and the cations to the cathode. The symmetry of the ionic atmosphere is therefore disturbed with respect to the surrounded ion, and the ions experience a retarding force due to the migration of the opposite-polarity ionic atmosphere in the opposite direction.

If high-frequency alternating current or a fast-pulse direct current is applied to the electrolyte, the symmetry is little disturbed and a higher conductivity of electrolyte results.

ionic bond A crystalline bond in which one or more electrons are transferred between the component atoms, allowing them to become ionized. The crystal retains overall charge neutrality, and the ions are arranged by electrostatic attraction in a regular array to minimize the energy of the system of ions and form the highest packing density. An example is sodium chloride, NaCl, in which the sodium atoms lose a ►valence electron to achieve a stable electronic configuration, and become positive ions, Na^+, and the chlorine atoms gain an extra valence electron to achieve a stable electronic configuration, and become negative ions, Cl^-; the crystal contains equal parts of sodium and chlorine to retain charge neutrality, and the ions are arranged in a regular array.

ionic conduction The movement of charges through a ►semiconductor due to the displacement of ions within the crystal lattice, such movement being maintained by a continuous supply of external energy.

ionic crystal A crystal that is in the form of an array of positively and negatively charged ions, the interatomic forces being of the Coulomb type. This electrostatic attraction between the ions is balanced by the repulsion experienced when the outer electronic shells of each ion approach too closely.

ionic semiconductor A type of semiconductor in which the movement of ►ions through the material contributes more to the conductivity than the movement of mobile charge carriers (►electrons and ►holes).

ion implantation A technique of introducing donor or acceptor impurities into a ►semiconductor to create n-type or p-type regions. The impurity atoms are ionized in a gaseous form, and then accelerated in a vacuum using a high potential and fired at the target semiconductor layer. The high-energy ions penetrate the surface of the semiconductor and rapidly come to rest in the solid. The average depth of the impurities in the semiconductor is a function of the accelerating potential, and so the depth of the doping can be carefully controlled. The number of ions implanted can be calculated from the current flow through the target, as the ions are neutralized, and the duration of the implantation process. Very carefully designed impurity profiles can be created using this process, which is used to make the very small transistors for high-speed, high-frequency, and ►VLSI integrated circuits.

ionization Any process in which ►ions are formed from neutral atoms. Ions are formed spontaneously when an electrolyte dissolves in a suitable solvent. The action of ionizing radiation (X-rays, alpha, beta, or gamma rays, fast electrons, etc.) is required for ions to be formed in a gas.

ionization chamber A ►gas-filled tube that is used to detect ionizing radiation. It is a very versatile radiation detector as it can be used to detect and measure a wide range of energies and intensities of radiation. A typical chamber has a sensitive volume defined using a ►guard ring, contained within it, and two electrodes with a potential difference maintained across them. When the gas is ionized by a beam of radiation the ions migrate to the electrodes under the influence of the applied voltage.

In conjunction with a suitable external circuit the ionization chamber can be used as a counter – an *ionization counter* – to count particles, such as alpha or beta particles, a pulse of current being produced by each particle of radiation. The most common application however is as a continuous measuring instrument in which an ionizing current is produced and the size of the current is proportional to the intensity of the ionizing radiation.

The sensitivity of the instrument is proportional to the mass of gas enclosed within the sensitive volume, and the size of instrument depends on the intensity of radiation to be measured. Extremely large chambers have been produced for measuring background radiation levels and very small chambers are used to calibrate high-output beams of X-rays or electrons.

ionization counter ➤ionization chamber.

ionization current Current that results from the movement of ➤ions through a conducting medium under the influence of an electric field.

ionization gauge A pressure gauge that is used to measure extremely low gas pressures and consists of a high-vacuum three-electrode thermionic valve fused to the gas system (see diagram). Electrons leaving the cathode are accelerated to the grid but fail to reach the anode because of its negative potential. Any gas molecules in the system are ionized by collision with some of the electrons that pass through the grid; the positive ions thus formed migrate to the anode producing an output current that is a function of the number of gas molecules present and hence of the gas pressure.

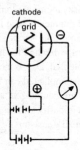

Ionization gauge

ionization potential *Syns.* electron binding energy; radiation potential. Symbol: I; unit: electronvolt. The minimum energy required to remove an electron from a given atom or molecule to infinity. Originally the ionization potential was defined as the minimum potential through which an electron would fall in order to ionize the atom, and was measured in volts.

The electron usually removed from the atom or molecule is in the outermost orbit, i.e. the least strongly bound electron. Some atoms and molecules may be ionized by the removal of an electron that is not in the outermost orbit, i.e. one that is more strongly bound. The resulting ion will be in an excited state. The ionization potential for removal of the least strongly bound electron is designated the *first ionization potential* (I_1) and results in an ion in the ground state. Ionization by the removal of the second (or subsequent) least strongly bound electron requires a greater ionization potential, termed the second (or third, etc.) ionization potential (I_2, I_3, etc.).

ionizing event Any physical process that produces an ion or group of ions. The physical agent involved in the process, such as a charged particle passing through a gas, is termed an *ionizing agent.*

ionizing radiation Any radiation, such as streams of energetic charged particles (electrons, protons, alpha particles, etc.) or energetic ultraviolet radiation, X-rays, or gamma rays, that produces ionization or excitation of the medium through which it passes.

ion microprobe ➤secondary-ion mass spectroscopy.

ion milling ➤etching.

ionosphere A region in the earth's atmosphere that has a high concentration of ions and free electrons and extends from about 50 km to over 1000 km in altitude. The ionosphere disturbs the propagation of radiowaves through it by reflecting and attenuating them. It does however allow long-distance radio transmission at frequencies up to about 30 megahertz by successive reflections between it and points on the earth's surface.

The ionosphere consists of several distinct layers or regions that can change in thickness between day and night and also show seasonal and latitude variations. The *D-layer,* extending from about 60 to 90 km, has a relatively low concentration of electrons and reflects low-frequency radiowaves. The *E-layer* extends from about 90 to 120 km, has a higher concentration of electrons than the D-layer, and reflects medium-frequency radiowaves.

The highest layers are the F_1-*layer*, which is a daytime feature centred on about 150 km, and the F_2-*layer*, which is centred on about 300 km. These layers contain the highest concentration of free electrons and reflect high-frequency radiowaves. At night the D- and E-layers become relatively inactive since there is no solar radiation to regenerate ion pairs lost by recombination. The F-layers have a lower density and hence a lower recombination rate of ions. The F_2-layer can be used for radio transmission at all times. It is thus the most useful region for long-range radio communication. There is a well-marked ionization maximum for the F-layers, the region above the maximum being the *topside ionosphere.*

Waves with wavelengths between about 6 millimetres and 20 metres lie within the ➤radio window and are not reflected by the ionosphere but pass straight through it. High-frequency ➤television transmissions fall within this band and require ➤communications satellites, usually in geostationary earth orbit, in order to achieve long-distance television links. Radioastronomy is restricted to wavelengths within the radio window.

ionospheric defocusing ➤ionospheric focusing.

ionospheric focusing A process that results in an enhancement of the field strength at a given receiver due to the focusing that arises from either small-scale or large-scale curvature of the ionospheric layers. *Ionospheric defocusing* results from a reduction of the field strength at a receiver due to the defocusing that arises from the curvature of the layers.

ionospheric wave *Syn.* sky wave. A radiowave deflected by the ionosphere.

ion source A device that provides ions, particularly for use in particle ➤accelerators. A common type consists of a minute jet of a suitable gas that is subjected to electron bombardment. The resulting ions (protons, alpha particles, etc.) are injected into the accelerator.

ion spot 1. A darker area appearing on the screen of a ➤cathode-ray tube and resulting from decreased luminescence of the screen (not necessarily of the phosphor) after bombardment by heavy negative ions in the beam. Use of an ➤ion trap can minimize this effect. **2.** A spurious signal appearing in the output of ➤camera tubes and ➤image converters due to alteration of the charge pattern on the target or cathode by ion bombardment.

ion trap A device used in ➤cathode-ray tubes that attracts heavy ions present in the electron beam and thus prevents them impinging on the phosphor coating of the screen and causing blemishes.

IP$_2$ (or **IP2**) *Abbrev. for* second-order intercept point (➤intermodulation). The IP$_2$ is used to predict ➤mixer performance in terms of the *half-IF response*, which is generated by mixing of internally generated ➤local oscillator and RF second harmonics:

$$\text{half-IF rejection} = \tfrac{1}{2}(IP_2 - S - CR) \quad \text{(dB)}$$

where S is the ➤sensitivity of the ➤radio receiver, and CR is the ➤cochannel rejection or capture ratio. Only the mixer IP_2 needs to be considered, as it is this stage that generates the half-IF response.

IP$_3$ (or **IP3**) *Abbrev. for* third-order intercept point (➤intermodulation). The 3rd-order nonlinearity will accommodate all the distortion products of interest in a ➤radio receiver or amplifier circuit, and so the IP$_3$ is an important measure of the circuit or system linearity. It is used to determine the amount of intermodulation (IM) distortion in the circuit at high signal levels. In a radio receiver, for example, the *IM rejection*, which is the difference between the receiver ➤sensitivity S and an input signal level sufficient to produce interference, is calculated from the system input intercept point:

$$\text{IM rejection} = \tfrac{1}{3}(2IP_3 - 2S - CR) \quad \text{(dB)}$$

where CR is the ➤cochannel rejection or capture ratio.

iris ➤waveguide.

I^{2}R loss *Syn.* copper loss. ➤dissipation; heating effect of a current.

iron Symbol: Fe. A metal, atomic number 26, that exhibits ➤ferromagnetism and has a high tensile strength. It is widely used in electronics as a magnetic material and for screening purposes. It is relatively abundant and therefore has a low cost.

iron loss ➤core loss.

irradiation The exposure of a body or substance to electromagnetic or corpuscular ➤ionizing radiation.

ISDN *Abbrev. for* integrated services digital network. A definition for global digital data communications. Its purpose is to ensure that people, computers, and other devices can communicate over standardized connection facilities. The criteria include the setting of standards in such a way that users have access through a limited set of multipurpose interfaces. ISDN has definitions for data-transmission speeds, or capacities, of channels and the number of channels in each service. ➤digital communications.

ISO *Abbrev. for* International Standards Organization. ➤standardization.

isoelectronic Denoting groups of atoms that have identical distributions of electrons in the outer orbits, and exhibit similar electronic properties.

isolating Disconnecting a circuit or device from an electric supply system, usually by making a circuit open at a time when it carries no current.

isolating transformer A transformer that is used to isolate any circuit or device from its power supply; thus the circuit derives power from the source without a continuous wire connection between them.

isolation diode 1. A ➤diode that is used in a circuit to allow signals to pass in one direction but to block those in the other direction and therefore prevent damage from surges in the reverse direction. **2.** The diodes formed by the collector-substrate junctions in ➤bipolar integrated circuits. In order to maintain isolation between parts of the integrated circuits these junctions must always be reverse biased. This is achieved by maintaining the potential of the substrate so that none of the diodes becomes forward biased. Similar diodes are formed in ➤MOS integrated circuits by the source-substrate junctions, drain-substrate junctions, and, once formed, the channel-substrate junctions. The substrate material must be held at a suitable potential to maintain the isolation between parts of the circuit, as with bipolar circuits.

isolator A device, usually made of a ferrite, that allows microwave energy to pass in one direction with little loss but absorbs power in the reverse direction.

isotropic Denoting a substance or medium in which physical properties, such as magnetic susceptibility, are the same in all directions.

isotropic radiator An ➤antenna that radiates equally in all directions. It is usually taken to be a point source, but is not in fact physically realizable. However it is useful as a reference for describing the directional properties of real antennas.

IT *Abbrev. for* information technology.

iterative impedance The impedance that when connected to one pair of terminals of a ➤two-port network produces a like impedance at the other pair of terminals. In general a two-port network has two iterative impedances, one for each pair of terminals. If the two iterative impedances are equal, their common value is termed the *characteristic impedance* of the network. ➤image impedance.

ITU *Abbrev. for* International Telecommunication Union. ➤standardization.

i-type semiconductor ➤intrinsic semiconductor.

J

j (or **i**) **1.** A mathematical operator equivalent to a rotation through 90°. **2.** Symbol for the square root of −1, √−1, which defines imaginary numbers.

If a current flows through a complex ➤impedance, comprising a real resistive component and a reactive component, then the voltage across the resistive component will be in phase with the current, whereas the voltage across the reactance will be shifted in phase from the current by 90°. The resistance can therefore be denoted by a real value R, and the reactance by jX, the j signifying the 90° phase shift.

jack plug and socket A type of plug-and-socket connector used when rapid and easy connections between circuits or devices are required. Insertion or removal of the plug can cause one or more switching functions to occur, such as the breaking of a short circuit. The basic construction is shown in the diagram. The plug slides into the sprung socket and is correctly located by means of a groove running round the plug. The contacts are arranged linearly along the length of the plug and socket and are insulated from each other. Two or more contacts may be used. The end of the plug forms one contact and the others are rings lying along the length of the metal casing. Contact is made by means of a central wire or conducting rod. More than one rod can be used if more than two contacts are required, the rods being insulated from the casing and from each other except at the connection point.

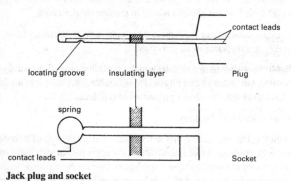

Jack plug and socket

jammer *Syn.* jamming transmitter. ➤jamming.

jamming Deliberate interference in communications and radar by means of unwanted signals that are intended to render unintelligible or to falsify the whole or part of the desired signal. A *jammer* is used to produce the jamming signal. Reducing the effects of jamming is termed *antijamming*.

Jansky noise *Syn. for* galactic noise. ➤radio noise.

JEDEC *Abbrev. for* Joint Electron Device Engineering Council. ➤standardization.

JFET *Abbrev. for* junction field-effect transistor. ➤field-effect transistor.

jitter Short-term instabilities in either the amplitude or phase of a signal, particularly the signal on a ➤cathode-ray tube. It has the effect of causing momentary displacements of the image on the screen, giving it a shaky or 'jittery' appearance. An oscilloscope that is used to measure the amount of jitter is termed a *jitter scope*. Momentary errors of synchronization between the scanner and receiver in ➤television or ➤facsimile transmission can cause jitter of the received images. This type of jitter is known as *jitters*.

In hi-fi sound reproduction systems that use a digital recording method, jitter can cause unwanted audible variations in the pitch of the sound output. These can be quite unpleasant. In ➤compact disc systems, for example, a noise similar to ignition interference is heard. Jitter in digital recording is the equivalent of ➤wow and ➤flutter in a system using analogue recording.

J-K (or **JK**) **flip-flop** ➤flip-flop.

Johnson–Lark–Harowitz effect The change in resistivity of a metal or a ➤degenerate semiconductor due to scattering of the charge carriers by impurity atoms.

Johnson noise *Syn. for* thermal noise. ➤noise.

Josephson effect An effect that occurs when a sufficiently thin layer of insulating material is introduced into a superconducting material (➤superconductivity). A superconducting current can flow across the junction, known as a *Josephson junction,* in the absence of an applied voltage. This is the *direct-current Josephson effect.* If the value of the current exceeds a critical value, I_c, determined by the properties of the insulating barrier, current can only flow when a finite voltage is applied. The current-voltage characteristic is shown in the diagram, in which the dashed curve is the current-voltage characteristic in the nonsuperconducting state.

The *alternating-current Josephson effect* occurs when a small direct voltage, V, is applied across a Josephson junction. The superconducting current across the junction becomes an alternating current given by

$$I_s = I_c \sin\omega t$$

where $\omega = 2\pi f = 2e/hV$: h is the Planck constant, f the frequency, and e the electron charge.

The direct-current Josephson effect is utilized in several devices, particularly the ➤Josephson memory. The alternating-current Josephson effect is utilized for radiofrequency detection, the determination of h/e, for accurate measurement of frequency, and as a monitor of voltage changes in standard cells or for the comparison of cells at different Standards laboratories.

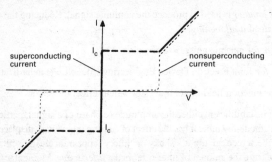

I-V characteristic of Josephson junction

Josephson junction ►Josephson effect.

Josephson memory A cryogenic ►memory that consists of an array of Josephson cells, i.e. memory cells containing ►Josephson junctions, held at a temperature very close to critical temperature. In the absence of an external magnetic field the cell is superconducting but the presence of a magnetic field destroys the ►superconductivity and hence the voltage across the device changes. Information is stored in the form of local variations of a magnetic field; the data is sensed by the voltage across the appropriate cell.

Josephson memories are inherently extremely fast in operation but because of the cryogenic requirements are extremely expensive to operate. The development of materials that exhibit superconductivity at higher temperatures is reducing the difficulties of using this type of memory.

joule Symbol: J. The ►SI unit of energy, including work and quantity of heat. It is defined as the work done when the point of application of a force of one newton is displaced one metre in the direction of the force. One joule equals one watt-second. ►kilowatt-hour.

Joule effect ►heating effect of a current.

Joule magnetostriction *Syn.* positive magnetostriction. ►magnetostriction.

jump A transition of an orbital electron from one atomic energy level to another.

jumper A direct electrical connection between two points on a printed circuit that is not a part of the interconnection pattern of the circuit.

junction 1. A contact between two different conducting materials such as two metals, as in a rectifier or thermocouple. **2.** A boundary between two semiconducting regions of differing electrical properties. ►p-n junction. **3.** A connection formed between two or more conductors of the same type or between sections of transmission lines.

junction field-effect transistor (JFET) ►field-effect transistor.

K

Ka band A band of microwave frequencies ranging from 27.00–40.00 gigahertz (IEEE designation). ➤frequency band.

Kaiser window ➤windowing.

Karnaugh map (K-map) A tabular form by which binary information may be represented and which allows functions to be simplified. For functions involving n variables, the data are in 2^n adjacent boxes and are grouped using the Boolean identity

$$ABC + ABC' = AB(C + C') = AB$$

where C' is the complement (or negative) of C. The coding of the boxes is such that any two adjacent boxes only differ by a one bit change (➤Gray code). A 3-variable Karnaugh map is shown in the diagram.

AB $\quad$ C	00	01	11	10
0	$A'B'C'$	$A'BC'$	ABC'	$AB'C'$
1	$A'B'C$	$A'BC$	ABC	$AB'C$

Karnaugh map for 3 variables

K band A band of microwave frequencies ranging from 18.00–27.00 gigahertz (IEEE designation). ➤frequency band.

keep-alive circuit ➤transmit-receive switch.

Kell factor ➤television.

kelvin Symbol: K. The basic ➤SI unit of ➤thermodynamic temperature, defined as 1/273.16 of the thermodynamic temperature of the triple point of water, i.e. the equilibrium point at which pure ice, air-free water, and water vapour can coexist in a sealed vacuum flask. It is related to the *degree Celsius* (°C):

$$t \,°C = T - 273.15 \text{ kelvin}$$

where T is the thermodynamic temperature and 273.15 K is the temperature of the ice point, which is the zero of the Celsius (or centigrade) temperature scale.

The kelvin is also used as a unit of temperature difference, in which case it is identical to the degree Celsius.

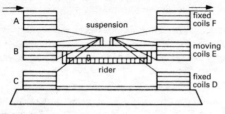

Kelvin balance

Kelvin balance *Syns.* current balance; ampere balance. A type of instrument in which the electromagnetic forces resulting from the passage of a current through a set of coils are balanced against the force of gravity. In one form two coils B, E at each end of a balanced rod are suspended between four fixed coils, A, F and C, D (see diagram). When current flows through all six coils, i.e. in the order A, B, C, D, E, F, the resulting electromagnetic forces cause each fixed coil to displace the balanced arm in the same direction. The arm is rebalanced by moving a rider along a scale on the arm, calibrated to give a reading of the current. If the current is reversed the direction of the displacement is unaltered since the current flow is reversed in all the coils. Thus both direct and alternating currents can be measured.

Kelvin contacts A means for testing or making measurements in electronic circuits or components, particularly when low values are being measured. Two sets of leads are used to each test point, similar with respect to thickness, material, and length; one set carries the test signal and the second set connects with the measuring instrument used. The effect of resistance in the leads is thus eliminated from the measurement.

Kelvin double bridge *Syn.* Thomson bridge. A bridge with six arms that has been developed from the ►Wheatstone bridge for measuring low resistances. The bridge is connected as shown in the diagram: X is the low resistance to be measured and S is a standard low resistance. Two sets of variable resistances, R_1, R_2 and r_1, r_2 are varied together until balance is achieved. At balance: $R_1/R_2 = r_1/r_2 = X/S$. The method eliminates errors due to contact resistance and the resistance of the leads.

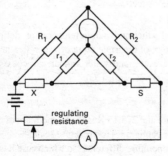

Kelvin double bridge

Kelvin effect *Syn.* Thomson effect. ➤thermoelectric effects.

Kelvin–Varley slide A device used in vernier potentiometers to reduce the effect of contact resistance. The device consists essentially of two or more sets of slide wires or resistance coils in cascade, each coil acting as a decade voltage divider for the preceding one (see diagram). If the total resistance of the first coil is $11R$, the resistance of the second coil is made $2R$ and it is connected to the first by a pair of sliding contacts that move together and shunt $2R$ of the first coil. The total resistance of the shunt and the shunted portion of the first coil is R, thus the first coil is effectively divided into 10 equal resistances. The second coil therefore acts as a vernier scale.

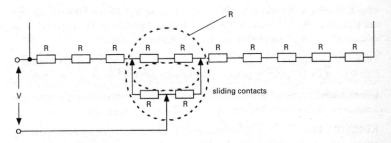

Kelvin–Varley slide

When a voltage V is required accurately the approximate voltage is set up by adjusting the contacts on the first coil, fine adjustment being achieved by a sliding contact on the second coil. A small error on the positioning of the contact on the second coil has much less effect on the total voltage tapped than the equivalent error on a single coil, thus reducing the effect of the contact resistance of the slider. Further subdivisions can be effected in a similar manner, by subdividing the second coil into 11 parts and bridging with a further coil of total resistance $4R/11$, and so on.

Kennelly–Heaviside layer *Syn. for* E-layer. ➤ionosphere.

Kerr cell *Syn.* electro-optical shutter. ➤Kerr effects.

Kerr effects Two effects in which the optical properties of transparent material are affected by electric or magnetic fields.

The *electro-optical effect* is the effect whereby the direction of polarization of plane-polarized light through a refractive medium is rotated by an electric field applied perpendicularly to the direction of propagation of the light. The *Kerr cell* utilizes this effect. It consists of two parallel plates immersed in a liquid that exhibits a marked Kerr effect. Polarized light passing through the cell can be interrupted by the application of an electric field. *Pockel's effect* is the Kerr effect when it occurs in a piezoelectric material. Pockel's effect can be used for the measurement of distance by a *mekometer.* Such an instrument can measure distance to an accuracy of 0.05 millimetres in 50 metres.

The *magneto-optical effect* occurs when plane-polarized light is reflected from a highly polished pole face of a strong electromagnet. Slight elliptical polarization of the light beam is produced.

keyboard 1. A computer input device that resembles a typewriter keyboard but with some additional keys, which may include a control key, function keys, and arrow keys. The keyboard sends electrical signals (called *scan codes*) to the CPU when a user presses a key. **2.** An electronic musical instrument that incorporates a piano-style keyboard.

keystone distortion ➤distortion.

kilo- 1. Symbol: k. A prefix to a unit, denoting a multiple of 10^3 (i.e. 1000) of that unit: one kilometre equals 10^3 metres. **2.** Symbol: k or K. A prefix used in computing to denote a multiple of 2^{10} (i.e. 1024): one kilobyte equals 2^{10} bytes.

kilogram Symbol: kg. The ➤SI unit of mass equal to the mass of the international prototype of the kilogram, which is a platinum-iridium bar kept at Sèvres, France.

kilowatt-hour Symbol: kWh. The energy produced when one kilowatt of power is expended for one hour. One kWh equals 3.6×10^6 joules.

Kirchhoff's laws (i) At any point in an electric circuit the algebraic sum of the currents meeting at that point is zero. (ii) In any closed electric circuit the algebraic sum of the products of current and resistance in each part of the network is equal to the algebraic sum of the electromotive forces in the circuit.

klystron An ➤electron tube that is used as a microwave amplifier or oscillator. It is a linear-beam ➤microwave tube in which ➤velocity modulation is applied to an electron beam in order to produce amplification of a microwave-frequency field.

Several variations of the basic klystron exist. A simple two-cavity klystron is shown in Fig. *a*. A beam of high-energy electrons produced from an electron gun is passed through a ➤cavity resonator excited by high-frequency radiowaves. The interaction between the high-frequency waves and the electron beam produces velocity modulation of the beam. The modulated beam leaving the cavity resonator (the *buncher*) traverses a field-free region (the *drift space*) where bunching occurs as the faster electrons catch up with the slower ones. The periodic current-density variations in the beam

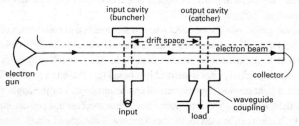

a Two-cavity **klystron** amplifier

due to the formation of the bunches are of the same frequency as the exciting radio-waves. The beam then passes through a second cavity resonator (the *catcher*) placed at a distance x away, where the current-density variations produce a voltage wave in the catcher, which is tuned to the exciting frequency or a harmonic of it.

The magnitude of the output waves depends on the velocity of the electrons; the phase is such that the negative maximum corresponds to the centre of the bunch. Most of the energy of the beam is given to the catcher since many more electrons are retarded by the induced field than are accelerated by it. Voltage amplification is obtained by conversion of the d.c. energy of the original beam into radiofrequency energy in the output circuit.

It can be shown that the optimum condition for power extraction from the beam occurs when

$$\omega t = 2\pi(n + \tfrac{3}{4})$$

where ω is the angular frequency, t the transit time between the two resonators, and n is an integer known as the *mode number*. Since $t = x/v_0$, where v_0 is the initial electron velocity, the transit time may be altered by adjusting the voltage of the electron gun. A collector electrode is used to collect that part of the electron beam leaving the second cavity. Two-cavity klystrons can be made to oscillate if positive feedback to the input cavity is employed.

The most important type of klystron is the *reflex klystron,* used as a low-power oscillator. This type of klystron has only one cavity, which acts as both buncher and catcher (Fig. *b*). Velocity modulation of the electron beam is caused by the input radiofrequency wave in the cavity, and the modulated beam leaving the cavity is reflected back by a *reflector electrode.* Bunching occurs because the faster electrons travel further towards the reflector before reversing their direction of travel than do the slower ones. The bunches of electrons returning to the cavity that experience the maximum positive field give up the most energy since the direction of motion is now reversed.

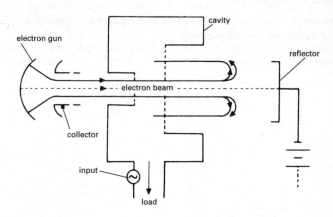

b Reflex **klystron**

As with the two-cavity klystron, optimum power transfer occurs when the transit time t of the electrons from and to the resonator is given by

$$\omega t = 2\pi(n + \tfrac{3}{4})$$

The klystron will only resonate around certain discrete values of the collector voltage, corresponding to the integers $n = 1, 2, 3$, etc. Oscillation is still possible for small excursions of collector voltage around these values, so that the reflex klystron is useful for providing automatic frequency control or in frequency-modulation transmission. This latter application requires a higher power output (up to about 10 watts) than for the more common low-power local-oscillator applications, where a typical power output of 10 milliwatts is needed.

Multicavity klystrons are used when either extremely high power pulses are required, as in the power source for a particle accelerator, or when continuous waves of moderate power are needed, as in UHF television transmitters. Three or more resonant cavities coupled to the electron beam are used to provide a high overall gain. The velocity-modulated beam leaving the first cavity interacts with the second and subsequent cavities in such a way that the induced amplified voltage in each cavity remodulates the beam received from the preceding cavity so that the beam becomes more strongly bunched and eventually excites a highly amplified wave in the output circuit.

The mutual electrostatic repulsion of the electrons tends to cause ➤debunching of the beam, particularly when very strong bunching is required. This limits the output of the device. Magnetic focusing may be used to minimize debunching.

K-map *Short for* Karnaugh map.

Koch resistance The resistance of a ➤photocell when light is incident on the active surface of the tube.

Kronig–Penney model (of energy band structure) ➤quantum mechanics.

k-space ➤momentum space.

Ku band A band of microwave frequencies ranging from 12.00–18.00 gigahertz (IEEE designation). ➤frequency band.

L

label In digital communications, *syn. for* header.

labyrinth loudspeaker ➤loudspeaker.

ladder network ➤filter.

LADT *Abbrev. for* local area data transport. The use of silent periods of a voice telephone connection for the transmission of digital information. The use of a voice channel means that only a low data rate can be achieved; however it avoids the need for an additional channel if the amount of information that needs to be sent is low.

lag 1. The amount, measured as a time interval or a proportion of the a.c. cycle, by which one periodically varying wave is delayed in phase with respect to the similar phase in another wave. ➤➤lead[1]. **2.** The time interval between the transmission of a signal and its detection by a receiver. **3.** The delay between a correcting signal of a control system and the response to it. **4.** The persistance of the electrical image in a television ➤camera tube; it may be several frames in duration.

lagging current An alternating current that has a ➤lag with respect to the applied electromotive force producing it. ➤➤leading current.

lagging load *Syn.* inductive load. A ➤reactive load in which the inductive ➤reactance exceeds the capacitive reactance and therefore carries a ➤lagging current with respect to the voltage across the terminals. A pure inductance introduces a lag of 90° or one quarter wavelength. ➤➤leading load.

Lagrange's equation A differential equation whose use provides a systematic unified approach for handling a broad class of physical systems (containing both electrical and mechanical components), no matter how complex their structure. Lagrange's equation is given by

$$\frac{\mathrm{d}}{\mathrm{d}t}\left(\frac{\partial T}{\partial \dot{q}_n}\right) - \frac{\partial T}{\partial q_n} + \frac{\partial D}{\partial \dot{q}_n} + \frac{\partial V}{\partial q_n} = Q_n \quad n = 1,2,3\ldots$$

where T is the total kinetic energy and V the total potential energy of a system, D is the dissipation function of the system, Q_n is the generalized applied force at the coordinate n, q_n is the generalized coordinate, and $\dot{q}_n = \mathrm{d}q_n/\mathrm{d}t$ (generalized velocity).

laminated core ➤core.

lamination A thin stamping of iron or steel, oxidized or lightly varnished on the surface, that is employed in the assembly of a laminated ➤core for use in transformers, transductors, relays, chokes, or similar apparatus. The use of laminations reduces ➤eddy currents in alternating-current applications.

LAN *Abbrev. for* local area network.

language *Short for* programming language.

Laplace operator Symbol: ∇^2. The differential operator

$$\partial^2/\partial x^2 + \partial^2/\partial y^2 + \partial^2/\partial z^2$$

Laplace transform A mathematical method of simplifying the transient analysis of a network or circuit. The Laplace transform $F(s)$ of a signal $f(t)$ is given by:

$$F(s) = \int_0^{+\infty} e^{-st} f(t)\, dt$$

where s is the *complex frequency* $\sigma + j\omega$. ➤s-domain circuit analysis.

lapping A method of reducing the thickness of a ➤wafer of semiconductor for applications where accuracy of the substrate thickness is critical to the operation of the device, or where the thermal resistance of the substrate must be reduced to a minimum. After processing of the front side of the substrate using planar process, a slurry of water and fine grit is used to wear down the back of the wafer. The slurry is placed between a flat plate and the back of the wafer, and the wafer moved with respect to the plate in order to remove substrate material by abrasion.

large-scale integration (LSI) ➤integrated circuit.

laryngophone *Syn. for* throat microphone. ➤microphone.

laser Acronym from *l*ight *a*mplification by *s*timulated *e*mission of *r*adiation. A source of ➤monochromatic ➤coherent radiation in the visible, ultraviolet, or infrared regions of the electromagnetic spectrum.

In an atom, an electron can jump from one allowed energy level E_2 to an empty level E_1, accompanied by the absorption or emission of a photon of electromagentic radiation of the appropriate wavelength, λ, given by ➤Planck's law:

$$|E_2 - E_1| = h\nu = hc/\lambda$$

If E_2 exceeds E_1, then emission of radiation occurs. *Spontaneous emission* occurs when an electron in the atom changes energy level without any specific external stimulus. *Stimulated emission* occurs in response to excitation by a photon of the same energy as the energy difference, $E_2 - E_1$. The incident and emitted photons are both in phase: coherent radiation. Laser action is based on the process of stimulated emission. For continued emission of radiation, the number of electrons in the higher energy level E_2 must be greater than the number in the lower level E_1. This state is known as *population inversion*. Population inversion is a nonequilibrium state, and internal relaxation processes will tend to restore the electron distribution to the thermal equilibrium state. External power must therefore be applied to maintain the inverted population in a laser;

for example, applied electromagnetic radiation of a shorter wavelength can excite other electrons into a third energy level above E_2, and this level is then the source of electrons for E_2. For continued stimulated emission, photons of the appropriate wavelength must continually pass through the laser medium. This can be achieved by placing two mirrors, facing each other at each end of the laser region, the distance between them being a whole number of wavelengths: this is a *Fabry–Perot cavity*, and is resonant at the laser wavelength. Spontaneous and stimulated photons then travel back and forth between the mirrors, stimulating further emission of photons, and building up a high intensity of light. If one of the mirrors is only partly reflecting, this light can be emitted in the form of a high-intensity monochromatic coherent beam or pulse.

Laser action can take place in solid, liquid, and gaseous media. The ruby laser was the first solid laser; population inversion was achieved by light excitation. Gas lasers are excited by a continual electric discharge (➤gas discharge tube); population inversion is a result of collisions between the gas particles and high-energy ions and electrons excited by the high voltage, creating the excited glow. Semiconductor lasers are made using forward-biased p-n junctions, the injection of carriers across the junction providing the inverted population.

➤semiconductor laser; maser.

latch A simple ➤register in a computer that holds values and places them on a ➤bus.

latching ➤locking.

lattice constant A parameter that describes the configuration of a crystal lattice. The lattice constant is given either as the lengths of the edges of a unit cell of a crystal or as the angle between the axes of the cell. The former description is also termed the *lattice parameter* or *lattice spacing*. The edge length of a cubic unit cell is usually given.

lattice filter ➤filter.

lattice network ➤network.

lattice parameter *Syn.* lattice spacing. ➤lattice constant.

Lauritzen's electroscope ➤electroscope.

lawnmower *Informal* A type of preamplifier used with radar receivers that reduces the level of ➤grass on the screen.

L band A band of microwave frequencies ranging from 1.00–2.00 gigahertz (IEEE designation). ➤frequency band.

LCC *Abbrev. for* leadless chip carrier.

LCD *Abbrev. for* liquid-crystal display.

L-C network (or **LC network**) A tuned circuit that contains both inductance and capacitance. The product, *LC,* of inductance and capacitance is constant for any given frequency.

lead[1] **1.** An electrical conductor, usually in the form of a wire or cable, that is used to make external connections between circuits or pieces of apparatus. ➤interconnection; intraconnection. **2.** The amount, measured as a time interval or as a proportion of the a.c. cycle, by which one periodically varying wave is advanced in phase with respect to the similar phase in another wave. ➤lag.

lead[2] Symbol: Pb. A heavy metal, atomic number 82, that is mainly used either as an alloy, with other metals such as tin, to form solders or in storage batteries.

leadframe The complete interconnection pattern of leads inside an ➤integrated-circuit package. It is used to connect the integrated circuit to the outside world. The leadframe is formed from thin copper sheet (➤wire bonding) or by plating the required pattern onto plastic tape (➤tape automated bonding). The frame is relatively rigid and robust. ➤dual in-line package; leadless chip carrier; pin grid array.

lead-in The cable that connects the active part of an antenna to the transmitter or receiver.

leading current An alternating current that has a ➤lead with respect to the applied electromotive force producing it. ➤lagging current.

leading edge (of a pulse) ➤pulse.

leading load *Syn.* capacitive load. A ➤reactive load usually containing resistance and capacitance that carries a ➤leading current with respect to the voltage across the terminals. A pure capacitance introduces a lead of 90° or one quarter wavelength. ➤lagging load.

leadless chip carrier (LCC) A common form of package used for ➤integrated circuits. It consists of a ceramic or plastic casing that contains a circuit using either ➤wire bonding or ➤tape automated bonding, and also forms the output contacts arranged around all four sides of the package. The metallic outer contacts are formed flush with the edges of the LCC package and can be flow-soldered to a printed circuit board or hybrid substrate. The number of contacts available varies depending on size of package and complexity of the circuit.

leakage 1. The passage of an electric current along a path other than that intended due to faulty insulation or isolation in a circuit, component, device, or other piece of apparatus. **2.** ➤discrete Fourier transform.

leakage current 1. A fault current occurring in any electronic device, circuit, etc., due to ➤leakage. It is small compared with a short-circuit current. **2.** The reverse saturation current in ➤p-n junctions and rectifying ➤metal-semiconductor contacts.

leakage flux Lost flux in any apparatus, such as a transformer, that contains a magnetic circuit. The leakage flux is flux that is outside the useful portion of the flux circuit.

leakage reactance Unwanted reactance in a transformer or alternator that is caused by ►leakage flux cutting one coil but not the other. A leakage inductance is produced by this effect and leads to losses in the system.

least significant bit (LSB) The rightmost bit in the binary representation of a number. It is the bit conveying the least amount of precision.

LEC *Abbrev. for* liquid-encapsulated Czochralski.

Leclanché cell A primary ►cell that contains a carbon-rod anode and a zinc cathode. The electrolyte is 10–20% ammonium chloride solution. ►Polarization is minimized by means of a depolarizer consisting of manganese dioxide mixed with crushed carbon, held in contact with the anode. This wet cell, devised by Leclanché in 1867, has an e.m.f. of about 1.5 volts. The present-day dry ►cell based on it uses an ammonium chloride paste as electrolyte with the zinc cathode as the container, protected by a plastic wrapping. The dry cell is used in torches, radios, etc.

LED *Abbrev. for* light-emitting diode.

Leduc effect ►Righi–Leduc effect.

LEED *Abbrev. for* low-energy electron diffraction. ►diffraction.

left-hand rule ►Fleming's rules.

Lenz's law ►electromagnetic induction.

LEOS *Abbrev. for* low earth orbit satellite. ►►GEOS.

level compensator A device or circuit that automatically compensates for the effects of amplitude variations in a received signal.

level shifter A circuit that changes the d.c. level of a signal without introducing signal attenuation. It is used instead of decoupling capacitors (►coupling) to interface one circuit with another while preserving the d.c. response of the combined circuit.

Leyden jar An early type of capacitor.

LF *Abbrev. for* low frequency. ►frequency band.

LFO *Abbrev. for* low-frequency oscillator.

life test A test in which a sample or population of items is subjected to stated stress conditions for specified times with stated failure or success criteria, in order to determine its ►reliability characteristics. The data from such tests will provide information giving ►failure rate and ►mean life of the item. The reliability of most semiconducor devices, etc., is so great that ►accelerated life tests and ►step-stress life tests are employed to avoid unnecessarily long tests. *Truncated tests* are those terminated after a predetermined time or a predetermined number of failures or a combination of these. A *screening test* is a test designed to remove unsatisfactory items or those likely to exhibit early failures. This is sometimes called *burn-in* (►failure rate).

lifetime The mean time interval between generation and recombination of a charge ➤carrier in a ➤semiconductor.

lifter A device that is equivalent to a filter in the spectral domain and can be used on the cepstrum of a signal. ➤cepstrum.

lift-off A technique used in processing semiconductor devices or wafers to produce a required metallization pattern (➤metallizing). The wafer is covered with ➤resist, which is then exposed and developed to produce the desired pattern. Metal is then applied, usually by evaporation, and deposited on top of the resist as well as the substrate. The resist is dissolved using a suitable solvent and the metal on the resist is removed or 'lifted off' with it, leaving the desired pattern on the substrate. This technique contrasts with the etch processes, where metal is applied to the wafer first. The metal is covered with resist, which is exposed and developed to produce a protective layer for the desired pattern. The metal is then etched away before the remaining resist is dissolved. These two processes, contrasting gallium arsenide and silicon, are shown in the diagram. For lift-off to be facilitated a small protruding lip at the top edge of the resist is desirable. *Assisted lift-off* is a multilevel resist technique used to produce such a protruding lip. Other methods of *edge profile modification* include baking the wafer after development of the resist, which produces a crust on the resist with a suitable lip. The resist may be soaked in chlorobenzene to remove residual solvents from the upper lay-

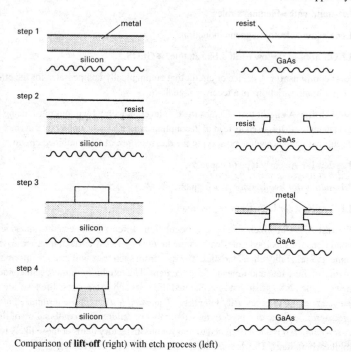

Comparison of **lift-off** (right) with etch process (left)

ers of the resist film. This retards development of the resist in the upper layer and produces a suitable undercut edge.

light-emitting diode (LED) A p-n junction diode that emits light as a result of direct radiative recombination of excess electron-hole pairs (see recombination processes). In ►direct-gap semiconductors, such as gallium arsenide, it is a major part of recombination and a significant amount of light will be emitted following ►injection of excess minority carriers. The quantity of light produced in a forward-biased p-n junction diode formed from such material will be proportional to the numbers of excess minority carriers, i.e. to the bias current. The useful light obtained from the diode is dependent on the optical quality of the crystal surfaces. The frequency, and hence the colour, is a property of the material used, since the energy of the emitted photon is determined by the band-gap energy. LEDs are useful for low-voltage display devices, such as calculators or digital watches. Single LEDs are widely used as on/off indicators.

lightning conductor A lightning protective system that consists of a single conductor providing a path between an air terminal and earth along which a lightning stroke can pass.

lightning stroke An electric discharge due to the discharge of one of the charged regions of a thunder cloud. The polarity of the lightning stroke is the polarity of the electric charge that comes to earth. A complete *lightning flash* is a complete discharge along a single path.

The path of the lightning flash is established by an initial discharge, the *leader stroke,* that can develop either downwards from the cloud to earth or upwards from earth to the cloud. A *dart leader stroke* is a leader stroke that develops continuously. One that develops in a series of relatively short steps is a *stepped leader stroke.* The *return stroke* consists of a high current discharge that flows upwards as soon as a downward leader stroke strikes the earth. If the flash is made up of more than one lightning stroke it is termed a *multiple-stroke lightning flash.*

A lightning stroke to any part of a power or communication system is described as a *direct stroke* and the ►surge produced in the system by it or by a flashover from it is a *direct lightning surge.* An *indirect stroke* induces a voltage in such a system without actually striking it and the surge induced by it is an *indirect lightning surge.*

A *lightning protective system* is a complete system of conductors that is designed to protect a building or equipment from the effects of a lightning stroke. ►►lightning conductor.

light-pen A penlike device that produces data, in the form of a visible image on the screen of a cathode-ray tube, by 'writing' in a manner similar to that with a pen on paper. Light-pens are almost invariably used in conjunction with an on-line ►visual display unit that inputs the data produced to a computer.

limb ►transformer.

limited space-charge accumulation mode (LSA mode). ►transferred electron device.

limited stability The property of any system, circuit, device, etc., that is stable only for a specific range of values of input signal and unstable outside this range.

limiter Any device that automatically sets a boundary value or values upon a signal. The term is usually applied to a device which, for inputs below a specified instantaneous value, gives an output proportional to the input, but for inputs above that value gives a constant peak output. A *base limiter* is one whose output comprises that part of an input signal exceeding a predetermined value. A *slicer* is a limiter having two boundary values, the portion of the signal between these values being passed on. A limiter designed to give a constant current whatever the applied voltage is a *current limiter.*

limiting amplifier ➤amplifier.

line 1. A physical communications medium, such as a telephone line or some other ➤transmission line, or a 'virtual' channel down which information can be routed. **2.** ➤television.

linear 1. Denoting any system, device, or apparatus that has its essential physical parts arranged in a line. **2.** Describing any device that has an output directly proportional to the value of the input and varies continuously with it, as in a linear ➤amplifier.

linear accelerator A particle ➤accelerator in which electrons or protons are accelerated as they travel along a straight evacuated chamber. The acceleration is provided by the electric field vector that is associated with the radiofrequency (r.f.) output from a ➤klystron or ➤magnetron.

In the *standing-wave accelerator* standing waves from an r.f. supply are established between a series of cylindrical electrodes coaxial with the chamber. The standing waves are established with the electric vector aligned axially. The electrons are only accelerated in the gaps between the electrodes; inside the electrodes they drift towards the next electrode. The electrodes are therefore termed *drift tubes.* The lengths of the drift tubes and the frequency of the r.f. supply is arranged so that the phase of the electric field vector is always in accelerating mode when the electrons emerge from a drift tube. As the energy of the particles increases, the lengths of successive drift tubes must be correspondingly increased in order to maintain the required phase relationship.

Higher particle energies are achieved using a *travelling-wave accelerator,* in which the accelerating chamber is a long ➤waveguide. An r.f. ➤travelling wave is established from a high-power source and the waveguide is excited so that a large-amplitude travelling wave is produced, travelling with a phase velocity equal to the local velocity of the electrons to be accelerated. Power is transferred from the r.f. wave to the accelerated electrons and the r.f. power is boosted at regular intervals along the length using klystrons.

linear amplifier ➤amplifier.

linear array antenna An antenna consisting of a series of elements of the same type arranged in a line. If the elements are fed with a signal of the same phase, the array operates in broadside mode (➤antenna array); if they are fed with a signal whose phase differs by an amount equivalent to the electric distance (►electric dimensions) between the elements, it operates in endfire mode.

A *phased linear array* is a linear array antenna in which the phase of the feed to each element is adjusted so that the maximum radiation lies in an arbitrary direction to form a scanning array. ►►planar array antenna; log-periodic antenna.

linear-beam microwave tube *Syn.* O-type microwave tube. ►microwave tube.

linear block codes ►digital codes.

linear circuit A circuit in which the output varies continuously as a given linear function of the input.

linear detector ►detector.

linear inverter ►inverter.

linearly graded junction A junction between two different-polarity semiconductors (p-n, p-i, or n-i) in which the concentration of impurities varies linearly across the junction. ►►abrupt junction.

linearly polarized wave *Syn.* plane-polarized wave. ►polarization.

linear motor ►MAGLEV.

linear network ►network.

linear phase response ►phase response.

linear prediction (LP) A widely employed method for predicting the output waveform from a ►linear system. It finds special application in modelling the acoustic waveform produced during speech, being used as a basis for speech coding, speech analysis, and speech synthesis; when used for speech coding it is known as *linear predictive coding* (*LPC*). Linear prediction relies on the fact that speech can be described in terms of an acoustic excitation waveform exciting the ►formants of the vocal tract. For those sounds in speech that have a ►pitch associated with them, such as the vowels in 'feed' and 'card', the formants are excited by a periodic acoustic waveform resulting from ►glottal airflow. This assumes a train of narrow pulses as the acoustic excitation, with an overall zero decibel per octave spectral shape, where each pulse results from a vocal-fold closure. The response of each formant to each individual pulse will be a sine wave at the formant frequency whose amplitude decays exponentially dependent on the bandwidth of the formant; each individual output sample from a formant can then be predicted mathematically from previous output samples.

In LP analysis of speech, a spectral estimate is made based on an all-pole filter (having only ►poles in its frequency response) that gives the minimum squared error when its output to a spectrally flat pulse waveform input is compared with the speech being analysed. LP speech analysis relies on five key assumptions: the ringing of the formants during voiced speech production is purely due to the most recent vocal fold excitation acoustic pressure pulse; the formant frequencies remain constant during each cycle; the formant bandwidths remain constant during each cycle; the vocal tract response can be completely modelled in terms of formants for all speech sounds; the acoustic excitation to the vocal tract can be modelled as being spectrally flat (0 dB per octave). In order to lessen the effect of these in practice, LP is usually carried out on input frames of 10–25 milliseconds in duration.

The error between the predicted speech and the input speech is known as the *residual*, which exhibits a large discontinuity at each excitation pulse as these are not predictable. When LPC is used as a means of coding speech for transmission, a more natural sounding output can be achieved by transmitting the residual as additional data to excite the LPC model in *residual-excited linear prediction*. Another method for improving the naturalness of resynthesized LPC-based speech is by coding the excitation signal as a series of pulses of varying amplitudes during each cycle; this is known as *multipulse linear predictive coding*. Alternatively, the excitation signal for resynthesis can be selected from a stored codebook of Gaussian sequence with zero mean in *code-excited linear prediction*. A variation is *vector sum excited linear prediction* in which the excitation signal is reconstructed from linear combinations of stored vectors.

linear predictive coding (LPC) ➤linear prediction.

linear scan 1. A sweep of the electron beam in a cathode-ray tube in which the beam scans the screen with constant velocity, usually by application of a sawtooth waveform to the deflection plates or coils. ➤➤timebase. **2.** A scan using a ➤radar beam that moves with constant angular velocity.

linear system A system that, for an input consisting of the sum of a number of signals, produces an output that is the sum, or *superposition*, of the system's response to each input considered separately. For example, if the input waveform $x_1[t]$ to a system produces the output $y_1[t]$ and the input $x_2[t]$ produces the output $y_2[t]$, then the system is linear if the output is $\{y_1[t] + y_2[t]\}$ when the input applied is $\{x_1[t] + x_2[t]\}$.

A linear system will also produce a weighted sum of outputs if the inputs are so weighted, such that an input of $\{ax_1[t] + bx_2[t]\}$ will produce an output of $\{ay_1[t] + by_2[t]\}$, where a and b are constants.

linear timebase oscillator A ➤relaxation oscillator that is used to generate a sawtooth waveform for use as a ➤timebase.

linear transducer ➤transducer.

line communication Communication, such as broadcasting or telephony, between two points by means of a physical path such as wire or waveguide.

line frequency ➤television.

line impedance stabilization network (LISN) ➤electromagnetic compatibility.

line of flux An imaginary line drawn in a magnetic field whose direction at any point along its length is that of the magnetic flux density, **B**. The number of lines of flux through unit area perpendicular to the direction of **B** is equal to the magnetic flux density at that point. ➤➤field.

line of force An imaginary line drawn in an electric field whose direction at any point along its length represents that of the field at that point. The number of lines of force through unit area perpendicular to the field is equal to the field strength at that point. ➤field.

line-of-sight path A straight-line path between two radio transmitters where there are no physical objects lying on the path.

line printer An output device used with computers and data-processing systems that prints an entire line of characters at a time. Typical operating speeds range from 200 to 3000 lines per minute.

line-sequential colour television A ➤colour-television system in which each of the video signals (red, blue, and green) is transmitted in turn for the duration of one entire scanning line.

line voltage ➤voltage between lines.

link 1. A communication channel or circuit that is used to connect other channels or circuits. **2.** A path between two switches that form part of a central control system in automatic switching.

link availability The proportion of time for which a telecommunication link's signal is above a certain level, usually that required for normal operational use.

link budget analysis Analysis carried out in a ➤telecommunication system to determine the change in signal along a particular link of that system associated with each component, including the effects of ➤antenna gain and ➤path loss, and the initial power and final required power. The aim is to determine whether there is sufficient signal at the receiving component of that link, usually with an additional component available (the *link margin*), to ensure successful transmission of the signal.

link margin ➤link budget analysis.

lin-log receiver A type of ➤radar receiver that has a linear output for small input signals and a logarithmic output for large ones. Such a receiver is useful over a large range of received signals.

lip microphone ➤microphone.

liquid crystal An organic liquid consisting of long-chain molecules that can change their crystal structure to allow them to flow like a liquid. When an electric field is applied, the orientation of the crystals is disturbed. A change in the applied field causes a change in the reflectivity of the liquid making them extremely suitable for use in display devices.

Liquid crystals are also temperature-sensitive. There is an apparent change of colour with increasing temperature that makes them suitable for use in temperature indicators.

liquid-crystal display (LCD) A type of display that uses ►liquid crystals. Examples range from the 7-segment displays in digital watches to screens in laptop computers. Controlling the electric field applied to each element of the display generates an image. Displays can be reflective, where the display is viewed from the same side as the incident light, or transmissive (backlit), where the light source is behind the display.

For computer screens, especially colour displays, a much higher performance may be obtained from *active-matrix LCDs*. In these LCDs, thin-film circuitry is deposited on a glass substrate in the screen and is used to control individual pixels (picture elements) in the image.

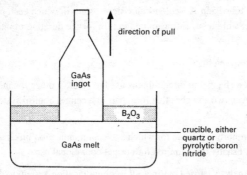

Schematic diagram of **liquid-encapsulated Czochralski** growth method

liquid-encapsulated Czochralski (LEC) A method of growing semiconductor crystals from molten form by slowly pulling the crystal vertically from the melt. A layer of liquid boric oxide (B_2O_3) is floated on the surface of the semiconductor to confine the melt – hence the description liquid-encapsulated. The growth is started using a seed crystal, which is introduced through the boric oxide when the appropriate temperature profile is achieved. The machines used to grow the crystals are referred to as *crystal pullers.* High-pressure LEC processes are performed under high external pressures – up to 50 atmospheres. Low-pressure LEC processes are carried out at pressures of about 1 atmosphere. The pullers required for each of these processes are different, and different heat-flow characteristics exist; the two types of growth machine are therefore not easily interchangeable. The diagram shows an example of gallium arsenide growth. The liquid encapsulation and high pressures originally used were required to contain the rather violent reactions that occur as the gallium and arsenic react exothermically to produce the gallium arsenide melt. The low-pressure technique was developed by introducing the arsenic in a controlled manner into the molten gallium below the boric oxide surface.

liquid-phase epitaxy A method of growing an ►epitaxial layer on a substrate from a molten material. The substrate crystal is placed in a slider and the material to be deposited is contained in molten form in a 'boat'. The melt is supercooled to just below the solidification temperature. As the slider containing the substrate material is moved slowly across the surface of the melt, atoms solidify onto the crystal substrate. This method of epitaxy is most useful for III-V or II-VI ➤compound semiconductors, particularly gallium arsenide substrates. It has limitations and is losing popularity, but is inexpensive and capable of growing many material compositions; it is therefore still used for some applications, such as light-emitting diodes, that do not require such thin uniform high-quality layers as are required for microwave devices.

LISN *Abbrev. for* line impedance stabilization network. ➤electromagnetic compatibility.

Lissajous' figure The displacement pattern traced out when two sinusoidally varying quantities are superimposed at right angles to each other. These patterns may be obtained in practice on the screen of a ➤cathode-ray oscilloscope by applying the two signals to the horizontal and vertical deflection plates.

The simplest pattern is a straight line, which occurs when two signals of equal frequency and in phase with each other are superimposed. Much more complicated figures result when the ratio of the frequencies is not a simple one. Typical figures are shown in the diagram for various frequency ratios and phase angles. The patterns can be used for accurate frequency matching of two signals or for identifying the phase relationship of two signals of the same frequency.

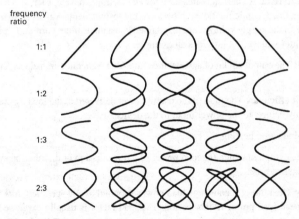

Lissajous' figures

literal In computing, an explicit version of one variable, usually an input, in either its true or its false (complemented) form.

lithography Patterning techniques used in the manufacture of ➤integrated circuits, ➤semiconductor components, ➤thin-film circuits, and ➤printed circuits. The techniques are used to create or transfer required patterns from masks to a substrate surface using energy-sensitive materials known as ➤resists.

The most common technique is ➤photolithography, but in VLSI applications where the dimensions of the pattern to be transferred are of the same order as the wavelengths of the light used other techniques have been developed. ►electron-beam lithography; ion-beam lithography; X-ray lithography.

Litzendraht wire A multistranded wire formed from many fine conducting filaments. It is employed for high-frequency applications, as in coils used in radio, in order to reduce the high-frequency resistance. ►skin effect.

live *Syn.* alive. Denoting a conductor or circuit that is not at earth potential.

L network *Short for* inductance network. ►network.

load 1. Any device or material that absorbs power from a source of electrical signals. Examples include loudspeakers, television and radio receivers, the material heated by dielectric or induction heating, logic circuits, and any driven circuit. **2.** The output power delivered by any electrical machine, generator, transducer, electronic circuit, or device. The machine, generator, etc., is *off-load* when it is operated under normal conditions but no absorbing load is connected at the output. It is *on-load* when connected to an absorbing load.

The maximum power that is absorbed by a load or delivered as a load in a predetermined time period is the *peak load.*

load characteristic A characteristic curve for an electronic device, such as a transistor, showing the dynamic relationship between the instantaneous values of two variables when all the supply voltages are maintained constant. Emitter current, I_e, plotted against emitter voltage, V_e, provides an example.

load circuit The output circuit of any electrical machine, generator, transducer, circuit, or device.

load-circuit efficiency The ratio of the useful power delivered to the load by the load circuit of any device, to the power input to the device.

load coil ►induction heating.

load curve A graph in which the load power of a transmission or distribution system is plotted against time.

load factor The ratio of the average load power supplied over a specified time interval to the peak load power in that period. The load factor is usually expressed as a percentage.

The *plant load factor* relates specifically to an electrical generator or group of generators, and is the ratio of the actual number of electrical units (kilowatt-hours) supplied by the generators in a given time interval to the number that would have been supplied had the plant operated at its maximum continuous rating for that interval. This factor also is usually expressed as a percentage.

load impedance The ►impedance presented by a load to the driver circuit that supplies power to it. The effect of variations in the load impedance on the performance of an oscillator is seen in a *load impedance diagram* in which the oscillator output is plotted against load impedance.

loading ➤transmission line.

loading coils ➤transmission line.

load leads The conductors or transmission lines that connect the power source for dielectric or induction heaters to the load, load coil, or applicator used.

load line A line drawn on the graph of a family of ➤characteristics of an electronic device, such as a transistor, that shows the relationship between current and voltage of the circuit under consideration for a given load.

load matching Adjustment of the output impedance of the load circuit of a dielectric or induction heater to optimize the energy transferred from the source to the load.

load regulator A circuit or device that maintains a load at a constant value or varies it in a predetermined manner.

load transfer switch A switch that is used to connect the power output from a generator or other power source to one or to other load circuits as required.

lobe ➤antenna pattern.

lobe switching *Syn.* beam switching. A form of scanning used in radar in which the direction of maximum radiation or reception is switched sequentially to each of a number of preferred directions, for example from one side to the other of the target area. Lobe switching is achieved by using a ➤steerable antenna and switching the output into the appropriate circuits, each of which has an optimum signal-to-noise response for the chosen direction.

local area network (LAN) A ➤network for linking computers together, typically personal computers, for sharing files, printers, and electronic mail. It usually provides high-speed data communication services to directly connected computers of a single organization, the typical overall size of the LAN being a few kilometres. LANs may be connected to each other or to longer-distance networks. ➤➤wide area network.

local exchange ➤telephony.

local feedback *Syn.* multiple-loop feedback. ➤feedback.

localizer beacon ➤beacon.

local oscillator ➤heterodyne reception.

locator beacon *Syn. for* homing beacon. ➤beacon.

lock-in detector *Syn.* lock-in amplifier. A detector that responds only to an input signal whose frequency is synchronous with the frequency of a locally generated control signal. It may be used as a null-point detector in a bridge circuit.

locking 1. Controlling the frequency of an oscillator by means of an applied signal of constant frequency from an external source. **2.** *Syn.* latching. Holding a circuit in position or in a certain state until previous operating circuits are ready to change the circuit.

locking-in Synchronizing the frequencies of two coupled oscillators, as in a frequency doubler, so that the two frequencies have a desired ratio, usually of two integers. One of the oscillators must be free-running and capable of being pulled to the desired frequency.

locking-on The following of a target, automatically, by a radar antenna.

locking relay ►relay.

lock range The frequency range within which a ►phase-lock loop (PLL) will capture and lock on to the input frequency. If the input signal range is greater than the PLL lock range, the input signal will need to be searched for by changing the frequency of the voltage-controlled oscillator before lock and tracking will result.

logarithmic amplifier A nonlinear ►amplifier whose response describes the mathematical logarithmic function. Computing operations such as multiplication, division, and the taking of powers or roots may all be performed using logarithmic amplifiers. They also provide a convenient method of compressing signals, termed *logarithmic compression*.

logarithmic compression ►logarithmic amplifier; data compression.

logarithmic decrement ►damped.

logarithmic potentiometer ►potentiometer.

logarithmic resistor A form of variable resistor designed so that the movement of the contact is proportional to the logarithm of the resistance.

logical address *Syn.* virtual address. The ►address of an instruction or item of data as seen by a computer program and independent of the hardware on which the program will execute. Logical addresses typically range between 0 and 2^N, where N is the number of address bits used by the program.

logical one The digit 1 used in ►binary notation. It is equivalent to the value 'true' of a logical statement. *Logical zero* is the digit 0 in binary notation. It is equivalent to the value 'false'.

logical operations Operations performed on binary data items, i.e. on data having the states '1' or '0'. These are usually implemented by logic gates, such as AND, OR, NOT, or exclusive OR gates (►logic circuit). The rules that define these operations are specified in ►Boolean algebra.

logic circuit A circuit designed to perform a particular logical function based on the concepts of 'and', 'either-or', 'neither-nor', etc. Normally these circuits operate between two discrete voltage levels, i.e. *high* and *low logic levels,* and are described as *binary logic circuits*. Logic using three or more logic levels is possible but not common.

The basic ►logic gates that implement the elementary logical functions are as follows.

AND gate: a circuit with two or more inputs and one output in which the output signal is high if and only if (sometimes written *iff*) all the inputs are high simultaneously;

NOT gate (or ►inverter): a circuit with one input whose output is high if the input is low and vice versa;

NAND gate: a circuit with two or more inputs and one output, whose output is high if any one or more of the inputs is low, and low if all the inputs are high;

NOR gate: a circuit with two or more inputs and one output, whose output is high if and only if all the inputs are low;

OR gate: a circuit with two or more inputs and one output whose output is high if any one or more of the inputs are high;

exclusive OR gate: a circuit with two or more inputs and one output whose output is low if all inputs are identical, otherwise it is high.

The graphical symbols for the logic gates are shown in the table. These circuits are for use with *positive logic*: that is, the high voltage level represents a ►logical 1 and low a logical 0. *Negative logic* has the high level representing a logical 0 and low a logical 1. The same circuits may be used in negative logic but become the complements of the positive logic circuits, i.e. a positive OR gate becomes a negative AND gate.

Binary logic circuits are extensively used in computers to carry out instructions and arithmetical processes. Any logical procedure may be effected by using a suitable combination of the basic gates. ►►truth table; Boolean algebra.

Binary circuits may be formed from discrete components or, more commonly, from ►integrated circuits. Families of integrated logic circuits exist based on ►bipolar junction transistors; these include ►emitter-coupled logic (ECL), ►I^2L, ►non-threshold logic (NTL), and ►transistor-transistor logic (TTL). ►MOS logic circuits are based on ►MOSFETs.

Popular (formerly BSI) symbol	Binary logic circuit	IEC approved symbol	Popular (formerly BSI) symbol	Binary logic circuit	IEC approved symbol
	AND gate	&		NOR gate, negated output	≥1
	NAND gate, negated output	&		NOR gate, negated inputs	≥1
	NAND gate, negated inputs	&		Exclusive-OR gate	=1
	OR gate	≥1		Inverter (NOT gate)	

Graphical symbols for logic gates

Bipolar logic circuits are capable of very high speed operation but have relatively complex structures compared to MOS logic circuits, and therefore a lower functional ►packing density. MOS logic circuits have thus been widely used for large-scale integration (LSI) despite their lower speed of operation, and bipolar logic circuits have been used for circuits demanding high performance and high speeds. Recent improvements in bipolar technology, however, have improved the packing densities that can be achieved with bipolar circuits. For VLSI (very large scale integration) applications demanding high speeds of operation, bipolar circuits have great potential. I^2L circuits offer the highest density and lowest power dissipation, approaching that of MOS circuits. ECL circuits have the highest performance at present.

logic diagram A diagram that shows the logic elements of a computer or data-processing system, or a function thereof, and their interconnections but does not usually show any constructional or engineering details. A logic diagram is useful when designing such systems or smaller networks to perform a specific mathematical operation such as integration. The logic diagram of a full ►adder is shown in the diagram. ►►logic circuit (table).

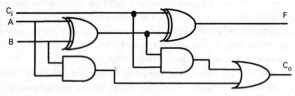

Logic diagram of a full adder

logic element A small part of a logic circuit, typically a ►logic gate, that may be represented by mathematical operators in symbolic logic. ►►logic circuit.

logic gate A digital circuit, such as an AND gate, that implements an elementary logical function. It has one or more inputs but only one output. The conditions applied to the input(s) determine the voltage levels at the output. The output usually has two possible output states: '1', '0'. ►logic circuit; tristate logic gate.

logic symbol A graphical symbol representing a logic gate. ►logic circuit (table).

log-periodic antenna A ►broadband antenna with almost frequency-independent characteristics within its designed band of operation. To achieve this, the structural dimensions increase in proportion to the distance from the origin of the structure, thus giving an approximately constant number of active elements within logarithmically changing frequency bands (hence the name log-periodic).

A common form is composed of ►dipoles arranged along a two-conductor tapered transmission line and is fed at one end of the transmission line. The dipoles are typically short in length and close together near the feed and become longer and more widely separated further from the feed. The radius of the dipoles should also increase with distance from the feed, but this is usually implemented in a series of steps. Where the elements of the dipoles on one side are arranged along the same transmission line, the antenna operates in broadside mode (➤antenna array); where they are arranged on alternate transmission lines, it operates in endfire mode.

long-line effect An effect sometimes observed when an oscillator is coupled to a load through a transmission line that is long compared with the wavelength of the oscillator output. The oscillator may jump from its desired frequency to a nearby unwanted frequency.

long-persistence screen A type of screen used in cathode-ray tubes that allows the image on the screen to persist for several seconds. This is usually achieved by mixing phosphorescent compounds with the usual fluorescent compounds of the screen.

long-tailed pair (LTP) A pair of matched ➤transistors, either two BJTs or two FETs, connected as an emitter-coupled or source-coupled pair, with a constant current source providing the bias at the common connection. The collector (drain) output current is related to the difference between the input voltages at the base (gate) terminals of the two transistors. This arrangement is the basis of a ➤differential amplifier, which amplifies differential-mode signals only. The ➤common-mode performance of the amplifier is related to the conductance of the constant current bias source. Any signals appearing across this source will be effectively common to both inputs, but will cause a change in the output current; this is unwanted common-mode gain.

This source can be realized using a large resistor; this requires a large voltage to maintain the bias current, and signal voltages will be much smaller than the steady bias voltage, giving good rejection of the common-mode signals. The larger this resistance, the longer the 'tail' of the differential pair. In integrated circuits the current source is more likely to be realized using active devices, in the form of a ➤current mirror.

long wave A radiowave that has a wavelength in the range 1–10 km. ➤frequency band.

loop ➤feedback control loop.

loop antenna *Syns.* frame antenna; coil antenna. A type of ➤directional antenna with moderate directivity that consists essentially of a coil with one or more turns of wire of very small axial length compared to the diameter. The direction of maximum sensitivity or transmission is the direction coplanar with the coil so that this type of antenna is very useful for radio direction finders. It is also simple and light in construction enabling it to be used with portable radio receivers.

loop direction finding *Syn. for* frame direction finding. ➤direction finding.

loop gain The product of the gains, in an amplifier with feedback, as the loop is traversed from the amplifier input, through the amplifier, and through the network creating the feedback path (with the summing junction, completing the loop, open).

loop signals ➤feedback control loop.

loose coupling *Syn. for* undercoupling. ➤coupling.

loosely coupled multiprocessor ➤multiprocessor.

Lorentz force The force experienced by an electron or other charged particle moving in a region of magnetic flux density, **B**. The force acts in a direction that is perpendicular both to the direction of particle motion and of the flux density and is given by the vector product $q\mathbf{B} \times \mathbf{v}$, where $\mathbf{v}$ is the particle velocity and q the charge.

loss ➤dissipation.

loss angle The angle by which the angle of ➤lead of the current is less than $90°$ when a capacitor or dielectric is subjected to sinusoidal alternating electric stress. It is mainly due to ➤electric hysteresis loss.

loss factor 1. The ratio of the average power dissipation in a line, circuit, or device to the power dissipated at peak load. **2.** The product of the ➤power factor and the relative ➤permittivity of a ➤dielectric. For a given alternating field the power factor is proportional to the heat generated in the material.

lossless line A hypothetical transmission line in which no attenuation occurs.

lossy Denoting an insulator that dissipates more energy than is considered normal for that class of material.

lossy line A type of ➤transmission line that is designed to produce a high degree of attenuation.

loudness The perceived position of a sound on a scale from soft to loud. Loudness varies primarily as the intensity of a sound is altered, but it also varies to a lesser extent when, for example, the spectral content or duration is changed.

loudspeaker An electroacoustic device that converts electrical energy into sound energy. It is the final unit of any sound reproducer or of the acoustic circuit of any broadcast receiver. Its action is the reverse of that of the ➤microphone but it is designed to handle far greater power, enabling the sound output to be audible over a large area.

Most types of loudspeaker use a coil and diaphragm arrangement in which a small coil is fixed at the centre of a diaphragm that is free to move in an annular gap. A strong magnetic field, produced by either a permanent magnet or an electromagnet, is applied across the gap. The audio signal is input to the coil as alternating current, causing it to move in the magnetic field as a result of ➤electromagnetic induction. The diaphragm is thus caused to vibrate at the same frequency as the alternating current and sound waves are produced by it. Any loudspeaker using this arrangement can be described as a *magnetic loudspeaker*.

For high efficiency (up to 50% of energy conversion) a *horn loudspeaker* is used. This type of speaker uses a small diaphragm at the mouth of a large exponential horn, and has marked directional properties. The horn speaker is impractical for most indoor uses as it is very large. A more convenient size is achieved using a large conical or ellipsoidal diaphragm, usually made of stiff paper, with the coil at its apex. The cone is supported round its edge by a metal frame, and the coil is maintained in position at the centre of the gap by thin flexible supports known as a *spider*. A large baffle, usually the cabinet housing the speaker, is used with this *cone loudspeaker* to prevent the direct passage of sound from front to back and thus improve the low-frequency response.

A simple cone loudspeaker has a uniform power output over a moderate range of frequencies but the output falls at the high- and low-frequency ends of the audible spectrum. For good reproduction at low frequencies a large cone is required in order to give a larger radiation resistance at these frequencies. At high frequencies the mass of the vibrating system sets an upper limit for good reproduction and a small cone is required to optimize the response at these frequencies. Two speakers are sometimes used together in order to overcome these requirements: a large cone for the low notes and a small one for the high notes. The large cone is designed to have a very low resonant frequency to improve its output and the strong resonant peaks, which would cause a boomy sound, are eliminated by using considerable damping (►damped). A frequency-selective network is required with this type of speaker in order to divide the spectrum between the two sound radiators.

Some speakers use an arrangement of multiple coils and cones designed to reduce the effective mass of the speaker at high frequencies. A well-designed speaker gives a uniform response between about 80 hertz and 10 000 hertz but its conversion efficiency is low. Cone loudspeakers are essentially omnidirectional but the use of a suitable cabinet can introduce directional properties if required.

Other methods of producing the sound vibrations are sometimes used. The *crystal loudspeaker* utilizes a ►piezoelectric crystal as the vibrating part. ►Magnetostriction is used in the *magnetostriction loudspeaker* and the vibrations of a magnetic ►armature produce the sound of a *magnetic-armature loudspeaker*. The action of electrostatic fields produces the mechanical movement in an *electrostatic loudspeaker*. Acoustic standing waves can be reduced by placing a loudspeaker in a special housing containing air chambers. Such an arrangement is termed a *labyrinth loudspeaker*.

A *loudspeaker microphone* is a dynamic loudspeaker that may also be used as a microphone. This arrangement is often used in an intercommunication (intercom) system enabling a single unit to be used for both speaking and listening. A manually operated switch connects the device to the appropriate circuit for the desired function. The most convenient arrangement is a push-button that returns automatically to the loudspeaker condition when pressure is removed from it.

loudspeaker microphone ►loudspeaker.

low earth orbit satellite (LEOS) A satellite that orbits the earth at a distance of a few hundred kilometres above the surface. The time of orbit is a few hours. When used for personal wireless communications or navigation purposes, a constellation of several tens of LEOS is required to provide continuous coverage over the globe, so that at least one satellite – or three or four for navigation – is in view all the time.

low electron velocity camera tube ►camera tube. ►►image orthicon; vidicon.

lower sideband ►carrier wave.

low frequency (LF) ►frequency band.

low-frequency compensation Compensation applied to an amplifier when it is used with low-frequency signals. The compensation is designed to avoid distortion due to signal attenuation and phase shift caused by the reactance of coupling capacitors.

low-frequency oscillator (LFO) An oscillator found in musical ►synthesizers operating at frequencies up to approximately 20 hertz. It is used to modulate the ►fundamental frequency or ►amplitude of the output sound, referred to as 'vibrato' or 'tremolo' respectively.

low-level injection ►injection.

low-level modulation A method of modulation whereby the modulation is produced at a stage in a system where the power level is low compared with the level of power that is output by the system.

low-level programming language ►programming language.

low logic level ►logic circuit.

low-loss line A type of ►transmission line that is designed to dissipate very little energy per unit length. The series resistance and shunt conductance of the line are therefore made low.

low-pass filter ►filter.

low-velocity scanning ►scanning.

LP *Abbrev. for* linear prediction.

LPC *Abbrev. for* linear predictive coding. ►linear prediction.

LPCVD *Abbrev. for* low pressure chemical vapour deposition. ►chemical vapour deposition.

LPI *Abbrev. for* low probability of intercept. In communication systems, an LPI system is one designed such that it cannot easily be detected by anybody other than the intended receiver.

LPPF *Abbrev. for* low probability of position fix. In communication or radar systems, an LPPF system is one in which, even if the presence of the signal is detected, the direction of the transmitter is difficult to pinpoint. ►Spread-spectrum signals can be used to achieve LPPF.

LSA mode *Short for* limited space-charge accumulation mode. ►transferred electron device.

LSB *Abbrev. for* least significant bit.

L-section ►two-port network.

LSI *Abbrev. for* large-scale integration. ►integrated circuit.

LTP *Abbrev. for* long-tailed pair.

luminance flicker ➤flicker.

luminance signal ➤colour television.

luminescence The emission of electromagnetic radiation from a substance due to a non-thermal process. The term is also used to describe the radiation itself, particularly when it falls within the visible spectrum. Luminescence occurs when atoms of the material are excited and then decay to their ➤ground state with the emission of radiant energy. If the luminescence ceases as soon as the source of excitation is removed, i.e. the persistence is less than about 10^{-8} second, it is termed *fluorescence*. If it persists for longer than about 10^{-8} second it is termed *phosphorescence*. A luminescent material is known as a *phosphor*.

The most common source of energy that results in luminescence is other electromagnetic radiation or electrons or other charged particles. *Stokes' law* states that the radiation emitted is usually of longer wavelength than that of the exciting radiation (although it may sometimes be shorter). Ultraviolet radiation therefore can produce visible light from a phosphor. The light produced has a characteristic colour for a particular fluorescent material: fluorescene, yellow-green; quinine sulphate, blue; chlorophyll, red.

Fluorescence is used for examining the spectrum in the ultraviolet region, in fluorescent lighting, and for display purposes, such as with the screen of a cathode-ray tube. Phosphorescence is used when longer persistence is required, such as with a long-persistence screen.

Thermoluminescence is an indirect effect of bombardment by ionizing radiation and is seen when the material is heated after subjection to radiation. The radiation releases electrons within the material and these are trapped at defects within the solid. These electrons are released on heating and the energy thus produced is emitted as visible radiation. Other energy sources that excite luminescence include friction (*triboluminescence*) and chemical reaction (*chemiluminescence*).

lumped parameter Any circuit parameter, such as inductance, capacitance, or resistance, that can be treated as a single parameter at a point in the circuit for the purposes of circuit analysis over a specified range of frequencies.

M

machine code The ➤binary code used to represent the ➤instruction set of a particular computer. The machine-code instruction set is defined by the designer of the computer and constrained by the architecture chosen.

macrocell (or **macro**) The definition of a design that occurs with a degree of regularity in a digital design, often in computer-aided design (CAD) tools. The macrocells can be simple gates, flip-flops, adders, CPUs, etc., and once defined can be called and placed in the required positions in a design.

MAGLEV *Short for* magnetic levitation, achieved by using an electromagnet to provide the magnetic field to support a load containing magnetic material, by repulsion of like magnetic poles. If the electromagnet is a linear structure of several electromagnets – a *linear motor* – then by switching each electromagnet on in turn the load can be transported by levitation, without friction. An example is the MAGLEV train at Birmingham International Airport, which transports passengers to and from the main hall to the flight departure and arrival terminal.

magnesium Symbol: Mg. A light metal, atomic number 12, used extensively for the construction of electronic components, most commonly as an alloy with aluminium.

magnet A body that possesses the property of ➤magnetism. The term is applied to those bodies that can produce an appreciable magnetic field external to themselves. Magnets are either temporary or permanent. ➤permanent magnet; electromagnet; ferromagnetism; magnetite.

magnetic-armature loudspeaker ➤loudspeaker.

magnetic balance A device that determines directly the force between two magnetic poles. A long magnet is balanced on a knife edge so that it takes up a horizontal position. One magnetic pole of a second long magnet is brought near to one end and the force (of attraction or repulsion) between the poles is balanced by the addition of weights or the action of a movable rider at the other end to restore the horizontal position. The magnets are made long in order to reduce interference by interaction between their second poles.

The magnetic balance can also be used to measure the strength of a magnetic field. In this type of balance the magnet that acts as the balance arm is replaced by a long conductor. A known current is passed through the conductor and the force exerted on one end by a pole of the magnet to be measured is balanced as above. The field due to the magnet may be calculated from the force between it and the known field due to the current. This type of balance can be calibrated to read magnetic field directly for a stated current and distance between the poles.

This method of measurement contains errors since the precise location of the magnetic poles is indeterminate. More accurate methods have been devised that measure the ➤magnetic moment or ➤magnetic flux density associated with permanent magnets and current-carrying conductors. ➤flip-coil; Cotton balance.

magnetic bias ➤magnetic recording.

magnetic blow-out ➤circuit-breaker.

magnetic circuit A completely closed path described by a given set of lines of ➤magnetic flux. The direction of the path at any point is that of the ➤magnetic flux density at that point.

magnetic constant *Syn. for* permeability of free space. ➤permeability.

magnetic contactor A ➤contactor that is operated by magnetic means, such as an alternating magnetic field.

magnetic controller ➤automatic control.

magnetic core *Syn. for* ferrite core. ➤core.

magnetic damping A method of ➤instrument damping in which the damper consists of a metal vane, connected to the pointer, that moves through a magnetic field. The induced currents in the vane are in such a direction as to oppose the motion (➤electromagnetic induction).

magnetic deflection ➤electromagnetic deflection.

magnetic dipole moment ➤magnetic moment (both definitions).

magnetic disk A computer storage medium usually in the form of a rigid circular aluminium plate with a metallic magnetic coating on both sides; this form is known as a *hard disk*. The flexible form, known as a *floppy disk*, has a magnetic coating of ferric oxide on both sides of a polyester base. Hard disks have a larger storage capacity than floppy disks; the latter provide a low-cost lightweight storage medium. In both forms data is stored in and retrieved from a set of concentric tracks in the magnetic coating. This is achieved by means of a *disk drive*, in which one or more disks may be rapidly rotated. *Read-write heads* in the disk drive are moved radially over each disk surface so that data may be written to (i.e. stored in) or read from the required track. Specific storage locations in the track may be accessed directly and in any order. Unlike floppy disks, hard disks are usually permanently mounted within the disk drive. ➤➤moving magnetic surface memory.

magnetic field The space surrounding a magnet or a current-carrying conductor and containing ➤magnetic flux. It may be represented by ➤lines of force whose direction at any point is the direction of the force exerted on a small coil (a search coil) placed in the field at that point. The direction of the force is normal to the magnetic flux density at that point. It is assumed that the dimensions of the coil are sufficiently small so as not to disturb the magnetic conditions.

magnetic field strength *Syn.* magnetizing force. Symbol: H; unit: ampere/metre. The strength of a magnetic field at a point in the direction of the line of force at that point. It is defined in a vacuum from the equation

$$B = \mu_0 H$$

where B is the ➤magnetic flux density and μ_0 is a constant, the ➤permeability of free space. ➤Ampere's law.

magnetic flux Symbol: Φ; unit: weber. The ➤flux through any area in the medium surrounding a magnet or current-carrying conductor, equal to the surface integral of the ➤magnetic flux density over the area. It is measured by the e.m.f. produced when a circuit linking the flux is removed from it. One weber of flux linking a circuit of one turn produces an e.m.f. of one volt in that circuit when the flux is reduced to zero.

magnetic flux density *Syn.* magnetic induction. Symbol: B; unit: tesla (weber/metre^2). The fundamental force vector in magnetism; the magnetic analogue of the electric field E. Both a magnet and a current-carrying coil exert forces on other coils or magnets. The magnetic flux density produced by such magnets or coils is a vector quantity and lines of flux can be drawn whose direction at any point is the direction of magnetic flux density. The value of B is given by the number of lines of flux per unit area and is expressed by the equation

$$dF = I(ds \times B)$$

where dF is the force exerted due to B on an element of length ds of wire carrying a current I. This defines the unit of magnetic flux density as that which exerts a force of one newton on a wire of length one metre carrying a current of one ampere.

magnetic focusing *Syn. for* electromagnetic focusing. ➤focusing.

magnetic head ➤magnetic recording.

magnetic hysteresis A phenomenon observed in ferromagnetic materials below the Curie point (➤ferromagnetism) where the ➤magnetization of the material varies nonlinearly with the ➤magnetic field strength and also lags behind it. The magnetic ➤susceptibility of such materials is large and positive, a large value of magnetization being produced for comparatively small fields. A characteristic plot of either magnetization, M, or ➤magnetic flux density, B, against magnetic field strength, H, demonstrates the hysteresis effect and is termed a *hysteresis loop* (see diagram). If an initially unmagnetized sample of iron is subjected to an increasing magnetic field the magnetization follows the curve shown by the dotted line OAS. This is known as the *magnetization curve*. If the specimen is then subjected to a complete magnetizing cycle with magnetic field varying symmetrically between +H and –H the curve shown by the solid line is followed.

The value of **B** at zero field is termed the *remanence*. The value of **H** at the point at which **B** (or **M**) falls to zero is known as the *coercive force*, and is the reverse field required to demagnetize the sample. The area enclosed by the hysteresis loop is proportional to the energy dissipated in each complete cycle when ferromagnetic material is subjected to an alternating magnetic field. This is known as *hysteresis loss*.

The effect of hysteresis in a magnetic core is to cause an increase in the effective resistance of the coil surrounding the core; the *hysteresis factor* is the increase in the effective resistance of a coil carrying a current of one ampere at a specified frequency.

The general form of the hysteresis curve is shown in the diagram. The area enclosed by the curve depends on the nature of the ferromagnetic sample: minimum area (and minimum coercive force) occur with soft iron, rising to a value some twenty times greater with tungsten steel. Any complete magnetizing cycle (say between the values of $H + h$ and $H - h$) will give rise to a hysteresis loop. ➤ferromagnetism.

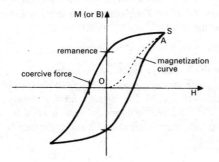

Magnetic hysteresis loop

magnetic induction ➤magnetic flux density.

magnetic intensity *Obsolete syn. for* magnetic field strength.

magnetic leakage The loss of magnetic flux from a magnetic circuit due to a portion of the total magnetic flux following a path that renders it ineffective for the desired function of the circuit. Magnetic leakage reduces the overall efficiency of the operation. In a transformer, for example, magnetic leakage occurs when some of the flux from the primary circuit does not link with the secondary circuit.

The *magnetic leakage coefficient,* σ, is defined by the ratio of the total magnetic flux to the effective (or useful) magnetic flux; i.e.

σ = 1 + leakage flux/useful flux

A typical value of σ is 1.2 in electrical machines.

magnetic lens ➤electromagnetic lens.

magnetic loudspeaker ➤loudspeaker.

magnetic memory A ➤memory device or medium, such as a magnetic disk, in which digital information is represented by the direction of magnetization of magnetic material. ➤moving magnetic surface memory.

magnetic microphone ➤microphone.

magnetic modulation ➤transducer.

magnetic moment 1. *Syn.* magnetic dipole moment. Symbol: m; unit: ampere metre^2. A measure of the strength of a ➤magnet. When a magnet is placed in a homogeneous magnetic flux of magnetic flux density B it experiences a torque T such that

$$T = m \times B$$

where m is the magnetic moment. If the magnet is a small coil of area dA carrying a current I, then the magnetic moment m is equal to IdA. **2.** *Syn.* (*obsolete*) magnetic dipole moment. Symbol: p_m; unit: weber metre. The product of magnetic moment m and the ➤permeability of free space, μ_0:

$$p_m = \mu_0 m$$

Historically magnets and small circulating currents were considered to be ➤dipoles consisting of two equal and opposite magnetic poles, analogous to electric charges, separated by a distance r (the magnetic length). The magnetic dipole moment p_m was the product of the magnetic pole strength and the magnetic length; it was determined by the torque, T, experienced when the magnet was placed in a magnetic field H. Thus p_m was determined by the relation:

$$T = p_m \times H$$

Modern usage strongly favours the first definition, the second definition being very rarely used. Small magnets, however, are frequently referred to as dipoles, particularly on an atomic scale, and the magnetic moment m termed the magnetic dipole moment. In modern texts these terms are defined by the first definition above. The term 'magnetic moment' has been used wherever possible throughout this dictionary as defined in **1.** above.

magnetic monopole A hypothetical magnetic particle with a single magnetic charge of either north or south. Such a particle would be analogous to the electrical particles the electron and proton. ➤Maxwell's equations would prove completely symmetrical if the existence of such particles were to be proved. Magnetic monopoles were postulated on conservation and symmetry principles and are thought to be more massive than nucleons. Their existence is not barred by quantum theory nor classical electromagnetic theory and it has been suggested that they occur in extremely high energy cosmic rays. No proof of their existence or otherwise has yet been found despite intensive searches for them.

magnetic pole A region of concentrated magnetism in a magnet analogous to an electrostatic point charge. Historically a magnet was considered as being formed from two magnetic poles of opposite types (north and south) located near its ends. Lines of magnetic force converge on or diverge from the magnetic poles. Use of the concept of magnetic poles allowed the theory of magnetostatics to be developed along similar lines to that of electrostatics, by applying the inverse square law of forces to these imaginary poles. This approach to magnetostatics however requires the use of the magnetic monopole for its development; also the precise location of the magnetic poles is indeterminate. Modern practice favours the use of ►magnetic moment instead.

magnetic potential *Obsolete syn. for* magnetomotive force.

magnetic recording A method of recording electrical signals on a magnetic medium. One application is the recording of sound on *magnetic tape* (*tape recording*). In magnetic sound recording the magnetic tape is moved at a uniform speed past the poles of an electromagnet and is longitudinally magnetized. Variations in the audiofrequency current supplying the electromagnet produce corresponding variations in the magnetization. During reproduction the process is reversed: the tape is fed past an electromagnet and the variations of magnetization induce currents in the coils corresponding to the original magnetizing currents. The recording medium is usually made from finely divided ferrous oxide particles deposited on a plastic (cellulose acetate) tape. Multitrack tapes are available containing two or more separate recording tracks.

The electromagnet used to record, reproduce, or erase the signal on the tape is called a *head* (or *magnetic head*). It is possible to use a single *read-write head* to perform each separate function but commercial tape recorders usually use separate heads for recording, reproducing, and erasing. A typical head consists of soft iron pole pieces wound with wire coils. The distance between the pole pieces is the *gap length* and a good recorder may have a gap length as small as 0.6 mm allowing a sharper record and thus more faithful reproduction. The magnetic tape completes the magnetic circuit (see diagram) and the magnetization produced represents the flux pattern in the gap at the moment the tape leaves the gap. During recording *magnetic bias* is applied to the *recording head*. This is an alternating current of frequency between 60 and 100 kilohertz superimposed on the audiofrequency signal. The frequency response, distortion, and signal-to-noise ratio characteristics of the system are improved by biasing in this way. The recording is erased by causing the tape to move past the *erasing head* to which a

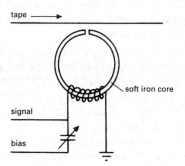

Magnetic tape recording

large direct current is applied. This produces uniform magnetization of the magnetic material.

Magnetic recording of information that contains both pictures and sound is achieved using ►videotape. Videotape is used for storage and broadcasting. In computer systems, magnetic recording of data is made on media such as ►magnetic disk and specially fabricated ►magnetic tape (►►moving magnetic surface memory).

magnetic residual loss ►ferromagnetism.

magnetic resistance ►reluctance.

magnetic saturation ►ferromagnetism.

magnetic screening The use of a screen of high ►permeability magnetic material in order to protect electric circuits, devices, or other apparatus from the effects of magnetic fields. Magnetic screening is often used to surround sensitive alternating-current measuring instruments and thus shield them from stray external magnetic fields.

magnetic shunt A means of varying the useful ►magnetic flux of a magnet in an electrical measuring instrument. It consists of a piece of magnetic material mounted near the magnet and adjustable in position relative to the magnet. It is often used to extend the range of the measuring instrument.

magnetic susceptibility ►susceptibility.

magnetic tape 1. ►magnetic recording; videotape. **2.** A computer storage medium consisting of specially fabricated magnetic tape. The tape is wound on a reel. Data is stored in and retrieved from a set of tracks in the magnetic coating, usually running along the tape. This is achieved by means of a device, known as a *tape unit*, which moves the tape at high speed over one or more *read-write heads*. Data may then be written to (i.e. stored in) or read from the required track. ►►DAT; moving magnetic surface memory.

magnetic transition temperature *Syn. for* Curie point. ►ferromagnetism.

magnetic tuning Tuning of a very high frequency (microwave) oscillator by means of a ferrite rod in the cavity resonator of the oscillator. The magnetization of the rod is determined by an external steady magnetic flux density, and is varied to provide a different frequency in the cavity by altering the flux density.

magnetism The phenomena associated with regions containing magnetic flux. Magnetic properties were first noticed in the naturally occurring oxide of iron, magnetite. Ampère discovered that a small coil carrying a current behaves like a magnet; he suggested that the origin of all magnetism lay in small circulating currents associated with each atom. Ampère's theory, which gave a natural explanation of the fact that no isolated magnetic pole had ever been observed, is essentially similar to modern atomic theory: his elementary current circuits (Amperean currents) are the motions of the negatively charged electrons in closed orbits around the positively charged atomic nucleus.

All materials exhibit magnetic properties, the nature of those properties depending on the distribution of electrons in the outer orbits of the atoms. ►Diamagnetism is a weak effect, common to all materials and resulting from the orbital motion of the atomic electrons. ►Paramagnetism occurs in certain materials that have a permanent molecular ►magnetic moment due to electron ►spin. It is a stronger effect than diamagnetism, opposed to it, and masks it in paramagnetic materials. Some paramagnetic materials, such as iron, also display ►ferromagnetism (►ferrimagnetism, antiferromagnetism) at temperatures below the Curie point. Ferromagnetic materials can produce a substantial magnetic flux density and some are suitable for use as ►permanent magnets. A magnetic field can also be produced by a flowing electric current (►electromagnet).

magnetization Symbol: M; unit: ampere per metre. A measure of the magnetic polarization that occurs when a material is placed in a magnetic flux. It is defined as the magnetic moment per unit volume and is the product of magnetic field strength, H, and magnetic ►susceptibility, χ_m:

$$M = H\chi_m$$

$$M = B/\mu_0 - H$$

where B is the magnetic flux density and μ_0 is the ►permeability of free space.

magnetization curve ►magnetic hysteresis; ferromagnetism.

magnetize To induce a magnetic flux density in a material; to cause a material to exhibit magnetic properties.

magnetizing force ►magnetic field strength.

magneto An alternating-current electrical ►generator, particularly of the ►synchronous a.c. generator type coupled to an induction coil, that uses a ►permanent magnet to provide the magnetic flux rather than an electromagnet.

magnetomotive force (m.m.f.) Symbol: F_m; unit: ampere; ampere-turn. The line integral of ►magnetic flux density, B, for a closed path, divided by the magnetic constant μ_0:

$$F_m = \mu_0^{-1} \oint B.\mathrm{d}l$$

It is equal to the total conduction current linked, during one traverse of the closed path. If the path encloses a current I the value of F_m increases by I for every traverse, i.e. F_m is multivalued unlike electromotive force, which is single-valued.

magneto-optical effect ►Kerr effects.

magnetoresistance A change, usually an increase, in the resistance of a conductor or semiconductor when placed in a magnetic field. This is due to a combination of the energy distribution of the electrons in the material in the magnetic field, and the scattering mechanisms that occur.

In some materials, held at very low temperatures, a large decrease in resistance is produced on applying a magnetic field. This effect is known as *colossal magneto-*

resistance; it is observed, for example, in a group of materials known as *perovskites*, all with similar crystal structures.

magnetostriction Mechanical deformation of a ferromagnetic material when subjected to a magnetic field. The effect arises because of internal stresses in the material due to the anisotropy energy required to magnetize it in certain directions relative to the crystal axes (►ferromagnetism). Conversely, when subjected to mechanical stress a change in the magnetization of the material is observed.

Joule magnetostriction (or *positive magnetostriction*) is an increase in the length of a rod or tube of ferromagnetic material when an axial magnetic flux is applied to it. *Negative magnetostriction* occurs in materials, such as nickel, that suffer a decrease in length with an increase in magnetic flux density. Magnetostriction also affects deformed ferromagnetic samples. A bent bar tends to straighten under the influence of a magnetic flux applied along its length (*Guillemin effect*); a twisted bar also tends to straighten under similar conditions (*Wiedemann effect*).

High-frequency alternating fields set up longitudinal vibrations in the sample, which thus acts as a source of sound waves; if the frequency of these vibrations corresponds to the natural frequency of the sample large values of amplitude occur. Magnetostrictive vibrations have many applications. At ultrasonic frequencies they form a source of ultrasonic energy that can be used commercially, for example for ultrasonic cleaning or for breaking the oxide film on aluminium for soldering. Magnetostriction oscillators utilize the magnetostrictive effect of alternating currents to produce frequency-controlled oscillations at frequencies in the range 25 000 hertz downwards.

Magnetostriction is also used to provide the vibrating part of a ►loudspeaker; conversely audio vibrations applied to a magnetostrictive rod produce corresponding flux changes, as in a magnetostriction ►microphone.

magnetron *Syn.* magnetron oscillator. A crossed-field ►microwave tube that produces radiofrequency (r.f.) oscillations in the microwave region. An early magnetron was used as a rectifier but all modern magnetrons are designed as oscillators.

The basic magnetron consists of a central cylindrical cathode surrounded by a cylindrical anode containing several ►cavity resonators (Fig. *a*). A steady electrostatic field is applied between the anode and cathode. A steady magnetic flux density is applied parallel to the cylindrical axis and orthogonal to the electrostatic field. The mag-

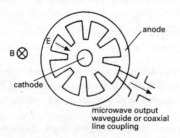

a Resonant-cavity **magnetron**

netic flux is usually provided by a permanent magnet or sometimes by an electromagnet. Electrons, emitted from the cathode, move under the influence of these two
fields. The interaction of the electrons with the gaps or the resonant cavities of the anode
produces radiofrequency oscillations. These oscillations are output through a coupled
waveguide or coaxial line.

As the electrons leave the cathode of a magnetron they are accelerated towards the
anode by the electrostatic field. In the absence of a magnetic field, described by a magnetic flux density, they travel radially towards the anode. When a magnetic field is applied it exerts a Lorentz force on them perpendicular to their direction of motion and
proportional to the velocity; this causes them to follow a cycloidal path. The distance
that an electron can travel towards the anode is a function of the anode voltage, V, and
the magnetic flux density $\boldsymbol{B}$.

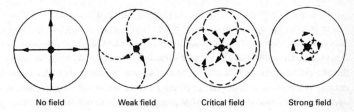

| No field | Weak field | Critical field | Strong field |

b Effect of magnetic field on electrons in **magnetron**

For a given value of anode voltage, the *critical field* is that value of $\boldsymbol{B}$ at which an
electron just fails to reach the anode and returns to the cathode with zero kinetic energy. The *critical voltage* is the maximum anode voltage, in the presence of a fixed magnetic flux density, at which an electron just fails to reach the anode. The effect of an
increasing magnetic field with fixed anode voltage is shown in Fig. b. Under the
strong field condition, when the value of $\boldsymbol{B}$ exceeds the critical field, an electron gains
kinetic energy; it returns to the cathode with a nonzero velocity following a cycloidal
path of relatively small radius. The effect of the strong field is thus to produce a narrow *sheath* of electrons rotating about the cathode with an angular velocity ω. If the
radii of the anode and cathode are b and a respectively, then provided that $(b - a)$ is
small compared to both b and a it can be shown that

$$\omega = 2V/[\boldsymbol{B}(b^2 - a^2)]$$

The rotating sheath of electrons interacts with the resonant cavities or gaps in the
anode structure to produce r.f. oscillations in them; the r.f. oscillations in turn interact
with the electrons in a complex manner: the electrons are either accelerated by the r.f.
field and turned back to the cathode or are decelerated by it and travel to the anode giving up energy to the r.f. field as they do so (Fig. c). On average the net power gained
by the r.f. fields when an electron loses kinetic energy is greater than that required to
return one to the cathode; r.f. oscillations are therefore sustained by the system. The
closed nature of the circuit effectively supplies the positive feedback required for the
oscillations to occur.

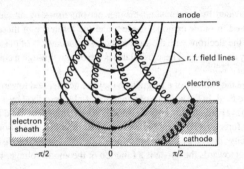

c Electron paths in the radiofrequency field

There are various possible modes of operation depending on the geometrical structure of the anode, the magnitudes of the fields, and the phase differences of the r.f. fields between successive cavities. For a particular design of anode the magnitudes of V and B must be adjusted so that the angular velocity of the electrons is synchronized with the alternation of the r.f. fields in the cavities so as to produce the optimum transfer of energy. When properly adjusted, an efficiency of about 70% is possible with *π-mode operation*; this is the simplest and most efficient mode of operation in which the phase difference between successive cavities is π. The electrons are retarded by several successive cavities when the proper phase relationship is maintained and travel towards the anode in 'spokes'.

Suitable design of the anode of a magnetron allows either standing waves or travelling waves to be output from the device. A *travelling-wave magnetron* has several possible modes of oscillation. The performance of a *standing-wave magnetron* is usually shown on a ►Rieke diagram.

Excessive heating of the anode by incident electrons is avoided by constructing it from a material, such as copper, that has good thermal conductivity. Electrons returning to the cathode produce *back heating*; this reduces the heater current required when the tube is running and also stimulates ►secondary emission of electrons, which provides a significant proportion of the total cathode emission.

Large power outputs are possible by running the tube in short pulses (*pulsed magnetrons*) rather than continuously; this gives an improvement of output power of up to 1000. A typical medium high power magnetron, pulsed for one microsecond at a repetition rate of 1000 hertz, can produce 0.1 mm wavelength waves with a power output during the pulse of 750 kilowatts. It requires an anode voltage of 31 kilovolts and magnetic flux density of 0.28 tesla.

Most magnetrons are *fixed-frequency magnetrons* but *tunable magnetrons* have been produced in which a variation of 10–20% in the frequency is achieved by means of plungers moved into the resonators from one end.

magnetron effect ►thermionic valve.

magnitude response Variation of the magnitude of a parameter with frequency; the parameter could, for example, be gain of an amplifier or filter circuit. Magnitude response is often expressed in decibels.

main lobe *Syn.* major lobe. ➤antenna pattern.

mains The source of domestic electrical power distributed nationally. The *mains frequency* is the frequency at which the electrical power is supplied. This is 50 hertz in Europe and usually 60 hertz in the US.

mains hum ➤radio noise; noise.

maintaining voltage ➤gas-discharge tube.

majority carrier The type of charge ➤carrier in an extrinsic ➤semiconductor that constitutes more than half of the total charge carrier concentration.

major lobe ➤antenna pattern.

make 1. To close a circuit by means of a switch, ➤circuit-breaker, or similar device. **2.** The maximum gap between the contacts of such a device for the circuit to be closed. ➤break.

make-and-break A type of switch that is automatically activated by the operation of the circuit in which it is incorporated, and that repetitively makes and breaks the circuit. Applications of make-and-break include electric bells and buzzers.

Manchester code The name given to biphase pulse code modulation. ➤pulse modulation. ➤digital codes.

manganese Symbol: Mn. A metal, atomic number 25, that is used in alloys such as ➤manganin and ➤Heusler alloys. The element is also used in some forms of primary ➤battery.

manganin An alloy that consists of 70–86% copper, 15–25% manganese, and 2–5% nickel and has a high coefficient of electrical ➤resistivity, low ➤temperature coefficient of resistance, and low contact potential (➤contact). It is used in the manufacture of precise wire-wound ➤resistors.

man-made noise ➤noise.

MAR *Abbrev. for* memory address register.

Marconi antenna Originally the simple vertical wire ➤antenna, earthed at its lower end, used by Marconi. The term is now applied to vertical antennas one quarter wavelength long. Usually the antenna is connected to earth by a series reactance, which may be either an inductance or a capacitance, of such a value that together with the antenna it is electrically equivalent to one quarter wavelength. Usually the series reactance is variable, allowing the antenna to be tuned to any resonant frequency within a designated range.

mark In analogue terms, the high state in pulse or square-wave signals, the low state being referred to as *space*. In digital terms, the high and low levels are commonly referred to as 1 and 0, respectively.

mark-space ratio The ratio in a pulse or square-wave signal of the pulse duration to the time between successive pulses. In a perfect square wave the mark-space ratio is unity. ➤mark.

M-ary FSK ➤frequency modulation.

M-ary PSK ➤phase modulation.

maser Acronym from *m*icrowave *a*mplification by *s*timulated *e*mission of *r*adiation. A source of intense coherent monochromatic radiation in the ➤microwave region of the electromagnetic spectrum. The maser can be used as a microwave amplifier or oscillator. The device operates on similar principles to the ➤laser, i.e. by means of population inversion and stimulated emission, but the transitions between energy levels are in a lower energy range.

mask 1. A device used to shield selected areas of a semiconductor ➤wafer during the manufacture of semiconductor components and integrated circuits. During the ➤photolithography process, a set of photographic masks is required to define the openings in the oxide layer through which the various ➤diffusions are made, the windows through which the metal contacts are formed, and the pattern in which the desired metal interconnections are formed. The photographic masks, or *photomasks*, are either emulsion on glass or an etched thin film of chromium or iron oxide on glass; they are produced by photographic reduction from large-scale layouts. **2.** A device made from metal foil and used during the manufacture of ➤thin-film circuits in order to define the pattern of material deposited as a thin film by vacuum evaporation on to the substrate.

mask register A ➤register in a computer whose bits specify the data required in an operation, i.e. they *mask* an operation. For example, in an ➤associative memory the bits mask the fields that will not participate in a match, in an interrupt controller the bits mask ➤interrupts, and in a DMA (➤direct memory access) controller the bits mask incoming requests for service.

Mason's rules *Syn.* Mason's nontouching loop rules. A technique for simplifying the analysis of ➤signal flowgraph descriptions of electrical ➤networks. Given a source node **a** and a destination node **b**, then the total ➤transfer function between the two nodes, accounting for all paths, is given by Mason's rules to be

$$\mathbf{b}/\mathbf{a} = (\Sigma_K T_K \Delta_K)/\Delta$$

where T_K is the path gain or total ➤transmittance of the Kth forward path from **a** to **b**; $\Delta = (1 - (\text{sum of all individual loop gains}) + (\text{sum of products of loop gains for all possible combinations of two nontouching loops}) - (\text{sum of products of loop gains for all possible combinations of three nontouching loops}) + \ldots)$;
Δ_K = sum of all terms in Δ not touching the Kth path.

A *path* is a continuous succession of individual branches passing any node in the network only once. A *loop* is a path that starts and ends on the same node in the network. Nontouching loops do not share any nodes.

master ►recording of sound.

master frequency meter ►integrating frequency meter.

master oscillator An ►oscillator of extremely high inherent frequency stability that is used to drive a power ►amplifier in order to supply a substantial power output to the load. This arrangement is used where substantial output power is required with a high degree of frequency stability, as in the generation of a ►carrier wave, since the frequency of an oscillator is partially dependent on the load supplied by it. The master oscillator 'sees' a constant load from the ►buffer between it and the multistage power amplifier and hence the frequency stability is maintained.

master trigger ►radar.

matched load ►matched termination.

matched termination A termination to a ►network or ►transmission line at which no reflected waves are produced. A load that absorbs all the power incident from a transmission line and forms a matched termination is known as a *matched load*.

matched waveguide ►waveguide.

matching ►impedance matching; load matching.

Mathiessen's rule In simple terms, a means of combining the carrier scattering rates due to the different scattering mechanisms that occur in a ►semiconductor; the individual scattering rates are added directly. This is the same as adding the reciprocals of the relaxation times (τ) associated with each scattering process:

$$1/\tau = \Sigma_{\text{scattering processes}} \ 1/\tau_i$$

The overall mobility (μ) is determined in a similar way:

$$1/\mu = \Sigma_{\text{scattering processes}} \ 1/\mu_i$$

►►drift mobility; Hall mobility.

maximally flat A term used to describe the pass-band magnitude characteristic of a ►filter that is as flat as possible near zero value of angular frequency; this occurs, for example, with the Butterworth ►filter.

maximum power theorem If a variable load is to be matched to a given power source in order to achieve the maximum possible power dissipation in the load, the resistance of the load must be made equal to the internal resistance of the source. The *available power* then obtainable is $V_o^2/4R_i$, where V_o is the open-circuit electromotive force of the source and R_i the internal resistance.

The converse problem, that of matching a power source to a given load in order to achieve maximum power dissipated in the load, is not solved by this theorem. In this case the power source with the lowest internal resistance gives the maximum power.

This theorem can be modified to apply to alternating-current linear networks. The impedances Z_S and Z_L of the voltage generator and load, respectively, contain an imaginary term:

$$Z_S = R_S + jX_S$$
$$Z_L = R_L + jX_L$$

For maximum power dissipation in the load it is necessary to satisfy the conditions given by

$$X_S + X_L = 0$$
$$R_S = R_L$$

When this applies the circuit is said to be *conjugate matched*.

maxterm In digital design, a sum (logical OR) that includes every ►literal in a ►truth table. The complement of any maxterm is a ►minterm.

maxwell Symbol: Mx. The unit of magnetic flux in the obsolete CGS electromagnetic system of units. One maxwell equals 10^{-8} weber.

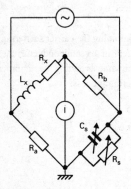

Maxwell bridge

Maxwell bridge A four-arm ►bridge for measuring inductance in terms of a capacitance and resistances (see diagram). At balance, as indicated by a null response on the instrument, I,

$$R_sR_x = R_bR_a$$
$$L_x = R_bR_aC_s$$

Maxwell's equations A set of classical equations relating the vector quantities applying at any point in a varying electric or magnetic field. The four basic equations are:

$$\text{curl } H = \partial D/\partial t + j$$

$$\text{div } B = 0$$

$$\text{curl } E = -\partial B/\partial t$$

$$\text{div } D = \rho$$

H is the magnetic field strength, D the electric displacement, t is time, j is the current density, B the magnetic flux density, E the electric field strength, and ρ the volume charge density.

From these equations Maxwell deduced that each field vector obeys a wave equation; he also showed that in free space, where $j = 0$ and $\rho = 0$, the solutions represent a transverse wave travelling through space with the speed of light, c. These waves are known as electromagnetic waves (►electromagnetic radiation). Further work showed that certain properties of these waves, i.e. reflection, refraction, and diffraction, are identical to the properties of light waves and that light waves are a form of electromagnetic radiation.

Maxwell's theory deals only with macroscopic phenomena and does not offer an explanation of phenomena arising from interactions on an atomic scale, such as dispersion and the photoelectric effect. On an atomic scale it has been found necessary to introduce the quantum mechanical theory of electromagnetic radiation.

From Maxwell's equations it is possible to deduce the wave velocity in a medium as

$$v = 1/\sqrt{(\mu\varepsilon)}$$

where ε is the ►permittivity of the medium and μ is its ►permeability, $\mu = B/H$. In a vacuum the wave velocity is given by

$$c = 1/\sqrt{(\mu_0\varepsilon_0)}$$

Thus in a nondispersive medium of refractive index n, where $n = c/v$,

$$n^2 = \mu_r\varepsilon_r$$

or in a nonferromagnetic material where $\mu_r \approx 1$,

$$n^2 = \varepsilon_r$$

where ε_r is the relative permittivity of the medium and μ_r the relative permeability. This is known as *Maxwell's formula*. In a dispersive medium the above formula applies provided that all measurements are carried out at the same frequency.

Maxwell's formula ►Maxwell's equations.

Maxwell's rule In an electric circuit linked by a magnetic flux each part of the circuit experiences a force causing it to tend to move in such a direction as to enclose the maximum possible magnetic flux.

MBE *Abbrev. for* molecular beam epitaxy.

MBR *Abbrev. for* memory buffer register.

MCPC *Abbrev. for* multichannel per carrier. In multiple-access systems (➤digital communications), the combination of a number of channels onto a single ➤wideband channel. An example is the multiplexing of a number of telephone channels onto a single microwave-frequency signal that is transmitted to a satellite for relay across the Atlantic.

Mealy circuit ➤finite-state machine.

mean current density ➤current density.

mean life The mean time to ➤failure of a device, component, or any part or subsystem that can be separately tested. For any particular item the mean life will be based on the results of ➤life tests and one or more of the following will be quoted.

The *observed mean life* is the mean value of the observed times to failure of all specimens in a sample of items under stated stress conditions. The criteria for what constitutes failure should be stated.

The *assessed mean life* is the mean life of an item determined as a limiting value of the confidence interval with a stated probability level based on the same data as the observed mean life of nominally identical items. The following conditions apply: the source of the data should be stated; results may be combined only when all conditions are similar; it should be stated whether one- or two-sided intervals are being used; the lower limiting value is usually used for mean life statements and the assumed underlying distribution should be stated.

The *extrapolated mean life* is the extension of the observed or assessed mean life by a defined extrapolation or interpolation for stress conditions different from those applying to the conditions of the assessed mean life.

The elapsed time at which a stated proportion (q per cent) of a sample or population of items has failed is the q-*percentile life*. This is quoted as the observed percentile life, assessed percentile life, and extrapolated percentile life under similar conditions to those quoted for mean life.

mean time between failures (MTBF) A measurement of system reliability for a fault-tolerant system. It is the average operating time between the start of normal operation and the system's first electronic or mechanical failure.

mean time to failure (MTTF) A measurement of system reliability for a fault-tolerant system when only one failure is possible, such as a component failure.

measurand ➤transducer.

medium frequency (MF) ➤frequency band.

medium-scale integration (MSI) ➤integrated circuit.

medium wave A radiowave that has a wavelength in the range 0.1 to 1 km, i.e. of frequency between 3000 and 300 kilohertz. ➤frequency band.

mega- Symbol: M. **1.** A prefix to a unit, denoting a multiple of 10^6 (i.e. 1 000 000) of that unit: one megavolt equals 10^6 volts. **2.** A prefix used in computing to denote a multiple of 2^{20} (i.e. 1 048 576): one megabyte equals 2^{20} bytes.

megaflops (MFLOPS) A measurement of the speed of a computer, usually applied to computers used in science rather than general-purpose computers, that focuses on floating-point operations rather than instruction executions. ('Flops' is an abbreviation of floating-point operations per second.) ➤mega-; floating-point representation.

megaphone A portable device used to amplify and direct sound. It contains a microphone, amplifier, and loudspeaker that is directional by virtue of a conical horn, together with a battery power supply.

Megger *Trademark* A portable insulation tester calibrated directly in megohms.

Meissner effects ➤superconductivity.

mekometer ➤Kerr effects.

mel A subjective unit of the relative ➤pitch of ➤sine waves. It is defined with respect to a 1 kHz sine wave at 40 dB above the listener's threshold, which is a pitch of 1000 mels. Pitches judged by a listener to be double or half this would be at 2000 and 500 mels respectively.

mel frequency cepstral coefficients ➤MFCC.

memory *Syn.* store. Any device or physical medium associated with a ➤computer and used to store information for subsequent retrieval. The information may, for example, be computer ➤programs or the data on which programs operate. The information is stored in digital form as sequences of ➤bits. The location of each item of information (usually in the form of a ➤word or ➤byte) can be identified by a unique ➤address, which allows a particular item to be stored (or *written*) and retrieved (or *read*). The time taken to retrieve an item of information from memory is known as the *access time*. The *memory capacity* is the total amount of information, usually in terms of the number of bits or bytes, that can be stored in any given memory, or in a computer system as a whole.

A computer system contains several types of memory that differ markedly in access time and capacity, and also in the amount of information that can be read or written on a given occasion and the cost of storing a given amount of information. For efficient and economical use of computer memory, the various types are organized into a hierarchy according to performance and cost. The highest performance and in general most expensive type is at the top level of the hierarchy, and is under the direct control of the ➤central processing unit (CPU). This memory is ➤RAM (random-access memory), composed of solid-state electronic circuitry with access times of tens of nanoseconds. Programs and data are stored here during execution of the program, part of the memory being used to store intermediate or partial results; the stored information can be readily altered. Various standard programs and sets of commonly used data are also stored in RAM in both read-only (➤ROM) and read-write sections of memory. A small-capacity solid-state ➤cache memory with extremely short access time is often inserted between the CPU and the main memory, and then the cache forms the top of the hierarchy.

Backing store is below solid-state memory in the hierarchy. It is ►nonvolatile memory on which information is held for reference but not for direct execution. Permanently connected (online) backing store is usually in the form of ►magnetic disk memory, and the information is transferred to and from the main memory by means of a disk drive. The capacity of disk memory is very much larger than solid-state memory and it is much less expensive, but the access time is reckoned in milliseconds. Information is also held offline on, for example, ►floppy disks, ►CD-ROM, or ►magnetic tape, and these storage devices are at the lowest levels of the hierarchy.

memory access time The average time the CPU of a computer requires to access memory, usually quoted in nanoseconds.

memory address register (MAR) The programmer-invisible ►register in which a computer holds the ►address of a reference during a memory access. It is not part of the register set.

memory bandwidth The number of bytes per second that a memory system can send to the processor, usually measured in megabytes per second.

memory buffer register (MBR) A programmer-invisible buffer in which a computer holds data for a store operation and into which it receives data during a load operation. It is not part of the register set.

memory capacity *Syn.* storage capacity. ►memory.

mercury-vapour lamp An arc-discharge tube used as a lamp. The gas in the tube is mercury vapour and the arc is struck between mercury electrodes. The light of the mercury arc is very rich in ultraviolet radiation and the tube can be used as a ►fluorescent lamp by coating the inside of the glass tube with a fluorescent powder, which converts the ultraviolet radiation to visible light (►luminescence).

merged CMOS/bipolar ►BiCMOS.

mesa A plateau of electrically active material on the surface of a semiconductor ►wafer, formed by etching away the electrically active material surrounding the location where a mesa is to be formed. Formation of mesas is a simple way of providing isolation between portions of a wafer.

mesa transistor A bipolar junction transistor or field-effect transistor in which the active region of the device is constructed as a ►mesa.

MESFET *Abbrev. for* metal-semiconductor field-effect transistor. *Syn.* Schottky-gate field-effect transistor. A type of junction ►field-effect transistor that has a ►Schottky barrier as the gate electrode rather than a semiconductor junction. The current-voltage characteristic is similar to that of a junction FET. The Schottky barrier gate electrode has the advantage that it can be made at much lower temperatures than are required to form a p-n junction.

mesh ➤network.

mesh contour ➤network.

mesh current ➤network.

mesh voltage *Syn. for* hexagon voltage. ➤voltage between lines.

metal Any of a class of chemical elements, including copper and iron, that are usually solids and are good ➤conductors of electricity and heat. Alloys, such as brass or steel, containing one or more metals are also considered as metals. The atoms in a metal or alloy crystal are held together by a covalent bond known as *metallic bond*: a regular lattice of positive ions held together by a cloud of ➤free electrons that move through the lattice. ➤➤energy bands.

metal-ceramic ➤cermet.

metal film resistor A ➤film resistor that uses a metal as the resistive element.

metallization ➤metallizing.

metallized film capacitors Capacitors that are constructed using a plastic film for the dielectric, which is coated in a metallic film to provide the parallel plate electrodes. The film is then usually tightly wound to minimize the volume.

metallized paper capacitors Capacitors that are constructed using a paper impregnated with a salt for the dielectric, which is coated in a metallic film to provide the parallel plate electrodes. The film is then usually tightly wound to minimize the volume. Such capacitors are often used for suppression of high-voltage burst interference, providing a self-healing breakdown phenomenon to prevent the transmission of the high-voltage interference.

metallizing Depositing thin films of metal (originally silver) on a glass, semiconductor, or other substrate in order to produce an electrically conductive layer. The metal is then etched into the required *metallization* pattern using a specially designed ➤mask. The technique is widely used in solid-state electronics for the formation of interconnections on ➤integrated circuits or ➤thin-film circuits and to form ➤bonding pads on integrated circuits or discrete components.

There are several processes used for metallizing. The chosen method depends on the substrate and the metal to be deposited. Techniques include ➤vacuum evaporation, ➤sputtering, ➤electroplating, and ➤chemical vapour deposition (CVD), plus variations such as ➤MOCVD. ➤➤multilevel metallization.

metal-semiconductor contact An intimate contact between a deposited metal and the underlying ➤semiconductor, used to make electrical contact between the solid-state electronic device and the external circuit. The metal-semiconductor contact can be either low-resistance, often termed an ➤ohmic contact, or a rectifying contact known as a ➤Schottky barrier. The difference is in principle due to the relative energy difference between the ➤Fermi levels in the metal and semiconductor. For an ohmic contact the Fermi levels should be in alignment, or their positions should be such that an accumulation of ➤majority carriers is created at the surface of the semiconductor; this is shown in Fig. *a* for a metal-to-➤n-type semiconductor. For a Schottky-barrier contact the Fermi levels are not in alignment, causing a ➤depletion of charge carriers at the surface of the semiconductor, shown in Fig. *b*, again for a metal-to-n-type semiconductor. This produces a potential barrier that the carriers must overcome in order for current to flow. The current-voltage relationship for a Schottky barrier is therefore very similar to that of a ➤p-n junction, except that only majority carriers are considered.

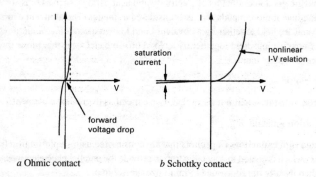

a Ohmic contact   *b* Schottky contact

Metal-semiconductor contact

    Many practical semiconductors, such as silicon or gallium arsenide, possess *defect* states at the surface of the crystal: these can be thought of as unsatisfied interatomic bonds. These surface states readily fill with electrons from the conduction band of the semiconductor, and cause a depletion region to form at the surface in n-type semiconductors, even in the absence of a metal. Thus there already exists a potential barrier. The deposition of a metal produces only a small change to this barrier height. Obtaining a rectifying or Schottky contact is therefore reasonably straightforward. To produce an ohmic contact, the surface region of the semiconductor must be very heavily ➤doped so that the resulting depletion region is narrow enough for conduction by ➤field emission, or *tunnelling*, to take place. The resulting current-voltage relation is not linear – i.e. not ohmic – but produces only a small voltage drop. The heavy doping is generally provided by the contact ➤metallization: this is often an alloy, a component of which is a ➤dopant for the underlying semiconductor. During the heat treatment that is part of the contact formation, the dopants ➤diffuse into the surface of the semiconductor to form the highly doped region for the ohmic contact.

metal-semiconductor diode *Syn. for* Schottky diode. ➤metal-semiconductor contact.

meter 1. Any measuring instrument, such as a voltmeter or ammeter. **2.** US spelling of ➤metre.

meter resistance The internal resistance of a meter, such as a voltmeter, measured at its terminals at a given temperature. The resistance of particular types of meter, such as rectifier instruments, varies with the frequency, magnitude, and waveform of the input signal. In such cases families of curves can be drawn at specified temperatures relating the resistance to these quantities.

method of images A method of deriving field distributions where a large perfectly reflecting surface is present in the problem. Any source in the problem is duplicated on the other side of the surface at the same distance from the surface, by reflection. The surface can then be removed and the field distribution from this combined collection of sources is then the same as if the surface had been left in place.

method of moments A method of solving field problems in which a self-consistent set of equations, usually in matrix form, is generated involving a set of unknown parameters, for example currents on a conducting surface, and a set of known excitations, for example voltage sources; the set of equations is solved through inversion of the matrix to derive the unknown parameters. The generation of these equations involves the sampling of the effects of the unknown parameters at a finite number of points within the problem, and weighting of these effects in order to satisfy a set of ➤boundary conditions at these points: the weighting or moment of these effects leads to the name method of moments.

In the case of an unknown set of currents, the boundary condition sampled is usually that of zero tangential ➤electric field on a conductor: at each point chosen to enforce this condition, an expression for the radiated field from the unknown currents on the whole structure is derived and set equal to any externally applied fields. When applied to all of the chosen points, this gives rise to the self-consistent matrix equation, which is then solved.

metre Symbol: m. The ➤SI unit of length defined (since 1983) as the length of the path travelled by light in vacuum during a time interval of $1/299\ 792\ 458$ of a second. It was previously defined as the length equal to $1\ 650\ 763.73$ wavelengths in a vacuum of the radiation associated with the transition between the levels $2p_{10}$ and $5d_5$ of the krypton-86 atom. The wavelength of this radiation is about 605.8 nanometres. The metre can be measured very accurately and, unlike the original *international prototype metre*, which was a bar of platinum, does not change with time.

metre bridge A form of ➤Wheatstone bridge with the ratio arms in the form of a tappable uniform wire of length one metre.

MF *Abbrev. for* medium frequency. ➤frequency band.

MFCC *Abbrev. for* mel frequency cepstral coefficients. The result of applying a speech analysis methodology, often as the input stage to a speech recognition system, that is designed to reflect the way we hear. The speech input is sampled, then windowed (➤windowing), and a fast Fourier transform is taken from which the power ➤spectrum is calculated. This is arranged into mel-spaced frequency bands (➤mel) and the logarithm of the amplitude of each band is taken. A discrete cosine transform is then taken to produce the MFCCs. Although this output is referred to as a ➤cepstrum, it should be noted that the two are not directly equivalent.

MFLOPS *Abbrev. for* megaflops.

mho Symbol: Ω^{-1}. The reciprocal ►ohm, formerly used to measure ►conductance but now replaced by (and equal to) the ►siemens.

mica A naturally occurring mineral that consists of complex aluminium-potassium silicates and has a monoclinic structure; it can therefore be readily cleaved into thin plates. It has a large dielectric constant virtually independent of frequency and retains its properties as an insulator even at very high temperatures. It also has low loss and high dielectric strength and is used as the dielectric material in ►mica capacitors. It is widely used for electrical insulation because of its unequalled combination of physical and electrical characteristics.

mica capacitor A ►capacitor that uses ►mica as the dielectric material. Mica capacitors have characteristics of low loss, low temperature coefficient of capacitance, and good frequency stability. There are several types.

Clamp-type mica capacitors (Fig. *a*) have the mica clamped between tin-foil electrodes. Alternate layers of foil are brought out on opposite sides. Metal lugs are soldered to the electrodes. This type of construction is used in the manufacture of standard capacitors but has been largely superseded for other applications.

Eyelet-construction mica capacitors consist of silvered mica plates fixed together with metal eyelets to form stacks. The possibility of relative movement and bowing of the plates in this construction leads to poor stability of capacitance and of temperature coefficient.

Bonded silvered mica capacitors (Fig. *b*) have mica plates that, except for the outer plates, are silvered on both sides, the silvering forming the appropriate electrode areas. The stacked plates are bonded together by firing. This arrangement gives a dimensionally stable stack with good stability of capacitance.

Button mica capacitors are circular capacitors with terminals in the form of a metal band round the perimeter and a metal eyelet in the centre. This type of capacitor also has good dimensional stability and is particularly suitable for high-frequency operation when mounted in a true coaxial arrangement.

The extremely good stability of most forms of mica capacitor both with temperature and frequency makes them highly suitable for use in ►filters.

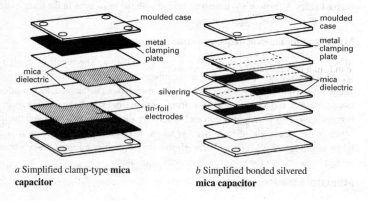

a Simplified clamp-type **mica capacitor**

b Simplified bonded silvered **mica capacitor**

micro- Symbol: μ. A prefix to a unit, denoting a submultiple of 10^{-6} (i.e. 0.000 001) of that unit: one microsecond equals 10^{-6} seconds.

microcircuit An ►integrated circuit, normally one that performs a very complex function.

microcode ►microprogram.

microcomputer ►computer.

microelectronics The branch of ►electronics concerned with or applied to the realization of electronic circuits or systems from extremely small electronic parts. It includes the design, production, and application of any technique to reduce the cost, size and weight of electronic parts, subassemblies, and assemblies and to replace vacuum-tube circuits with solid-state compatible parts. Increased miniaturization is particularly desirable in the field of computers and therefore ►integrated circuits of ever increasing packing density are being designed. ►►nanoelectronics.

micromanipulator probe ►probe.

micron Symbol: μ. A unit of length equal to 10^{-6} metre; it has been renamed as the micrometre (μm).

microphone A device that converts sound energy into electrical energy. It forms the first element in a telephone, a broadcast transmitter, and all forms of electrical sound recording, sound reinforcement, and public address systems. It is the converse of the ►loudspeaker.

There are many different types of microphone: the most common types are the ►carbon, ►capacitor, ►crystal, ►electret, moving-coil, and ►ribbon microphones. Most types of microphone operate by converting the sound waves into mechanical vibrations that in turn produce electrical energy. The most common method is to use a thin diaphragm mechanically coupled to a suitable device. The force exerted is usually proportional to the sound pressure but in the case of the ribbon microphone it is proportional to the particle velocity.

In the *moving-coil microphone* a small coil is attached to the centre of the diaphragm; when the diaphragm is caused to move by sound waves in a steady magnetic flux, an e.m.f. is produced in the coil by electromagnetic induction. The e.m.f. is a function of the incident sound pressure. The *moving-iron microphone* operates in a similar manner but a small piece of iron is moved by the diaphragm. This induces an e.m.f. in a current-carrying coil surrounding it (cf. ►induction microphone). Microphones in which the electrical energy is produced by the motion of a coil or conductor in a magnetic flux density are described as *magnetic microphones*.

In the *magnetostriction* and *crystal microphones* the sound pressure is converted into electrical pressure by direct deformation of suitable magnetostrictive or piezoelectric crystals. In the *glow-discharge microphone* the current of the glow discharge is modulated by the sound waves.

Specially shaped microphones have been designed. The *ear microphone* is specially shaped to fit into the human ear; the *lip microphone* is designed to be held close to the lips and thus cut down on extraneous external sounds. The *throat microphone* is shaped to be worn on the throat and responds directly to vibrations of the larynx, thus cutting out background noise.

Most microphones have strong directional properties that provide improved rejection of extraneous noise (►cardioid microphone; figure-eight microphone; hypercardioid microphone; omnidirectional microphone). Their usual lack of sensitivity is not a great disadvantage since the output from the microphone is usually amplified and the directional properties tend to minimize background noise. Good quality sound reproduction is usually achieved but resonance in the mechanical system must be avoided. This is usually done by making the resonant frequency of the moving parts either much higher or much lower than the required operating sound-frequency range.

microphony ►noise.

microprocessor The physical realization of the ►central processing unit of a given computer system on either a single ►chip of semiconductor or on a small number of chips.

microprogram A ►program that is used to define the ►instruction set of a computer. Computers that can have their instruction sets entered in this way are *microprogrammable*. The languages in which the microprograms are written are called *microcodes*.

microstrip line A transmission line formed on a ►microwave integrated circuit. The impedance of the microstrip line is determined by the ratio of conductor width to substrate thickness, dielectric constant of the substrate, and, to a lesser extent, the thickness of the conductor.

microwave An electromagnetic wave with a frequency in the range of approximately 0.3 gigahertz to 300 gigahertz or more, corresponding to a wavelength of roughly one millimetre or below to about one metre. The microwave frequency range has been divided into various ►frequency bands. Microwaves are used in radar and telecommunications and are also used commercially for extremely rapid cooking.

microwave generator A device, such as a ►klystron or ►magnetron, used to generate microwaves.

microwave integrated circuit (MIC) An ►integrated circuit that operates at microwave frequencies. Strictly, the term refers to a hybrid circuit; a monolithic microwave circuit is referred to as an MMIC.

microwave landing system (MLS) A guidance system to enable civil aircraft to land in conditions of poor visibility. The MLS uses a highly directional microwave signal that the aeroplane can receive and 'home in' on, thereby guiding it to the runway.

microwave tube A ►vacuum tube that is suitable for use as an amplifier or oscillator at ►microwave frequencies. These tubes usually employ ►velocity modulation of the electron beam rather than density modulation as in the valves used in ►audiofrequency valve amplifiers or oscillators.

Microwave tubes may be classified into two main types: *linear-beam tubes,* in which the electron beam travels in an essentially linear direction, and *crossed-field tubes*, in which the electron beam follows a curved path under the influence of orthogonal electric and magnetic fields.

The ►klystron and most forms of ►travelling-wave tubes are linear-beam tubes; the ►magnetron is a crossed-field tube.

MIDI *Abbrev. for* musical instrument digital interface. An industry internationally agreed standard for the serial interconnection of control signals between electronic musical instruments. It is mainly used for the transfer of musical performance data, such as which notes are played and how rapidly they are depressed and released (coded as note-on and note-off messages that incorporate note-velocity information), and the state of performance controllers, such as volume controls, ►pitch bend, and ►modulation wheels. MIDI data can be sent to and from computer ►sequencer systems for storage and modification. MIDI also enables instrument-specific, or 'system-exclusive', data to be communicated relating to internal instrument settings and digital audio samples.

The MIDI standard defines a unidirectional asynchronous serial interface at 31.25 kilobaud, opto-isolated at the MIDI input of every MIDI device. All devices equipped with MIDI will have a MIDI input (IN), some will have a MIDI output (OUT) if MIDI data is generated by the device, and many provide a buffered version of the MIDI input at a MIDI THRU socket for connection to other MIDI devices in the chain.

mike *Informal* A microphone.

mil *US syn. for* thou.

milking generator *Obsolete* A low-voltage d.c. generator that is used to charge one or more cells of an accumulator independently of the others.

Miller effect The phenomenon by which an effective feedback path between the input and output of an electronic device is provided by the ►interelectrode capacitance of the device. If the device is used in a voltage amplifier, this can increase the input admittance of the circuit and result in reduced operating bandwidth. ►cascode.

Miller indices Triplets of numbers corresponding to the three spatial directions and used to specify particular directions and planes within a crystal structure orientated with respect to the unit cube of the crystal.

Miller integrator An ►integrator that contains an active device, such as a ►transistor or ►operational amplifier, in order to improve the linearity of the output from a pulse generator. Miller integrators are used particularly with sawtooth pulse generators, such as those used to generate a ►timebase.

Miller sweep generator ►timebase.

milli- Symbol: m. A prefix to a unit, denoting a submultiple of 10^{-3} (i.e. 0.001) of that unit: one millimetre equals 10^{-3} metre.

millimetre waves Electromagnetic radiation at frequencies between 30 and 300 gigahertz, which result in wavelengths of millimetre dimensions.

MIM capacitor *Short for* metal-insulator-metal capacitor. A thin-film capacitor consisting of two metal plates separated by a dielectric material and used in integrated circuits. MIM capacitors are the most commonly used type of ➤monolithic capacitor in
MMICs (➤monolithic microwave integrated circuits).

minimum discernible signal (mds) *Syn.* threshold signal. The smallest value of input
signal to any circuit or device that just produces a discernible change in the output.
➤radar.

minimum sampling frequency ➤sampling; pulse modulation.

minimum shift keying (MSK) ➤amplitude modulation.

minority carrier The type of charge ➤carrier in an extrinsic ➤semiconductor that constitutes less than half of the total charge carrier concentration.

minterm In digital design, a product (logical AND) that includes every input (i.e. every
➤literal) in a ➤truth table. The complement of any minterm is a ➤maxterm.

mirror galvanometer ➤galvanometer.

mismatch A condition that occurs when the impedance of a ➤load does not equal the
output impedance of the source to which it is connected.

misuse failure ➤failure.

mixed coupling ➤coupling.

mixer *Syn.* frequency changer. A device that is used in ➤heterodyne reception to produce an output signal of different frequency from the input signal. The received carrier wave is mixed with a locally generated signal from the local oscillator to produce
an *intermediate-frequency* (IF) signal that retains the modulation characteristics of the
original signal. The amplitude of the output signal has a fixed relationship to the
input-signal amplitude and is usually a linear function of it.

mixer noise balance A measure of ➤insertion loss for wideband local-oscillator
➤noise converted by the ➤mixer to the output port at the intermediate frequency (IF).

MKS system A system of ➤units, now obsolete, in which the fundamental units of
mass, length, and time are the kilogram, metre, and second and the fourth fundamental quantity necessary to define completely electric and magnetic quantities is the
➤permeability of free space, μ_0. The permeability was orginally defined as having the
value 10^{-7} henry/metre (➤CGS system).

It was later shown that if μ_0 were given the value $4\pi \times 10^{-7}$ henry/metre then electric and magnetic equations would appear more rational: equations concerned with
spherical systems would contain the factor 4π, which characterizes any concept of
sphericity; those concerned with circular or cylindrical systems, such as a coil of
wire, would contain the characteristic factor 2π; those concerned with a linear system,
such as a straight wire, would not contain the factor π. Adoption of the value $4\pi \times 10^{-7}$

H/m for μ_0 leads to the *rationalized MKS system*, which is the basis of the system of ➤SI units.

MLS *Abbrev. for* microwave landing system.

MMIC *Abbrev. for* monolithic microwave integrated circuit.

mobile multimedia system ➤cellular communications.

mobile phone ➤cellular communications.

mobility ➤drift mobility; Hall mobility.

mode *Syn.* transmission mode. Any one of the several different states of oscillation of an electromagnetic wave of given frequency. The mode of an electromagnetic wave depends upon the configurations of the vectors describing the wave; three main types of wave exist:

TE waves (or *H waves*) are transverse electric waves in which the electric field vector (*E*-vector) is always perpendicular to the direction of propagation, *z*, i.e.

$$E_z = 0$$

TM waves (or *E waves*) are transverse magnetic waves in which the magnetic vector (*H*-vector) is always perpendicular to the direction of propagation, *z*, i.e.

$$H_z = 0$$

TEM waves are transverse electromagnetic waves in which both the *E*-vector and the *H*-vector are perpendicular to the direction of propagation, i.e.

$$E_z = H_z = 0$$

The TEM mode is the one most commonly excited in coaxial lines. It cannot be propagated in a waveguide.

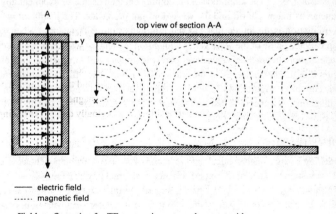

—— electric field
······ magnetic field

a Field configuration for $TE_{1,0}$ wave in rectangular waveguide

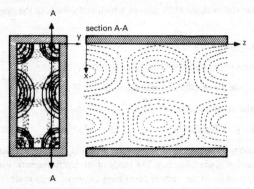

b Field configuration for TE$_{2,1}$ wave in rectangular waveguide

Solving Maxwell's equations for the particular conditions existing, it is commonly found that the solutions are characterized by the presence of one or more integers (*m*, *n*) that can take values from zero to infinity; this allows several possible modes for each type of wave. The physical constraints and the frequency of the radiation usually limit the number of permitted values of *m* and *n* and therefore the number of possible modes.

A particular example may be considered. In a ►waveguide each component of the wave contains a factor given by

$$\exp(j\omega t - \gamma_{m,n}z)$$

where ω is 2π times the frequency and j is $\sqrt{-1}$; $\gamma_{m,n}$ is the ►propagation coefficient and determines the phase and amplitude of the wave components. For each mode a lower limit exists where the complex quantity $\gamma_{m,n}$ is purely real and is equal to the ►attenuation constant, $\alpha_{m,n}$. The amplitude of the component then decreases exponentially. The frequency is below cut-off and the wave does not propagate. The *dominant mode* of the oscillation is the mode that has the minimum cut-off frequency. When $\gamma_{m,n}$ is imaginary the phase of the wave varies with distance *z* and the wave is propagated with no attenuation. In practice $\gamma_{m,n}$ is never purely imaginary and some attenuation always occurs due to energy losses in the transmission line. The effects of different *m,n* modes on the configuration of a wave travelling along a waveguide can be seen from the diagrams. A waveguide that is designed to separate different modes of oscillation of a wave is termed a *mode filter.*

mode filter ►mode.

modelling Using a mathematical representation of an electrical component, device, circuit, or system to determine the performance of the real physical entity.

Equivalent-circuit modelling uses standard circuit elements – resistors, capacitors, voltage and current sources, etc. – to represent the electrical behaviour of ►transistors and other ►active devices and circuits, the behaviour being described by the

➤semiconductor device physics. Thus straightforward circuit theory is used to model the physical behaviour of these devices and circuits.

Physical modelling uses the semiconductor equations describing ➤drift and ➤diffusion current flows, ➤continuity equation, ➤Poisson's equation, etc., to describe the movement of the charge ➤carriers within the semiconductor device, and hence describe the external current-voltage behaviour. Further detail can be incorporated at the ➤quantum mechanical level for those devices where such phenomena play a role in their electrical behaviour (in the ➤high electron mobility transistor or in ➤nanoelectronics, for example).

The modelling itself can be analytical, though nowadays greater emphasis is placed on *numerical modelling*, which can be incorporated into ➤computer-aided design (CAD) software.

modem Acronym from *mo*dulator-*dem*odulator. A device that is used to convert signals output from one type of equipment into a form suitable for input to another type. One of the best-known uses for modems is to connect digital devices, such as computers, across analogue transmission lines: a modem can convert a computer's digital signal into an analogue signal suitable for transmission over an analogue telephone line, and can convert an incoming analogue signal back into digital form.

MODFET *Abbrev. for* modulation-doped field-effect transistor. ➤high electron mobility transistor.

modulated wave ➤modulation.

modulating wave ➤modulation.

modulation In general, the alteration or modification of one electronic parameter, the ➤carrier wave, by another wave or signal, the *modulating wave.* The resultant composite signal is the *modulated wave.* The reverse process is *demodulation,* by which an output wave is obtained having the characteristics of the original modulating wave. If the receiver uses knowledge of the carrier's phase to detect the modulating signals, the demodulation process is called *coherent detection* (or *coherent demodulation*); when the receiver does not utilize the carrier-phase information, the demodulation process is called *noncoherent detection* (or *noncoherent demodulation*).

The characteristics of the carrier that may be modulated are the amplitude (➤amplitude modulation) or phase angle (*angle modulation*), particular forms being ➤phase modulation and ➤frequency modulation. Modulation by an undesired signal is *cross modulation. Multiple modulation* is a succession of processes of modulation in which the whole or part of the modulated wave from one process becomes the modulating wave for the next.

➤➤pulse modulation; intermodulation.

modulation-doped FET (MODFET) ➤high electron mobility transistor.

modulation index *Syn.* modulation factor. ➤amplitude modulation; frequency modulation.

modulation wheel A wheel found on electronic musical instruments that enables the degree of modulation by a ➤low-frequency oscillator of an aspect of the output sound to be varied. When the ➤fundamental frequency or ➤amplitude is modulated, these are perceived as ➤pitch or ➤loudness changes and usually known as *vibrato* or *tremolo* respectively.

modulator 1. Any device that effects ➤modulation. In radio transmitters, for example, where the signal to be transmitted is usually modulated onto a carrier, one method of creating a modulator is to vary the reactance of the tuned circuit of an oscillator, such as by a ➤varactor, thereby tuning the frequency. Where such an arrangement is used the modulator is referred to as a *reactance modulator*. **2.** *Syn.* master trigger. A device, usually a ➤multivibrator, that is used in ➤radar to generate a train of short pulses, each of which acts as a trigger for the oscillator.

modulator electrode An electrode that is used to modulate the flow of current in an electrode device. In a ➤field-effect transistor it is the gate electrode(s), used to modulate the channel conductivity. In a ➤cathode-ray tube it is the electrode used to control the electron-beam intensity.

molecular-beam epitaxy (MBE) A method of ➤epitaxy that is carried out in ultrahigh vacuum. It is a thermal evaporation technique in which the elements to be deposited are evaporated from ovens and impinge on the substrate where they are deposited in crystalline order. The substrate wafer is rotated to ensure a uniform growth rate across the wafer. Under suitable conditions the process can be controlled to produce almost any required epitaxial layer composition, thickness, and doping level with a resolution of virtually one atomic layer, to a high degree of accuracy and uniformity across the wafer. Disadvantages are the high-vacuum requirements, complex and costly equipment, and the slow growth rate of the epitaxial layer.

moment method ➤method of moments.

momentum space The motion of electrons in a crystalline solid can in principle be found from the solution of Schrödinger's equation in ➤quantum mechanics. This results in the electron ➤energy band structure, which describes the relationship between the electron energy E and momentum p in the crystal. The motion of the electron is described by this energy–momentum (E–p) relation, which is known as momentum space.

From ➤de Broglie's relationship, the momentum p and wavelength λ of the electron are related:

$$p = h/\lambda = \hbar k$$

where k is the *wavevector*; k and p are in direct proportion, related by the rationalized ➤Planck constant. An alternative way of expressing the energy–momentum relation is thus as an energy–wavevector (E–k) relation; this is known as *k-space*.

monochromatic radiation Electromagnetic radiation of a single frequency (or, equivalently, of a single wavelength). In practice radiation of a single frequency is never achieved and the term is applied to a narrow range of frequencies. The term is also applied to particulate radiation when the particles are all of the same type and energy but in this case the description *homogeneous* or *monoenergetic* is usually preferred. ➤polychromatic radiation.

monochrome Displaying a picture in only one colour. A monochrome TV displays a black and white picture. A monochrome display may have a white, amber, or green screen. ➤television.

monolithic capacitor A capacitor that is formed as part of a monolithic ➤integrated circuit. It can be used as a ➤blocking capacitor, a ➤bypass capacitor, or a tuning capacitor as part of a filter or ➤tuned circuit. The main types of monolithic capacitor are as follows: *interdigitated capacitors* (Fig. *a*), in which the capacitance is produced by capacitive coupling between adjacent conductors separated by the substrate dielectric material; ➤MIM (metal-insulator-metal) capacitors (Fig. *b*), which are parallel plate capacitors and the most common type used; and ➤Schottky diodes used as ➤varactors in voltage-controlled oscillator circuits.

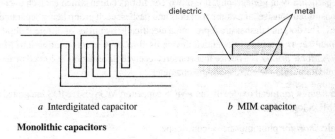

a Interdigitated capacitor   *b* MIM capacitor

Monolithic capacitors

monolithic integrated circuit ➤integrated circuit.

monolithic microwave integrated circuit (MMIC) An ➤integrated circuit operating at ➤microwave frequencies and built on a single crystal of semiconductor. This is usually gallium arsenide though silicon can be used at low microwave frequencies, below about 2 gigahertz.

monophonic Involving a sound system with only one channel. ➤reproduction of sound; aftertouch.

monopole antenna One half of a ➤dipole antenna. Some antennas, particularly wire and helical antennas mounted on vehicles and communications devices, are often referred to as monopole antennas although they are actually dipole antennas where the body of the vehicle or device acts as the other half of the dipole.

monoscope A type of ➤electron tube that is used to produce single images, such as a test pattern, suitable for television broadcasting. It contains an aluminium plate electrode with the desired image printed on it in carbon. The plate is scanned by an electron beam and video signals are produced because of the difference in ➤secondary emission of electrons of aluminium and carbon. ➤camera tube.

monostable *Syn.* one-shot; univibrator. A type of circuit that has only one stable state but on the application of a ➤trigger pulse it can take up a second quasi-stable state. A common form of monostable consists of a ➤multivibrator with resistive-capacitive coupling. Monostables can be used to provide pulses of fixed duration, for pulse stretching or shortening, or as a delay element.

Moog synthesizer An early electronic musical instrument.

Moore circuit ➤finite-state machine.

Moore's law An observation made by Gordon Moore in 1965 that the number of transistors that could be fabricated on the most cost-effective ➤integrated circuits had roughly doubled every year since their commercial introduction in 1958, and that this would continue into the near future. In the late 1970s the integration rate slowed to a doubling of transistors every 18 months, and has since held to this rate.

Morse code An internationally agreed code formerly used for the transmission of signals, particularly in ➤telegraphy. It is a two-condition code in which each character to be transmitted consists of a number of dots and dashes, each group being separated by spaces. The dots and dashes are represented electrically by pulses of different duration. A *Morse key* is a manually operated device used to convert information into Morse code. A *Morse printer* is a device that converts received information, recorded on a suitable medium, into the corresponding characters.

MOS *Abbrev. for* metal oxide silicon. ➤MOS capacitor; MOSFET; MOS integrated circuit; MOS logic circuit.

mosaic *Short for* photomosaic. ➤iconoscope.

mosaic crystal An imperfect crystal composed of a number of smaller crystals that have been grown together so that the corresponding crystal planes are nearly or exactly parallel but have discontinuities at their mutual surfaces. Such crystals are effectively single crystals for most practical purposes but the discontinuities form crystal defects that can affect their properties, for example by providing additional energy levels in ➤semiconductors.

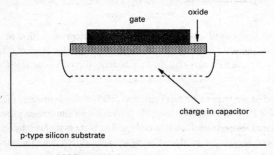

a n-channel **MOS capacitor** structure

mosaic electrode The light-sensitive electrode of a television ➤camera tube on which the image is formed.

MOS capacitor A two-terminal capacitor structure made in silicon technology: the top electrode – the *gate* – is a metal contact, the insulating or dielectric layer is silicon dioxide, and the counter electrode is provided by the silicon itself. An example of this structure for an n-channel capacitor is shown in Fig. *a*.

The electrical behaviour of the MOS capacitor is governed by the behaviour of the silicon beneath the oxide layer, and can be described by four modes of operation: *flat band*, *accumulation*, *depletion*, and *inversion*; the schematic electron ➤energy band diagrams for these operation modes are shown in Fig. *b*.

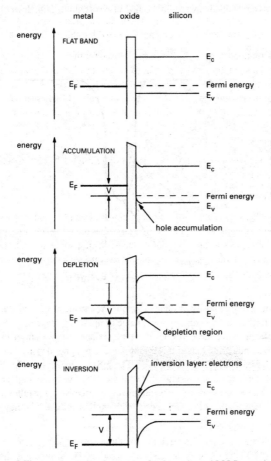

b Electron energy band structures for the n-channel **MOS capacitor** in four operation modes

In flat-band operation mode, the ➤Fermi levels E_F in the metal and the silicon are in alignment so the energy bands are completely flat. This is an equilibrium situation, and is defined as a reference point in MOS structures.

In accumulation mode, a negative voltage is applied to the metal, attracting free ➤holes in the silicon to the surface beneath the oxide. There is a greater density of holes at the surface than the equilibrium p-type doping density, so there is effectively an *accumulation* of holes at the surface. The energy bands are bent upwards at the silicon surface, indicating higher hole density than deep in the silicon.

In depletion mode, a positive voltage is applied to the metal, repelling free holes from the silicon surface beneath the oxide. The hole density is now less than the equilibrium doping density, and the region beneath the oxide is *depleted* of charge carriers. The energy bands are bent downwards at the silicon surface, indicating lower hole density.

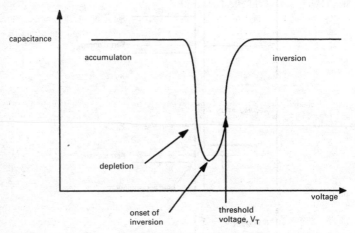

c Variation of capacitance with applied voltage for a **MOS capacitor**

In inversion mode, a further positive voltage is applied to the metal, causing the energy bands to bend further downwards. At some value of applied voltage, the energy bands will bend sufficiently for the Fermi levels at the surface to move closer to the conduction band edge, indicating that in this region there is a higher density of electrons than holes: the surface layer appears ➤n-type. The surface region is said to be *inverted*. This is the normal region of operation for ➤MOSFETS. The inversion layer of electrons forms the counter electrode to the metal in what is effectively a parallel-plate capacitor structure. The variation of capacitance C with applied voltage V is shown in Fig. *c*.

MOSFET *Abbrev. for* metal-oxide-silicon field-effect transistor. A ➤field-effect transistor that uses the ➤MOS capacitor as the basis of operation. The structure is shown in the diagram. The *inversion* layer forms a conducting bridge or *channel* between two

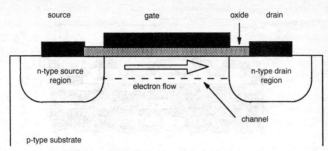

n-channel **MOSFET** structure

electrodes – *source* and *drain*; the *gate* forms the third electrode. (For details of the electrical behaviour, ➤field-effect transistor.)

Both p-channel and n-channel MOSFET technologies are available, known as *PMOS* and *NMOS* respectively. Depletion-mode and enhancement-mode transistors can be made in both technologies. PMOS was the first MOSFET technology to be realized practically, followed by NMOS, which offered higher switching speeds for digital ➤logic circuits. Both PMOS and NMOS technologies are combined on a single silicon wafer in one technology, called *complementary MOS* or *CMOS*. This offers lower power consumption than other MOS technologies, since power is consumed only during the logic switching cycle and not in static conditions. MOS technologies are used in the largest ➤microprocessor and ➤memory ICs.

MOS integrated circuit A type of ➤integrated circuit based on ➤MOSFETs. MOS circuits have several advantages compared to ➤bipolar integrated circuits for many low-frequency applications, and account for the majority of all semiconductor devices produced.

MOSFETs are self-isolating and no area-consuming isolation diffusions are required; this enables a very high functional ➤packing density to be obtained. MOSFETs may be used as active load devices and a separate process is not therefore required to form resistors. When used as load devices ➤pulse operation of the circuit is easily achieved using the gate electrodes to activate the device: power dissipation is greatly reduced, involving less complicated heating problems. A characteristic of MOSFETs is their exceptionally high input impedance. This allows the gate electrodes to be used as temporary storage capacitors thus keeping the circuits comparatively simple. This is called *dynamic operation*. Usually these circuits operate above a minimum specified frequency.

The relatively few processing steps required in manufacture, compared to bipolar integrated circuits, enables large chips to be made thus further increasing the functional compactness and reducing the costs. MOS integrated circuits tend however to be slower than their bipolar counterparts due to their inherently lower ➤transconductance and to the fact that their speed is extremely dependent on the load capacitance.

Modern MOS ICs are made using a *self-aligned gate* process. In the self-aligned gate circuit the gate electrodes are formed before the source/drain diffusions are made;

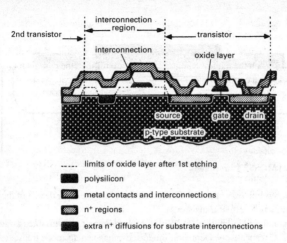

interconnection region

2nd transistor

interconnection

transistor

oxide layer

source | gate | drain

p-type substrate

- - - - limits of oxide layer after 1st etching

polysilicon

metal contacts and interconnections

n⁺ regions

extra n⁺ diffusions for substrate interconnections

MOS integrated circuit: cross section of n-channel self-aligned circuit

the most widely used method is known as the *silicon gate technology* (see diagram). Openings are made in the initial thick oxide layer and a thin layer of oxide grown to form the ►gate insulators. This is immediately covered with a layer of polycrystalline silicon (polysilicon) using a vapour-phase reaction. The polysilicon is then etched to form the gate electrodes together with some interconnections. The gate oxide is then removed from the regions not covered with polysilicon leaving openings through which the source/drain diffusions are made. The edges of these diffused regions are defined by the previously etched gate regions thus providing the required precision of alignment. The diffusing material also enters the polysilicon regions and dopes them, which has the desirable effect of reducing their resistivity. The whole wafer is covered with a further oxide layer, contact windows etched, and a metal layer is deposited and etched to form the interconnections. Several layers of metal interconnections are possible using this technique.

MOS logic circuit A ►logic circuit constructed in a ►MOS integrated circuit. It consists of a combination of ►MOSFETs in series or in parallel that perform the logic functions, i.e. act as AND or OR gates, etc. (Fig. *a*), coupled to other MOSFETs that

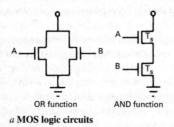

OR function   AND function

a **MOS logic circuits**

determine the output voltages of the circuit. MOS logic circuits are classified according to the method of determining the output voltage, i.e. into *ratio* or *ratioless circuits*. The logic gates effectively act as switches when used with a suitable choice of high and low logic levels. When the required input conditions are fulfilled the combination switch is 'on' and provides a conducting path. If the switch is 'off' the gate does not conduct. The high logic level is chosen to be greater than the threshold voltage, V_T, of the MOSFET; the low level is lower.

In a *ratio circuit* the logic gate, represented as a single switch transistor, T_S, is connected in series with a load transistor T_L. The drain of the load transistor is connected to the power supply and the source of the switch transistor to earth (Fig. *b*). The output is taken from the ►node, A, between the transistors. The circuit will usually be driving similar MOS logic gates. These have a very high input impedance. A voltage of magnitude greater than or equal to the drain voltage, V_{DD}, is applied to the gate of T_L. In static operation the voltage is applied continuously; in dynamic operation the voltage is applied on the application of a ►clock pulse in order to reduce dissipation.

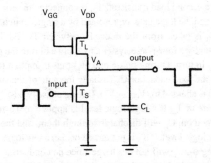

b Ratio circuit

A low logic level input to the gate of the switch transistor T_S results in T_S being 'off'; C_L is then charged by T_L until the output voltage, V_A, reaches a value sufficient to cause T_L to turn off, i.e. until V_A reaches $(V_{GG} - V_T)$ or V_{DD}, whichever is lower. Application of a high logic level to the gate of T_S causes T_S to be 'on' and C_L discharges through T_S. The output voltage V_A falls to a level determined by the relative impedances of the two transistors.

It can be shown that the voltage V_A at the node depends on the ratio of the ►aspect ratios of the devices, and these are manufactured to ensure an output voltage suitable for a low logic level, i.e. less than the threshold voltage of the following gate. The circuit provides inversion of the logic function; thus an AND function in T_S provides a NAND output, etc.

If the dynamic version of the circuit (with the gate voltage of the load transistor clocked) is used a minimum rate of clocking must be specified to prevent loss of information at the output due to leakage paths causing the charge on the load capacitor to decay.

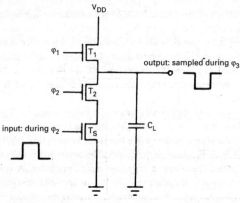

c Ratioless circuit

In a *ratioless circuit* (Fig. *c*) a second load transistor, T_2, is connected in series between the first load transistor T_1 and the logic gate, represented by a single switch transistor T_S; the output voltage V_B is taken from the node, B, between T_1 and T_2. A clocking system is employed, usually a four-phase system, to apply a bias to the gates of the load transistors T_1 and T_2 in turn. During phase one (ϕ_1) bias is applied to the gate of T_1, T_1 is turned on, and the load capacitor C_L (usually the gate of the switch transistor of the following stage) is charged to ($V_{GG} - V_T$). During phase two (ϕ_2) bias is applied not to T_1 but to the gate of T_2. If a high logic level is applied to T_S at this time, both T_2 and T_S will be turned on, C_L will discharge through them, and the output voltage V_B will fall to the low logic level. If T_S is not turned on, i.e. a low logic level is applied to the gate of T_S during ϕ_2, C_L will not discharge since no conducting path to earth exists and V_B will remain at the high logic level. The output of the circuit is sampled by the following circuit during phases ϕ_3 and ϕ_4; information may thus only be supplied to T_S once in every four clock phases.

Operation of this circuit does not depend on the impedances of the devices and it is therefore termed ratioless. Power dissipation is very low since no conducting path ever exists directly between the power supply and earth and the circuits depend solely on charge storage in the load capacitance. The circuit is inverting and two gates are frequently combined to provide a noninverting circuit. If used in a dynamic ►shift register, for example, six transistors are needed for each ►bit of information.

A *CMOS logic circuit* uses ►complementary MOS transistors to provide the basic logic functions. The basic NAND gate is shown in Fig. *d*. CMOS circuits have the advantage that the power required is extremely low and they are suitable for applications where very little power consumption is a condition. They have a lower packing density than ratio circuits since every transistor requires its complement and therefore isolation of p-channel devices from n-channel devices is required. For convenience, groups of n-channel devices (and p-channel devices) are formed in the same area of the chip. The speed of operation is relatively slow, compared to ►transistor-transistor

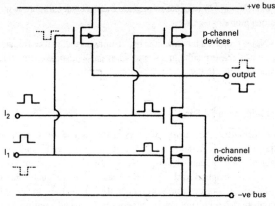

d CMOS NAND gate

logic, because of the relatively large bulk capacitance of the substrate. CMOS circuits are however very resistant to stray noise pulses. Faster versions of CMOS circuits have been designed; the fastest version is the silicon-on-sapphire type of circuit (➤silicon-on-insulator).

MOSRAM RAM fabricated in MOS technology.

MOST *Abbrev. for* MOS transistor. ➤MOSFET.

MOS transistor ➤MOSFET.

most significant bit (MSB) The leftmost bit in the binary representation of a number. It is the bit conveying the greatest amount of precision.

motherboard A printed circuit board (PCB) that holds and supports all the other PCBs in a computer. These PCBs may include one to control access to the disk drives, one to provide high-resolution graphics, and one to support the input/output user ports.

motional impedance *Syns.* driving impedance; driving-point impedance. A component of the impedance of an electromechanical or acoustic ➤transducer that results from the counter e.m.f. generated by the motion of the transducer. It is equal to the vector difference between the input impedance measured under specified load conditions and the ➤blocked impedance.

motor A machine that converts electrical energy into mechanical motion. The motion is produced by the torque due to the magnetic fields associated with the currents in the windings (➤electromagnetic induction).

An alternating-current (a.c.) motor operates with an alternating-current power supply. The rotating parts are termed the *armature* (or *rotor*) and the stationary windings the *stator*. A direct-current (d.c.) motor operates with a direct-current power supply and usually contains a *commutator* that connects each of the sections of the primary winding in turn to the power supply in order to provide the necessary torque. ➤induction motor; synchronous motor.

motorboating ➤Noise produced by unwanted oscillations arising in low-frequency and audiofrequency amplifiers. It results in a sound from the loudspeaker that resembles an outboard motor running slowly.

mount ➤waveguide.

mountain effect ➤direction finding.

moving-coil instrument A measuring instrument that is usually used as a ➤galvanometer. It depends for its operation on the interaction between a steady magnetic field and the magnetic field induced in a movable coil carrying the current to be measured. The instrument is designed so that the coil rotates in the magnetic flux density and the angular deflection is directly proportional to the current to be measured. Moving-coil instruments have a relatively low power consumption and a uniformly divided scale. The magnetic flux density is usually provided by the suitably shaped pole pieces of a permanent magnet although an electromagnet may be used. The instrument is only suitable for use with direct current but may be adapted for alternating-current measurements by means of a suitable rectifier (➤rectifier instrument). It can be used as a ➤voltmeter with a large series resistance.

moving-coil microphone ➤microphone.

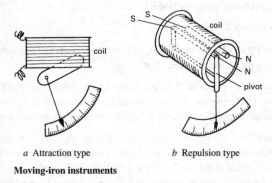

a Attraction type   *b* Repulsion type

Moving-iron instruments

moving-iron instrument A measuring instrument that depends for its operation on the interaction between current-carrying fixed coils and soft-iron pieces in the moving system. The attraction-type of instrument depends on the attraction of the coil for a soft-iron armature when the magnetic flux due to the current in the coil changes (Fig. *a*). The repulsion-type of instrument depends on the mutual repulsion between two soft-iron rods located inside the coil (Fig. *b*). Since both are contained within the coil similar poles are induced in adjacent ends.

moving-iron microphone ➤microphone.

moving magnetic surface memory A ➤memory device in which the storage medium moves during operation and in which digital information is stored in magnetic material thinly coated on a nonmagnetic substrate; the direction of magnetization of small localized areas of the magnetic material represents the stored information. Information is stored and accessed to and from the material by means of small electromagnets – read and write heads (➤magnetic recording) – past which the storage medium is moved or rotated at high speed. The same electromagnet can be used for both reading and writing – a read/write head – using time-division multiplexing.

The nature of the device and the speed of operation depend on the substrate material, which is either flexible or rigid, and the physical geometry used. ➤Magnetic disk and ➤magnetic tape are the major forms of moving magnetic surface memory. They are nonvolatile memory and are used as backing store (➤memory). Interchange of disks or reels of tape greatly increases the volume of information available to a given computer system.

moving-magnet instrument A measuring instrument that depends for its operation on the interaction between the magnetic flux density induced in a fixed coil by the current to be measured and a small permanent magnet suspended in the plane of the coil. It is essentially the converse of the ➤moving-coil instrument and has the disadvantage that it must be accurately aligned relative to the magnetic meridian.

MSB *Abbrev. for* most significant bit.

MSK *Abbrev. for* minimum shift keying. ➤amplitude modulation.

MTBF *Abbrev. for* mean time between failures.

MTTF *Abbrev. for* mean time to failure.

M-type microwave tube *Syn. for* crossed-field microwave tube. ➤microwave tube.

mu circuit ➤feedback.

multicavity microwave tube ➤klystron; magnetron.

multichannel analyser An instrument that assigns an input waveform to a number of channels according to a specific parameter of the input. A device that sorts a number of pulses into selected ranges of amplitude is known as a *pulse-height analyser.*

multielectrode valve 1. A ➤thermionic valve that contains more than three electrodes.
2. A ➤thermionic valve that contains within one envelope two or more sets of electrodes each having its own independent stream of electrons.

multilayer PCB A printed circuit board (PCB) that comprises several alternating layers of metal and insulating medium, usually a glass-fibre weave. This enables complex circuits to be created in a small area by allowing several interconnection layers between the components. Metal layers can also be configured as 'ground planes' to provide shielding between different signal pathways; for example one metal level could be used for analogue signals, another for digital signals, with a ground plane between them.

multilevel metallization ➤Metallization on a monolithic ➤integrated circuit that allows the various active and passive components to be connected together by a highly conductive path. On simple circuits a single layer of metallization, often ➤aluminium, can be used to interconnect the components to form the circuit. In more complex ICs, several layers of metal pathways may be needed: these layers are isolated from one another using insulating film deposited onto the IC, and connections between the metal layers are effected using ➤via holes through the films, allowing the two metal films to come into contact.

multilevel resist *Syn.* portable conformable mask. A technique in ➤photolithography that uses two or more layers of resist to produce a desired pattern. The bottom layer of resist is relatively thick and produces a very planar topography. The topmost layer is very thin and is used for the optical exposure. The exposed pattern in the top layer is then replicated in the lower layer or layers, i.e. the topmost layer acts as a ➤mask for the lower layers. *Bilevel resist* consists of only two layers, one thick and one thin. Intermixing between the two layers of resist can be a problem, and to prevent this *trilevel resists* can be used. In this case a very thin *transfer layer* of metal or dielectric film is used to completely separate the two layers of resist. Multilevel resist techniques are rather complex but offer several advantages, particularly the planar surface produced and initial exposure in a thin planar resist which provides the optimum conditions for photolithography.

multilevel signalling ➤signal.

multimedia ➤CD-ROM.

multipath A term used to refer to the multiple paths taken by a signal from a radio transmitter to a fixed or moving receiver. These are the result of scattering from objects in the environment between the transmitter and receiver; the objects themselves may be fixed, such as hills or buildings, or moving, such as vehicles. The signals may add or cancel at the receiver leading to variations of up to several orders of magnitude in the received signal over very short distances. If the receiver is moving this leads to ➤fading of the received signal.

multiple access 1. ➤digital communications. **2.** ➤communications satellite.

multiple folded dipole ➤dipole.

multiple-loop feedback *Syn. for* local feedback. ➤feedback.

multiple modulation ➤modulation.

multiplexer (MUX) ➤multiplex operation.

multiplex operation The simultaneous transmission of several signals along a single path without any loss of identity of an individual signal. The various signals are input to a *multiplexer* that is used to allocate a transmission path to the input according to a particular parameter of the signal (➤frequency-division multiplexing; time-division multiplexing). The original signals are reconstructed at the receiver by a *demulti-*

plexer that is operated so as to synchronize with the multiplexer. The transmission path may use any suitable medium, such as wire, waveguide, optical fibres, or radiowaves. The communication channel chosen is termed a *multiplex channel.*

multiplier 1. ➤electron multiplier; photomultiplier. **2.** A device that has two or more inputs and that produces an output of magnitude equal to the product of the magnitudes of the input signals. **3.** A device or circuit, such as a ➤frequency multiplier, that produces an output equal to a specified multiple of the input signal.

multiprocessor A ➤parallel-processor architecture with two or more processors that share the same ➤logical-address space. A *loosely coupled multiprocessor* is a multiprocessor in which the individual processors have their own local memories and usually execute programs out of them but also share the memories of other processors. A *tightly coupled multiprocessor* is one in which the CPUs share a common main memory; it is approximately equivalent to the group of multiprocessors with global memory.

multistable A circuit or device that has more than one stable state.

multistage amplifier ➤amplifier.

multistage feedback *Syn.* single-loop feedback. ➤feedback.

multivibrator An oscillator that contains two linear ➤inverters coupled in such a way that the output of one provides the input for the other. There are several types of multivibrator, the action of which depends on the type of coupling used.

Capacitive coupling produces an *astable multivibrator* that has two quasi-stable states; once the oscillations are established the device is free-running, i.e. a continuous waveform is generated without the application of a ➤trigger.

Resistive-capacitive coupling produces a ➤monostable multivibrator.

Resistive coupling (see diagram) produces a bistable circuit that has two stable states and can change state on the application of a trigger pulse. ➤flip-flop.

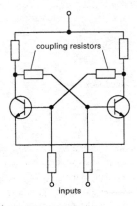

Bistable **multivibrator**

musa ➤steerable antenna.

mush area ➤service area.

mushroom gate ➤T-gate.

musical instrument digital interface ➤MIDI.

muting ➤automatic gain control.

muting switch A manual or automatic switch that activates a noise-suppression circuit in a particular piece of equipment when required, as when the ➤noise in the system exceeds a predetermined level.

mutual branch *Syn.* common branch. ➤network.

mutual capacitance An indication of the extent to which two capacitors can interact. It is expressed as the ratio of the electric charge transferred to one to the corresponding potential difference of the other.

mutual conductance *Syn. for* transconductance, now rarely used.

mutual impedance *Syn. for* transimpedance, now rarely used.

mutual inductance ➤inductance; electromagnetic induction.

mutual-inductance coupling ➤coupling.

MUX *Short for* multiplexer.

Mylar *Trademark* A polyester, usually produced in sheets of various thicknesses, that is used for a variety of applications such as insulation, the base material of a magnetic tape, or as the dielectric in certain types of capacitor.

N

NAK signal *Short for* negative acknowledgment signal. ►digital communications.

NAND circuit (or **gate**) ►logic circuit.

nano- Symbol: n. A prefix to a unit, denoting a submultiple of 10^{-9} of that unit: one nanometre equals 10^{-9} metre.

nanoelectronics A branch of electronics concerned with components and devices whose dimensions are significantly less than one micrometre, and with associated electrical phenomena. Many of these devices and components rely on ►quantum mechanical phenomena by virtue of their small size, and as a result demonstrate electrical behaviour that cannot be observed in larger structures. ►quantum dot; quantum wire. ►►nanotechnology.

nanotechnology The fabrication techniques used to produce the very submicrometre features required for the quantum-confined devices and miniature structures characteristic of ►nanoelectronics. High-resolution ►lithographic techniques, such as ►electron-beam lithography, are used to define the features on the semiconductor surface. ►Epitaxial growth techniques, such as ►molecular-beam epitaxy, are used to provide control of the semiconductor composition on an atomic layer-by-layer basis. Highly anisotropic dry ►etching techniques, for example plasma etching and ion milling, are used to cut the very fine surface features.

narrowband Denoting a system or circuit that is operational over a frequency range which is small compared to its ►centre frequency. A common measure of narrowband is where the bandwidth is up to a few percent of the centre frequency.

narrowband FM ►frequency modulation.

narrowband FSK ►frequency modulation.

National Institute of Standards and Technology (NIST) ►standardization.

natural frequency The frequency at which ►free oscillations occur in an electrical or mechanical system. It is the frequency at which ►resonance occurs, or at which ►formant in speech occurs, in such a system in response to a periodic driving force.

n-channel Denoting a junction FET or MOSFET in which the conducting channel is formed as n-type semiconductor. ►field-effect transistor. ►►p-channel.

NDR *Abbrev. for* negative differential resistance. ►negative resistance.

NDR effect *Syn. for* Gunn effect.

near-end crosstalk ►crosstalk.

near-field region ►far-field region.

Néel temperature ►antiferromagnetism; ferrimagnetism.

negative bias A voltage that is applied to an electrode of an electronic device and is negative with respect to some fixed reference potential, usually earth potential.

negative booster ►booster.

negative charge ►charge.

negative differential resistance (NDR) ►negative resistance.

negative feedback *Syns.* inverse feedback; degeneration. ►feedback.

negative ion ►ion.

negative logic ►logic circuit.

negative phase sequence ►phase sequence.

negative photoresist ►photoresist.

negative resistance (or strictly **negative differential resistance**) A property of certain devices whereby a portion of the current-voltage characteristic has a negative slope, i.e. the current decreases with increasing applied voltage. Devices that exhibit negative resistance include the ►thyristor, ►tunnel diode, ►Gunn diode, ►magnetron, and the output port of any ►oscillator.

negative-resistance oscillator ►oscillator.

negative sequence ►phase sequence.

negative transmission ►television.

neon Symbol: Ne. An inert gas, atomic number 10, that exhibits a characteristic red glow when ionized and is extensively used as the gas in ►neon lamps.

neon lamp A small lamp that uses a glow discharge in low-pressure neon as the source of light (►gas-discharge tube). The light is red. Neon lamps consume very little power. They can be made extremely large for use in illuminated signs or extremely small for use as indicators and voltage regulators.

neper Symbol: Np. A dimensionless unit used in telecommunications to express the ratio of two powers. It is the natural logarithm of the square root of the power ratio. Thus if two values of power, P_1 and P_2, differ by n nepers then

$$n = \log_e[\sqrt{(P_2/P_1)}] = \frac{1}{2} \log_e(P_2/P_1)$$

$$P_2/P_1 = e^{2n}$$

In a single transmission line or other transmission network the neper can be used to compare two currents, usually the input and output currents, the current being proportional to the square root of the power. Thus if two currents, I_1 and I_2, differ by N nepers then

$$N = \log_e(I_2/I_1)$$

One neper equals 8.686 ►decibels.

Nernst effect If a conductor or semiconductor is placed in a magnetic flux density and a temperature gradient maintained across it at right angles to the magnetic flux density, an e.m.f. is developed in the material orthogonal to both the magnetic flux density and the temperature gradient. This is the converse of the ►Ettinghausen effect and is related to the ►Righi–Leduc effect.

Net *Short for* Internet.

net loss The difference between the attenuation and the gain in any circuit, device, network, or transmission line.

network 1. In electronics, a number of impedances connected together to form a system that consists of a set of interrelated circuits and that performs specific functions. The behaviour of the network depends on the values of the components, such as the resistances, capacitances, and inductances, from which it is formed and the manner in which they are interconnected. The values of the components are termed the *network parameters* or *network constants*. The nomenclature of networks describes either the type of component, the method of interconnection, or the expected behaviour of the network.

Networks are described as *resistive, resistance-capacitance (R-C), inductance-capacitance (L-C), inductance (L) networks,* etc., depending on their components.

Lattice networks have the input and output terminals at a junction between two or more conductors (Fig. *a*); a *bridge network* is a particular type of lattice network (Fig. *b*).

Series networks and *parallel networks* have their elements connected in series and in parallel, respectively.

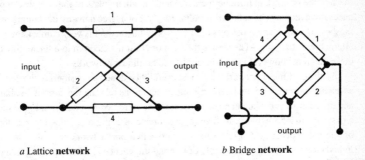

a Lattice **network**   *b* Bridge **network**

Linear networks have a linear relationship between the voltages and currents; otherwise they are *nonlinear.*

Bilateral networks conduct in both directions whereas those that conduct in only one direction are *unilateral.*

Passive networks contain no energy source or sink other than normal ohmic losses; those that do contain an energy source or sink are *active.*

All-pass networks attenuate all frequencies equally; other networks are described according to their frequency response ($\blacktriangleright$filter).

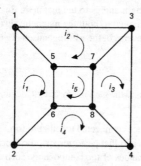

c Conducting paths of a **network**

A point within a network at which three or more of the elements are joined is termed a *node* (or *branch point*); points 1–8 in Fig. *c* are nodes. A conducting path between two such points is termed a *branch* (1–2, 3–4, etc., in Fig. *c*). A voltage at a point in the network measured relative to the voltage at a designated node is termed a *node voltage.* A closed conducting loop in the network (e.g. 1, 3, 7, 5, 1) forms a *mesh contour*, and the portion of the network bounded by it is termed a *mesh.* Any branch that is common to two or more meshes is a *mutual branch* (e.g. 5–6). Two branches of a network are said to be *conjugate* if an e.m.f. in one of them does not produce a current in the other. The currents circulating in the meshes are known as *mesh currents.*

The behaviour of a network may be analysed by applying $\blacktriangleright$Kirchhoff's laws to each mesh in the network in turn; both the real and imaginary parts of the complex impedances involved must be satisfied simultaneously. For a large network containing many meshes, as with many types of $\blacktriangleright$filter network, this method is very cumbersome. An alternative is to apply $\blacktriangleright$Thévenin's theorem or Norton's theorem to a linear network; these theorems however cannot be applied to nonlinear networks.

Analysis of linear networks can most usefully be done by considering the network as a $\blacktriangleright$two-port network and deriving sets of equations relating the currents, voltages, and impedances at the input and output; this is known as *two-port analysis.* Fig. *d* shows a passive two-port network with an input source consisting of a voltage generator V_s of internal impedance Z_s. Fig. *e* shows an active two-port network presenting an input impedance z_1 to the input source V_s of internal impedance Z_s, and appearing as a cur-

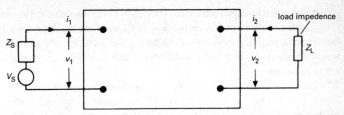

d Passive two-port **network**

rent generator $g_m v_1$ shunted by a resistance r_0 and producing a voltage v_2 in the output circuit (g_m being the transconductance).

Three different sets of equations can be written down: the *impedance equations*, the *admittance equations*, and derived from these the *hybrid equations*. The impedance equations can be written in the form of a matrix:

$$\begin{vmatrix} v_1 \\ v_2 \end{vmatrix} = \begin{vmatrix} z_{11} & z_{12} \\ z_{21} & z_{22} \end{vmatrix} \begin{vmatrix} i_1 \\ i_2 \end{vmatrix}$$

Equivalent matrices can be written for the hybrid and admittance equations. The constants in these equations are known as *z*, *h*, and *y parameters*, respectively, or collectively as *two-port parameters*. Three-terminal devices, such as transistors, can be represented as two-port networks that have two terminals joined together (►transistor parameters).

In the case of nonlinear networks the matrix equations are only true for small changes of current and voltage. In such cases the two-port parameters are termed *small-signal parameters* and are quantities that change value according to the operating conditions of the device.

The input and output impedances of a network, v_1/i_1 and v_2/i_2, can be calculated from the matrix equations; it can be shown that v_1/i_1 depends on the load impedance Z_L connected to the output and conversely that v_2/i_2 depends on the impedance Z_s of the source connected to the input.

The ►driving-point impedance is the impedance presented at a pair of terminals of a network of four or more terminals, under designated conditions at the other pair(s) of terminals. In the limiting case, for a two-port network, if the input (or output) is open

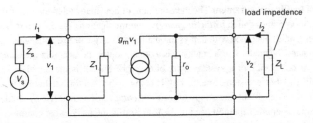

e Active two-port **network**

circuit, the output (or input) impedance is the *open-circuit impedance*. The other limiting case is when the input (or output) is a short circuit in which case the output (or input) impedance is the *short-circuit impedance*. The quantities v_2/i_1 and v_1/i_2 are the ►transimpedances of the network under open-circuit conditions, i.e. when $i_2 = 0$ and $i_1 = 0$, respectively.

2. In communications, a collection of resources used by a group of users to exchange information. In a ►local area network (LAN), users generally belong to a single organization located on a single site or a small number of nearby sites. A ►wide area network (WAN) is also usually operated by a single organization but communications are over large distances. Communication paths in a network are established and switched between computer terminals following agreed procedures known as ►protocols. The communication lines may include cables, optical fibres, phone lines, or radio links. These communication lines are interconnected at points known as *nodes*. The nodal device may be an electrical interface or a computer. ►►digital communications; bus network; ring network; star network.

network access In communication systems, the equipment or ►protocol required to interface to or make use of a digital communications ►network.

network analyser An instrument used for measuring the incident, reflected, and transmitted power at a ►two-port network, particularly at radio and microwave frequencies. This will provide the ►scattering parameters of the network under test. Both scalar (magnitude only) and vector (magnitude and phase) measurements can be made.

network analysis The analysis of a given network in order to determine the output signal, given the input signal. ►network.

network constants *Syn. for* network parameters. ►network.

network parameters ►network.

network synthesis *Syn.* filter synthesis. A systematic technique of determining the network ►transfer function that will produce a given output signal for a given input signal. This is in contrast to the normal analysis of a given network to determine the output signal given the input signal. The answer provided by the process of network synthesis is never unique, and with some combinations of input signal and required output signal an answer may not be possible.

network traffic measurement The measurement of how many telephone calls or connections (sometimes called *traffic intensity*) a communications system can make. The measurement is concerned with delays in obtaining a service or lost calls where a connection cannot be made. Traffic intensity is a dimensionless number relating to the congestion or occupancy of a system, line, or network link. It is measured in *erlangs* (symbol: E): when a system, line, or link is being fully used for one hour it is said to have a traffic intensity of one erlang of traffic. *Centum call second (CCS)*, or hundred call second, is another method used to measure traffic intensity, most common in the US: one CCS is one thirty-sixth of an erlang.

neural network 1. A modelling technique based on the observed structure and function of biological neurons and used to mimic the performance of a computer system. **2.** An electronic network simulating this.

neuroelectricity Electricity generated in the nervous systems of humans and other live animals.

neutral 1. Having no net positive or negative electric charge; at ➤earth potential. **2.** Denoting the line that completes the domestic ➤mains supply and is connected to earth at the power station.

neutralization The provision in an amplifier of negative ➤feedback of such amplitude and phase that it counteracts any inherent positive feedback. If the positive feedback is not neutralized, unwanted oscillations may occur.

Unwanted positive feedback in an amplifier can arise from the ➤Miller effect and is counteracted by neutralization: a circuit is used to provide a 180° phase shift in the voltage fed back to the input (base) circuit. ➤Parasitic oscillations produced during ➤push-pull operation can be counteracted by *cross neutralization,* in which a portion of the output voltage of each device is fed back by means of a neutralizing capacitor to the input (base) circuit of the other device.

The voltage fed back to the input is termed the *neutralizing voltage.* The degree of neutralization in an amplifier may be observed using an indicating device termed a *neutralizing indicator.*

neutral temperature ➤thermocouple.

newton Symbol: N. The ➤SI unit of force defined as the force that, when applied to a mass of one kilogram, gives it an acceleration of one metre per second per second.

NICAM *Abbrev. for* near instantaneously companded audio multiplex. A system used by the BBC and ITV whereby an extra digital code is added to the TV signal to carry digital stereo sound, so improving sound quality. To receive the signal, a TV or video cassette recorder has to have a NICAM decoder, and also a stereo amplifier and a pair of loudspeakers.

Nichrome *Trademark* An alloy of approximately 62% nickel, 15% chromium, and 23% iron. It has a very high resistivity and can operate at extremely high temperatures; it is therefore suitable for use in making thin-film resistors, wire-wound resistors, and heating elements.

nickel Symbol: Ni. A metal, atomic number 28, that is widely used in electronics. It exhibits strong ➤ferromagnetism and is used in magnetic alloys. It is also used as a conductor and in electrolytic cells.

nickel-cadmium cell A secondary ➤cell that has a nickel hydroxide/nickel oxide mixture as the anode, a cadmium cathode, and potassium hydroxide as the electrolyte.

nickel-iron cell *Syn.* Ni-Fe cell. A secondary ►cell that has a nickel hydroxide/nickel oxide anode, an iron cathode, and potassium hydroxide as the electrolyte. The e.m.f. is about 1.2 volts.

Ni-Fe cell ►nickel-iron cell.

NIST *Abbrev. for* National Institute of Standards and Technology. ►standardization.

NMOS ►MOSFET.

noctovision A television system that views an infrared image rather than a visible light image and is sometimes used at night. A specially designed infrared-sensitive ►camera tube produces a video signal that can be detected by a standard receiver.

nodal analysis A network or circuit analysis technique used to find all the unknown node voltages (►network) using ►Kirchhoff's current laws.

node 1. Any point, line, or surface in a distributed field at which some specified variable of a standing wave, such as voltage or current, attains a minimum value, usually zero. A *partial node* has a nonzero minimum. A point at which maximum magnitude is attained is an *antinode*. **2.** *Syn.* branch point. A point within an electrical network at which three or more elements are joined. ►network. **3.** An interconnection point in a communications ►network.

node voltage ►network.

noise Unwanted electrical signals that occur within electrical devices, circuits, or systems and result in a spurious signal at the output. The source of the noise can be *man-made*, the following being examples.

Mains hum, caused by electromagnetic induction from power cabling, transformers, etc.: this is a periodic signal (at the mains frequency), but is an unwanted signal.

Spark interference, such as from a car ignition system, fluorescent light starter, or commutating motor: this is high-frequency noise seen as a burst of high-voltage spikes.

Microphony: acoustic interference with sensitive apparatus can introduce unwanted electrical signals, particularly with ►vacuum-tube apparatus.

Radiofrequency interference: this may be on the desired receiving channel, on an adjacent channel to the desired channel, or may be an out-of-band or spurious response of the receiver; it is an inherent problem with heterodyne systems and requires careful design for spurious response rejection (►cochannel rejection; spurious rejection; electromagnetic compatibility).

Quantization noise, resulting from the inaccurate representation of analogue signals in a digital system of limited resolution.

These man-made sources of noise can generally be minimized by careful component choice, circuit and system design, and appropriate filtering.

The second category of noise sources is naturally occurring noise due to the particle-like nature of electrical conduction, and includes the following:

Thermal noise (or *Johnson noise*), due to the random motion of charge ►carriers in a conductor or semiconductor, giving rise to a fluctuating voltage: this voltage increases with temperature as the carriers move more rapidly.

Shot noise, due to charge carriers crossing a potential barrier, such as a ➤p-n junction or ➤metal-semiconductor contact.

Flicker noise, due to the random trapping and release of charge carriers in many electrical and naturally occurring systems: it has an inverse frequency response and is often called l/f (one over f) noise.

Radiation noise, caused by cosmic radiation, and similar short-wavelength radiation such as gamma rays or X-rays: this is ➤impulse noise and can result in errors in digital systems.

Contact noise, occurring in discontinuous conductors such as carbon resistors, which are made from compressed particles: this is an excess noise source over the normal thermal noise.

noise factor Symbol: F. In a given electrical component, circuit, or system, a measure of the additional noise contributed by that component, circuit, or system. It can be defined as the ratio of the ➤signal-to-noise ratio at input to the signal-to-noise ratio at output:

$$F = (S_i/N_i)/(S_o/N_o)$$

If the ➤gain of the system is given by G, this can be written

noise figure Symbol: F_{dB}. The ➤noise factor F expressed in decibels:

$$F = \frac{S_i/N_i}{GS_i/(GN_i + N_{add})} = \frac{N_i + (N_{add}/G)}{N_i}$$

$$F_{dB} = 10 \log F$$

noise margin The degree to which a ➤logic gate can withstand variations on logic levels at the input without causing the output state of the gate to change significantly.

noise temperature The equivalent temperature that is often used as a figure of merit for the power associated with noise in a device or system. If there is power P associated with the noise, and the system is operating with bandwidth B, then

$$P = kTB$$

where T is the noise temperature and k is the Boltzmann constant.

no-load Operation of any electronic or electrical machine, apparatus, circuit, or device under rated operating conditions but in the absence of a ➤load.

noncoherent detection *Syn.* noncoherent demodulation. ➤modulation.

nondestructive read operation ➤read.

noninverting amplifier A configuration of an ➤operational amplifier (see diagram) where the gain G of the amplifier is given by $1 + (R_f/R_c)$. ➤➤inverting amplifier.

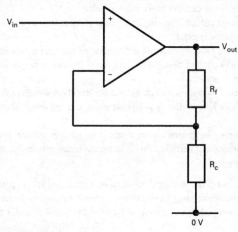

Noninverting amplifier

nonlinear amplifier ➤amplifier.

nonlinear distortion ➤distortion.

nonlinear network ➤network.

nonperiodic signal ➤aperiodic signal.

nonreactive Denoting any electric circuit, device, or winding that for its desired function has negligible ➤reactance.

nonreactive load A ➤load in which the alternating current is in phase with the alternating voltage at the terminals. ➤➤reactive load.

nonrecursive filters ➤digital filter.

nonresonant line ➤transmission line.

nonsaturated mode ➤saturated mode.

nonthreshold logic (NTL) A family of integrated ➤logic circuits that operates at a high speed at relatively low power. The basic NTL gate is shown in the diagram. It may be considered as a mixture of ➤resistor-transistor logic (RTL) and a simplified form of ➤emitter-coupled logic (ECL). The input transistors A and B form a ➤long-tailed pair with resistor R_1. The voltage gain is determined by the ratio R_2/R_1 and is adjusted to be slightly more than unity. Unlike other logic circuits the transistors conduct a relatively large current in the 'off' state, i.e. a logical 0 at the input. If a logical 1 is applied to the base of either transistor, that transistor saturates and the collector voltage falls. If both inputs are high the output is low, but if either input is low (or if both inputs are low) then the output voltage is high. The voltage swing is low, thus avoiding deep saturation and optimizing the switching time. A capacitor C may also be used,

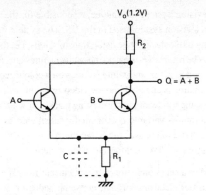

Nonthreshold logic: NAND gate

in parallel with resistor R_1, to further improve the switching speed. The relative simplicity of the basic gate gives a good potential for VLSI applications, but the low voltage gain employed makes the circuits susceptible to stray noise.

nonvolatile memory ➤Memory that retains information stored in it when the power supply is switched off.

NOR circuit (or **gate**) ➤logic circuit.

normal distribution ➤Gaussian distribution.

normalized filter *Syn.* prototype filter. A low-pass ➤filter with a normalized ➤cut-off frequency of one radian per second. From this normalized prototype filter, conversions to other frequencies or to other types (such as high-pass or band-pass) is achieved through ➤denormalization and ➤frequency transformation respectively.

Norton's theorem ➤Thévenin's theorem.

notch filter A band-stop ➤filter, usually one that only attenuates a narrow band of frequencies.

NOT circuit (or **gate**) *Syn. for* inverter. ➤logic circuit.

note frequency *Syn. for* beat frequency. ➤beats.

n-p junction A ➤p-n junction.

n-p-n transistor (or **NPN transistor**) ➤bipolar junction transistor.

NRZ codes *Short for* nonreturn to zero codes. ➤digital codes.

NRZ PCM *Abbrev. for* NRZ pulse code modulation. ➤pulse modulation.

NTL *Abbrev. for* nonthreshold logic.

NTSC *Abbrev. for* National Television System Committee, responsible for the specification, in 1953, of the colour television system used in the US. The system needed to be compatible with the existing monochrome television transmissions, i.e. the colour transmissions were required to be receivable as black-and-white pictures on a monochrome receiver and allow reception of black-and-white pictures on a colour receiver. In the NTSC system colour information is encoded into the video signal in such a way that the transmissions occupy the same bandwidth as the existing black-and-white transmissions. The system is susceptible to variable colour quality attributed to small changes of differential gain and differential phase within the transmission chain. This problem was overcome in the European ➤PAL colour television system.

n-type conductivity Conduction in a semiconductor in which current flow is caused by the movement of majority carrier electrons through the semiconductor. ➤semiconductor; p-type conductivity.

n-type semiconductor An extrinsic semiconductor that contains a higher density of conduction ➤electrons than of mobile ➤holes, i.e. electrons are the majority carriers. ➤semiconductor; p-type semiconductor.

nucleus The central and most massive part of an atom. It carries a positive charge Ze, where Z is the atomic number of the atom and e the electronic charge. A nucleus consists of tightly bound protons and neutrons, the total number of which is called the mass number, A. The number of neutrons associated with a given number of protons can vary within limits, giving rise to various isotopes of an element.

null A region of low ➤gain in an ➤antenna pattern. It is desirable to avoid positioning nulls along directions in which an antenna is required to perform well.

null method *Syn.* balance method. An accurate method of measurement in which the quantity being measured, such as resistance or capacitance, is balanced by another of a similar kind: voltages in different circuits are adjusted so that the response of an indicating instrument falls to zero. The best known example of the null method is in the ➤Wheatstone bridge although it is used in many other bridge circuits.

null-point detector An instrument that is used with ➤bridge circuits to give a zero response when the balance point is achieved. Direct-current bridge circuits employ a sensitive galvanometer: balance is achieved when a zero indication is obtained. Alternating-current bridge circuits frequently employ a sensitive loudspeaker, usually as a headset: the sound becomes inaudible at the balance point.

number density The number of electrons, holes, atoms, etc. per unit volume.

number of poles The number of different conducting paths that may be simultaneously opened or closed by a switch, circuit-breaker, or other similar apparatus. The device is described as single pole, double pole, triple pole, or multi-pole depending on the number of poles that it can operate.

numerical control A type of automatic control in which a number generated by the controlling device, such as a digital computer, is used to control another device, particularly automatic machines such as machine tools.

Nyquist diagram A diagram that can be used as a criterion of stability in a system. It provides a simple graphical procedure for determining closed-loop stability from the frequency-response curves of the open-loop transfer function of a system.

Nyquist frequency ➤Nyquist rate.

Nyquist noise theorem The law relating the power P dissipated in a resistor due to thermal ➤noise with the frequency f of the signal. At ordinary temperatures the law is expressed as:

$$dP = kT \, df$$

where k is the Boltzmann constant and T the thermodynamic temperature.

Nyquist point ➤gain margin.

Nyquist rate The minimum sampling rate at which the maximum frequency component in an analogue signal (the *Nyquist frequency*) can be recovered following ➤sampling. The Nyquist rate equals twice the Nyquist frequency.

O

observed failure rate ➤failure rate.

observed mean life ➤mean life.

OCR *Abbrev. for* optical character reader.

octal A symbolic notation that uses only the numerals 0 to 7 to represent numbers, which are expressed in base 8.

octave An interval that has the frequency ratio 2:1.

oersted Symbol: Oe. The unit of magnetic field strength in the obsolete CGS electromagnetic system of units. One oersted equals 79.58 (i.e. $1000/4\pi$) amperes per metre.

OFC *Abbrev. for* oxygen-free copper.

offline Denoting a device, possibly an ➤online device, that is switched off, broken, or disconnected from a computer. A device may be physically connected but offline if the computer system has been instructed not to use it. ➤online.

offset QPSK (OQPSK) ➤amplitude modulation.

ohm Symbol: Ω. The ➤SI unit of electrical ➤resistance, ➤reactance, and ➤impedance. It is defined as the resistance between two points on a conductor when a constant current of one ampere flows as a result of an external potential difference of one volt applied between the points. This unit replaced the *international ohm* (Ω_{int}) as the standard of resistance: one Ω_{int} equals $1.000\ 49\ \Omega$.

ohmic ➤Ohm's law.

ohmic contact Strictly, an electrical contact in which the potential difference across it is linearly proportional to the current flowing through it. In practice, any contact that allows a current flow with minimal voltage drop is termed ohmic. Practical ohmic contacts between a metal and a semiconductor can be formed by having a highly doped surface layer on the semiconductor to ensure that the dominant method of transfer of carriers (electrons in an n-type semiconductor) is due to the ➤tunnel effect. ➤metal-semiconductor contact.

ohmic loss Power dissipation in an electrical circuit, network, or device that is due to the resistance present rather than to other causes such as eddy currents or back e.m.f.

ohmmeter An instrument that measures the electrical ➤resistance of conductors or insulators. The indicating scale is calibrated in ohms or suitable multiples or submultiples of ohms.

ohm metre Symbol: Ω m. The ➤SI unit of electric ➤resistivity.

Ohm's law The electric current, I, flowing in a conductor or resistor is linearly proportional to the applied potential difference, V, across it. From the definition of ➤resistance, R, Ohm's law can be written:

$$V = IR$$

Any electrical component, circuit, or device that maintains such a linear relationship between current and voltage can be described as *ohmic*.

The form of Ohm's law at a particular point is given by

$$J = \sigma E$$

which relates the current density J at a point to the electric field E at that point and the ➤conductivity σ of the material.

ohms per square ➤sheet resistance.

omnidirectional antenna An ➤antenna whose radiative or receiving properties are the same around the azimuthal direction (i.e. pivoting around a vertical axis) but vary over elevational directions. An example is a ➤dipole.

omnidirectional microphone A ➤microphone that picks up sound equally from all directions (see diagram). Practically, such a response is difficult to achieve for frequencies above which the wavelength becomes comparable with the physical dimensions of the microphone itself.

OMVPE *Abbrev. for* organo-metallic vapour phase epitaxy. ➤chemical vapour deposition.

ones' complement notation An integer representation that uses absolute ➤binary notation for positive values, and forms negative values by complementing the bits (converting zeros to ones and vice versa) of positive values, including all leading zeros. For example, the 10-bit representation of the number 14 is

$$00\ 0000\ 1110_2$$

and the 10-bit ones' complement representation of -14 is

$$11\ 1111\ 0001_2$$

one-shot multivibrator *Syn.* single-shot multivibrator. ➤monostable.

one-sided abrupt junction ➤abrupt junction.

O-network ➤two-port network.

online Connected to a computer system and usable. ➤➤interactive; offline.

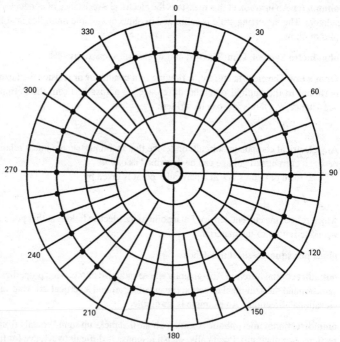

Idealized directional response of an **omnidirectional microphone** (scale in degrees)

on-off keying (OOK) ➤amplitude modulation.

OOK *Abbrev. for* on-off keying. ➤amplitude modulation.

op-amp *Short for* operational amplifier.

op-code *Short for* operation code.

open-area test site *Syn.* open-field test site. A region of open ground used for ➤electromagnetic compatibility testing. The surface of the ground is usually covered with a solid conducting sheet or conducting mesh to provide a well-defined electrical surface or ➤ground plane. The product under test and a receiving antenna are placed above the ground plane and the ➤radiated emissions from the product are monitored.

open circuit ➤circuit.

open-circuit impedance ➤network.

open-circuit voltage The voltage across the output terminals of any electrical or electronic network, device, machine, or other apparatus at ➤no-load when designated operating conditions are carried out.

open-ended Denoting any circuit, system, or process that can be readily augmented by the addition of further stages, parts, or steps.

open-field test site ►open-area test site.

open-loop gain The gain of an amplifier without a feedback loop created by an external network. ►►closed-loop gain.

open systems interconnection ►OSI.

operand One of the inputs to a mathematical or logical function or to a computer instruction.

operating point The point on the family of ►characteristic curves of an active electronic device, such as a transistor, that represents the magnitudes of voltage and current when designated operating conditions are applied to the device.

operational amplifier A high-gain direct-coupled ►amplifier originally intended for performing mathematical operations such as integration or differentiation. An operational amplifier also has a wide range of applications outside computation, including signal-conditioning functions such as filtering.

operational transconductance amplifier (OTA) ►dependent sources.

operation code The part of a computer instruction (usually the first field of the instruction) that specifies the operation.

optical ammeter An instrument that measures the current flowing through the filament of an incandescent lamp by comparing, photometrically, the illumination produced with that produced by a current of known magnitude in the same filament.

optical character reader (OCR) A device that is used to produce coded signals suitable for a given computer from information in the form of characters, numbers, or other symbols printed on paper.

optical fibre A fine filament of glass or clear plastic that transmits light along its length. To reduce light loss, the fibre is made of very pure material and the refractive index of the material is made to vary across the fibre. The refractive index variation causes light rays at less than a certain angle to the axis of the fibre to be totally internally reflected back into the fibre, and this continues along the transmission path. Optical fibres are used primarily for communication purposes. ►fibre-optics system.

optical image ►camera tube.

optical stepper ►photolithography.

optical switch A means of diverting an optical signal, such as one transmitted down an ►optical fibre, between two or more paths. Two common methods of achieving an optical switch are by the physical movement of the output fibres by means of a solenoid, or by using a mirror to reflect the optical signal between the two outputs.

OQPSK *Abbrev. for* offset quadrature phase shift keying. ►modulation.

Oracle *Trademark* ➤teletext.

OR circuit (or **gate**) ➤logic circuit.

order (of a filter) A means of classifying a ➤filter. The order of a filter is an integer number, also called the ➤number of poles; for example, a *second-order filter* is said to have two poles. In general, the higher the order of a filter, the more closely it approximates an ideal filter and the more complex the circuitry required to construct it.

orthicon ➤iconoscope.

orthogonal Mutually perpendicular; at right angles.

orthogonal codes ➤digital codes.

oscillating current (or **voltage**) A current (or voltage) waveform that periodically increases and decreases in amplitude with respect to time. Oscillating waveforms can, for example, be sinusoidal, sawtooth, or square.

oscillation 1. A periodic variation of an electrical quantity, such as current or voltage. **2.** A phenomenon that occurs in an electrical circuit if the values of self-inductance and capacitance in the circuit are such that an oscillating current results from a disturbance of the electrical equilibrium of the circuit.

A circuit that produces oscillations freely is termed an *oscillatory circuit*. Oscillations that result from the application of a direct-voltage input to the circuit and continue until the direct voltage is removed are termed *self-sustaining oscillations*. Oscillations that tend to decrease in amplitude with respect to time are known as *stable oscillations*; *unstable oscillations* tend to increase in amplitude with respect to time and soon exceed the rated operating conditions of the circuit. ➤free oscillations; forced oscillations; damped; parasitic oscillations.

oscillator A circuit that converts direct-current power into alternating-current power at a frequency that is usually greater than can be achieved by rotating electromechanical alternating-current generators. Application of the direct-voltage supply to the circuit is usually sufficient to cause it to oscillate and for the oscillations to be maintained until the direct voltage is switched off.

There are two broad categories of oscillator: *harmonic oscillators* generate essentially sinusoidal waveforms and contain one or more active circuit elements continuously supplying power to the passive components; ➤relaxation oscillators are characterized by nonsinusoidal waveforms, such as sawtooth waveforms, and the switched exchange of electrical energy between the active and passive circuit elements.

A simple harmonic oscillator consists essentially of a frequency-determining device, such as a resonant circuit, and an active element that supplies direct power to the resonant circuit and also compensates for damping due to resistive losses. In the case of a simple L-C circuit, application of a direct voltage causes ➤free oscillations in the circuit that decay because of the inevitable resistance in the circuit (➤damped). In the absence of the resistance no damping would occur and the free oscillations would continue at a constant amplitude until the direct voltage was removed. The active element in an oscillator can be considered as supplying a ►negative resistance of sufficient value to compensate for the positive resistance; consequently the complete oscillator contains effectively zero resistance and when shocked will oscillate continuously.

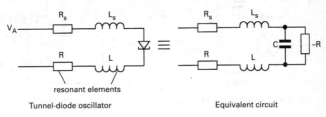

a Negative-resistance **oscillator**

The effective negative resistance is provided either by a device, such as a unijunction transistor, that exhibits a negative-resistance portion of its characteristic or by employing positive ►feedback of power in order to overcome the damping. Any particular oscillator may be studied from the negative-resistance approach or from a feedback approach. In the latter case internal positive feedback is considered to be present in the negative-resistance device. Usually *negative-resistance oscillators* are those that contain a device such as a unijunction transistor or tunnel diode (Fig. *a*), operated in the negative-resistance portion of the characteristic determined by the applied voltage, V_A, and external source resistance, R_s.

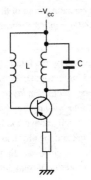

b Common-emitter **oscillator** with transformer feedback

Feedback oscillators are those that employ external positive feedback. An inherent phase shift of 180° occurs between the base and collector of the common-emitter connection shown in Fig. *b*. Various types of feedback circuit are used in order to provide the necessary counterbalancing phase-shift. Transformer coupling is shown in the diagram; the resonant circuit is formed by the transformer primary L and the capacitor C.

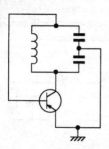
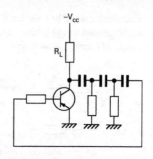

c Colpitt's **oscillator**: common emitter   *d* Phase-shift **oscillator**

The frequency-determining device may consist of a component, such as a piezo-electric or magnetostrictive crystal, that converts mechanical stress into electrical impulses; alternatively such a device may be coupled to the resonant circuit to prevent frequency drift. *Colpitt's oscillator* and the *phase-shift oscillator* are shown in Figs. *c* and *d*.

oscillatory circuit ➤oscillation.

oscilloscope An instrument that produces a visible image of one or more rapidly varying electrical quantities. An image can be produced showing the variation of a signal with respect to time or with respect to another electrical quantity. The most usual type of oscilloscope is the ➤cathode-ray oscilloscope.

An oscilloscope provided with the means to produce a permanent record of the signal on film or magnetically sensitive paper is known as an *oscillograph*; the record thus produced is an *oscillogram*.

OSI *Abbrev. for* open systems interconnection. A model devised by the International Standards Organization for network communication involving computer hardware and software applications. The model is defined in seven layers: *application layer*, *presentation layer*, *session layer*, *transport layer*, *network layer*, *data link layer*, and *physical layer* (see diagram). Each layer passes data to the layers directly above and below it. The application layer is furthest from the communication link and the physical layer is nearest the link. Data passes over the communication link to the corresponding lowest layer in the remote system. Dashed lines in the diagram represent virtual circuits.

OTA *Abbrev. for* operational transconductance amplifier. ➤dependent sources.

OTDR *Abbrev. for* optical time delay reflectometer. ➤reflectometer.

O-type microwave tube *Syn. for* linear-beam microwave tube. ➤microwave tube.

out of phase ➤phase.

output 1. The power, voltage, or current delivered by any circuit, device, or apparatus. **2.** The part of any circuit, device, or apparatus, usually in the form of terminals, at which the power, voltage, or current is delivered. ➤➤input/output. **3.** To deliver as an output signal.

output gap An ➤interaction space in a ➤microwave tube where power is extracted from the electron beam; it hence constitutes the output section of the tube.

output impedance The ➤impedance presented at the output of an electronic circuit or device.

output transformer A transformer that is used for coupling an output circuit, particularly that of an amplifier, to the ➤load.

overall efficiency The ratio of the power absorbed by the ➤load of a device to the power supplied by the source.

overbunching ➤velocity modulation.

overcoupling ➤coupling.

overcurrent ➤overcurrent release.

overcurrent release *Syn.* overload release. A switch, circuit-breaker, or other tripping device that operates when the current in a circuit exceeds a predetermined value. A current that causes the release to operate is an *overcurrent*.

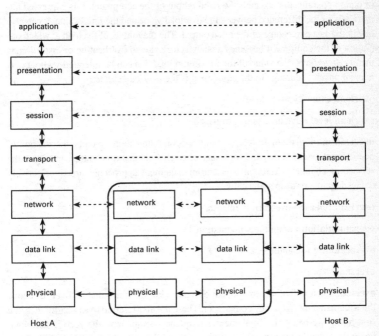

OSI model

The device is often designed so that a delay occurs after the overcurrent is sensed and before the device trips. Several different delay conditions can be used: *definite time lag* overcurrent release has a predetermined delay independent of the magnitude of the overcurrent; *inverse time lag* overcurrent release has a delay that is an inverse function of the magnitude of the overcurrent; *inverse and definite minimum time lag* overcurrent release occurs when the delay is an inverse function of the magnitude of the overcurrent until a minimum value of the delay is reached. ➤undercurrent release.

overdamping *Syn.* periodic damping. ➤damped.

overdriven amplifier An ➤amplifier that is operated with an input voltage greater than that for which the circuit was designed. Overdriving results in ➤distortion being introduced into the output waveform.

overflow The state in which a number is too large for the intended representation (for example, any value larger than 255 for an 8-bit binary representation).

overlay network In telephone or telecommunication systems, a network that is laid over the same geographic area as the basic network and provides a specific service or fulfils a specific function.

overload Any ➤load delivered at the output of an electrical device, circuit, machine, or other apparatus that exceeds the rated output of the equipment. It is expressed numerically as the difference between the overload value and the rated value; it can also be quoted as a percentage of the rated output. The magnitude of the load at which operation of the equipment becomes unsatisfactory, due to overheating or distortion, is the *overload level;* the value of the maximum load that can be tolerated without permanent physical damage to the equipment is the *overload capacity.*

overload capacity ➤overload.

overload level ➤dynamic range; overload.

overload management In telephone or telecommunication systems, the method of dealing with peaks in demand for a service or channel that exceed the capacity of the service or channel. Common overload-management approaches are to selectively delay, degrade, or drop some calls.

overload release ➤overcurrent release.

overmodulation ➤amplitude modulation.

overscanning ➤scanning.

overshoot ➤pulse.

overtone Each individual component of a periodic waveform that is 'over' or above the ➤fundamental frequency component. These are also referred to as ➤harmonics, the first harmonic being the fundamental frequency component, the second harmonic being the first overtone, the third harmonic being the second overtone, etc.

overvoltage 1. A voltage that exceeds the normal voltage applied between two conductors or between a conductor and earth. **2.** (at a given electrode in an electrolytic cell) The amount by which the applied e.m.f. necessary to release hydrogen on that electrode from a particular electrolyte exceeds the e.m.f. required to liberate hydrogen from the same electrolyte on a standard platinum electrode.

overvoltage release A switch, circuit-breaker, or other tripping device that operates when the voltage in a circuit exceeds a predetermined value. A voltage that causes the release to operate is an overvoltage. ➤undervoltage release.

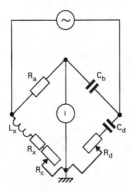

Owen bridge

Owen bridge A four-arm ➤bridge used for the measurement of inductance in terms of known resistance and capacitance (see diagram). At balance, as shown by the null response of the indicating instrument, I,

$$L_x = C_b R_a R_d$$

$$R_x = (C_b R_a / C_d) - R_c$$

oxidation A chemical reaction used in electronics in which a thin portion of the surface of a silicon chip is converted to silicon dioxide.

oxide In general, a chemical compound of an element with oxygen. In particular, *short for* silicon dioxide (➤silica). It is the most widely used insulating and/or passivating material in the construction of devices, components, and integrated circuits fabricated in silicon. Silicon dioxide can be readily grown on the surface of silicon and layers of varying thicknesses are grown during the fabrication of ➤silicon devices and circuits.

Oxide acts as a barrier to the diffusion of impurities into the substrate material and is used as an 'on-chip' ➤mask during the ➤planar process. A thin oxide layer is used to form the insulator in the manufacture of MOS devices, such as ➤MOSFETs and ➤charge-coupled devices. An extra thick oxide layer can be formed by means of the ➤coplanar process in order to prevent spurious MOST formation. A layer of oxide is formed over both bipolar and MOS integrated ciruits and devices for ➤passivation of the surface.

oxide masking The use of the silicon dioxide layer (➤oxidation) on the surface of a silicon chip to provide a mask for selective processing of the chip. The oxide formed by oxidation of the surface of the silicon is selectively etched to expose the underlying silicon prior to ➤diffusion or ➤metallizing.

oxygen-free copper (OFC) Single-crystal or large-crystal copper wire formed from highly pure copper that contains no oxides between the crystal grains. It is said to offer superior sound quality in audio applications.

P

PABX *Abbrev. for* private automatic branch exchange. ➤telephony.

packet A segment of a ➤digital signal – i.e. a group of ➤bits – that travels through a communications network as an information unit. Each packet has a fixed maximum size, and includes information about its source and destination. ➤➤packet switching; digital communications.

packet switching In digital communications, a technique that provides point to point transmission of ➤packets through a switching network. When a packet is received by a digital switching point such as a digital exchange, the packet is stored, has its destination read, and is then directed towards the destination in the most effective way: the exchange can choose from a number of channels, the selection depending on how busy each channels is. ➤digital communications.

packing density 1. *Syn.* functional packing density. The number of devices or logic gates per unit area of an ➤integrated circuit. **2.** The amount of information that is contained in a given dimension of a storage system of a digital computer, e.g. the number of ➤bits per inch of magnetic tape.

pad 1. ➤attenuator. **2.** ➤bonding pad.

PAD *Abbrev. for* packet assembler/disassembler. In digital communication systems, the interface circuit that lies between the user's information and the format required for transmission over the network. The PAD converts the data between the two formats.

pair A ➤transmission line consisting of two similar conductors that are insulated from each other but are associated to form part of a communication channel or channels. If the two conductors are twisted around each other they form a *twisted pair*. Many pairs can be further twisted together. The twisted pair or pairs can be inside an earthed conducting shield, usually formed from braided fine wire mesh, to minimize ➤electromagnetic interference; this is a *shielded twisted pair* (STP) construction. In an *unshielded twisted pair* (UTP) construction, the twisted pair(s) are uncovered or bundled within an insulating PVC covering. A *coaxial pair* is a pair of cylindrical conductors that are coaxial and may be used to form a ➤coaxial cable.

paired cable *Syn* twin cable. A type of cable composed of several ➤pairs of conductors: each pair is twisted together but no two sets of pairs are twisted.

pairing A fault occurring in television picture tubes that employ interlaced scanning
(➤television). Lines of alternate fields tend to coincide instead of interlacing with
each other and the vertical resolution is halved.

PAL 1. *Abbrev. for* phase alternation line. A colour-television system developed in Germany. The chrominance signal (➤colour television) is resolved into two components
in ➤quadrature that are used for ➤amplitude modulation of the chrominance subcarriers. In the PAL system the relative phase of the quadrature components is reversed on
alternate lines in order to minimize phase errors. This system has been generally
adopted in Europe. ➤SECAM. **2.** *Abbrev. for* programmable array logic. A logic device that has a number of input variables that are programmable to inputs of AND gates,
the outputs of the AND gates forming fixed inputs to an OR gate whose output is the
output of the PAL (see diagram).

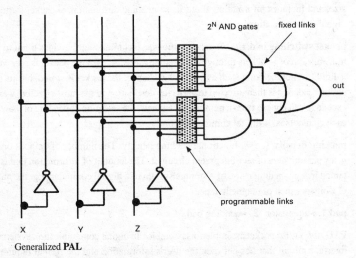

Generalized **PAL**

PAM *Abbrev. for* pulse-amplitude modulation ➤pulse modulation.

panoramic radar indicator ➤radar indicator.

panoramic receiver A ➤radio receiver that is automatically tuned so as to receive certain frequency bands, each for a preselected frequency. The period of the tuning variation is preselected. In this way a range of radiofrequencies can be regularly monitored,
for example, when listening for distress calls.

paper capacitor A capacitor of medium loss and medium capacitance-stability that
is used in high-voltage a.c. and d.c. applications. These capacitors are manufactured
by winding together aluminium foils interleaved with layers of tissue paper. The moisture content of the paper is removed by impregnation with a suitable oil or wax.
➤Metallized paper capacitors use an evaporated metal film as the electrode instead
of aluminium.

PAR *Abbrev. for* precision approach radar.

parabolic reflector *Syns.* paraboloid reflector; dish. A radiofrequency or microwave-frequency reflector (►directional antenna) that has a hollow concave paraboloid shape so that all waves passing through the focus will be reflected parallel to the axis of rotation.

parallel 1. Circuit elements are said to be *in parallel* if they are connected so that the current divides between them and later reunites (see diagram).

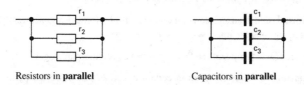

Resistors in **parallel**   Capacitors in **parallel**

For *n* resistors in parallel the total resistance, *R*, is given by:

$$1/R = 1/r_1 + 1/r_2 + \ldots 1/r_n$$

where $r_1, r_2, \ldots r_n$ are the values of the individual resistors.

For *n* capacitors in parallel the total capacitance, *C*, is given by:

$$C = c_1 + c_2 + c_3 + \ldots c_n$$

where $c_1, c_2, \ldots c_n$ are the individual capacitances. The capacitors thus behave collectively as a large capacitor having the total plate area of the component capacitors.

Machines, transformers, and cells are said to be in parallel when terminals of the same polarity are connected together. Several cells connected together in parallel have a lower total internal resistance than a single cell and can therefore supply a larger maximum current. ►series; shunt. **2.** (in computing) Involving the simultaneous transfer or processing of the individual parts of a whole.

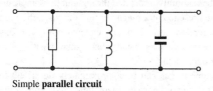

Simple **parallel circuit**

parallel circuit A circuit containing two or more elements connected in parallel across a pair of lines or terminals (see diagram).

parallel network *Syn.* shunt network. ►network.

parallel-plate capacitor A ►capacitor formed from two parallel metal electrodes separated by dielectric material.

parallel processing A computer system consisting of a number of processors or processing elements that can operate in parallel on more than one item of data at a time. Parallel processing is usually used to increase the speed of processing.

parallel resistance ►resonant frequency.

parallel resonant circuit *Syn.* rejector. ►resonant circuit; resonant frequency.

parallel supply ►series supply.

parallel-T network *Syn. for* twin-T network. ►two-port network.

parallel transmission Transmission of multiple units of information concurrently, i.e. in ►parallel along separate lines. ►►serial transmission.

paramagnetic Curie temperature *Syn. for* Weiss constant. ►paramagnetism.

paramagnetism An effect observed in certain materials that possess a permanent atomic or molecular ►magnetic moment. Each orbital electron in an atom constitutes an individual current and hence has a magnetic moment; however the possible ►energy levels available are such that only unfilled shells of electrons contribute to the magnetic moment of the atom as a whole. The ►spin of the atomic electrons also has a magnetic moment associated with it but in an atom only unpaired spins contribute to the magnetic moment of the atom as a whole. Most free atoms would have a magnetic moment due to orbital electrons in unfilled outer shells but practical substances combine in general so as to complete the outer shells; most gaseous molecules and ionic or homopolar liquids or solids have no overall magnetic moment and are diamagnetic (►diamagnetism). Permanent magnetic moments are only possessed by molecules or ions containing unpaired electron spins (oxygen, O_2, for example contains two unpaired spins) or by particular ions of multivalent transition elements in which there is an unfilled inner shell. Paramagnetism is most commonly associated with electron spin but a few compounds also have an orbital contribution.

In the absence of an applied magnetic flux density, thermal motion causes the individual magnetic moments to be randomly orientated throughout the sample and the net magnetization is zero. In the presence of a magnetic flux the magnetic moments tend to align themselves in the direction of the field; this tendency however is opposed by the thermal agitations and a paramagnetic substance has a small positive ►susceptibility, χ, which is temperature dependent.

The behaviour of a paramagnetic gas can be approximately described by the Langevin function, which shows that at ordinary magnetic fields and temperatures the gas obeys *Curie's law:*

$$\chi = C/T$$

where C is a constant and T the thermodynamic temperature. However, at sufficiently high flux density and low temperature a saturation point is achieved with all the molecules aligned along the field with neglibible thermal effects. Very dilute paramagnetic liquids also obey Curie's law.

The magnetic properties of paramagnetic solids and liquids depend on the complex intra-atomic and interatomic forces operating within them and the behaviour cannot always be described by a simple equation. Nonhydrated liquids and many paramagnetic solids at ordinary temperatures and fields obey the *Curie–Weiss law*:

$$\chi = C/(T - \theta),$$

where θ, the *Weiss constant,* can be either positive or negative. The Curie–Weiss law is only obeyed at temperatures $T > |\theta|$ and is a modification of Curie's law arising from the mutual interactions of the ions or molecules.

Certain metals, such as sodium or potassium, exhibit *free-electron paramagnetism* or *Pauli paramagnetism* in which a small positive susceptibility with only slight temperature dependence is observed. Both effects are due to the conduction electrons in the metals. The individual atoms in the solid are left as diamagnetic ions and the conduction electrons exhibit both diamagnetism and paramagnetism. In most metals these effects are of the same order of magnitude but Pauli paramagnetism occurs when the paramagnetism effect is greater than the diamagnetism.

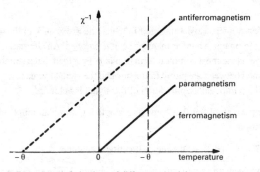

a Paramagnetic behaviour of different materials

At temperatures below certain critical temperatures and approximately equal to the modulus of the Weiss constant, the intermolecular forces of many paramagnetic solids become much greater than the thermal agitations. The magnetic moments are no longer randomly orientated but take up an appropriate ordered state and the materials become either ferromagnetic, antiferromagnetic, or ferrimagnetic. The paramagnetic behaviour of different materials can be compared by plotting the inverse susceptibility χ^{-1} against thermodynamic temperature in the paramagnetic region (Fig. *a*).

Paramagnetism causes an increase of magnetic flux density within a sample; this is represented schematically by a concentration of the lines of magnetic flux density passing through it (Fig. *b*). If a paramagnetic substance is placed in a nonuniform magnetic field it tends to move from the weaker to the stronger region of the field; a bar of paramagnetic material placed in a uniform magnetic flux tends to orientate itself with the longer axis parallel to the flux.

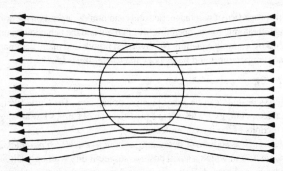

b Change in magnetic flux density in a paramagnetic substance

parameter A quantity that is constant in a given case but has a particular value for each different case considered. Examples include the values of the resistances, capacitances, etc., that form an electrical ►network or the constants appearing in the equations connecting currents and voltages at the terminals of a network. ►►transistor parameters.

parametric amplifier A microwave ►amplifier in which the ►reactance of the device is varied with respect to time in a regular manner, i.e. it is pumped. An alternating voltage, the pump voltage, is the most common form of *pump*. Provided that a suitable relationship is maintained between the pump frequency and the signal frequency, energy is transferred from the pump and amplification of the signal is achieved.

paraphase amplifier An amplifier that converts a single input into an input suitable for ►push-pull operation.

parasitic antenna *Syn. for* passive antenna. ►directional antenna.

parasitic capacitance Any unwanted capacitance in a circuit, device, or system, or in another form any ►stray capacitance between circuit ►nodes and earth. These stray capacitances are especially detrimental at high frequencies and with high-impedance nodes.

parasitic inductance Any unwanted inductance in a circuit, device, or system. It is especially detrimental at low frequencies.

parasitic oscillations Unwanted oscillations that can occur in ►amplifier or ►oscillator circuits. The frequency of such oscillations is usually very much higher than the frequencies for which the circuit has been designed since it is mainly determined by stray inductances and capacitances, such as in connecting leads, and by interelectrode capacitances.

Parasitic stoppers are devices incorporated into a circuit to prevent parasitic oscillations being generated. They are commonly resistors used in the input and output circuits of the device. They take their name from the particular part of the circuit in

which they are incorporated. For example, an *anode stopper* is connected to an
►anode, a *grid stopper* to a ►grid, and a *base stopper* to a ►base.

parasitic radiator *Syn. for* passive antenna. ►directional antenna.

parasitic stopper ►parasitic oscillations.

parity A count of the number of 1_2 bits (►binary notation) in a data unit, such as a word
or message. The parity is *odd* if there is an odd number of 1_2 bits and *even* if there is
an even number of 1_2 bits. For example, the parity of 101110_2 is even. ►parity check.

parity check A process used to detect errors in data in binary form. The ►parity is computed for a data unit before and after it has been transferred or otherwise manipulated;
comparison of the two values will determine if the prescribed even or odd parity condition is present, and hence will detect all single bit errors and most simple bit errors.
►digital codes; digital communications.

Parseval's theorem The proposition stating that for periodic signals the average
power dissipated by the signal into a one ohm resistance is the sum of the squares of
its d.c. value and the r.m.s. values of its harmonic components.

partial node ►node.

partition noise ►Noise that occurs in active electronic devices due to random fluctuations in the distribution of the currents between the electrodes.

pascal Symbol: Pa. The ►SI unit of pressure, defined as the pressure that results when
a force of one newton acts uniformly over an area of one square metre.

Paschen's law The ►breakdown voltage that initiates a discharge between electrodes
in a gas is a function of the product of pressure and distance. For example, if the distance between the electrodes is doubled breakdown will only occur at the same potential
difference if the gas pressure is halved.

pass band ►filter.

passivation Protection of the junctions and surfaces of solid-state electronic components and integrated circuits from harmful environments. Passivation is most commonly
achieved by forming a layer of silicon dioxide or silicon nitride on the surface of the
silicon chip. An alternative method is *glassivation,* in which the passivating layer is a
melt of vitreous material that is deposited on the surface of the semiconductor and allowed to harden.

passive Denoting any device, component, or circuit that does not introduce ►gain. In
practice only pure resistance, capacitance, inductance, or combinations of these are passive. ►active.

passive antenna *Syns.* secondary radiator; parasitic antenna; parasitic radiator.
►directional antenna.

passive filter ►filter.

passive network ➤network.

passive substrate A substrate, such as glass or ceramic, that is used in microelectronics for its lack of semiconductor behaviour.

passive transducer ➤transducer.

patch 1. In receiving or transmitting antennas, a physical area of conductor that, when excited with a signal of appropriate frequency, will radiate that frequency into free space in a direction at right angles to the plane of the conducting area. The frequency at which radiation occurs is dictated by the physical dimensions of the conductor. **2.** A sound configuration on a music ➤synthesizer, often selected by means of *patch buttons* according to the *patch number*.

patchbay A means of interconnecting equipment in a nonpermanent manner where various source and destination connections can be made rapidly and easily. Patchbays are often found in studios to enable recording and signal-processing equipment to be configured as desired.

path loss ➤Friis transmission equation.

Pauli exclusion principle No two electrons in an atom can exist in the same quantum energy state, i.e. they cannot be described by the same four quantum numbers, $l, m, n,$ and s.

Pauli paramagnetism ➤paramagnetism.

PBT *Abbrev. for* permeable base transistor. ➤vertical FET.

PBX *Abbrev. for* private branch exchange. ➤telephony.

PC 1. *Abbrev. for* personal computer, usually an IBM-compatible one. **2.** *Abbrev. for* program counter.

PCB *Abbrev. for* printed circuit board. The supporting sheet plus circuit of a ➤printed circuit. ➤➤multilayer PCB.

p-channel Denoting a junction FET or MOSFET in which the conducting channel is formed as p-type semiconductor. Usually n-channel devices are preferred. ➤field-effect transistor. ➤➤n-channel.

PCM 1. (or **pcm**) *Abbrev. for* pulse code modulation. ➤pulse modulation. **2.** *Abbrev. for* portable conformable mask. ➤multilevel resist.

p.d. (or **pd**) *Abbrev. for* potential difference.

PDM *Abbrev. for* pulse-duration modulation. ➤pulse modulation.

peak factor *Syn.* crest factor. The ratio of the ➤peak value of a periodically varying quantity to the ➤root-mean-square value. If the quantity varies sinusoidally the peak factor is $\sqrt{2}$.

peak forward voltage The maximum instantaneous voltage applied to a device in the forward direction, i.e. in the direction in which the device is designed to pass current with the minimum resistance.

peak inverse voltage The maximum instantaneous voltage applied to a device in the ►reverse direction, i.e. in the direction of maximum resistance. If the peak inverse voltage becomes too great, ►breakdown of the device will occur. The breakdown is avalanche breakdown in a semiconductor device and arc formation in a valve. A rated value of peak inverse voltage for a device specifies the maximum inverse voltage that the device can tolerate without breakdown.

peak limiter *Syn. for* clipper. ►limiter.

peak load ►load.

peak picker A circuit that locates either the positive or negative peaks of a waveform.

peak point ►tunnel diode.

peak pulse amplitude ►peak value.

peak-riding clipper A ►limiter that automatically adjusts the value of voltage at which it operates according to the peak value of the pulse train input to the circuit.

peak-to-peak amplitude ►amplitude.

peak value *Syn.* amplitude; crest value. **1.** The maximum positive or negative value of any alternating quantity, such as current or voltage, during a given time interval. The positive and negative values are not necessarily equal in magnitude. **2.** *Syn.* peak pulse amplitude. The maximum value of an ►impulse voltage or current.

PECVD *Abbrev. for* plasma-enhanced chemical vapour deposition. ►chemical vapour deposition.

pedestal A ►pulse waveform that is 'flat-topped', i.e. the amplitude is made constant for a pulse interval large compared to the rise and fall times (►pulse). A pedestal is combined with a second waveform in order to increase the magnitude of the second waveform by a constant amount.

Peltier effect ►thermoelectric effects.

pentode A ►thermionic valve containing five electrodes. It is equivalent to a tetrode containing an additional electrode, the *suppressor grid,* between the screen and the anode. The suppressor grid is at a negative potential relative to both anode and screen and is used to prevent secondary electrons from the anode reaching the screen. The suppressor grid must be of an open mesh design otherwise the passage of the primary electron beam would be impeded by it.

PEP *Abbrev. for* peak envelope power. A measure of the power rating of ►single-sideband transmitters. It is calculated by multiplying the maximum (peak) envelope voltage by 0.707, squaring the result, and dividing by the load resistance.

percentage modulation ➤amplitude modulation.

percentile life ➤mean life.

perfect dielectric ➤dielectric.

perfect transformer *Syn. for* ideal transformer. ➤transformer.

period *Syn.* periodic time. Symbol: *T*. The time required to complete a single cycle of regularly recurring events. The period of an oscillation is related to the frequency, *f*, and the angular frequency, ω, by

$$T = 1/f = 2\pi/\omega$$

periodic Denoting any variable quantity that has regularly recurring values with respect to equal increments of some independent variable, such as time. The interval between two successive repetitions is the ➤period.

periodic damping *Syn. for* overdamping. ➤damped.

periodic duty ➤duty.

periodic signal A signal that repeats itself at regular intervals; a periodic waveform.

periodic table The classification of chemical elements, first introduced by Mendeléev, that demonstrates a periodicity of chemical properties when the elements are arranged in order of their atomic numbers. One form of periodic table is given in Table 11, in the back matter. Elements of similar chemical properties occur in the same vertical group in the periodic table. The periodic table was used to predict the existence of undetected elements.

periodic waveform A waveform (➤wave) where the variation of a quantity repeats regularly with time. ➤➤aperiodic waveform.

peripheral devices Devices that are connected to a ➤computer, form part of the computer system, and whose operation is controlled by the ➤central processing unit of the computer. Terminals, visual display units, printers, and backing store are examples of peripherals.

permanent magnet A magnetized sample of a ferromagnetic material, such as steel, that possesses high retentivity and is stable against reasonable handling. It requires a definite demagnetizing flux in order to destroy the residual magnetism. A *simple magnet* consists of a single bar, which can be horseshoe shaped, of the material. A *compound magnet* has several suitably shaped bars or laminations fastened together. ➤➤ferromagnetism; magnetic hysteresis.

permeability Symbol: μ; unit: henry per metre. The ratio of the magnetic flux density, *B*, in a medium to the external magnetic field strength, *H*, i.e.

$$\mu = B/H$$

The *permeability of free space* is designated μ_0 and is termed the *magnetic constant*. In the system of ►SI units it has the value $4\pi \times 10^{-7}$ henry/metre. In other systems, such as the ►CGS and ►MKS systems, it has been given different values. Using ►Maxwell's equations it can be shown that

$$\mu_0 \varepsilon_0 = 1/c^2$$

where ε_0 is the ►permittivity of free space and c is the speed of light.

The *relative permeability,* μ_r, is the ratio of the magnetic flux density in a medium to the magnetic flux density in free space for the same value of external magnetic field strength, i.e.

$$\mu_r = \mu/\mu_0$$

For most materials μ_r is a constant. Diamagnetic materials have a value of μ_r less than unity. Paramagnetic materials have a value just greater than unity (►diamagnetism; paramagnetism). Ferromagnetic materials have values of μ_r that are very much greater than unity and that depend on the magnetic flux density (►ferromagnetism; magnetic hysteresis).

The *incremental permeability* is the permeability measured when a small alternating magnetic field is superimposed on a large steady one.

permeability of free space *Syn.* magnetic constant. ►permeability.

permeability tuning *Syn. for* slug tuning. ►tuned circuit.

permeable base transistor (PBT) ►vertical FET.

permeameter An instrument that measures the magnetic properties of a ferromagnetic material, particularly its permeability.

permeance Symbol: Λ; unit: henry. The reciprocal of ►reluctance.

permittivity Symbol: ε; unit: farad per metre. The ratio of the electric ►displacement, D, in a dielectric medium to the applied electric field strength, E, i.e.

$$\varepsilon = D/E$$

The *permittivity of free space* is designated ε_0 and is termed the *electric constant*. It is related to the ►permeability of free space, μ_0, by the equation

$$\varepsilon_0 \mu_0 = 1/c^2$$

where c is the speed of light. Thus since μ_0 has the value $4\pi \times 10^{-7}$ henry per metre in the system of ►SI units, then

$$\varepsilon_0 = (1/4\pi c^2) \times 10^7 \text{ farad/metre}$$

$$= 8.854\ 187\ 817 \times 10^{-12} \text{ farad/metre}$$

The *relative permittivity*, ε_r, is the ratio of the electric displacement in a medium to the electric displacement in free space for the same value of applied electric field strength, i.e.

$$\varepsilon_r = \varepsilon/\varepsilon_0$$

The dimensionless quantity ε_r is also termed the *dielectric constant* when it is independent of electric field strength and refers to the dielectric medium of a capacitor. It can then be defined as the ratio of the capacitance of the capacitor containing the dielectric medium to the capacitance it would have were the dielectric removed.

permittivity of free space *Syn.* electric constant. ➤permittivity.

persistence *Syn.* afterglow. **1.** The time interval after excitation during which a phosphor continues to emit light (➤luminescence), particularly the phosphor on the screen of a ➤cathode-ray tube. The magnitude of the luminance with respect to time after excitation of a luminescent screen is a *persistence characteristic.* The persistence depends on the nature of the phosphor and for luminescent screens is commonly chosen to be less than the persistence of the image on the human retina (about 0.1 seconds); it can however vary from fractions of a second to several years. **2.** A faint luminosity observed after the passage of an electric discharge through certain gases; it can last for several seconds.

persistor A device, commonly in the form of a miniature bimetallic printed circuit, that depends for its operation on the sharp changes of resistance in a metal as it passes from a state of superconductivity to its normal resistive state. It can be used as a low-temperature storage element or very fast switch.

personal communications device ➤cellular communications.

personal computer A single-user computer with a base price of less than about £3000, examples being the IBM PC, the IBM PS/2, the Apple Macintosh.

perveance The space-charge-limited characteristic between the electrodes in an ➤electron tube. It is a function of the current density, j, and the collector voltage, V, being equal to $j/V^{3/2}$.

PFM *Abbrev. for* pulse-frequency modulation. ➤pulse modulation.

PGA *Abbrev. for* pin grid array.

phase 1. The stage or state of development of a regularly recurring quantity; it is the fraction of the ➤period that has elapsed with respect to a fixed datum point.

The amplitude variations of a sinusoidally varying quantity are similar to simple harmonic motion; such a quantity may be represented as a rotating vector *OA* (see diagram) of length equal to the maximum amplitude and rotating through an angle 2π during the ➤period, T, of the waveform. The vector has an angular velocity, ω, equal to $2\pi/T$ and related to the frequency, f, of the waveform by

$$f = \omega/2\pi$$

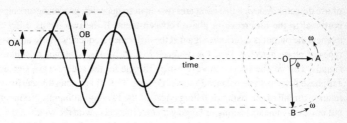

Phase angle between two quantities of the same frequency

The phase of the quantity **OA** with respect to another quantity, represented by the vector **OB,** is given by the angle, ϕ, between them (see diagram). This is the ►phase angle (➤phase difference), which is constant if the two quantities have the same frequency.

Particles in a travelling wavefront moving in the same direction with the same relative displacement are said to be in the same phase of vibration. The wavelength is equal to the distance travelled between two points, in the direction of propagation of a wavefront, at which the same phase recurs.

Periodic quantities that have the same frequency and wave shape and that reach corresponding values simultaneously are said to be *in phase;* otherwise they are *out of phase.* **2.** One of the separate circuits or windings of a ►polyphase system or apparatus. **3.** One of the lines or terminals of a ►polyphase system or apparatus.

phase angle Symbol: ϕ. The angle between two vectors that represent two sinusoidally varying quantities of the same frequency (►phase). If the two quantities are nonsinusoidal but have the same ►fundamental frequency, the phase angle is the angle between the two vectors that represent the fundamental components. Waveforms that have a phase angle of $\pi/2$ are said to be in quadrature. If the phase angle is equal to π they are in antiphase.

phase centre A point on or near an antenna from which the radiated fields form spherical waves. In practice there may be no single point for radiation in all directions but there is usually a single point for the main radiated signal. In the case of a ►broadband antenna, the point is likely to move around depending on the frequency.

phase-change coefficient *Syns.* phase constant; wavelength constant. ►propagation coefficient.

phase constant *Syn. for* phase-change coefficient. ►propagation coefficient.

phase corrector A network that restores the original phase of a waveform that has suffered phase ►distortion.

phase delay The ratio of the inserted ►phase shift undergone by a periodic quantity to the frequency.

phase detector A device that compares two input frequencies, producing an output proportional to the difference in phase between them. If the two inputs differ in frequency, the output is a periodic signal at the difference frequency. There are two basic types of phase detector. Type I is designed to be driven by analogue or square-wave signals. This can be simply an ➤exclusive-OR gate for the square-wave signals. Linear phase detectors use balanced ➤mixer circuits. Type II is a detector sensitive to the relative timing of the rising or falling edges of the two digital inputs, producing output pulses to indicate leading or lagging phase difference, with the width of the pulses being equal to the time between the compared edges. The output pulses will disappear entirely when the two input signals are exactly in phase.

phase deviation ➤phase modulation.

phase difference 1. Symbol: ϕ. The difference in phase between two sinusoidally varying quantities of the same frequency. It may be expressed as an angle – the ➤phase angle – or as a time. **2.** The angle between the reversed secondary vector of an ➤instrument transformer and the corresponding primary vector. The vectors represent current in a current transformer and voltage in a voltage transformer. The phase difference is positive if the reversed secondary vector ➤leads the primary and negative if it ➤lags. The term *phase error* has been used in this application but this is deprecated.

phase discriminator A ➤detector circuit that produces an output wave in which the amplitude is a function of the ➤phase of the input.

phase distortion ➤distortion.

phased linear array ➤linear array antenna.

phase error ➤phase difference.

phase inverter A circuit that changes the phase of an input signal by π. A common application is for driving one side of a ➤push-pull amplifier.

phase lag ➤lag.

phase-lock loop (PLL) A circuit comprising a ➤phase detector, low-pass ➤filter, ➤amplifier, and voltage-controlled ➤oscillator (VCO). The phase detector compares the frequencies of an input signal and the output of the VCO. If the two frequencies are different, the phase detector produces a phase error signal at the difference frequency, which, after low-pass filtering and amplification, is used to drive the VCO in the direction of the input frequency. When the PLL is 'locked', the VCO frequency is the same as the input frequency, maintaining a constant phase difference. Under these conditions, the phase detector output is then a DC voltage driving the VCO at a constant frequency. This DC voltage is therefore a measure of the input frequency. Further, any ➤frequency modulation present on the input signal will be present on the control voltage, so the PLL is acting as a demodulator for FM signals.

The VCO output is a locally generated frequency equal to the input signal frequency, and therefore can provide a lower noise replica of the input signal. By placing a divider between the VCO output and the phase detector, a multiple of the input frequency can be generated in the VCO. This is the basic technique of frequency ►synthesizers.

phase modulation (PM) A type of ►modulation in which the phase of the ►carrier wave is varied about its unmodulated value by an amount proportional to the amplitude of the signal wave and at a frequency of the modulating signal, the amplitude of the carrier wave remaining constant. If the modulating signal is sinusoidal the instantaneous amplitude, e, of the phase-modulated wave may be written:

$$e = E_m \sin(2\pi Ft + \beta \sin 2\pi ft)$$

where E_m is the amplitude of the carrier wave, F the unmodulated carrier wave frequency, β the peak variation in the phase of the carrier wave due to modulation, and f is the modulating signal frequency. The peak difference between the instantaneous phase angle of the modulated wave and the phase angle of the carrier is the *phase deviation*.

If the modulating signal is not sinusoidal but is made up of discrete levels, the resulting modulation is known as *phase shift keying* (PSK). Any number of discrete-amplitude signal levels can be used. In the general case where M discrete levels of the modulating signal are used, the modulation is called *M-ary PSK*. If only two levels are used, the result is *binary PSK* (BPSK).

Modulation of a phase-modulated signal may or may not make use of a phase reference to determine the phase of the modulating signal. Instead of a reference just the difference between the phases can be used to modulate the carrier. This form of phase modulation is called *differential phase modulation* or, in the case of phase shift keying, it is abbreviated to *DPSK*.

phase response The variation of ►phase difference between input and output signals of a circuit or network, hence the phase of the ►transfer function, as a function of the applied signal frequency. The circuit or network is said to have a *linear phase response* if its phase shift changes in a linear fashion with respect to frequency.

phase sequence The order in which the three ►phases of a three-phase system (►polyphase system) reach a maximum potential of given polarity. The normal order in a particular system is termed the *positive sequence*; the reverse order is the *negative sequence*. A *phase-sequence indicator* is an instrument used to indicate the phase-sequence of such a system.

phase shift Any change that occurs or is introduced into the phase of a periodic quantity or in the phase difference between two or more such quantities. Phase shift can occur as a result of errors introduced by a particular device or circuit or can be deliberately inserted by a phase-shifting network.

phase shift keying (PSK) ►phase modulation.

phase-shift oscillator ►oscillator.

phase splitter A circuit that has a single input signal and produces two separate outputs with a predetermined phase difference. An example is the driver for a ➤push-pull amplifier.

phase velocity The velocity at which an equiphase surface of a travelling wave is propagated through a medium, i.e. the velocity at which the crests and troughs travel. It is equal to λ/T where λ is the wavelength and T the period of the wave. If the frequency is f and the number of wavelengths per unit distance is σ (the wavenumber) then

$$\lambda/T = f/\sigma$$

phase voltage The voltage in one ➤phase of a polyphase system. The term is sometimes used to indicate the voltage relative to neutral but such usage can give rise to ambiguities and is deprecated.

phasing Adjusting the position of the picture transmitted in television or facsimile transmission along the scanning line.

phonon One quantum of thermal energy associated with the vibrations of the atoms in a crystal lattice. If the frequency of vibration is ν and h the Planck constant, the phonon is equal to $h\nu$.

phosphor A luminescent material. ➤luminescence.

phosphor-bronze Bronze that contains at least 0.18% of added phosphorus. The addition of the phosphorus enhances the tensile strength, ductility, and shock resistance of the alloy. Phosphor-bronze in strip form has been widely used for galvanometer suspensions and other similar applications, including bearing surfaces.

phosphorescence ➤luminescence.

photocathode A cathode that emits electrons as a result of the ➤photoelectric effect.

photocell A light–electric transducer. The term was originally short for *photoelectric cell,* which is a vacuum ➤diode containing a ➤photocathode and an anode. When the photocathode is illuminated electrons are emitted. The electrons are driven to the anode by a positive potential, the *driving potential,* applied to the anode and a photocurrent flows in the external circuit. The small current flowing when the device is not illuminated, but the driving potential is on, is the dark current.

The term is now most commonly used to designate a *photoconductive cell,* which consists of ➤semiconductor material sandwiched between two ➤ohmic contacts (see diagram). The semiconductor may be either bulk material in the form of a rod or bar or a thin polycrystalline film on a glass substrate. The conductivity of the sample increases markedly when it is subjected to light or other radiation of suitable wavelength (➤photoconductivity) and a photocurrent, superimposed on the small dark current, flows in an external circuit. The gain of such a device is defined as:

$$\text{gain} = \Delta I/eG_{\text{pair}}$$

where ΔI is the incremental current due to photoconductivity (the photocurrent), e is the electron charge, and G_{pair} the number of ►electron-hole pairs created per second.

The term photocell is also sometimes applied to a ►photodiode or a ►photovoltaic cell. The resistance of photocells drops markedly upon illumination and the *dark resistance* – i.e. the resistance when the devices are not illuminated – is much higher than the ►dynamic resistance of the devices.

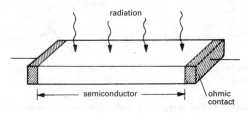

Photoconductive cell

photoconductive camera tube ►camera tube. ►►vidicon.

photoconductive cell ►photocell.

photoconductivity An enhancement of the conductivity of certain semiconductors due to the absorption of ►electromagnetic radiation. Photoconductivity can result from the action of radiation in the visible portion of the spectrum in some materials.

Radiation, of frequency v, can be considered as a stream of ►photons of energy hv, where h is the Planck constant. If a semiconductor is exposed to radiation an electron in the ►valence band can be excited by the radiation. If the photon energy is sufficiently great the energy absorbed by the electron causes it to be excited across the forbidden band into the ►conduction band and a hole remains in the valence band. Thus when the photon energy, hv, exceeds the energy gap, E_g, there is a sudden marked change in the conductivity of the material due to the creation of excess charge carriers. For small intensities of illumination the increase in conductivity is approximately proportional to the intensity. If the photon energy exceeds the ►work function, Φ, of the material, the electron is liberated from the solid by the ►photoelectric effect. Photoconductivity resulting under the condition

$$\Phi > hv > E_g$$

is sometimes termed the *internal photoelectric effect*. Band-to-band transitions as described above result in *intrinsic photoconductivity*.

Measurement of the absorption spectra of semiconductors can yield valuable information on the magnitude of E_g. ►Direct-gap semiconductors have an absorption edge corresponding exactly to E_g (Fig. *a*). ►Indirect-gap semiconductors yield a value larger than E_g (Fig. *b*). Even if the photon energy is only equal to E_g so that a direct transition is not possible, *indirect photoconductivity* can occur in indirect-gap semiconductors provided that a ►phonon is simultaneously created or destroyed. If k is the wavevector and thus represents the momentum of the electron in the crystal lattice

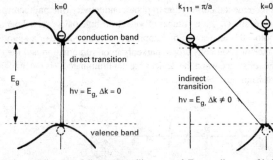

a Energy diagram of direct-gap gallium arsenide semiconductor

b Energy diagram of indirect-gap germanium semiconductor

($\blacktriangleright$momentum space), transitions of energy E_g can only take place if there is a change in momentum so that $\Delta k \neq 0$. This results in a small absorption edge in the spectrum corresponding to E_g. The momentum change required for the occurrence of indirect transitions has to be absorbed or released by the assisting phonon in order to conserve momentum.

Extrinsic photoconductivity is an effect observed in some semiconductors when the photon energy, $h\nu$, of the incident radiation is not sufficiently large to cause band-to-band transitions; instead the energy corresponds to the energy gap required to excite an electron from the valence band, energy E_v, into an acceptor level or from a donor level into the conduction band, energy E_c (Fig. *c*). E_a and E_d are the acceptor and donor energy levels. In this case electron-hole pairs are not created: an increase in the p- or n-type carriers results, respectively.

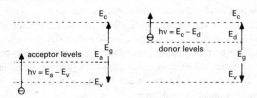

c Extrinsic **photoconductivity** transitions

Photoconductive materials have a short response time to the incident radiation because the excess carriers generated disappear very quickly due to recombination. They are very useful as switches and photodetectors and as $\blacktriangleright$photocells can produce an a.c. signal if the incoming radiation is suitably modulated, as by a mechanical chopping device.

photoconverter A photoelectric $\blacktriangleright$transducer that produces digital electrical signals from an optical pattern. It consists of an array of photocells each of which produces a current proportional to the intensity of light falling on it. The illumination may be due to an optical image produced on a transparent screen or a pattern resulting from an illuminated photographic ➤mask or cut-out of opaque material.

photocurrent An electric current that is produced in a device by the effect of incident electromagnetic radiation. ➤photocell; photoconductivity; photoelectric effect; photoionization.

photodetachment The interaction of electromagnetic radiation with a negative ion to detach an electron from the ion and leave a neutral atom or molecule, i.e.

$$M^- + h\nu \rightarrow M + e^-$$

where $h\nu$ represents a photon of frequency ν and h is the Planck constant. The mechanism in this process is exactly the same as for ➤photoionization of a neutral species. The ➤ionization potential of the negative ion is equal to the ➤electron affinity of the atom or molecule.

photodetector Any electronic device that detects (➤detector) or responds to light energy. ➤photocell; photodiode; phototransistor.

photodiode A semiconductor ➤diode that produces a significant ➤photocurrent when illuminated. There are two main classes of photodiode: *depletion-layer photodiodes* and *avalanche photodiodes*.

  A common form of depletion-layer photodiode consists of a reverse-biased ➤p-n junction operated below the ➤breakdown voltage. When exposed to electromagnetic

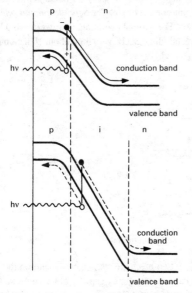

a Energy bands of reverse-biased p-n junction (top) and p-i-n **photodiodes**

radiation of suitable frequency, excess charge carriers are produced as a result of ►photoconductivity; these carriers are in the form of ►electron-hole pairs. They usually recombine very quickly but those generated in or near the ►depletion layer present at the junction cross the junction and produce a photocurrent (Fig. *a*). The photocurrent is superimposed on the normally very small reverse saturation current, or dark current. The p-n junction can be formed by any of the usual methods and illumination can be either normal to the plane of the junction or parallel to it.

The *p-i-n* (or *PIN*) *photodiode* (Figs. *a, b*) contains a layer of intrinsic (i-type) semiconductor material sandwiched between the p- and n-regions. The depletion layer is wholly contained within the i-region. The thickness of the intrinsic region can be adjusted to produce devices with optimum sensitivity and frequency response. The p-i-n photodiode is the most common type of depletion-layer photodiode.

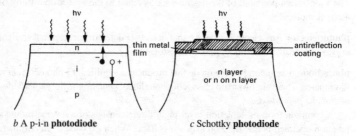

b A p-i-n **photodiode**   *c* Schottky **photodiode**

Depletion-layer photodiodes may also be realized using a ►metal-semiconductor junction, a ►heterojunction, or a ►point-contact diode. The *Schottky photodiode* (Fig. *c*) uses a metal-semiconductor junction. For optimal operation, in order to avoid losses due to absorption and reflection the metal film must be very thin (about 10 nanometres) and an antireflection coating must be used.

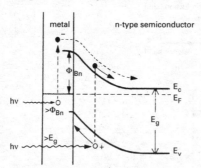

d Simplified energy diagram for Schottky **photodiode**

The photocurrent can be caused by two mechanisms, which depend on the value of the incident photon energy, $h\nu$, relative to the energy gap, E_g, of the semiconductor and the reduced work function, Φ_{Bn}, of the metal (Fig. *d*):

if $E_g > h\nu > \Phi_{Bn}$

photoelectric emission of electrons from metal to semiconductor occurs;

if $h\nu > E_g$

photoconductive electron-hole pairs are produced in the semiconductor.

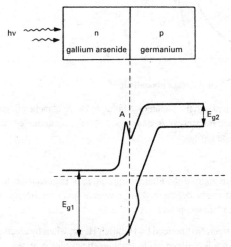

e Heterojunction **photodiode**

The *heterojunction photodiode* is formed from two semiconductors of different band gaps (Fig. *e*). The frequency response depends on the relative absorption of the two materials, suitable choice of material causing most of the radiation of a particular frequency to be absorbed close to the junction. Such diodes have a high speed of response and are highly frequency-selective. The small barrier, A, in the energy diagram is due to the discontinuity in the conduction bands and can be crossed by tunnelling (►tunnel effect) or surmounted by electrons with sufficiently high energy.

Modulation of the intensity of illumination falling on a photodiode produces a modulated photocurrent that is dependent on the incident illumination.

The other class of photodiodes, avalanche photodiodes, are reverse-biased p-n junction diodes that are operated at voltages above breakdown voltage. Current multiplication of electron-hole pairs, generated by the incident electromagnetic radiation, occurs due to the ►avalanche process. The photomultiplication factor, M_{ph}, is defined as the ratio of the multiplied photocurrent, I_{ph}, to the photocurrent, I_{pho}, at voltages below breakdown where no avalanche multiplication takes place. Static current-voltage characteristics are shown in Fig. *f*. The current I is the sum $I_{ph} + I_{do}$, where I_{do} is the dark current. The variation of I_{do} is also shown. The maximum photomultiplication depends inversely on the square root of I_{do}, and in order to achieve optimum operation I_{do} must be kept as small as possible. The device is thus operated with the

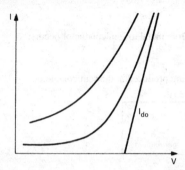

f I–V characteristics of an avalanche **photodiode**

voltage, *V,* approximately equal to the breakdown voltage, V_B. Avalanche photodiodes provide a substantial gain at microwave frequencies. Schottky photodiodes may also be operated in the avalanche region.

photoelectric cell ►photocell.

photoelectric constant The ratio of the Planck constant, *h,* to the electron charge, *e.* The photoelectric constant can be determined by photoemission experiments and has been used to calculate the Planck constant.

photoelectric effect An effect, first noticed by Heinrich Hertz, whereby electrons are liberated from matter when it is exposed to electromagnetic radiation of certain energies. In solids electrons are only liberated when the frequency of the exciting radiation is greater than a characteristic value – the *photoelectric threshold* of the material. The photoelectric threshold is usually in the mid-ultraviolet region of the electromagnetic spectrum for most solids, although some metals exhibit photoelectric emission with visible or near-ultraviolet radiation. It was found that the numbers of electrons ejected from a solid is not dependent on the frequency of the radiation but on the intensity; the maximum velocity however is directly proportional to the frequency.

Einstein explained this phenomenon by assuming that the radiant energy could only be transferred in discrete amounts, i.e. as ►photons. The energy of each photon is given by *h*ν, where *h* is the Planck constant and ν the frequency of the incident radiation. Provided that *h*ν exceeds the ►work function of the material, Φ, an electron in the material absorbing a photon is ejected from the surface. The maximum kinetic energy, *E,* of the electrons is given by the *Einstein photoelectric equation*:

$$E = h\nu - \Phi$$

Photoelectric emission from a material may be prevented by applying a negative potential to the surface. The minimum potential required to prevent photoelectric emission is the *stopping potential.*

The photoelectric effect is utilized in some types of ►photocell and in ►photomultipliers. ►►photoconductivity.

photoelectric galvanometer ➤galvanometer.

photoelectric threshold ➤photoelectric effect.

photoelectron spectroscopy A method of analysis of the composition of a semiconductor by detecting photoelectrons generated by photons impinging on the surface of the material (➤photoelectric effect). The method is nondestructive and only photoelectrons generated at or very near to the surface are detected. The photoelectrons have an energy typical of the parent atom, but also characterized by neighbouring atoms: chemical bonding between the parent atom and its neighbours causes perturbations of the energies detected.

X-ray photoelectron spectroscopy (XPS) uses photons in the X-ray range of energy and *ultraviolet photoelectron spectroscopy* (UPS) uses photons in the ultraviolet energy range. This technique contrasts with ➤Auger electron spectroscopy in that the electrons result from transitions between the outer atomic orbitals (XPS) or the valence and conduction bands (UPS) as opposed to electrons in the inner orbitals (Auger ES). XPS yields sharper more easily identifiable energy peaks than UPS but since focusing is not easy the lateral resolution is poor.

photoemission Emission of electrons from a material as a result of exposure to electromagnetic radiation. ➤photoionization; photoelectric effect.

photoemissive camera tube ➤camera tube. ➤➤iconoscope.

photogeneration The generation of charge carriers in a semiconductor by means of pulses of light or other electromagnetic radiation.

photoglow tube ➤phototube.

photoionization Ionization of an atom or molecule that results from exposure to electromagnetic radiation. The mechanism of photoionization is the same as that operating in the ➤photoelectric effect, but in photoionization an electron is ejected from a single atom or molecule, such as occur in the gaseous state. The incident radiation, of frequency v, can be considered to consist of discrete ➤photons of energy hv, where h is the Planck constant. Photoionization can only occur when the photon energy, hv, exceeds the first ➤ionization potential, I_1; there is thus a *photoionization threshold* of frequency below which the effect does not occur.

photolithography *Syn.* (*informal*) printing. A technique used during the manufacture of ➤integrated circuits, ➤semiconductor components, ➤thin-film circuits, and ➤printed circuits. Photolithography is used in order to produce a desired pattern from a photographic ➤mask – a *photomask* – on a substrate material preparatory to a particular processing step.

The clean substrate is covered with a solution of ➤photoresist by spincoating, spraying, or immersion. The solution is allowed to dry and is then exposed to light or near ultraviolet radiation through the mask. *Deep ultraviolet exposure* can be used where greater resolution is required because of the shorter wavelength. Quartz masks rather than glass ones must then be used together with different resist material. The depolymerized portions of the photoresist are removed using a suitable solvent, and the polymerized portion remains and acts as a barrier to etching substances or as a mask for deposition processes. When the processing step is completed the remaining photoresist is removed using another suitable solvent.

Different methods are used to expose the resist through the mask. *Proximity printing* places the mask close to the slice but not in actual contact with it (Fig. *a*). The diffraction that occurs at the edges of the patterns on the mask causes divergence of the light and this method is only suitable for applications not requiring a high lateral resolution. *Contact photolithography* has the mask in contact with the surface of the slice. After alignment it is vacuum clamped to the slice for exposure (Fig. *b*). The resolution and uniformity of the technique depend critically on the mask being undamaged and the degree of contact that can be achieved. Any slight curvature of the slice or the mask causes *runout*, i.e. errors in the exposed pattern. The actual contact tends to damage the masks and they therefore have a limited life. Contact lithography however is a cheap and rapid technique and suitable for use with smaller slices and small-scale integrated circuits.

For large slices and VLSI circuits other methods are used. *Projection photolithography* uses an optical system to produce an image of the mask on the slice. The mask and slice are moved in synchronism to scan across the entire area (Fig. *c*). The depth of focus is very small and the slice therefore must have an extremely flat surface for accurate reproduction of the pattern. *Optical stepping* is the most commonly used technique for large-scale integration (LSI) applications, where good resolution and high yield are required. A mask, known as a reticle, contains the pattern for a portion of the slice (Fig. *d*). The pattern is imaged onto the slice, then the slice is moved and the exposure repeated. The step and repeat process continues until the entire slice has been exposed. The technique is particularly suitable for digital circuits where tens of thousands of identical devices are required. The pattern on the reticle can be up to ten times as large as the final pattern on the slice, allowing very accurate masks to be made. Optical steppers require extremely accurate optical systems and their complexity makes them very expensive.

All photolithographic techniques require excellent collimation of the light source, uniform and constant intensity over the whole of the mask area, and a vibration-free environment.

Positive photoresist is used to produce a positive image of the photographic mask. In this case the exposed portion is depolymerized and removed during development (Fig. *e*). Negative photoresist produces a negative image of the mask and it is the exposed portion that is polymerized and remains after development (Fig. *f*).

Photolithography is an important part of the ➤planar process for the manufacture of integrated circuits. In the case of very complex integrated circuits the size of the components forming the circuits is nearly comparable to the wavelengths of the radiation used to produce the masks and large geometric errors can easily arise. Shorter wavelengths are therefore required and other lithography systems have been developed. ➤electron-beam lithography; ion-beam lithography; X-ray lithography.

425

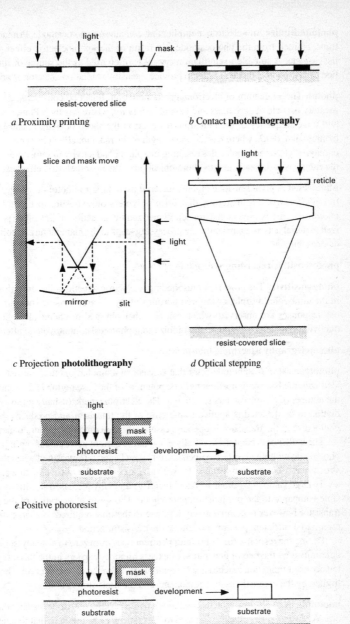

light

mask

resist-covered slice

a Proximity printing

b Contact **photolithography**

slice and mask move

light

reticle

light

mirror   slit

resist-covered slice

c Projection **photolithography**

d Optical stepping

light

mask

photoresist

development

substrate

substrate

e Positive photoresist

light

mask

photoresist

development

substrate

substrate

f Negative photoresist

photomultiplier An ►electron multiplier that contains a ►photocathode. Primary electrons, emitted from the photocathode as a result of the ►photoelectric effect, initiate the cascade. A suitable ►scintillation crystal is often used as the source of illumination for the photocathode in order to provide a sensitive radiation detector or ►counter.

photon The ►quantum of electromagnetic radiation. It can be considered as an elementary particle of zero mass and having energy $h\nu$, where h is the ►Planck constant and ν the frequency of the radiation. It travels at the speed of light, c, and has momentum $h\nu/c$ or $h\lambda$, where λ is the wavelength of the radiation. Photons can cause ►excitation of atoms and molecules resulting in ►photoconductivity in semiconductors; if the energy is sufficiently great ►photoionization or the ►photoelectric effect can result.

photoresist A photosensitive organic material used during ►photolithography. *Negative photoresists* are materials that form polymers on exposure to light. *Positive photoresists* are polymers that are depolymerized by the action of light. The polymerized material acts as a barrier during processing steps in the manufacture of solid-state devices, etc.

photosensitive recording ►recording of sound.

photosensitivity The property of responding to electromagnetic radiation, particularly in the ultraviolet, visible, or infrared portions of the electromagnetic spectrum. Various responses are observed, which can be either physical or chemical. ►photoconductivity; photoelectric effect; photoionization; photoresist; photovoltaic effect.

phototelegraphy ►facsimile transmission.

phototransistor A ►photodetector that consists of a bipolar junction transistor operated with the base region ►floating. The potential of the base region is determined by the number of charge carriers stored in it. The electromagnetic (usually ultraviolet) radiation to be detected is applied to the base of the transistor and produces the base photocurrent; the transistor is operated essentially in ►common-emitter connection.

The collector current is essentially equal to the current generated in a p-n junction photodiode multiplied by β, the ►beta-current gain factor. β is measured when the same structure is used as a simple transistor with a base contact. A typical structure can have a very large value of β (about 100) and therefore a greatly increased sensitivity is possible compared to the p-n junction photodiode. The speed of operation of the phototransistor however is comparatively less due to the time required to charge the base region to a sufficient potential to realize the transistor action.

During the response time the emitter current rises from zero to a steady-state value determined by the rate of generation of excess minority carriers in the base. If the intensity of illumination reaches a sufficiently high value the collector current becomes limited by the external circuit components and the transistor saturates.

phototube An ►electron tube that contains a photosensitive electrode, usually the cathode. A *vacuum phototube* is evacuated to a sufficiently low pressure that ionization of the residual gas in the tube does not affect the characteristics. A *gas phototube* is one that contains a gas, such as argon, at very low pressure in order to minimize the space charge effect of the photoelectrons in the tube. If the gas pressure is such that a glow discharge (►gas-discharge tube) occurs across the tube, then the tube is termed a *photoglow tube.* The sensitivity of such a tube is increased by the presence of the glow discharge.

The *dynamic sensitivity* of a phototube used as a ►photodetector is defined as the ratio of the alternating component of the anode current to the alternating component of the incident radiant flux. Such tubes respond only to the proportion of the incident radiant flux falling on the photocathode; the *acceptance angle* is the solid angle at the photocathode within which all the exciting flux reaches the cathode. This is determined by the geometry of the tube. Like other types of photodetector, phototubes exhibit a definite ►dark current on which the photocurrent is superimposed and which contributes to the total current flowing through the device. ►►photocell; photomultiplier.

photovoltaic cell A cell that utilizes the ►photovoltaic effect in order to produce an e.m.f. An example is the ►solar cell, the basis of which is an unbiased p-n junction.

photovoltaic effect An effect arising when a junction between two dissimilar materials, such as a metal and a semiconductor or two opposite polarity semiconductors, is exposed to electromagnetic radiation, usually in the range near-ultraviolet to infrared. A forward voltage appears across the illuminated junction and power can be delivered from it to an external circuit. The effect results from the depletion region and resulting potential barrier invariably associated with an unbiased junction (►►p-n junction; metal-semiconductor contact).

The energy bands are shown for a p-n junction and a metal-semiconductor contact (Figs. *a, b* overleaf). The incident radiation imparts energy to electrons in the valence band and electron-hole pairs are produced in the depletion region around the p-n junction or in the barrier layer at the metal-semiconductor contact. As the electron-hole pairs are produced they cross the junction due to the inherent field (Figs. *a, b*) and produce the forward bias: an excess of holes migrating into the p-type semiconductor or the metal produces a positive bias; electrons migrating into the n-type semiconductor produces a negative bias.

The photovoltaic effect is utilized in ►photovoltaic cells, such as ►solar cells.

pi circuit (or **π circuit**) ►delta circuit.

picket-fence effect ►discrete Fourier transform.

pick-up A ►transducer that converts information, usually recorded, into electrical signals. The term is particularly applied to the electromechanical transducers used to reproduce the signals recorded in the grooves of a gramophone record.

Ceramic pick-ups are in common use. They are constructed from a suitable ceramic material, such as barium titanate, that exhibits the ►piezoelectric effect. The mechanical vibrations produced by the grooves of the rotating record stress the ceramic material and result in a corresponding e.m.f. Ceramic materials are mechanically reliable and relatively stable under ambient conditions.

In a ►compact disc system, the pick-up is an assembly of ►semiconductor laser and light sensor. No mechanical vibrations are involved.

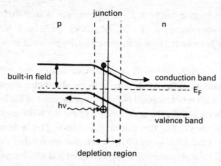

a **Photovoltaic effect** in unbiased p-n junction

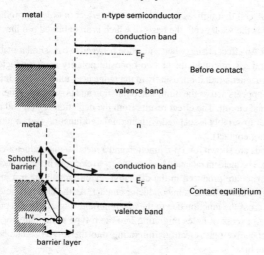

b **Photovoltaic effect** in metal-semiconductor contact

pico- Symbol: p. A prefix to a unit, denoting a submultiple of 10^{-12} of that unit: one picofarad is 10^{-12} farads.

picture carrier ►television.

picture element 1. The smallest portion of a picture area in a ►television system that is resolved by the scanning processes. In transmission it effectively corresponds to the smallest area of the optical image that produces a discernible video signal. In reception it effectively corresponds to the area of smallest detail that can be resolved on the screen of the picture tube. **2.** ►pixel.

picture frequency *Syn. for* frame frequency. ►television.

picture noise ►grass.

picture signal *Syn. for* video signal. ➤television.

picture tube A ➤cathode-ray tube used in a ➤television receiver to reproduce the transmitted picture. The electron beam is intensity-modulated by the transmitted video signal in order to reproduce the transmitted luminance; it is then caused to traverse the screen by ➤sawtooth waveforms applied to deflection coils around the tube. Automatic focusing is applied to convergence coils to maintain a clear image at the edges of the screen. The scanning process (➤television) must be performed in synchronism with the transmitted information in order to maintain a satisfactory picture. ➤➤colour picture tube.

picture white *Syn. for* white peak. ➤television.

Pierce crystal oscillator ➤piezoelectric oscillator (Figs. *b, c*).

piezoelectric crystal A crystal that exhibits the ➤piezoelectric effect. All ➤ferroelectric crystals are piezoelectric as well as certain nonferroelectric crystals and some ceramics. The best-known examples of piezoelectric crystals include quartz crystal, Rochelle salt, and barium titanate. ➤➤piezoelectric oscillator.

piezoelectric effect An effect that occurs when certain materials are subjected to mechanical stress. A ➤dielectric polarization is set up in the crystal and the faces of the crystal became electrically charged. The polarity of the charges reverses if the compression is changed to tension. Conversely, an electric field applied across the material causes it to contract or expand according to the sign of the electric field.

The piezoelectric effect is reversible with an approximately linear relation between deformation and electric field strength. The *piezoelectric strain constant, d,* is defined as

$$d_{i,k} = \delta e_k/\delta E_i$$

where δe_k is the incremental stress and δE_i the change in electric field strength along defined axes in the crystal ($i \equiv x, y, z$ and $k \equiv xx, yy, zz, yz, zx, xy$).

The piezoelectric effect is observed in all ➤ferroelectric crystals and in nonferroelectric crystals that are asymmetric and have one or more polar axes. The magnitude of the piezoelectric effect depends on the direction of the stress relative to the crystal axes. The maximum effect is obtained when the electrical and mechanical stresses are applied along the X-axis (the electric axis) and the Y-axis (the mechanical axis), respectively. The third major axis of a piezoelectric crystal is the Z-axis (the optical axis).

The piezoelectric effect is important because it couples electrical and mechanical energy and thus has many applications for electromechanical ➤transducers.

piezoelectricity Electrical signals generated as a result of the ➤piezoelectric effect.

piezoelectric oscillator An oscillator that utilizes a piezoelectric crystal in order to determine the frequency. Such oscillators are very stable. If an alternating electric field is applied across a suitable direction of a piezoelectric crystal, mechanical vibrations result (➤piezoelectric effect). If the frequency corresponds to a natural frequency of vibration of the crystal, substantial mechanical vibrations result. These in turn produce an alternating electric field across the crystal. The mechanical vibrations suffer little from damping and have a sharp resonance peak; piezoelectric crystals are therefore suitable for use as frequency standards.

A suitably cut piezoelectric crystal is mounted between the plates of a capacitor in order to apply the alternating voltage. The capacitor is usually formed by ►sputtering a metallic film on the large faces of the crystal in order to minimize the mechanical loading. The crystal is supported by lightweight supports that touch it at a mechanical node. In order to produce extremely high frequency stability the crystal can be supported in vacuo; for the highest frequency stability, required for the control of powerful frequency transmitters, the crystal is placed in an electrically heated oven, thermostatically controlled to within 0.1 kelvin. In the latter case, where the temperature coefficient of frequency of oscillation is required to be substantially zero, a T-cut crystal is usually used. This is cut as a thin plate whose faces contain the X-axis and a line in the YZ-plane inclined at an angle to the Z-axis. This cut exhibits lower piezoelectric activity than X- or Y-cuts.

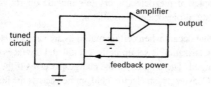

a Basic **piezoelectric oscillator**

Piezoelectric crystals can be connected in various ways to an oscillator circuit. The circuits used may be classified into two main types.

In the *crystal oscillator* the crystal replaces the ►tuned circuit in the oscillator and thus provides the resonant frequency (Fig. *a*); the *Pierce crystal oscillator* is an example (Figs. *b*, *c*).

In the *crystal-controlled oscillator* the crystal is coupled to the oscillator circuit, which is tuned approximately to the crystal frequency. The crystal controls the oscillator frequency by ►pulling the frequency to its own natural frequency and thus preventing frequency drift. The *Hartley crystal-controlled oscillator* is an example (Fig. *d*).

piezoelectric strain constant ►piezoelectric effect.

piezoelectric strain gauge ►strain gauge.

pigtails *Informal* The conductor connecting the shield on a piece of cable to the earth on an enclosure or connector. The conducting shield typically becomes imperfect in these regions and is often a curved piece of wire, hence 'pigtail'. The consequence of using these rather than a continuous shield may be severe degradation of the EMC performance of the equipment.

π-mode operation ►magnetron.

pincushion distortion ►distortion.

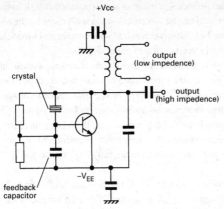

b Pierce crystal oscillator with crystal between base and collector

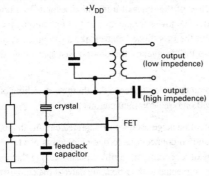

c Pierce crystal oscillator using a FET

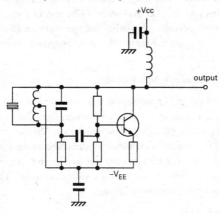

d Hartley crystal-controlled oscillator

PIN diode (or **p-i-n diode**) A diode that exhibits an ➤impedance at high frequency that is almost entirely resistive, and can be controlled over a wide range by the applied d.c. bias. These devices are used as voltage-controlled ➤attenuators and ➤modulators, and as switches in signal control circuits, such as phase shifters.

The PIN diode structure consists of a layer of lightly doped high-resistivity ➤semiconductor (the 'I' or intrinsic layer), sandwiched between heavily doped (highly conductive) p-type and n-type regions. With zero or negative applied voltage bias, the I-layer is depleted of charge ➤carriers and exhibits a very high resistance. In forward bias, carriers are injected from the P and N regions into the I-layer, increasing its conductivity rapidly. In these conditions, the diode behaves essentially as a pure resistor at radio frequencies. The resistance ratio from the fully 'on' state to the 'off' state can be as large as 5000 to 1.

The ➤capacitance associated with the diode can be thought of as that of a parallel-plate capacitor, with the P and N regions as the two plates. This capacitance is usually very low as a result of the thickness of the I-layer. A major parameter of the PIN diode is the carrier lifetime τ, which is of the order of a few nanoseconds to a few microseconds. It determines the effective lower frequency limit of operation, $f_0 = 1/(2\pi\tau)$. At frequencies below this limit the PIN diode behaves like a p-n junction diode, but with higher distortion. Above this frequency the diode resistance is the dominant feature.

ping-pong *Informal* In telecommunication systems, the alternate switching of parts of the incoming and outgoing call onto the communications line.

pin grid array (PGA) A form of package used for ➤integrated circuits that is capable of providing up to several hundred connections to one chip and is used for LSI and VLSI applications. It consists of a ceramic or moulded plastic casing that contains a ➤leadframe. The leadframe is connected to the ➤bonding pads of the chip using either ➤wire bonding or ➤tape automated bonding, and is connected to an array of output pins around the edges of the PGA package. The grid array of pins may be formed as several parallel rows of pins at two opposite sides of the package or around all four sides of the package, depending on the size and complexity of the integrated circuit. ➤➤dual in-line package; leadless chip carrier.

PIN photodiode (or **p-i-n photodiode**) ➤photodiode.

Pirani gauge *Syn.* hot-wire gauge. An instrument that measures low pressure by means of the variation of resistance of a conductor with pressure. An electrically heated wire loses heat by conduction through the low-pressure gas. The rate of heat loss depends on the resistance of the wire and if a constant potential difference is maintained across a wire, the change in resistance with pressure can be measured. Alternatively, the applied p.d. can be varied in order to maintain the resistance constant. The gauge must be calibrated against known pressures before use.

π-section ➤two-port network; filter.

pitch The perceived position of a sound on a scale from low to high. Pitch varies primarily as the ➤fundamental frequency of a sound is altered, but it can also vary to a much lesser extent when, for example, the intensity, spectral content, or duration is changed.

pitch bend A wheel or lever found on electronic musical instruments that enables the ➤pitch of the sound to be varied up or down by altering the fundamental frequency of the sound. The control is sprung such that it returns to the centre when released.

pixel A picture element, i.e. one element in a digitized image or a video display.

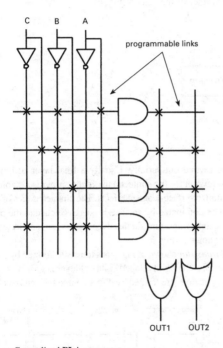

Generalized **PLA**

PLA *Abbrev. for* programmable logic array. A ➤logic circuit that is a sum-of-products device – input devices AND gates, output devices OR gates – where the input variables to the AND gates are programmable (as in ➤PAL devices), and the inputs to the OR gates are also programmable (see diagram). Crosses in the diagram indicate connections of input variables (A, B, C) to inputs of AND and OR gates.

planar array antenna An antenna consisting of an array of elements filling a rectangular area. Depending on the phasing of the feed to the elements, the array can be made to scan a wide range of angles in both horizontal and vertical directions. ➤linear array antenna.

planar process The most commonly used method of producing junctions during the manufacture of ►semiconductor devices. All the semiconductor devices are formed at the same time on one side of a semiconductor wafer. The device and IC structure is then planar rather than bulk three-dimensional. A layer of silicon dioxide is thermally grown on the surface of a silicon substrate of the desired conductivity type. ►Photo-lithography is used to etch holes in the oxide layer, which then acts as a ►mask for the ►diffusion of suitable impurities into the substrate in order to produce a region of opposite polarity. The junction between the two semiconductor types actually meets the surface of the substrate below the oxide since the diffusion occurs in directions both normal to and parallel to the surface of the silicon (Fig. *a*).

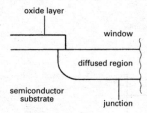

a **Planar process**

Several diffusions can be carried out serially. Usually a final layer of oxide is grown to cover the entire chip (except for the contacts) in order to provide a stable sur-face for the silicon and to minimize surface-leakage effects. The characteristics of early junction transistors tended to be dominated by surface-leakage effects and the planar process proved to be one of the most important single advances in semiconductor tech-nology. A planar transistor is shown in Fig. *b*.

The n-p-n planar *epitaxial transistor* shown in Fig. *c* is a transistor formed by a com-bination of ►epitaxy and diffusion. The lightly doped epitaxial layer is grown on to the highly doped substrate and the junctions are formed by diffusion into the epitaxial layer.

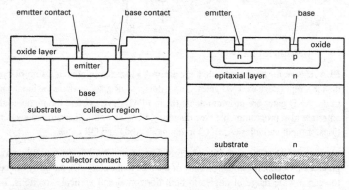

b Planar transistor   *c* Planar epitaxial transistor

In this technique the highly doped substrate forms the bulk of the collector and the collector series resistance is therefore reduced while the lightly doped epitaxial layer maintains the collector-base breakdown characteristics.

Planck constant Symbol: *h*. A universal constant arising from *Planck's law,* which states that the energy of ►electromagnetic radiation is confined to discrete packets or ►photons. The energy of each photon is given by *h*ν, where ν is the frequency of the radiation. Planck's law is fundamental to quantum theory. The value of *h* is

$$6.626\ 0755 \times 10^{-34} \text{ joule second}$$

The *rationalized* Planck constant, $h/2\pi$, symbol $\hbar$, has the value

$$1.054\ 5726 \times 10^{-34} \text{ J s}$$

Planck's law ►Planck constant.

plane-polarized wave *Syn.* linearly polarized wave. An ►electromagnetic wave in which the vibrations are rectilinear and parallel to a plane that is transverse to the direction of propagation of the wave. ►polarization.

plan position indicator (PPI) A form of presentation of information used in a ►radar receiver, where the receiving antenna is continuously performing a circular scan. The signal from the target appears as a bright spot on a phosphor screen, and is refreshed on each revolution of the radar antenna. The distance and bearing of the target are given by the polar coordinates of the spot with respect to the centre of the screen, which is the location of the receiver.

Planté cell The first secondary cell to be constructed. It consisted of rolled lead sheets that were dipped into dilute sulphuric acid.

plant load factor ►load factor.

plasma 1. A region of ionized gas in an arc-discharge tube (►gas-discharge tube) that contains approximately equal numbers of electrons and positive ►ions and provides a conducting path for the arc discharge. **2.** *Syn. for* positive glow. ►gas-discharge tube. **3.** A fully ionized gas in which electrons and ions move freely, with electric and magnetic fields affecting and arising from the moving charged particles. A plasma can be formed at high temperatures, as in stars and experimental thermonuclear reactors, or by photoionization.

plasma-enhanced CVD (PECVD) ►chemical vapour deposition.

plasma etching ►etching.

plasma screen *Syn.* plasma display. ►flat-panel display.

plastic-film capacitor A capacitor the dielectric of which is a plastic film. The electrical properties depend on the molecular structure. Material made from polar (asymmetrical) molecules will have an increased dielectric constant that is frequency

plate   436

dependent; nonpolar (symmetrical) molecules give a material the properties of which are independent of frequency.

Two main types of plastic-film capacitor exist: *polystyrene film* and *polyester film capacitors*. The former contains a nonpolar plastic with excellent properties that is used with metal-foil electrodes to produce a low-loss capacitor with good capacitance stability. Polyester-film capacitors are slightly polar plastic-film capacitors with either foil or metallized electrodes and are useful for d.c. applications and for operation up to 125 °C.

plate 1. *US syn. for* anode. **2.** An electrode in an electrolytic ►cell or in a ►capacitor.

plateau A region of the current-voltage characteristic of an electronic device that exhibits a substantially constant value of current for a significant voltage range.

plated heat sink A ►heat sink formed on a power device by plating a thick layer of metal on the back of the wafer. Plated heat sinks are often used in conjunction with ►via holes in integrated circuits containing power devices.

plated magnetic wire A magnetic wire (►magnetic recording) that has a surface of ferromagnetic material plated on to a nonmagnetic core.

plated-through-hole (PTH) A printed circuit board (PCB) technique where holes in the board for insertion of the components make metallic contact to both sides (and all layers in a multilayer PCB), so that the component is electrically connected to all the appropriate interconnection layers. Early double-sided PCBs without this technology could only make contact to the components on one side of the board, and connections to the other side had to made explicitly using ►via (through) pins.

plating ►electroplating.

platinum Symbol: Pt. A metal, atomic number 78, that is extremely stable and noncorroding and is used as an electrical contact or conductor at high temperatures or when chemical attack is likely. It is also used as a diffusion barrier in multilayer metal contacts, preventing interactions between the constituent metals.

platinum resistance thermometer ►resistance thermometer.

PLB *Abbrev. for* personal locator beacon. A radio transmitter carried by an individual for use in emergency situations to alert the search and rescue services to the emergency. The transmitted signal is usually an amplitude modulated signal at 121.5 megahertz, one of the international emergency distress frequencies. ►►EPIRB.

PLD *Abbrev. for* programmable logic device. A logic device where the designer must program the function required into the device. Such devices include ►ROM, ►PALs, ►PLAs, and ►FPGAs.

PLL *Abbrev. for* phase-lock loop.

PLM *Abbrev. for* pulse-length modulation. ►pulse modulation.

plug and socket A device that enables electrical apparatus to be connected or disconnected from a source of supply or other equipment. It consists of two separable portions, the 'male' plug and the 'female' socket, with metal contacts that engage each other when connection is effected. A *polarized plug* is constructed so that engagement with the socket is only possible in one position.

plumbicon ➤vidicon.

PM *Abbrev. for* phase modulation.

PMBX *Abbrev. for* private manual branch exchange. ➤telephony.

PMOS ➤MOSFET.

p-n junction The region at which two ➤semiconductors of opposite polarity meet, i.e. at which a p-type and n-type semiconductor meet. A simple p-n junction is formed from the same material in which approximately equal doping levels lead to two different conductivity types; this is known as a *homojunction*. A ➤heterojunction is formed from two dissimilar materials. The properties of p-n junctions are used in many semiconductor devices, such as ➤diodes and ➤transistors.

The simplified energy diagrams of an unbiased junction, a forward-biased junction (positive voltage applied to the p-region), and a reverse-biased junction are shown in Fig. *a*. In the unbiased state equilibrium considerations demand that the Fermi level, E_F (➤energy bands), is constant throughout the bulk of the material. This causes distortion of the energy bands at the junction and results in an electric field across the junction. This field is known as the *built-in field*. At equilibrium there is a small ➤depletion layer containing fixed ionized atoms and substantially no mobile charge carriers.

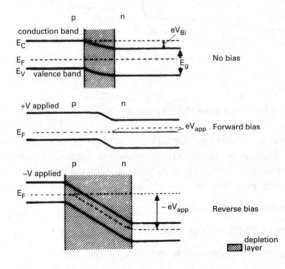

a Energy diagrams of a **p-n junction** (a homojunction)

Under reverse-bias conditions the depletion layer is increased as is the electric field across the junction, which thus acts as a barrier to current flow. Thermally generated charge carriers diffusing into the depletion layer are swept across the junction by the electric field and thus produce a small *reverse saturation current, I_0*.

Under forward-bias conditions the built-in field virtually disappears and charge carriers are attracted across the junction into the opposite polarity type (where they become ►minority carriers) and cause a current to flow in an external circuit.

The current-voltage relationship of an ideal p-n homojunction under forward-bias conditions obeys an exponential relationship:

$$I = I_0(e^{eV/kT} - 1)$$

where e is the electronic charge, V the applied voltage, k the Boltzmann constant, and T the thermodynamic temperature. This is the *Shockley equation* (or *ideal diode equation*) and assumes that the current is due to the diffusion of charge carriers across the junction, that the injected minority carrier density is small compared to the majority carrier density, and that there is no generation or recombination of charge carriers in the depletion region. In practice these conditions are not always fulfilled, particularly in silicon where the intrinsic carrier concentration is low, and it is found that the Shockley equation may be modified so that the term eV/kT is replaced by eV/nkT, where n has a value between one and two.

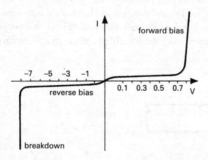

b Current-voltage characteristic of a **p-n junction**

The characteristic curve of a p-n junction depends on the geometry, the bias conditions, and the doping level on each side of the junction. A typical characteristic curve for a p-n junction diode is shown in Fig. *b*. Applications of simple p-n junctions include rectification, voltage regulation, and use as a varistor or varactor and as a switch.

pnpn (or **p-n-p-n**) **device** ►thyristor.

p-n-p transistor (or **PNP transistor**) ►bipolar junction transistor.

PN sequence *Short for* pseudonoise sequence. ►digital codes.

Pockel's effect ►Kerr effects.

point contact A non-ohmic contact between a pointed metallic wire (a *cat's whisker*) and the surface of a ➤semiconductor.

point-contact diode A diode rectifier formed by making a point contact between a small metal wire with a sharp point and a semiconductor. The point-contact diode has a very small area, which results in a very small capacitance, and it is therefore suitable for microwave applications. It has however a large spreading resistance, a large leakage current, and poor reverse breakdown characteristics. Its characteristics are difficult to predict since the device is subject to wide variations in its physical properties, including the pressure of the wire whisker, the contact area, and the crystal structure.

point-contact transistor A packaged version of the original ➤transistor that was demonstrated in 1947 and from which the ➤bipolar junction transistor was developed. Two ➤point contacts – 'emitter' and 'collector' – rested very close together on a germanium (semiconductor) crystal, which in turn was soldered to a metal disc – the 'base'. The components were housed within a metal cylinder connected electrically to the metal disc and crystal, forming the earth terminal, while the point contacts connected to pins that could be plugged into a socket. When a small signal was passed between the two point contacts, the current flowing in the device was amplified, largely due to the great difference in resistance between input and output ends. The larger output was modulated in accordance with the input signal.

Poisson's equation If the electric potential is V at a point (x, y, z) and the charge density is ρ, then Poisson's equation is given as

$$\partial^2 V/\partial x^2 + \partial^2 V/\partial y^2 + \partial^2 V/\partial z^2 = -\rho/\varepsilon$$

where ε is the permittivity. This can be rewritten as

$$\nabla^2 V = -\rho/\varepsilon$$

polar Denoting a component, such as an ➤electrolytic capacitor, that can only operate normally in one direction of applied voltage.

polarity 1. (magnetic) The manifestation of two types of regions in a magnet at which the inherent magnetism appears to be concentrated (poles). There are two types of magnetic pole: north-seeking (N) and south-seeking (S). **2.** (electrical) The manifestation of positive or negative parameters in an electrical circuit or device. The parameters include voltage, charge, current, and majority-carrier type.

polarization 1. A phenomenon occurring in a simple electrolytic ➤cell containing dissimilar electrodes. The current obtained from the cell decreases substantially soon after commencing the operation of the cell. This is due to the accumulation of bubbles of hydrogen gas, released from the electrolyte, around one of the electrodes. The bubbles partially cover the plate and thus increase the internal resistance of the cell and also create an e.m.f. of opposite polarity to the cell e.m.f. Cells such as the ➤Leclanché cell are designed to minimize polarization. **2.** The property of a radiated ➤electromagnetic wave that describes the time-varying direction and relative magnitude of the electric field vector; specifically, it is the figure traced as a function of time by the end-point of the vector at a fixed point in space (and the sense – clockwise or anticlockwise – in which it is traced), as observed along the direction of wave propagation. Polarization may be classified as *linear*, *circular*, or *elliptical*, depending on the shape of the figure traced by the electric field vector. In general the field is elliptically polarized. If the vector is always directed along a line that is perpendicular to the propagation direction, then the field is linearly polarized (or *plane-polarized*). **3.** ➤antenna polarization. **4.** *Short for* dielectric polarization.

polarization diversity ➤diversity system.

polarized plug ➤plug and socket.

polarized relay *Syn.* directional relay. ➤relay.

polarizing angle ➤Brewster angle.

pole 1. Each of the terminals or lines to a piece of electrical apparatus, a circuit, or network between which the circuit voltages are applied or produced. ➤➤number of poles. **2.** ➤magnetic pole. **3.** An electrode of an electrolytic ➤cell.

poles (and zeros) ➤s-domain circuit analysis.

polling Sequential interrogation of a number of difference sources of data in a digital system, each data source having its own line or digital address. The information gathering part of the circuit will interrogate each source to see if it has data it wishes to send.

polychromatic radiation Electromagnetic radiation that contains more than one frequency. The term is also applied to particulate radiation when the particles are all the same type but of different energies. The description *inhomogeneous* is usually preferred in this latter case. ➤➤monochromatic radiation.

polycrystalline silicon ➤polysilicon.

polyester-film capacitor ➤plastic-film capacitor.

polyphase system An electrical system or apparatus that has two or more alternating supply voltages displaced in phase relative to each other. In a symmetrical polyphase system each voltage is of the same magnitude and frequency and is displaced by an equal amount. If there are *n* sinusoidal voltages the mutual phase displacement is $2\pi/n$ radians and the system requires *n* lines at least. Thus in the *three-phase system* there is a phase difference of $2\pi/3$ between three voltage lines. An exception is the *two-phase system* that has a phase difference of $\pi/2$ between the two voltages.

polyphase transformer A ➤transformer that is used with a ➤polyphase system. The magnetic circuits required for each of the phase windings usually have portions in common with each other in order to retain the correct voltages.

polyphonic Involving a sound system with more than one channel.

polysilicon *Short for* polycrystalline silicon. Silicon in polycrystalline form is most often used to form the gate electrodes in silicon-gate ➤MOS integrated circuits and ➤charge-coupled devices. In this application the silicon is doped with a sufficiently high doping concentration so that it becomes degenerate and exhibits metallic properties.

polystyrene-film capacitor ➤plastic-film capacitor.

population inversion ➤laser.

port An access point in an electronic circuit, device, network, or other apparatus where signals can be input or output or where the variables of the system may be observed or measured. ➤➤two-port network.

portable conformable mask (PCM) ➤multilevel resist.

positive booster ➤booster.

positive charge ➤charge.

positive column *Syn. for* positive glow. ➤gas-discharge tube.

positive feedback *Syns.* direct feedback; regeneration. ➤feedback.

positive glow *Syns.* positive column; plasma. ➤gas-discharge tube.

positive ion ➤ion.

positive logic ➤logic circuit.

positive phase sequence ➤phase sequence.

positive photoresist ➤photoresist.

positive sequence ➤phase sequence.

positive transmission ➤television.

positron The antiparticle of the ➤electron.

pot *Informal* A potentiometer.

potential ➤electric potential.

potential barrier A region in an electric or magnetic field of force in which the potential is of such polarity as to oppose the motion of a particle subject to the field.

potential difference (p.d.) Symbol: $\Delta V, U$; unit: volt. The difference in ➤electric potential between two points, equal to the line integral of the electric field strength between the points. If a charge is moved from one to the other of the points by any path, the work done is equal to the product of the potential difference and the charge. The *potential gradient* at a point is the potential difference per unit length. ➤➤electromotive force.

potential divider *Syn.* voltage divider. A chain of resistors, inductors, or capacitors arranged in series. It is tapped at one or more points along the chain in order to obtain one or more predetermined fractions of the total voltage across the chain.

potential gradient ➤potential difference.

potential transformer *Syn. for* voltage transformer. ➤transformer.

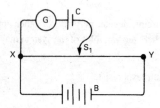

Potentiometer

potentiometer 1. A form of ➤potential divider that uses a uniform wire as the resistive chain. A movable sliding contact is used to tap off any potential difference less than that between the ends of the wire.

A typical use is for the measurement of potential difference or e.m.f. by balancing the unknown e.m.f. with that of a standard. For example, to measure the e.m.f. of a cell, C, the cell is arranged as shown in the diagram. The slider, S, is moved along XY until the null position is found on the galvanometer, G. The e.m.f. due to C just balances the potential across XS_1 (of length l_1). The cell is replaced by a standard cell, C_S, and the new balance point found at XS_2 (length l_2). Then

$$E_C/E_S = l_1/l_2$$

where E_C and E_S are the e.m.fs. of the unknown and standard cells, respectively.

More elaborate forms of potentiometer are available for precision applications, such as the ➤Kelvin–Varley slide. **2.** Any variable resistor, usually wire-wound, used in electronic circuits that has a third movable contact. The geometry of the device can be arranged so that the output voltage is a particular function of the applied voltage. The uniform wire can be arranged as a single coil or a spiral with the movable contact rotating about the axis through the centre of the coil. A *sine, cosine,* or *logarithmic potentiometer* produces an output proportional to the sine, cosine, or logarithm of the angular displacement of the shaft, respectively.

powdered-iron core A magnetic ➤core that is constructed of finely divided particles of iron embedded in a plastic or ceramic binding material. The low dissipation of such a core makes it very useful for high-frequency applications.

power Symbol: *P*; unit: watt. The rate at which energy is expended or work is done. In a direct-current circuit or device the power developed is equal to *VI*, where *V* is the potential difference in volts and *I* the current in amperes. In an alternating-current circuit the power developed is equal to *VI* cosϕ, where *V* and *I* are the root-mean-square values of voltage and current and ϕ is the ►phase angle between them. Cosϕ is the ►power factor of the circuit or device and the *apparent power* is the product *VI*, measured in volt-amperes. The product *VI* sinϕ is the *reactive power*. ►var.

power amplifier ►amplifier.

power component 1. (of a current or voltage) ►active current; active voltage. **2.** (of the volt-amperes) ►active volt-amperes.

power density *Syn.* Poynting vector. Symbol: *S*. Of an electromagnetic wave, the power passing through unit area carried by the wave. If the electric and magnetic fields in the wave are *E* and *H* respectively (both vector quantities), then the power density is equal to the vector product of the fields:

$$S = E \times H$$

It is measured in watts per square metre. For a ►harmonic wave (with *E* and *H* both harmonic), a more often quoted value is the average power density, S_{av}, which is given by

$$S_{av} = \tfrac{1}{2}\mathcal{R}(E \times H*)$$

where $\mathcal{R}$ indicates the real rather than imaginary value, and *H** is the complex conjugate of *H*.

power efficiency The ratio of the energy output to energy input under a specified set of operating conditions. The term is especially applied to an electroacoustic transducer, such as a loudspeaker. The inverse of the power efficiency is the *power loss* of the device.

power factor The ratio of the actual power in watts developed by an a.c. system, as measured by a ►wattmeter, to the apparent power indicated by ►voltmeter and ►ammeter readings and usually measured in ►volt-amperes (VA) rather than watts. If the voltage and current are sinusoidal the power factor, *P/VI*, is equal to the cosine of the ►phase angle between them. The power factor is also equal to the ratio of the ►resistance, *R*, to the ►impedance, *Z*, and thus indicates the dissipation in an ►insulator, ►inductor, or ►capacitor.

power frequency The frequency at which domestic and industrial mains electricity is supplied and distributed. In Europe the standard value is 50 hertz; in the US it is 60 hertz.

power line ►transmission line.

power loss ►power efficiency.

power pack A device that converts power from an a.c. or d.c. supply, usually the ►mains, into a form that is suitable for operating electronic devices.

power station *Syn.* generating station. A complete assembly consisting of all necessary plant, equipment, and buildings at a suitable site for the conversion of energy of one type, such as thermal or nuclear energy, into electrical power.

power supply Any source of electrical power in a form suitable for operating electronic circuits. Alternating-current power may be derived from the mains either directly or by means of a suitable ➤transformer. Direct-current power may be supplied from batteries, suitable rectifier/filter circuits, or from a converter.

A ➤bus is frequently employed to supply power to several circuits or to several different points in one circuit. Suitable values of voltage are derived from the common supply by ➤coupling through dropping resistors or by capacitive coupling.

power supply rail *Syn.* power rail. ➤rails.

power transformer ➤transformer.

power transistor A transistor, such as an IGBT, that is designed to operate at relatively high values of power or to produce a relatively high power gain. Power transistors are used for switching and amplification. They usually require some form of temperature control since the power dissipation in them ranges from 1 watt to 100 watts.

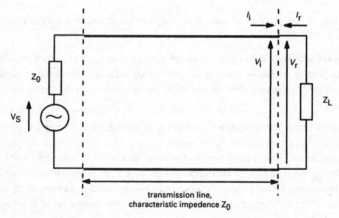

Power waves: transmission line connecting load to source

power waves A signal representation used in ➤distributed circuits and ➤transmission lines. In such a circuit or network, the voltage and current vary spatially, therefore the indicated impedance also varies with the location of the measurement. To allow an understanding and analysis of the signal flow in the circuit, the signals are represented by power waves, which describe the incident and reflected signals in the transmission line. Given a simple transmission line circuit (see diagram) of characteristic impedance Z_0, with V_i and V_r as the incident and reflected voltages at the load,

incident power $P_i = |V_i|^2/Z_0$

reflected power $P_r = |V_r|^2/Z_0$

power delivered to load $P_L = (|V_i|^2 - |V_r|^2)/Z_0$

load ►reflection coefficient $\Gamma = (Z_L - Z_0)/(Z_L + Z_0)$

If this circuit is viewed in terms of power flow, then if the power available from the source is given by

$$P_A = |a|^2$$

and the power delivered to the load is

$$P_L = |a|^2 - |b|^2$$

then $|b|^2$ can be thought of as the power that is reflected, or *scattered* by the load. It then follows that

$$a = (V + Z_0I)/(2\sqrt{Z_0}) \text{ and } b = (V - Z_0I)/(2\sqrt{Z_0})$$

are the incident and reflected power waves, respectively. They have dimensions of $\sqrt{(\text{power})}$. ►scattering parameters.

power winding ►transductor.

Poynting vector ►power density.

PPI *Abbrev. for* plan position indicator.

PPM *Abbrev. for* pulse-position modulation. ►pulse modulation.

preamp *Short for* preamplifier.

preamplifier An amplifier that is used in a system, such as a ►radio receiver, in order to amplify the received signals before they are input to the main part of the system.

precision approach radar (PAR) A ►radar system that is used at an airport to present accurate information about the location of incoming aircraft in the vicinity of the airport and is used as an aid for air traffic control. An associated *airport surveillance radar* system (ASR) is usually employed separately to scan the surrounding area and presents continuous information to the air traffic controller about the distance and bearing of all aircraft within a given radius of the airport.

precision rectifier A rectifier circuit in which the ►p-n junction diode turn-on voltage is eliminated, allowing small-amplitude signals to be rectified accurately. This is

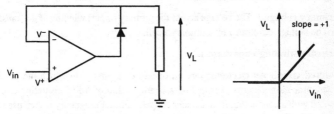

Precision rectifier

important for measurement instrumentation. An example of a precision rectifier circuit is shown in the diagram: when the input is positive, the diode is forward biased and behaves like a ➤voltage follower, keeping $V^+ = V^-$.

pre-emphasis A technique used to improve the ➤signal-to-noise ratio in a radio-communication system that employs ➤frequency modulation or ➤phase modulation. A network that increases the modulation index of the higher modulation frequencies relative to the lower ones is inserted at the transmitter. *De-emphasis* is used at the receiver in order to restore the relative strengths of the audiofrequency signals, i.e. a network is inserted that reduces the relative strength of the higher frequencies. Amplitude-modulation systems rarely use pre-emphasis and de-emphasis as the resulting improvement in signal-to-noise ratio is only slight.

The technique is also used for magnetic tape recordings, gramophone records, and speech analysis systems to counter the average −6 dB per octave spectral tilt found in speech and music.

preferred values ➤standardization.

preselector A low-pass filter placed at the input of a ➤heterodyne receiver to limit the bandwidth of the received input signal. This improves the noise performance and also prevents harmonics of the desired input frequency from being captured by the receiver.

PRF *Abbrev. for* pulse repetition frequency. ➤pulse.

primary cell ➤cell.

primary electrons Electrons that impinge on a surface and cause ➤secondary emission of electrons from the surface. The term is also sometimes used to describe electrons released from atoms by one of the processes of ➤electron emission other than secondary emission. Such electrons however are more commonly described in terms of the relevant process of emission, as with thermionic electrons.

primary emission ➤Electron emission other than ➤secondary emission.

primary failure ➤failure.

primary radiator *Syn. for* active antenna. ➤directional antenna.

primary standard A standard that is used nationally or internationally as the basis for a given unit. ➤secondary standard.

primary voltage 1. The voltage across the primary (input) winding of a ➤transformer. **2.** The voltage developed by a primary ➤cell.

primary winding ➤transformer.

printed circuit An electronic circuit, or part of a circuit, in which the conducting interconnection pattern is formed on a board. A thin board of insulating material is coated with a conducting film, usually copper. ➤Photolithography is then used to coat part of the film with protective material. The unprotected metal is removed by etching, leaving the desired pattern of interconnections. Discrete components or packaged ➤integrated circuits may then be added to complete the circuit.

Double-sided printed circuits are commonly produced in which both sides of the board have a circuit formed on them, with ➤feedthroughs to connect the two sides as required. Printed circuits have been produced with several alternating layers of metal film and thin insulating film mounted on a single board. Boards with up to 12 layers of interconnections can be produced, commonly for digital systems.

The use of plug-in printed circuits in electric or electronic equipment facilitates maintenance and repair. Printed circuits are reasonably robust when subjected to careful handling.

printer A device that produces printed characters, numbers, pictures, or symbols on paper from a source of information, such as a source of experimental data, the output of a computer, or information transmitted by a telecommunication system.

printing *Informal* Photolithography.

print-through A form of ➤distortion arising during magnetic tape recording. It is caused by a region of strongly magnetized tape affecting adjacent layers, and is a temperature-sensitive effect.

privacy In communications, protection against unauthorized access to communicated data. There is often the desire to communicate such that others cannot understand the message being sent. Privacy may be acquired by the coding of an information message such that unauthorized users cannot gain access to the information.

private branch exchange (PBX) ➤telephony.

private exchange ➤telephony.

probe 1. An electric lead that connects to a measuring or monitoring circuit, or contains such a circuit at its end or along its length, and that is used for testing purposes. The measuring or monitoring circuit may be formed from either active or passive components. **2.** A means of testing integrated circuits or devices while they are still on the complete semiconductor wafer, prior to dicing. The probes provide mechanical and electrical connection to the bond pads that are usually found around the periphery of the IC. The probes can be moved individually to the correct locations on the wafer – these are known as *micromanipulator probes* – or they can be mounted onto a *probe card* in a fixed location that reflects the arrangement of the bond pad connections around the particular IC. **3.** A resonant conductor that is inserted into a ➤waveguide or ➤cavity resonator in order to inject or extract energy.

probe card ➤probe.

process-control system *Syn.* regulator. ➤control system.

processor An electronic device that performs some sort of calculations or complex manipulations on information to obtain the desired result. A computer contains one or more processors; when a processor is designed to perform a wide variety of functions, making it possible to use just one processor instead of several, it is referred to as a ➤central processing unit or central processor.

program 1. A complete set of instructions written in a particular ➤programming language that after translation by a compiler into executable form causes a computer to perform a set of defined operations. A given program includes all the necessary instructions that cause the computer to input or output data or results, to perform mathematical operations, to store data in designated locations, to transfer data or instructions from one part of the system to another, and to perform any necessary operations required for the successful completion of the program.

A *procedure* (or *subroutine*) is a section of a program to which control may be transferred from a number of points throughout the program. When the instructions in the subroutine have been obeyed, control is returned to the point from which the transfer was made. This saves the repetition of identical sections of code in different places in the program. A *diagnostic routine* is used to check automatically the operation of a program or part of a program and to detect errors (➤bug). **2.** To cause a logic circuit, logic device, etc., to realize a particular function.

program counter (PC) *Syns.* instruction pointer; instruction counter. The ➤register in a CPU that points to the next instruction to be executed.

programmable array logic ➤PAL.

programmable logic array ➤PLA.

programmable logic device ➤PLD.

programmable read-only memory (PROM) ➤ROM.

programme signal The complex wave containing the information corresponding to the sound information (audio signal) and vision information (video signal) during a specific radio or television broadcast.

programming language A notation for the precise description of computer ➤programs or ➤algorithms. It is designed to be mutually comprehensible to computers and humans. A particular digital computer operates in ➤binary notation and all instructions must ultimately be produced in the appropiate binary code, known as ➤machine code.

A *high-level programming language* is a language that resembles natural language or mathematical notation more closely than machine code. It is designed to reflect the requirements of particular problems. A *compiler* is a program that is used to convert a high-level language to machine code. There are many different high-level languages suitable for different applications. Examples include *Pascal, Fortran, C, C++*, and *occam*.

A *low-level programming language* is one that resembles machine code more closely that natural language, and has structures that directly reflect the architecture of the computer. ➤assembly language.

projection lithography ➤photolithography.

PROM *Abbrev. for* programmable ROM. ➤ROM.

propagation coefficient *Syn.* propagation constant. Symbol: γ, P. A complex quantity that expresses the effect of a ➤transmission line on a sinusoidal progressive wave. The propagation coefficient is defined for a uniform transmission line of infinite length supplied with a sinusoidal current of specified frequency at its sending end.

Under steady-state conditions, if the currents at two points along the line, separated by unit length, are I_1 and I_2, where I_1 is nearer the sending end of the line, then

$$\gamma = \log_e(I_1/I_2)$$

at the specified frequency; I_1/I_2 is the vector ratio of the currents. γ is a complex quantity and may be written

$$\gamma = \alpha + j\beta$$

where $j = \sqrt{-1}$. The real part, α, is the ➤attenuation constant and is measured in nepers per unit length of line. It measures the transmission losses in the line. The imaginary part, β, is the *phase-change coefficient* and is measured in radians per unit length of line. It is the ➤phase difference between I_1 and I_2 introduced by the transmission line. Thus

$$I_1/I_2 = \exp(\alpha + \beta j) = \exp(\alpha).\exp(j\beta)$$

If the displacement of the vibration is a maximum at a given point and equal to p_1, then at the same instant the displacement, p_2, at a distance x along the transmission line is given by

$$p_2 = p_1 \exp(-\alpha - j\beta)x$$

An infinite transmission line is not physically possible but conditions simulating those in an infinite line are realized when a transmission line of finite length is terminated by its characteristic impedance. ➤image transfer constant.

propagation constant ➤propagation coefficient.

propagation delay The time taken for an electrical signal to move from one point to another. The concept is often associated with digital circuits and the time to pass information through logic gates.

propagation loss Energy loss from a beam of ➤electromagnetic radiation as a result of absorption, scattering, and spreading of the beam.

proportional control A control system that operates by first determining the difference between the actual value of the quantity to be controlled and the desired value, and then applying a correction proportional to this difference (see diagram).

protective horn ➤arcing horn.

protocol In the transfer of digital signals across a switched system, the structure to which a digital message must conform if it is to be acceptable to the switching network: the message must be able to be identified as a valid message and must have a sender and recipient identifier. A typical example would be where the message has a header block of digits giving information about the sender and the recipient and possibly the length of the message. It will also have a terminating block of digits that indicate the end of the message and, optionally, further information to verify the accuracy of the message.

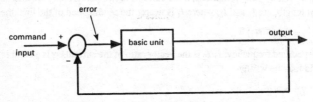

Proportional control

proton An elementary particle having a positive charge equal to that of the electron, mass 1.6726×10^{-27}kg, which is about 1836 times the electron mass, and spin ½. It also has an intrinsic ►magnetic moment. The proton forms the nucleus of the hydrogen atom and is a constituent part of all atomic nuclei. The number of protons in an atomic nucleus is equal to the number of orbital electrons and is the atomic number of the atom.

prototype filter *Syn. for* normalized filter.

proximity effect 1. An effect observed when two or more conductors carrying alternating current are placed close to one another. The distribution of current across the cross section of any one conductor is changed under the influence of magnetic fields due to the other(s). The ►effective resistance of the conductor is modified by this effect and it is particularly significant in coils used at high (radio) frequencies. **2.** In ►electron-beam lithography, unwanted exposure of the resist between two lines or patterns if these lines or patterns are exposed very close to each other.

proximity printing ►photolithography.

PSK *Abbrev. for* phase shift keying. ►phase modulation.

psophometer An instrument that measures the amount of ►noise in a transmission system, particularly a telephone system. It includes a device for measuring noise power through a weighting network in order to produce objective results that approximately parallel subjective results with human observers. Standard psophometric weighting characteristics have been produced by the ►CCITT.

Crosstalk and other noise may be measured by passing a suitable noise signal down one telephone channel and measuring the amount of this noise that appears on an adjacent line with a psophometer with standard weighting characteristics.

PSPICE A version of SPICE especially adapted to run on PC-compatible computers. ►transistor parameters.

PSTN *Abbrev. for* public switched telephone network.

PTFE *Abbrev. for* polytetrafluoroethylene. ➤Teflon.

PTH *Abbrev. for* plated-through hole.

PTM *Abbrev. for* pulse-time modulation. ➤pulse modulation.

p-type conductivity Conduction in a semiconductor in which current flow is caused by the effective movement of mobile ➤holes through the semiconductor. ➤semiconductor; n-type conductivity.

p-type semiconductor An extrinsic semiconductor that contains a higher density of mobile ➤holes than of conduction ➤electrons, i.e. holes are the majority carriers. ➤semiconductor; n-type semiconductor.

puff *Informal syn. for* picofarad.

pulling A change of frequency observed in an electronic oscillator when it is coupled to a circuit containing another independent oscillation. The frequency of the oscillator tends to change towards that of the independent oscillation; the tendency is particularly marked if the difference in frequency is small. Complete synchronization can sometimes be achieved. Pulling is used to control the frequency of an oscillator, as in crystal-controlled oscillators (➤piezoelectric oscillator).

pulsating current A current that exhibits regularly recurring variations of magnitude. The term implies that the current is unidirectional. A pulsating current or other pulsating quantity can be considered as the sum of a steady component and a superimposed alternating component whose average value cannot be zero.

pulse A single transient disturbance, one of a series of transient disturbances recurring at regular intervals, or a short train of high-frequency waves as are used in echo sounding or radar. A pulse consists of a voltage or current that increases from zero (or a constant value) to a maximum and then decreases to zero (or the constant value), both in a comparatively short time. The zero or constant value is termed the *base level*. A pulse is described according to its geometric shape when the instantaneous value is plotted as a function of time: it can be rectangular, square, triangular, etc. Unless otherwise specified a pulse is assumed to be *rectangular*.

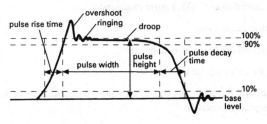

Practical rectangular **pulse**

In practice a perfect shape is never achieved and a practical rectangular pulse is shown in the diagram. The portion of the pulse that first increases in amplitude is the *leading edge*. The time interval during which the leading edge increases between specified limits, usually between 10% and 90% of the pulse height, is termed the *rise time*. The pulse decays back to the base level with a finite *decay time*, usually taken between the same limits as the rise time. The major portion of the decay time is termed the *trailing edge* of the pulse.

The time interval between the end of the rise time and the start of the decay time is the *pulse width*. The magnitude of the pulse taken over the pulse width is normally substantially constant, and is the *pulse height*. The pulse height can be quoted either as the maximum value, the average value, or the root-mean-square value, all measured over the pulse width and ignoring any spikes or ripples. The height of a nonrectangular pulse, e.g. a triangular one, is normally the maximum amplitude.

A practical rectangular pulse can suffer from *droop*, which occurs when the pulse height falls slightly below the nominal value. The degree of droop is given by the *pulse flatness deviation;* this is the ratio of the difference between the maximum and minimum values of the pulse amplitude to the maximum amplitude, during the pulse width. A *valley* occurs when the droop exists only over a portion of the pulse width and the pulse height then recovers. The pulse *crest factor* is the ratio of the peak amplitude of the pulse to the root-mean-square amplitude.

A *spike* is an unwanted pulse of relatively short duration superimposed on the main pulse; *ripple* is unwanted small periodic variations in amplitude. A practical pulse frequently rises to a value above the pulse height and then decays to it with damped oscillations. These phenomena are known as *overshoot* and *ringing*. A similar effect occurs as the pulse decays to the base level. Pulse circuits may frequently incorporate a *smearer,* which is a circuit designed to minimize overshoot.

A group of regularly recurring pulses of similar characteristics is called a *pulse train*. The time interval between corresponding portions of the pulses in the train is the *pulse spacing* (or *pulse-repetition period*), *T*, and its reciprocal is the *pulse-repetition frequency* (or *pulse rate*), which is measured in hertz. Minor variations in the pulse spacing in a pulse train are known as pulse *jitter.* The *duty factor* of a pulse train is the ratio of the average pulse width to the average pulse spacing of pulses in the train.

pulse amplitude *Syn. for* pulse height. ➤pulse.

pulse-amplitude modulation (PAM) ➤pulse modulation.

pulse carrier ➤pulse modulation.

pulse code modulation (PCM) ➤pulse modulation.

pulse coder *Syn. for* coder. ➤pulse modulation.

pulse communications Telecommunications involving the transmission of information by means of ➤pulse modulation.

pulse detector *Syn. for* decoder. ➤pulse modulation.

pulse discriminator A ➤discriminator that selects and responds only to a ➤pulse with a particular characteristic of amplitude, period, etc., and that is used in ➤pulse operation.

pulse duration *Syn. for* pulse width. ➤pulse.

pulse-duration modulation (PDM) *Syns.* pulse-width modulation; pulse-length modulation. ➤pulse modulation.

pulse-flatness deviation *Syn.* pulse tilt. ➤pulse.

pulse-forming line An ➤artificial line that contains inductances and capacitors in series and is used to produce fast high-voltage pulses in ➤radar.

pulse-frequency modulation (PFM) ➤pulse modulation.

pulse generator An electronic circuit or device that generates voltage or current ➤pulses of a desired waveform. It can be designed to produce either single pulses or a pulse train. A *pulser* is a particular type of pulse generator that produces fast (short-duration) pulses of high voltage.

pulse height *Syn.* pulse amplitude. ➤pulse.

pulse-height analyser ➤multichannel analyser.

pulse interval *Syn. for* pulse spacing. ➤pulse.

pulse jamming In radar systems, the transmission of band-limited pulses of wideband noise in an attempt to prevent a receiver from receiving a signal.

pulse length *Syn. for* pulse width. ➤pulse.

pulse-length modulation (PLM) ➤pulse modulation.

pulse modulation A form of modulation in which ➤pulses are used to modulate the ➤carrier wave or, more commonly, in which a pulse train is used as the carrier (the *pulse carrier*). Information is conveyed by modulating some parameter of the pulses with a set of discrete instantaneous samples of the message signal. The *minimum sampling frequency* is the minimum frequency at which the modulating waveform can be sampled to provide the set of discrete values without a significant loss of information.

Different forms of pulse modulation are shown in Fig. *a*. In *pulse-amplitude modulation* (PAM), the amplitude of the pulses is modulated by the corresponding samples of the modulating wave. In *pulse-time modulation* (PTM), the samples are used to vary the time of occurrence of some parameter of the pulses. Particular forms of pulse-time modulation are *pulse-duration modulation* (PDM), also known as *pulse-length modulation* (PLM) or *pulse-width modulation* (PWM), in which the time of occurrence of the leading edge or trailing edge is varied from its unmodulated position, *pulse-frequency modulation* (PFM), in which the ➤pulse repetition frequency of the carrier pulses is varied from its unmodulated value, and *pulse-position modulation* (PPM), in which the time of occurrence of a pulse is modulated from its unmodulated time of

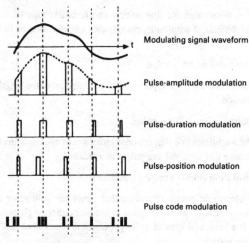

a Forms of **pulse modulation**

occurrence, i.e. the pulse repetition period is varied. All these types of pulse modulation are examples of uncoded modulation.

Pulse code modulation (PCM) is a form of ►digital modulation: only certain discrete values are allowed for the modulating signals. The modulating signal is sampled, as in other forms of pulse modulation, but any sample falling within a specified range of values is assigned a discrete value. Each value is assigned a pattern of pulses and the signal transmitted by means of this code. There are a number of variations on the way in which this code is transmitted, including NRZ (nonreturn to zero) PCM, RZ (return to zero) PCM, return to zero AMI (alternate mark inversion), and biphase PCM, also called Manchester code. These different schemes are shown in Fig. *b*. The electronic circuit or device that produces the coded pulse train from the modulating waveform is termed a *coder* (or pulse coder). A suitable *decoder* must be used at the receiver in order to extract the original information from the transmitted pulse train. ►Morse code is a very well known example of a pulse code.

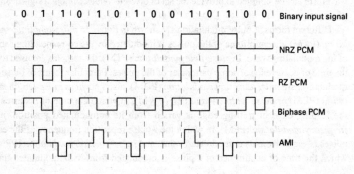

b Forms of pulse code modulation

Delta modulation (DM), also called *slope modulation*, is another form of digital modulation in which the transmitted information only indicates whether the signal to be transmitted has been encoded to have a rising or falling transition.

Pulse modulation is commonly used for ➤time-division multiplexing.

pulse operation Any method of operation of an electronic circuit or device that transfers electrical energy in the form of pulses.

pulse-position modulation (PPM) *Syn.* pulse-phase modulation. ➤pulse modulation.

pulser ➤pulse generator.

pulse radar ➤radar.

pulse rate ➤pulse repetition frequency.

pulse regeneration The process of restoring a ➤pulse or pulse train to its original form, timing, and magnitude. Pulse regeneration is required in most forms of pulse operation since the circuits or circuit elements used can introduce ➤distortion.

pulse repetition frequency (PRF) *Syn.* pulse rate. The rate at which pulses are transmitted in a pulse train. It is measured in hertz. ➤pulse.

pulse repetition period *Syn. for* pulse spacing. ➤pulse.

pulse separation *Syn. for* pulse spacing. ➤pulse.

pulse shaper Any circuit or device that is used to alter any of the characteristics of a ➤pulse or pulse train. ➤Pulse regeneration is a special case of pulse shaping.

pulse spacing *Syns.* pulse separation; pulse interval; pulse repetition period. ➤pulse.

pulse tilt *Syn. for* pulse flatness deviation. ➤pulse.

pulse-time modulation (PTM) ➤pulse modulation.

pulse train ➤pulse.

pulse width *Syns.* pulse duration; pulse length. ➤pulse.

pulse-width modulation (PWM) ➤pulse modulation.

pump ➤parametric amplifier.

punch-through A type of ➤breakdown that can occur in both ➤bipolar junction transistors and ➤field-effect transistors. If the collector-base voltage, V_{CB}, applied to a bipolar junction transistor is increased, the depletion layer associated with the collector-base junction spreads across the base region. At a sufficiently high collector voltage, known as the punch-through voltage, V_{PT}, the depletion layer spreads through the entire base region and reaches the emitter junction. A direct conducting path is therefore formed from emitter to collector and charge carriers from the emitter 'punch through' to the collector. The associated energy diagram is shown in Fig. *a*.

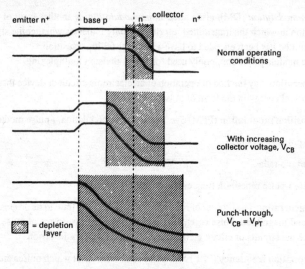

a **Punch-through**: energy diagrams for an n-p-n transistor at different collector voltages

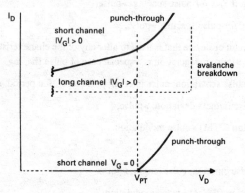

b FET characteristics above pinch-off

In a field-effect transistor the effect operates in a similar manner. When the drain voltage, V_D, reaches a sufficiently large value (V_{PT}) the depletion layer associated with the drain spreads across the substrate and reaches the source. Charge carriers then punch through the substrate.

Punch-through can be a problem in a short-channel device used as a switch. Due to the relatively high doping level of the drain of a FET the drain-substrate depletion layer spreads readily across the substrate. Punch-through therefore must be avoided in the 'off' state of the switch, when the gate voltage, V_G, is zero. Punch-through can also occur in the 'on' state ($|V_G| > 0$) when the device is operated in the saturated region above pinch-off (►field-effect transistor); charge carriers punch through the substrate from source to drain at sufficiently large values of V_D. In FETs that have relatively long channel lengths and high doping levels, punch-through does not occur; ►avalanche breakdown occurs first (Fig. *b*).

puncture voltage The value cf voltage that causes an insulator to be punctured when it is subjected to a gradually increasing voltage. ►►impulse voltage.

purple plague The formation of purple-coloured areas on the bonds of a silicon integrated circuit. The purple colour is caused by the formation of an unwanted aluminium-gold ►eutectic mixture ($AuA1_2$) at the bond between the gold connecting wire and the aluminium ►bonding pads. The resulting bond is mechanically weak, since the eutectic is very brittle, and is therefore susceptible to failure.

push-pull amplifier *Syn.* balanced amplifier. ►push-pull operation.

push-pull operation The use of two matched devices in such a way that they operate with a 180° phase difference. The output circuits combine the separate outputs in phase (Fig. *a*). One common means of achieving the desired 180° phase shift in the inputs is a transformer-coupled input circuit (Fig. *b*). ►Complementary transistors may also be used (Fig. *c*), in which case no phase shift is required in the inputs.

Push-pull circuits are frequently used for ►class A and ►class B amplification and are then termed *push-pull amplifiers*. A push-pull amplifier that is suitably biased to

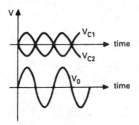

a Output of transformer-coupled **push-pull operation**

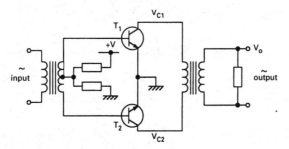

b Transformer-coupled class A **push-pull operation**

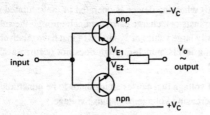

c Complementary transistor **push-pull operation**

give negligible output current when no input signal is present is termed a *quiescent push-pull amplifier.*

PWM *Abbrev. for* pulse-width modulation. ➤pulse modulation.

Q

QAM *Abbrev. for* quadrature amplitude modulation. ➤amplitude modulation.

Q channel ➤I channel.

Q factor *Syn.* quality factor. Symbol: Q. A factor that is associated with a ➤resonant circuit and describes both the ability of the circuit to produce a large output at the resonant frequency and the ➤selectivity of the circuit. The Q factor is defined as

$(2\pi \times$ energy stored$)$/(energy dissipated per cycle$)$

or alternatively as

(energy stored)/(energy dissipated per radian)

In the case of a simple series resonant circuit at the resonant frequency,

$\omega_0 L = 1/\omega_0 C$

and the Q factor is given by

$Q = \omega_0 L/R$ or $Q = 1/\omega_0 CR$

where ω_0 is 2π times the resonant frequency f_0 and L, C, and R are respectively the inductance, capacitance, and resistance of the circuit.

The selectivity of a resonant circuit can be written as

$Q = f_0/\text{BW}$

where BW is the difference in frequency between the two 3 dB points.

A single reactive component may be capable of resonance without any other components, i.e. if the self-capacitance of an inductance coil or the self-inductance of a capacitor is sufficiently large. The Q factor of a single component is defined as the ratio of the reactance to the effective series resistance of the component, hence for an inductance

$Q = \omega L/R$

and for a capacitor

$Q = 1/\omega CR$

Q point ➤quiescent point.

QPSK *Abbrev. for* quadrature phase shift keying. ➤amplitude modulation.

quadrant electrometer An ►electrometer that consists of a flat cylindrical metal box formed from individually isolated quadrantal segments and containing a light foil-covered vane that is supported by a quartz fibre and is free to move (see diagram). Opposite quadrants are connected together. The instrument is usually arranged so that the vane hangs symmetrically within the quadrants when it and both pairs of quadrants are at zero potential. In addition the potential, V_c, applied to the vane must be large compared to the potentials V_A and V_B applied to the quadrants. Under these circumstances the deflection, θ, of the vane is given by

$$\theta = k_1(V_A - V_B)$$

or if one pair of quadrants is at earth potential by

$$\theta = k_2 V$$

where k_1 and k_2 are constants and characteristic of the instrument. The deflection of the vane is observed by using a small mirror that reflects a spot of light attached to the torsion thread.

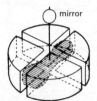

Quadrant electrometer

quadraphony A four-channel sound-reproduction system that is an extension of stereophonic sound reproduction (►reproduction of sound) using four separate loudspeakers. The sound may be recorded either using four coincident directional microphones at right angles to each other and placed at the position of the listener, or using separated microphones whose outputs are divided between the channels and combined in the correct proportions to achieve the desired effect.

Various methods exist for recording and reproducing quadraphonic signals, all of which combine the signals to produce two channels that may be recorded by stereophonic techniques. *Compatible discrete four-channel recording* has four separate channels. Each stereo channel contains signals from two of the quadraphonic channels, multiplexed together. The sum signal of the two channels occupies one frequency band and the difference signal another frequency band. On playback, the sum and difference signals in each of the stereo channels are detected and the original information extracted. *Matrix recording* systems combine the outputs of the four channels together in fixed proportions and phase relationships in order to produce two channels of information. The two channels are then decoded during reproduction in order to restore the original four channels.

quadrature Two periodic quantities that have the same frequency and waveform are *in quadrature* when the phase difference between them is $\pi/2$ (90°). They therefore differ by one quarter of a period, one wave reaching its ►peak value when the other passes through zero.

quadrature amplitude modulation (QAM) ►amplitude modulation.

quadrature component 1. (of current) ►reactive current. **2.** (of voltage) ►reactive voltage. **3.** (of volt-amperes) ►reactive volt-amperes.

quadrature phase shift keying (QPSK) ►amplitude modulation.

quadripole ►two-port network.

quality factor ►Q factor.

quantization A method of producing a set of discrete or quantized values that represents a continuous quantity. In the case of a voltage waveform, it is divided into a finite number of subranges each of which is represented by an assigned value within the subrange. This technique is used whenever a set of discrete values is required, for example to produce data suitable for a digital computer or in ►pulse modulation. Some loss of information of the original waveform is inherent in this process, leading to a certain amount of noise, or *quantization distortion.*

quantization noise ►noise.

quantizer A device that can convert an analogue signal into a signal having values that are identical to the analogue signal only at discrete instants. The output signal then consists of a number of steps between specified levels.

quantum ►quantum theory.

quantum dot A ►semiconductor structure in which the electron population is constrained in a three-dimensional ►quantum well, producing an electronic system in which the electrons can only have discrete energy values and are physically confined within the dot. The quantum dot therefore closely resembles the energy structure of an atom. This provides a practical device for studying the electronic behaviour in atoms. A quantum dot can be manufactured using ►nanotechnology: ►molecular-beam epitaxy can be used to produce quantum confinement in one dimension, and ►etching is used to create a small island of semiconductor, producing *size quantization* in two physical dimensions. ►►quantum wire; nanoelectronics.

quantum efficiency *Syn.* quantum yield. The number of reactions of a particular type induced per photon of absorbed electromagnetic radiation. It can, for example, be the average number of photoelectrons produced per photon at a specified frequency in a photoelectric ►photocell or ►phototube.

quantum mechanics *Syn.* wave mechanics. The theory of atomic and nuclear systems, providing a mathematical framework for ►quantum theory. In 1926 Erwin Schrödinger developed a wave equation to describe the behaviour of elementary particles in materials by treating them as *matter waves* or ➤de Broglie waves, following the hypothesis by de Broglie that ➤electrons, etc., could be described both in terms of particle-like and wave-like behaviour. This wave equation, now known as the *Schrödinger equation*, described the motion of de Broglie waves in a given potential energy, for instance in a potential well, or inside a crystalline solid. The Schrödinger equation for the electron, for example, is:

$$(-\hbar^2/2m)\nabla^2\Psi + V\Psi = j\hbar(\partial\Psi/\partial t)$$

where $\hbar$ is the rationalized ➤Planck constant, m is the mass of the electron, V is the potential in which it moves, and ∇^2 is the ➤Laplace operator. This is a differential equation in the quality Ψ, which is known as the *wavefunction*; this is a complex quality that can be thought of as a measure of the amplitude of the de Broglie wave describing the electron. Hence the square of the wavefunction is a measure of the intensity of the electron wave, or

$$|\Psi(x, y, z, t)|^2 \, dxdydz$$

gives the probability of finding an electron at time t in the incremental volume element $dxdydz$ located at (x, y, z).

Solutions of the Schrödinger equation can be found analytically only for simple examples, usually by eliminating the time dependence by assuming the de Broglie waves have a typical $\exp(j\omega t)$ time dependence and solving the *time-independent* Schrödinger equation:

$$(-\hbar^2/2m)\nabla^2\psi + V\psi = E\psi$$

Now ψ is the time-independent component of the wavefunction and E is the total energy of the electron, and the equation describes the physical variation of the electron wavefunction in space. Analytical solutions of the time-independent Schrödinger equation can be used to provide valuable insight into the behaviour of electrons in atoms and solids. For example, the solution of a particle in a potential well leads to the concept of ➤quantum numbers describing the allowed energy states in the well. This structure is known as a ➤quantum well. By using a periodic value for the potential energy V, solutions of the Schrödinger equation are found to describe the origins of ➤energy-band structure in crystalline solids; this is the *Kronig–Penney model* of energy-band structure.

quantum numbers A set of numbers that are used to label the various possible values of certain physical properties. The property is restricted to certain discrete values (quantized) as a result of applying the principles of ➤quantum mechanics to a physical system. A set of quantum numbers is usually in the form of a set of integers and half-integers.

quantum theory A theory introduced by Max Planck in 1900 to explain the emission and absorption of radiation (black-body radiation) by material bodies. Planck postulated that the atoms in the body behaved as tiny oscillators, which could radiate or absorb energy only in discrete packages. He called these packages *light quanta*, known today as ➤photons, whose energy is given by

$$E = h\nu = \hbar\omega$$

where $\hbar$ is the rationalized ➤Planck constant, equal to $h/2\pi$, ν is the frequency of the absorbed or emitted radiation, and ω is the angular frequency.

The quantum theory was a profound departure from classical physics, which presumed energy to be continuous and hence infinitely divisible into smaller and smaller quantities. Quantum theory was used successfully by Einstein in 1905 to explain the ➤photoelectric effect: if the frequency of the irradiating light is less than some threshold frequency ν_0, then no matter how intense the light no photoelectric effect would be observed; for incident light of frequency greater than ν_0, photoelectrons are emitted in numbers directly proportional to the intensity of the light – in other words, proportional to the number of incident light quanta or photons.

Quantum theory is used to describe physical behaviour on the atomic scale of matter – at the level of individual atoms, electrons, or photons – or in solids where the dimensions approach those of a few atomic diameters, say less than about 10 nanometres. The mathematical framework describing quantum theory is found in ➤quantum mechanics.

quantum well The one-dimensional infinitely deep quantum well is often used as an example for a simple solution of Schrödinger's equation in ➤quantum mechanics to illustrate a *quantized system*. Such a quantum well has a finite width W and in this region the local potential V is zero; outside this region the potential is infinite, thereby constraining the electrons to be inside the well. The solution of Schrödinger's equation (time-independent) for the quantum well is:

$$\psi = A\cos(2mEx/\hbar^2) + B\sin(2mEx/\hbar^2)$$

The wavefunction ψ must fall to zero at the edges of the well, $x = 0, W$, which gives

$$A = 0; \qquad \sin(2mEW/\hbar^2) = 0, \text{ or } 2mEW/\hbar^2 = n\pi;$$

n is an integer and is a ➤quantum number. The wavefunction ψ is an integral number of half-wavelengths in W, as shown in the diagram. The energies that the electron can have in this system are discrete values – i.e. are quantized; these values of energy are the ➤eigenvalues of the system.

Practical one-dimensional quantum wells can be realized by sandwiching a ➤semiconductor with a narrow ➤energy band gap between wider band gap semiconductors, using ➤molecular beam epitaxy. The energy step at the band edges effectively constrains the electrons (or holes) to the narrow band gap material. Solving Schrödinger's equation again yields a quantized system, where the electron population is quantized in one dimension, but remains free in the remaining two dimensions, forming a *two-dimensional electron gas*. This type of well is used as the basis for ➤high electron mobility transistors, and for quantum well ➤semiconductor lasers, where the quantization of electron and hole energy levels produces a narrow line width.

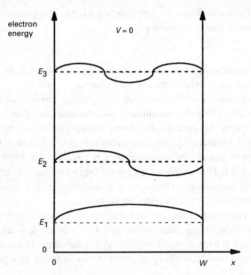

Quantum well, one-dimensional and infinitely deep, showing solutions of
Schrödinger's equation

quantum wire A ➤semiconductor structure in which the electron population is constrained in a two-dimensional ➤quantum well, resulting in only one degree of freedom for the electrons. Such structures can be made using ➤nanotechnology: ➤molecular beam epitaxy can be used to produce quantum confinement in one dimension, and ➤etching is used to create a stripe of semiconductor narrow enough to produce *size quantization*. Quantum wires display interesting electron-transport properties characteristic of highly quantized systems. ➤quantum dot; nanoelectronics.

quantum yield ➤quantum efficiency.

quarter-phase system *Syn. for* two-phase system. ➤polyphase system.

quarter-wavelength line *Syn.* quarter-wavelength transformer. A ➤transmission line of length equal to one quarter of the wavelength of the ➤fundamental frequency. It is used for ➤impedance matching, particularly in systems designed to operate at high radiofrequencies, for the suppression of even-order harmonics in filter networks, and for coupling and feeders used with antennas.

quarter-wavelength transformer ➤quarter-wavelength line.

quartz Naturally occurring crystalline silicon dioxide (SiO_2). It exhibits marked piezoelectric properties and is frequently used as the piezoelectric crystal in ➤piezoelectric oscillators. It also exhibits marked ➤dielectric strength.

 Quartz may be readily drawn into extremely fine uniform filaments that are very strong, elastic, and physically and chemically stable. Such *quartz fibres* are frequently used as torsion threads in delicate measuring instruments, such as ➤electrometers.

quartz-crystal oscillator ➤piezoelectric oscillator.

quasi-bistable circuit An astable ➤multivibrator that is caused to operate by a ➤trigger (i.e. it is ➤clocked) rather than being allowed to run freely. Provided that the frequency of application of the trigger is high compared to the natural frequency of vibration of the free-running circuit it operates as a bistable circuit, i.e. a ➤flip-flop.

quasi-complementary push-pull amplifier A configuration of ➤push-pull amplifier output stage for a power amplifier but instead of using ➤complementary output power transistors, two n-p-n ➤bipolar junction transistors are employed.

quasi-peak detector A detector designed to record the maximum received signal over a particular time span. However, rather than just detecting the overall maximum, the device does not record single very fast signals but only those that last for a significant length of time; in this way it is similar to a resistor-capacitor circuit with a finite time constant that will not respond to fast pulses. It is used for ➤electromagnetic compatibility testing, where it is employed to detect only those signals that pose a significant long-term probability of causing interference.

quefrency The X-axis of the ➤cepstrum of a signal, which has the dimension of time.

quench A capacitor, resistor, or combination of the two, that is used in parallel with a contact to an inductive circuit and that inhibits ➤spark discharge across the contact when the current ceases. A quench is commonly employed across the make-and-break contacts of an induction coil.

quench frequency ➤super-regenerative reception.

quiescent-carrier telephony ➤telephony.

quiescent component A component of an electronic circuit that, at a specific instant, is not in operation but at a short time following the specified instant will become operative.

quiescent current The current that flows in any circuit under specified normal operating conditions but in the absence of an applied signal.

quiescent period The period between transmissions in a pulse transmission system.

quiescent point *Syn.* Q point. The region on the characteristic curve of an active device, such as a transistor, during which the device is not operating.

quiescent push-pull amplifier ➤push-pull operation.

quieting circuit ➤squelch circuit.

quieting sensitivity ➤radio receiver.

Quine–McCluskey An algorithm to find a minimum ➤sum-of-products for a particular function.

R

radar Acronym from *ra*dio *d*irection *a*nd *r*anging. A system that locates distant objects using reflected radiowaves of microwave frequencies. Modern radar systems are highly sophisticated and can produce detailed information about both stationary and moving objects and can be used for navigation and guidance of ships, aircraft, and other vehicles and systems.

A complete *radar system* contains a source of microwave power, such as a ➤magnetron, a modulator to produce pulses of microwave energy where necessary, transmitting and receiving antennas, a receiver that detects the echo, and a cathode-ray tube (CRT) that displays the output in a suitable form. Several types of radar system are in common use.

Pulse radar systems transmit short bursts of high-frequency radiowaves and the reflected pulse is received during the time interval between the transmitted pulses.

Continuous-wave systems transmit energy continuously and a small proportion is reflected by the target and returned to the transmitter.

Doppler radar utilizes the ➤Doppler effect in order to distinguish between stationary and moving objects. The change in frequency between the transmitted and received waves is measured and hence the velocity of moving targets deduced.

Frequency-modulated radar is a system that transmits a frequency-modulated radar wave. The reflected echo ➤beats with the transmitted wave and the range of the target is deduced from the beat frequency produced.

Volumetric radar systems can produce three-dimensional positional information about one or more targets. Two transmitters used simultaneously are commonly employed. *V-beam radar* is a volumetric system using two fan-shaped beams.

In any of the above systems the direction and distance of the target is given by the direction of the receiving antenna and the time interval between transmission of the radar signal and reception of the echo.

The direction of the transmitting and receiving antennas can be periodically varied in order to scan a given area. *Coarse scanning* is often used to obtain an approximate target location before repeating the scan more accurately. A common arrangement is to rotate the antennas in a horizontal plane, and produce a synchronous circular scan on the CRT, in order to display any targets within the vicinity of the transmitter. Such a presentation is termed a ➤plan position indicator (PPI). Scanning in a vertical plane can also be used.

Pulsed radar systems frequently use the same antenna as both a transmitting and receiving antenna. The appropriate transmitting and receiver circuits are connected to it using a ➤transmit-receive (TR) switch. The pulse repetition frequency of the transmitted signals is determined by a ➤multivibrator known as the *master trigger.*

The *radar range* is the maximum distance at which a particular radar system is effective in detecting a target. It is usually defined as the distance at which a designated target is distinguished for at least 50% of the transmitted pulses. The range is dependent on the ➤minimum discernible signal that the radar receiver can accept, i.e. the minimum power input to the receiver that produces a discernible signal on the radar indicator. The power in the return echo is dependent on the peak power of the transmitted pulse. In general the larger the output power of the transmitter, the greater the range of the system. A given radar system is characterized by the performance figure; this is the ratio of the peak power of the transmitted pulse to the minimum discernible signal of the receiver. The ability of a radar system to differentiate objects along the same bearing is usually defined as the minimum radial distance separating targets at which they can be separately resolved.

Radar systems are used for the detection and control of aircraft (➤precision approach radar), guiding of ships in fog, and for locating distant storm centres when an echo is produced by the associated heavy rainfall. Radar is used in astronomy and also has an extremely wide range of military applications.

radar beacon ➤beacon.

radar indicator *Syns.* radar screen; radarscope. A ➤cathode-ray tube that provides a visual display of the return echoes in a ➤radar system. A *panoramic radar indicator* simultaneously displays all the received echoes of different frequencies.

radar range ➤radar.

radarscope ➤radar indicator.

radar screen ➤radar indicator.

radial-beam tube An ➤electron tube in which the electron beam travels in a radial path from a central cathode to one of a set of circumferential anodes. The beam is rotated between the anodes by means of a rotating magnetic field. The tube can be used as a high-speed switch or a commutation switch.

radiated emissions ➤electromagnetic compatibility.

radiated interference ➤electromagnetic compatibility

radiated susceptibility ➤electromagnetic compatibility.

radiation Any form of energy that is propagated as waves or streams of charged particles. ➤electromagnetic radiation.

radiation counter ➤counter.

radiation efficiency ➤antenna efficiency.

radiation noise ➤noise.

radiation pattern ➤antenna pattern.

radiation potential ➤ionization potential.

radiation resistance ➤antenna radiation resistance.

radiative recombination ➤recombination processes.

radio 1. The use of ➤electromagnetic radiation of frequency within the radiofrequency portion of the electromagnetic spectrum (➤frequency band) for the transmission and reception of electrical impulses or signals without connecting wires or waveguides. It is also the process of transmitting or receiving such signals.

The unqualified term usually denotes the ➤telecommunication system that transmits audio information. Any communication channel, circuit, or link in which information is transmitted by radio is described as a radiocommunication or radio channel, radio circuit, or radio link. **2.** *Short for* radiofrequency. Denoting electromagnetic radiation in the ➤radiofrequency range or any device, component, or other apparatus used to transmit or receive information at frequencies within this range, as in radio telephony, radio telemetry, or radio astronomy. **3.** A ➤radio receiver.

radio astronomy The study of astronomical bodies and events by means of the radio signals associated with them. ➤radio telescope.

radio beacon ➤beacon.

radio compass A navigational aid carried on board aircraft and ships. It consists of a radio receiver together with a directional antenna and is essentially a ➤direction finder. The antenna is rotated in order to find the direction of a specific radio transmitter relative to the craft; the information is presented as the heading of the craft relative to the transmitter.

radio device *Short for* radiofrequency device. ➤radiofrequency.

radio direction finding ➤direction finding.

radio effect *Short for* radiofrequency effect. ➤radiofrequency.

radiofrequency (r.f. or RF) Any frequency of ➤electromagnetic radiation or alternating currents in the range 3 kilohertz to 300 gigahertz (➤frequency band). An electronic device, such as an ➤amplifier, ➤choke, or ➤transformer, that operates in this range is known as a *radiofrequency device* or *radio device;* similarly any associated effect such as ➤distortion is termed a *radiofrequency effect* or *radio effect.*

radiofrequency choke (RFC) A type of ➤inductor whose inductance presents a high impedance to high-frequency signals. It is used in the biasing of high-frequency transistor amplifiers thereby preventing the d.c. bias source acting as a ➤short to the a.c. signal.

radiofrequency heating ➤Dielectric heating or ➤induction heating that is carried out using an alternating field of frequency greater than about 25 kilohertz.

radiofrequency interference (RFI) ►electromagnetic compatibility; noise.

radiography ►X-rays.

radio horizon In radio signal transmission, the line that includes the path taken by direct rays. Direct transmission of a signal from a transmitter to a receiver occurs when the receiver can see the transmitter, i.e. they are in line-of-sight of one another. In reality ►diffraction occurs and the radio waves are slightly bent around the curvature of the earth. The actual radio horizon is about $^4/_3$ greater than the geometric line-of-sight.

radio interferometer ►radio telescope.

radiolocation *Obsolete name for* ►radar.

radio noise Any unwanted sound or distortion appearing at the loudspeaker of a ►radio receiver. A portion of the noise is due to ►interference from various sources: disturbances in the atmosphere or discharges in the ionosphere (*atmospheric noise*), extraterrestrial sources such as the sun (*galactic noise*) and signals from other radio transmitters and other man-made sources. The remaining portion of the noise is an inherent property of the electronic circuits and devices and is caused by the random motions of electrons in the circuits.

   A very common form of noise is *mains hum,* which is caused by harmonics of the mains frequency being detected and amplified by the radio receiver, and results in a humming noise. It tends to worsen as the components in the receiver age. ►►interference; noise.

radio receiver A circuit or system that converts radiofrequency signals into audio or video signals so that the information can be accessed. This is generally accomplished by mixing the ►modulated radio signal containing the information down to the audio or video frequency band using a nonlinear device or circuit; the process can be carried out directly or by means of an intermediate frequency (►heterodyne reception; direct conversion receiver).

   Domestic radios can detect either ►amplitude-modulated signals or ►frequency-modulated signals. They are described as *AM receivers* or *FM receivers,* respectively. A receiver that has the facility to detect both types of signal is an *AM/FM receiver.* High-fidelity (hi-fi) radios usually contain additional circuits that are associated with the audiofrequency amplifier and are used to restore the *bass response* and *treble response* of the output to that of the original audible source. The *bass boost* circuit restores the lower audiofrequency signals; the *treble compensation* circuit acts on the higher audiofrequencies.

   Stereophonic radio receivers contain suitable detecting circuits that demodulate stereophonic radio transmissions and produce two outputs, each of which is separately amplified and output to a loudspeaker.

   ►►reproduction of sound.

radio set 1. A radio receiver. **2.** A combined radio transmitter and receiver such as is used by amateur radio operators, in aircraft, or in ships.

radiosonde A small radio transmitter together with suitable transducers that is carried by a balloon or kite into the upper atmosphere and transmits meteorological and other scientific data to the ground.

radio spectroscope A device that analyses and displays the total radiofrequency signals received at an antenna. The signals are usually displayed on the screen of a cathode-ray tube and some indication of the modulation and field strength at the transmitted carrier frequency can be obtained from the height and spread of the trace.

radio telegraphy ►telegraphy.

radio telemetry The capture of data at a distance where the data is transmitted back to the receiving point by a radio link. The data may be obtained, for example, from aircraft or satellites. ►telemetry.

radio telephone ►telephony.

radio telescope A telescope used in ►radio astronomy to record and measure extraterrestrial radio signals. It consists of an ►antenna, or system of antennas, connected by ►feeders to one or more ►receivers, where the signals are amplified and analysed, usually by a computer. The antennas may be in the form of *dishes*, linear ►dipoles, or ►Yagi antennas. A dish is a large generally parabolic metal reflector that brings radiowaves to a focus above the dish centre. The focused waves are collected by a secondary feed antenna that is connected to the receiver. Dishes can usually be steered to point to different regions of the sky.

The antenna system may form a *radio interferometer*, in which the electrical signals from two separated antenna units are fed to a common receiver. The antenna units are often mounted on an E–W line, pointing in the same direction, so that the earth's rotation causes an extraterrestrial radio source to move through the antenna beam. The two identical signals from the antenna units travel a different pathlength; when combined, the amplitude of the summed signal changes periodically producing *interference fringes* at the receiver output. An interferometer is generally used to improve the resolution of the telescope, i.e. to allow finer detail to be distinguished. Analysis of the fringes allows the structure of a radio source to be determined.

radiotherapy ►X-rays.

radiowave An electromagnetic wave that has a frequency lying in the ►radiofrequency range.

radio window The range of ►radiofrequencies that are not reflected by the ►ionosphere but pass straight through it. The radio window extends from about 300 gigahertz to about 15 megahertz (approximate wavelength range: 1 mm to 20 m). The effect of the ionosphere is still noticeable up to 100 MHz but decreases as the radiofrequency is increased. At frequencies above 10 GHz heavy rain can severely affect transmission.

High-frequency ►television broadcasts fall within the radio window and long-distance television communication therefore requires the use of ►communications satellites as reflectors. ►Radio astronomy is also restricted to this frequency range.

radome A dielectric sheet used to cover a radar or other transmitting antenna. It is usually curved, particularly if on an aircraft, and is designed such that the geometry, shape and thickness, and dielectric characteristics allow for maximum transmitted signal with minimum reflection.

rahmonic A component in a line ➤cepstrum.

rails *Syns.* power supply rail; power rail. The common power supply connection within a circuit and between circuits.

raised cosine window *Syn. for* Hanning window. ➤windowing.

R-ALOHA *Short for* reservation ALOHA. ➤ALOHA system.

RAM *Abbrev. for* random-access memory. A ➤solid-state memory device which allows reading and writing of data and to which there is ➤random access to the individual ➤memory locations. The memory is arranged as a rectangular array of memory cells forming rows and columns. Each memory cell in the array forms an intersection between the rows and columns. A simple array of 16 storage cells arranged as a 4 × 4 matrix is shown in the diagram. Any individual cell in the array, such as the cell indicated, is defined by the address of one row and one column, as shown, since each row and column intersect once only. Each cell can store one ➤bit of information.

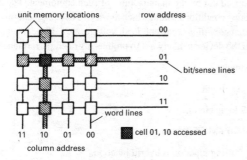

RAM: array of 4 × 4 storage cells

In order to retrieve information from a particular location, the address codes of the row and column are specified. The output is sampled by suitable sensing devices that are attached to each row and the rows are therefore termed *bit lines*. The columns are known as *word lines*.

random access A form of access to computer ➤memory in which individual memory locations can be directly accessed in any order to allow reading or writing of data. The ➤access time to any location is substantially constant and is independent of the ➤address of the location or of the previously accessed location.

random logic A ➤logic circuit that contains an arrangement of different interconnected logic gates rather than linked arrays of similar gates.

range 1. The maximum distance from a ➤radio or ➤television transmitter at which reception of the signal is possible. **2.** ➤radar.

range tracking Operation of a ➤radar system when viewing a moving target so that the ➤transmit-receive switch is automatically adjusted to switch to the receive mode at the correct instant for reception of the return echo; i.e. the ➤gate is adjusted to account for the alteration in distance of the target from the system.

raster ➤raster scanning.

raster scanning 1. A method of presenting images on the screen of a TV or VDU in which the electron beam of the ➤cathode-ray tube is swept across the screen in a horizontal pattern from top to bottom. The pattern of scanning lines is called a *raster*. ➤television. **2.** A method of sweeping a radar antenna beam either horizontally or vertically. Scanning is achieved mechanically by moving the entire antenna, or electronically using an array of antennas, a combination of which is selected to produce the desired beam shape and direction. **3.** ➤electron-beam lithography.

rated conditions ➤rating.

rating Stipulating or the stipulation of operating conditions for a machine, transformer, or other device or circuit and stating the performance limitations of such equipment. Rating is carried out by the manufacturer of such equipment. The designated limits to the operating conditions within which the device or equipment functions satisfactorily are the *rated conditions* (current, load, voltage, etc.). If the rated conditions are not adhered to the device is likely not to produce its rated performance.

ratio adjuster ➤tap changer.

ratio circuit ➤MOS logic circuit.

ratioless circuit ➤MOS logic circuit.

rationalized MKS system ➤MKS system.

rat-race *Informal syn. for* ring junction. ➤hybrid junction.

Rayleigh fading An effect occurring in radio signal propagation when the signal is received by a mobile receiver. The signal can take many paths from the transmitter, especially in urban areas where the signal will reflect off buildings and other obstructions. The signals from these different paths will arrive at the receiver with different ➤phases resulting from the different path lengths. The signals combine and in some cases will cancel one another out, resulting in nulls in the signal strength – Rayleigh fading.

RBS *Abbrev. for* Rutherford back scattering.

R-C (or **RC**) *Abbrev. for* resistance-capacitance. It is used as a prefix to describe circuits or devices that depend critically for their operation on their resistance and capacitance or that employ resistance-capacitance ➤coupling.

R-C network *Short for* resistance-capacitance network. ➤network.

RCT *Abbrev. for* reverse conducting thryristor. ►thyristor.

reactance Symbol: X; unit: ohm. The part of the total ►impedance of a circuit not due to pure ►resistance. It is the imaginary part of the complex impedance, Z, i.e.

$$Z = R + jX$$

where R is the resistance and j is equal to $\sqrt{-1}$. Reactance is due to the presence of ►capacitance or ►inductance in a circuit. The effect of reactance is to cause the voltage and current to become out of ►phase.

If an alternating voltage, given by

$$V = V_o \cos\omega t$$

where ω is the angular frequency, is applied to a circuit containing capacitance the impedance of the circuit is given by

$$Z = R - j/\omega C$$

where $1/\omega C$ is the *capacitive reactance*, X_C, which decreases with frequency. The current ►leads the voltage: the phase angle is 90° in a purely capacitive circuit.

In a circuit containing inductance the impedance is given by

$$Z = R + j\omega L$$

where ωL is the *inductive reactance*, X_L, which increases with frequency. The current ►lags the voltage: the phase angle is 90° in a purely inductive circuit.

reactance chart A chart that is presented in a form that enables the user to read directly the ►reactance of any given ►capacitor or ►inductor at a specified frequency, and conversely to deduce the capacitance or inductance of a given reactance at a particular frequency.

reactance drop ►voltage drop.

reactance modulator ►modulator.

reactance transformer A device that consists of pure reactances arranged in a suitable circuit and commonly used for impedance matching at radiofrequencies.

reactivation A process used with a thoriated tungsten filament cathode in order to improve the emission of electrons from the surface. An abnormally high voltage is applied to the filament and this causes a layer of thorium atoms to migrate to the surface.

reactive current *Syns.* reactive component, idle component, quadrature component of the current. The component of an alternating current vector that is in ►quadrature with the voltage vector.

reactive factor The ratio of the ►reactive volt-amperes of any ►load, circuit, or device to the total ►volt-amperes.

reactive ion beam etching ►etching.

reactive ion etching ►etching.

reactive load A ►load in which the current and voltage at the terminals are out of phase with each other. ►nonreactive load.

reactive power ►power.

reactive sputtering ►sputtering.

reactive voltage *Syns.* reactive component, idle component, quadrature component of the voltage. The component of an alternating voltage vector that is in ►quadrature with the current vector.

reactive volt-amperes *Syns.* reactive component, idle component, quadrature component of the volt-amperes. The product of the current and the ►reactive voltage or the product of the voltage and the ►reactive current. ►var; active volt-amperes.

reactor A device or apparatus, particularly a capacitor or inductor, that possesses ►reactance and is used because of that property.

read To remove information from a computer storage device or memory. *Destructive read operation* (DRO) is a read operation that leaves no information in the storage device. If the information is to be preserved DRO must be immediately followed by a write operation to restore it. *Nondestructive read operation* does not destroy the information. The type of read operation depends on the nature of the memory.

Read diode ►IMPATT diode.

read-only memory ►ROM.

read-out pulse A pulse applied to a word line of a ►RAM in order to enable a particular storage location on that line to become available to the sense amplifier connected to the appropriate bit line.

read-write head ►magnetic recording; magnetic disk.

read-write memory A memory used in computing in which the stored information can be readily altered. ►solid-state memory.

real-time operation Operation of a ►computer during the actual time in which a physical process occurs. Data generated by the physical process is input to the computer and the results produced can be used to control the process. ►interactive.

real-time system A system where correct operation is dependent on the output values and the time at which they appear. The term is often applied to ►control systems, where the controlling outputs must be sent at regular intervals in order to have correct system operation.

receiver The part of a telecommunication system that converts transmitted waves into a desired form of output. The range of frequencies over which a receiver operates with a selected performance, i.e. a known sensitivity, is the *bandwidth* of the receiver. The process of limiting reception to a desired portion of a cycle of operation is termed *gating.* The ►minimum discernible signal is the smallest value of input power that produces an output. ►radio receiver; television; heterodyne reception.

receiving antenna ►antenna.

recessed gate FET A power FET (►field-effect transistor) that has the gate electrode formed in a slot etched in the substrate between the source and drain electrodes (see diagram). Use of a recessed gate structure has the advantage that the extra thickness on each side of the gate reduces parasitic resistances between the gate and the source and drain.

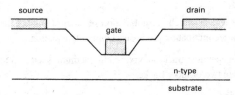

Recessed gate FET

rechargeable battery ►battery.

reciprocity ►reciprocity theorem.

reciprocity theorem (for electric circuits) For any physical linear system, the positions of an ideal voltage source and an ideal ammeter can be interchanged without affecting their readings. This can be generalized to situations involving electric fields and current sources. A system is said to possess *reciprocity* if it obeys the reciprocity theorem. Reciprocity can be useful in solving the electrical properties of that system.

recombination processes Various processes by which excess electrons and holes in a ►semiconductor recombine and tend to restore the system to the thermal equilibrium condition given by

$$pn = n_i{}^2$$

p is the number of holes, n the number of electrons, and n_i the number of holes or electrons in the intrinsic semiconductor at the same temperature.

The basic recombination processes are *band-to-band recombination,* when an electron in the conduction band recombines with a hole in the valence band, and *trapping recombination,* when *electron* or *hole capture* by a suitable acceptor or donor impurity occurs in the semiconductor (see diagram overleaf).

The energy lost by the conduction electron involved in band-to-band recombination may be emitted as a photon of radiation (*radiative recombination*) or may be transferred as kinetic energy to a free electron or hole (*Auger process*). Radiative recombination is the inverse process to ►photoconductivity and forms a significant proportion of the total recombination in ►direct-gap semiconductors. The Auger process is the inverse process to ►impact ionization.

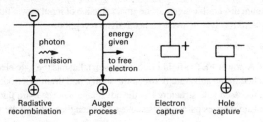

Recombination processes

recombination rate The rate at which recombination of electrons and holes in a ➤semiconductor occurs. ➤recombination processes; continuity equation.

recording channel An independent track on a recording medium, such as magnetic tape, that can accommodate two or more tracks.

recording head ➤magnetic recording.

recording of sound The process of producing a permanent or semipermanent record of sounds that may be used in suitable replay or reproducing apparatus (➤reproduction of sound) in order to reproduce the original sounds. Sounds may be recorded in analogue or digital formats, and the three main methods of sound recording are *electromechanical sound recording, photosensitive recording* (or *sound-on-film*), and ➤magnetic recording.

Electromechanical sound recording is used to produce gramophone records. The sound to be recorded is detected with one or more suitable microphones and the signals are amplified. The amplified signals are then used to drive a suitable *cutter* that produces an undulating groove in the surface of a wax or cellulose *master* disc. The undulations of the groove depend on the magnitude of the signals. In ➤compact disc systems the sound to be recorded is coded using ➤pulse code modulation, and the resulting coded signals are used to cut a series of tiny bumps on the wax or cellulose master disc with a laser. A copper-plated positive *mother* is usually produced from the master before mass production so that further copies of the master may be made as required.

Photosensitive or sound-on-film recording is used mainly in the cinema or for broadcast entertainment when film is used. The amplified audiosignals are used to modulate a light source, to which the recording film is exposed. The film is then developed and the record appears as a strip of varying density known as the *soundtrack*. The frequency of the light source is usually such that for convenience a ➤carrier wave is commonly used that is either amplitude-modulated or frequency-modulated by the audiosignal. In *white recording*, the minimum density of the developed film corresponds to the maximum received power of an amplitude-modulated signal or the lowest received frequency of a frequency-modulated system; the opposite applies for a *black recording*.

Magnetic recording is widely used for a variety of modern applications. The overall performance is as good as, or better than, the above two systems and it is very convenient in use.

rectangular pulse ➤pulse.

rectangular window ➤windowing.

rectification efficiency ➤rectifier.

rectifier A device that passes current only in the forward direction and can therefore be used as an a.c. to d.c. converter. A single device usually suppresses or attenuates alternate half-cycles of the alternating-current input (➤half-wave rectifier circuit). ➤Full-wave rectifiers usually contain two devices in a back-to-back arrangement. The most common type of rectifier is the semiconductor ➤diode. The *rectification efficiency* of any rectifier is the ratio of the direct-current output power to the alternating-current input power.

The output of a rectifier consists of a unidirectional current that rises to a maximum value periodically; this value corresponds to the peak value of the alternating-current input. The output is usually smoothed by a ➤smoothing circuit before being applied to the load in order to reduce the amount of ➤ripple. The fluctuating output can be considered as a steady d.c. component with an a.c. component superimposed on it. A portion of the ripple is sometimes due to the *rectifier leakage current,* i.e. alternating current that flows through the rectifier without being rectified.

rectifier filter ➤smoothing circuit.

rectifier instrument A d.c. instrument that can be made suitable for a.c. measurements by using a ➤rectifier to convert the alternating current to be measured into a unidirectional current. A common arrangement is a ➤bridge circuit formed from four ➤diodes and a ➤moving-coil instrument, M (see diagram). The indicated value on the instrument is usually the ➤root-mean-square value of a sinusoidal a.c. input waveform; rectifier instruments are subject to waveform error if used with nonsinusoidal input waveforms.

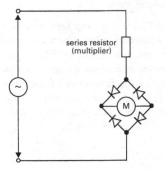

Rectifier instrument used as an a.c. voltmeter

rectifier leakage current ➤rectifier.

rectifier voltmeter A voltmeter containing a ➤rectifier circuit in the input that converts alternating voltage into an essentially unidirectional voltage, which is then measured.

rectilinear scanner ➤scanner.

rectilinear scanning ➤scanning.

recurrent-surge oscilloscope An instrument used to investigate electrical ➤surges. A surge generator is used in conjunction with a ➤cathode-ray oscilloscope (CRO). The repetition rate of the surges produced by the surge generator is made to synchronize with the timebase of the CRO so that a steady picture is obtained on the screen, suitable for visual or photographic inspection.

recursive filters ➤digital filter.

reduced instruction set computer ➤RISC.

redundancy 1. The fraction of the information in a transmission system that may be eliminated without loss of essential information. The excess information is often deliberately included to allow for loss in the transmission system. **2.** The extra components, devices, or circuits included in an electronic circuit or apparatus to increase the reliability of the system. If therefore a fault develops in one portion of the system the redundant circuits or components provided can take over the function of the faulty part. Redundancy is very important in systems, such as aircraft systems, in which high reliability is essential.

Reed–Solomon code Any of a family of error-correcting codes. ➤error correction; compact disc system.

reflected current *Syn. for* return current. ➤reflection coefficient.

reflected impedance ➤transformer.

reflected power Power that is returned from the ➤load back to a generator.

reflected wave 1. ➤travelling wave. **2.** ➤indirect wave.

reflection coefficient *Syn.* return-current coefficient. The vector ratio of the *return current*, I_R, to the *incident current*, I_0, when a ➤transmission line is incorrectly terminated with an impedance, Z_R, not equal to the characteristic impedance, Z_0, of the line. The incident current is that current flowing in the line at termination when Z_R is made equal to Z_0; the return current is the portion of the current flowing back along the line due to incorrect matching. The actual current in the line at termination is the vector sum of $\boldsymbol{I}_R$ and $\boldsymbol{I}_0$. In this case the reflection coefficient may be expressed in terms of the impedances Z_0 and Z_R, i.e. the reflection coefficient can be given by

$$(Z_0 - Z_R)/(Z_0 + Z_R)$$

➤reflection factor.

reflection error An error arising in radio navigation systems, ➤direction finding, and ➤radar systems due to undesired reflections of the transmitted energy.

reflection factor The vector ratio of the current, I, delivered to a load of impedance Z_B by a source of impedance Z_A when the impedances are not matched, to the current, I_0, delivered to the load when the impedances are matched by an impedance matching network of ➤image transfer constant, θ, equal to zero (see diagram). The reflection factor is thus the ratio of the current delivered to an unmatched load to the current that would be delivered to a perfectly matched load. It is given by

$$I/I_0 = \sqrt{(4Z_A Z_B)}/(Z_A + Z_B)$$

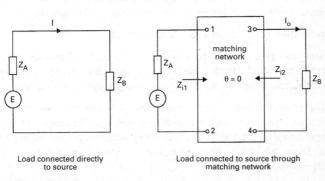

Load connected directly to source

Load connected to source through matching network

Reflection factor

The ratio of the electrical powers in Z_B for these matched and unmatched cases is also given by I/I_0. The *reflection loss* between Z_A and Z_B is the ratio of the powers in Z_B expressed in decibels, i.e. it is given by

$$10 \log_{10}(I/I_0)$$

If the reflection loss is negative it represents a reflection gain in the system. ➤➤reflection coefficient.

reflection loss ➤reflection factor.

reflectometer An instrument for comparing the incident and reflected waves at a device or equipment under test. It is generally used at radio or microwave frequencies in a ➤distributed circuit or ➤transmission line system, and compares the incident and reflected electrical ➤power waves. The comparison can be either magnitude only, yielding a scalar measurement, or of magnitude and phase, a vector measurement. The reflectometer is generally used in a ➤CW mode, but can be used with pulsed waveforms, where the reflected signal is received after a time delay, and provides a measure of the location in space of discontinuities in the network. When used in this manner, the instrument is called a *time-domain reflectometer*. Such measurement techniques find application in optical fibre systems, where discontinuities and branches in the fibre network can be characterized.

reflector ►directional antenna.

reflex bunching ►velocity modulation.

reflex circuit A circuit that is used to amplify a signal at one frequency and also after it has been converted to a second frequency.

reflex klystron ►klystron.

refresh To restore the condition of a memory cell in a computer to its original state in order to maintain the integrity of the information stored in it. A refresh is required following destructive read operation (►read). In dynamic memories periodic refresh must be carried out to prevent loss of information during standby intervals.

regeneration *Syn. for* positive feedback. ►feedback.

regenerative receiver A ►radio receiver used with amplitude-modulated radiowaves in which positive ►feedback is used in order to increase the sensitivity and ►selectivity of the receiver by reducing the damping. ►super-regenerative reception.

register One of a number of word-sized locations (►word) in the ►central processing unit of a computer in which the arithmetic and logic operations required by a program are performed on data obtained from ►memory, ►input/output devices, or other registers. The access time of processor registers is extremely brief, the registers being composed of ►flip-flops.

register set The set of operational ►registers of a computer, hence those registers that a programmer can access using the machine's instruction set.

regulator An electronic device that is used to maintain the voltage (*voltage regulator*) or current (*current regulator*) constant at a given point in a circuit or to vary it in a controlled manner. Regulators are used to control the output of a load or to regulate the voltage or current of an electronic device, despite fluctuations in the circuit conditions, particularly variations in the supply voltage.

reignition voltage The voltage required to re-establish the discharge in a ►gas-discharge tube after the tube has ceased conducting. The reignition voltage is often applied during the period of deionization of the tube, i.e. the interval during which the ions in the tube recombine to form gas molecules, and can therefore be less than the voltage required to initiate the first discharge.

rejector *Syn. for* parallel resonant circuit. ►resonant circuit; resonant frequency.

relative permeability ►permeability.

relative permittivity ►permittivity.

relaxation oscillator An oscillator in which one or more voltages or currents change suddenly at least once during each cycle. The circuit is arranged so that during each cycle energy is stored in and then discharged from a reactive element (e.g. a capacitor

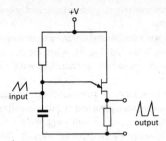

Unijunction transistor **relaxation oscillator**

or inductance), the two processes occupying very different time intervals. An oscillator of this type has an asymmetrical output waveform that is far from being sinusoidal.

A commonly produced output waveform is a ►sawtooth waveform; square or triangular waveforms can be easily produced when required by means of a suitable circuit. Sawtooth waveforms are particularly useful as the internal ►timebase of a ►cathode-ray tube.

The output waveform is very rich in harmonics and for some purposes this is particularly useful. Common types of relaxation oscillator include the ►multivibrator and ►unijunction transistor (see diagram) but many other circuit arrangements are possible.

relaxation time Symbol: τ. **1.** The time interval during which the ►dielectric polarization of a point in a dielectric falls to $1/e$ of its original value due to the electric conductivity of the dielectric. **2.** The travel time of a charge carrier in a conductor or semiconductor before it is scattered and loses its momentum.

relay An electrical device in which one electrical phenomenon (current, voltage, etc.) controls the switching on or off of an independent electrical phenomenon. There are many types of relay, most of which are either electromagnetic or solid-state relays.

The *armature relay* is an electromagnetic relay in which a coil wound on a soft-iron core attracts a pivoted armature that operates contacts or tilts a mercury switch (Fig. *a*). There are several different designs for the armature: *differential relays* have two coils and only operate when the currents in the coils are additive not subtractive. The armature may be split so that a small section of the metal operates with small currents independently of the main contacts, which require large currents to move the whole

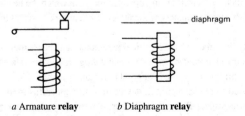

a Armature **relay**   *b* Diaphragm **relay**

armature; *polarized relays* have a central permanently magnetized core and operate differently with currents in different directions.

The electromagnetic *diaphragm relay* has a coil wound around a central core with a thin metal diaphragm plate mounted close to its end (Fig. *b*). When the coil is energized the central portion of the diaphragm moves towards the core and makes contact with it.

A true solid-state relay has all its components made from solid-state devices and involves no mechanical movement. Isolation between input and output terminals is provided using a ➤light-emitting diode (LED) in conjunction with a ➤photodetector. The switching is achieved using a ➤thyristor or bidirectional thyristor (a triac). This type of relay is compatible with digital circuitry and has a wide variety of uses with such circuits. The relay cannot normally be formed on a single chip since the LED is usually formed in gallium arsenide and the photodetector in silicon. Isolation may also be achieved by transformer-coupling on the input. Again a single chip may not be used. Examples of solid-state relays are shown in Figs. *c* and *d*.

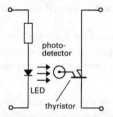

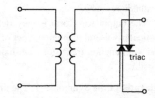

c LED-coupled solid-state **relay**   *d* Transformer-coupled solid-state **relay**

Solid-state relays have advantages over electromechanical relays because of increased lifetime, particularly at a high rate of switching, decreased electrical noise, compatibility with digital circuitry, and ability to be used in explosive environments since there are no contacts across which ➤arcs can form; the lack of physical contacts and moving elements also gives increased resistance to corrosion. No mechanical noise is associated with them. Disadvantages include the substantial amount of heat generated at a current above several amperes, necessitating some form of cooling, and greatly increased production costs for multipole devices compared to single pole devices; in certain applications a physical disconnection may be required for safety purposes and this is not available in solid-state relays.

Other types of relay include thermionically operated relays in which the heating effect of a current is used to operate contacts or the effect of a heating coil on a bimetallic strip is employed.

Relays are frequently described by the electrical parameter that causes them to operate or according to their function. *Current relays* and *voltage relays* operate when a predetermined value of current or voltage is applied to the input circuit.

A *locking relay* is used to render a circuit, device, or other apparatus inoperative under particular conditions, especially fault conditions. A *slow-operating relay* has an intentional delay between the energizing input and operation of the contacts; it includes some form of time-delay mechanism to achieve this.

reliability The ability of any device, component, or circuit to perform a required function under stated conditions for a stated period of time. This may be expressed as a probability. ('Time' may be considered as distance, cycles, or other appropriate units.) *Reliability characteristics* are those quantities used to express reliability in numerical terms. Electronic items are usually approached from the viewpoint of ►failure, because of their extremely high reliability.

reluctance *Syn.* magnetic resistance. Symbol: R; unit: henry^{-1}. The ratio of the magnetomotive force, F_m, to the total magnetic flux, Φ,

$$R = F_m/\Phi$$

reluctivity The reciprocal of magnetic ►permeability.

remanence *Syn.* retentivity. ►magnetic hysteresis.

repeater A device that receives signals in one circuit and automatically delivers corresponding signals to one or more other circuits. A repeater is most often used with telephonic or telegraph circuits or in radio systems, usually amplifying the signal, and in pulse telegraphy performs pulse regeneration on the transmitted pulses. Repeaters can operate either on signals in one direction only or on two-way signals; telephone repeaters operate on four-wire circuits or two-wire circuits. A *terminal repeater* is a repeater that is used at the end of a trunk feeder or transmission line.

repeating coil An audiofrequency transformer that is used to couple two sections of telephone line.

reproduction of sound The reproduction of sound information from a source of audiofrequency electrical signals. A complete sound reproduction system contains the original source of audio information, preamplifier and control circuits, audiofrequency power amplifier(s), and loudspeaker(s). The source of sound may be a compact disc, gramophone record, magnetic tape, a broadcast transmission, or a sound-on-film recording.

  Monophonic sound reproduction uses only a single audiofrequency channel. One or more loudspeakers may be used in parallel at the output. *Stereophonic sound reproduction* uses two channels to carry the audio information and at least two loudspeakers. Stereophony utilizes the human binaural (two-ear) processing ability to detect the direction of sounds by assessing the difference in time taken by the sound to reach each ear and the difference in volume caused by the screening effect of the head.

  Stereophonic signals require at least two microphones. A coincident pair of matched directional microphones may be used at the same location or a pair of separated microphones may be used. In the latter case the outputs are divided in the correct proportion between the two channels using a suitable potential divider – a panoramic potentiometer. In order to provide a signal that is compatible with monophonic sound reproduction the stereophonic signals are combined into sum and difference signals. If the two channels are A and B, the sum $(A + B)$ and the difference $(A - B)$ of the signals are used. A monophonic system produces an output using only the $(A + B)$ signal. The stereo system combines them to produce two signals that correspond to the original A and B information. In *stereophonic broadcast transmission* the sum signal is used to modulate the main carrier and the difference signal is used to modulate a subcarrier, separated from the main carrier frequency.

A sound-reproduction system that is composed of high-quality expensive parts and reproduces the original audio information faithfully and with very low noise levels is referred to as a *high-fidelity* (hi-fi) system. In such a system several different inputs may be available, which share the use of the power amplifiers and loudspeakers. Domestic systems usually have facilities for reproducing compact discs, tape recordings, gramophone records, and broadcast radio transmissions. Suitable impedance matching circuits are required for each input. The system may be modular, i.e. each unit is separately boxed and interconnections made between units by means of wires and plugs; alternatively all the units may be combined in a single housing.

reset To restore an electrical or electronic device or apparatus to its original state following operation of the equipment. ➤clear.

residual charge The portion of the charge stored in a capacitor that is retained when the capacitor is discharged rapidly and may be withdrawn from it subsequently. It results from viscous movement of the dielectric under charge causing some of the charge to penetrate the dielectric and hence become relatively remote from the plates. Only the charge near to the plates is removed by rapid discharge.

residual current Current that flows for a short time in the external circuit of an active electronic device after the power supply to the device has been switched off. The residual current results from the finite velocity of the charge carriers passing through the device.

residual resistance The inherent resistance of a conductor that is independent of temperature variations. It is usually ascribed to irregularities in the molecular structure of the material.

resist An energy-sensitive material used in ➤lithography. Resists are applied to the substrate material as a thin film and selectively exposed to an energy beam (light, electrons, etc.), which causes chemical changes in portions of the resist. The exposed film is then developed to selectively remove either the exposed portions (positive resist) or the unexposed portions (negative resist). ➤photoresist; electron-beam lithography; X-ray lithography; ion-beam lithography.

resistance 1. Symbol: R; unit: ohm. The tendency of a material to resist the passage of an electric current and to convert electrical energy into heat energy. It is the ratio of the applied potential difference across a conductor to the current flowing through it (➤Ohm's law). If the current is an alternating current the resistance is the real part – the *resistive component* – of the electrical ➤impedance, Z:

$$Z = R + jX$$

where j is equal to $\sqrt{-1}$ and X is the ➤reactance. **2.** ➤resistor.

resistance-capacitance coupling (RC coupling) ➤coupling.

resistance coupling ➤direct coupling.

resistance drop ➤voltage drop.

resistance gauge A gauge that is used to measure high fluid pressures by measuring the change in electrical ➤resistance of a sample of manganin or mercury when the sample is subjected to the pressure. The gauge must be calibrated against known pressures before use.

resistance strain gauge ➤strain gauge.

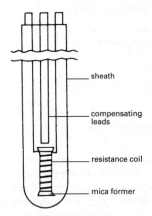

Resistance thermometer

resistance thermometer An electrical thermometer that utilizes the change in electrical ➤resistance with temperature (➤temperature coefficient of resistance) of a wire to measure the temperature of its surroundings.

It consists of a small coil of wire (usually platinum but other metals may be used at low temperatures) wound on a mica former and enclosed in a sheath of silica or porcelain (see diagram). The change in resistance is determined by placing the coil in one arm of a ➤Wheatstone bridge. Compensating leads are usually added to the other arm of the bridge to compensate for temperature variations in the leads as the coil is usually remote from the measuring instrument. Resistance thermometers can be used over a wide range of temperatures from –200 °C to over 1200 °C.

resistance wire Wire constructed from a material, such as nichrome or constantan, that has a high resistivity and low temperature coefficient of resistance. It is used for accurate wire-wound resistors.

resistive component (of a complex ➤impedance) ➤resistance.

resistive coupling *Syn. for* resistance coupling. ➤coupling.

resistivity *Syn.* volume resistivity. Symbol: ρ; unit: ohm metres. An intrinsic property of a material equal to the resistance per metre of material with cross-sectional area of one square metre:

$$\rho = RA/L$$

where R is the resistance, A the cross-sectional area, and L the length. Resistivity depends only on the nature of the material whereas resistance depends not only on the material but on its length and cross-sectional area.

Resistivity is the reciprocal of ➤conductivity: the lower the resistivity of a material the better conductor it is. Materials can be classified as conductors, semiconductors, or insulators according to their resistivities (see table). In semiconductors, the higher the ➤doping level the lower the resistivity. ➤➤surface resistivity.

Material	Resistivity (ohm metres)
conductors	$10^{-8} - 10^{-6}$
semiconductors	$10^{-6} - 10^7$
insulators	$10^7 - 10^{23}$

resistor *Syn.* resistance. An electronic device that posseses ➤resistance and is selected for use because of that property. There are several different types of resistor in common use; the type of resistor chosen depends on the particular application for which it is designed. The three main types of resistance element are *carbon, wire-wound,* and ➤film resistors. These may be produced as fixed-value resistors or adjustable resistors.

Carbon resistors consist of finely ground particles of carbon mixed with a ceramic material and encapsulated in an insulating material. The encapsulation has a set of coloured stripes or dots – the ➤colour code – denoting the value of the resistance (➤Table 2 in the back matter). Carbon resistors are compact, robust, and relatively cheap to manufacture and are widely used in electronic circuits in which the resistance value is not critical. The value of the resistance however is a function of the operating voltage and of temperature and close tolerances cannot be maintained over a wide range of load and ambient conditions. When used at power levels above one megohm the level of thermal ➤noise becomes high, precluding their use at high levels of power. The relatively short length of these resistors causes them to have a noticeable shunted capacitance and therefore at high (VHF and higher) operating frequencies the effective resistance is reduced as a result of the dielectric losses involved.

Adjustable carbon composition resistors may be formed on an insulating base or moulded at a high temperature onto a moulded plastic base. The composition is formed with a linear rotating contact or it may be tapered to produce a nonlinear characteristic. The resistance change is continuous. These resistors, particularly the thinner ones, tend to be noisy and are subject to mechanical wear in frequent use.

Wire-wound resistors are formed from a wire of uniform cross section wound on a suitable former. The value of resistance may be determined very accurately and for uses where the resistance value is critical, wire-wound resistors are usually preferred. Wire-wound resistors however are unsuited to use above 50 kilohertz as they have marked inductive and capacitive effects even when specially wound. At high frequencies the ►skin effect causes an increase in the effective resistance.

Adjustable wire-wound resistors are produced in a wide range of values and types. Linear or circular types are made with single-turn or multiturn contacts. By tapering or producing other shapes of the former the contact rotation versus resistance curve may be altered to generate various functions: logarithmic, sine wave, or other characteristics are available. Wire-wound resistors almost invariably change resistance values in steps. This can sometimes provide an unwanted pulse in the output. The motion of the contact across the turns also tends to generate noise.

Resistive ►film resistors are the most suitable type of resistor for high-frequency applications since the inductance, even in high-value spiral films, is very much less than with wire-wound resistors. Special high-frequency film resistors with very low values of inductance are available.

Metal film resistors are the most widely used type for precision requirements. They are manufactured by depositing nichrome alloys on a substrate. Extremely good characteristics may be obtained with normal tolerances of ±1% achievable; ±0.5%, ±0.25%, and ±0.1% are also available.

resistor-transistor logic (RTL) A family of integrated ►logic circuits that was the first to be developed but is now little used. The input is through a resistor into the ►base of an inverting transistor. The basic NOR gate is shown. The output is high (corresponding to a logical 1) only if both the inputs are low (corresponding to a logical 0). If either input is high the transistor conducts and saturates and the voltage at the output is low. RTL circuits tend to be slow low ►fan-out circuits that are susceptible to noise but the power dissipated is low compared to emitter-coupled, diode-transistor, and transistor-transistor logic circuits.

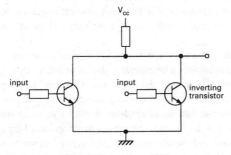

Resistor-transistor logic: two-input NOR gate

resolving time ►counter.

resonance A condition existing when an oscillatory circuit responds with maximum amplitude to a periodic driving force so that a relatively small amplitude of the driving force produces a large amplitude of oscillation. Resonance is achieved when the frequency of the driving force coincides with the natural undamped frequency of the oscillatory system. ►forced oscillations; resonant frequency; tuned circuit.

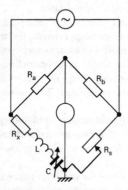

Resonance bridge

resonance bridge A four-arm bridge that has a tuned circuit in one of the arms (see diagram). The bridge will only balance at the resonant frequency of the circuit and at balance the frequency and resistance of the circuit are given by

$$\omega^2 LC = 1$$

$$R_x = R_s R_a / R_b$$

where ω is the angular frequency.

resonant cavity ►cavity resonator.

resonant circuit A circuit that contains both ►inductance and ►capacitance so arranged that the circuit is capable of ►resonance. The frequency at which resonance occurs – the ►resonant frequency – depends on the value of the circuit elements and their arrangement.

A *series resonant circuit* contains the inductance and capacitance in series. Resonance occurs at the minimum combined impedance of the circuit and a very large current is produced at the resonant frequency; the circuit is said to accept that frequency. A *parallel resonant circuit* contains the circuit elements arranged in parallel. Resonance occurs at or near the maximum combined impedance of the circuit (►resonant frequency). The overall current in the circuit is a minimum at the resonant frequency, a voltage maximum being produced; the circuit is said to reject that frequency. ►tuned circuit.

resonant frequency Symbols: ω_0, f_0. The frequency at which ►resonance occurs in a particular circuit or network. Resonance occurs in a circuit containing both capacitance and inductance when the imaginary component of the complex combined impedance of the circuit is zero, i.e. when the supply current and voltage are in ►phase and the circuit has unit ►power factor.

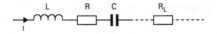

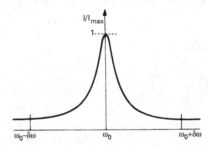

a Series resonant circuit and frequency response

In a *series resonant circuit,* which contains the capacitive and inductive elements in series (Fig. *a*), the combined impedance, Z, is given by

$$Z = R + j\omega L - j/\omega C$$

where R is the ohmic resistance, ω the angular frequency ($\omega = 2\pi \times$ frequency), L the inductance and C the capacitance, and j equals $\sqrt{-1}$. The resonance condition is fulfilled when

$$\omega_0 L = 1/\omega_0 C$$

i.e. when

$$\omega_0 = 1/\sqrt{(LC)}$$

In a series resonant circuit resonance occurs therefore when the combined impedance is purely resistive and is a minimum. The resistance can be low even for large values of L and C. In this case the current flowing through the circuit will be high, and although large voltages are developed across the individual elements these are out of phase with each other so that the total voltage developed across the circuit is relatively low; maximum current will then flow in a load resistor, R_L, in series with the circuit.

In the case of a *parallel resonant circuit,* in which the circuit elements are in parallel (Fig. *b*), it is convenient to consider the combined admittance, Y, of the circuit ($Y = Z^{-1}$), given by

$$Y = j\omega C + (R - j\omega L)/(R^2 + \omega^2 L^2)$$

The resonance condition is fulfilled when

$$R^2 + \omega_0^2 L^2 = L/C$$

i.e. when

$$\omega_0^2 = [1 - (R^2 C/L)][1/LC]$$

Since the term $R^2 C/L$ is usually very small this approximates to

$$\omega_0 = 1/\sqrt{(LC)}$$

which is the value of the series resonant frequency; the term however cannot always be neglected.

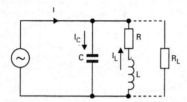

b Parallel resonant circuit

In a parallel resonant circuit therefore resonance occurs when the combined admittance is low. At resonance $Z = 1/Y$ is high and is termed the *parallel resistance* of the circuit. The overall current is low but the voltage developed across the circuit, and therefore across a load resistor, R_L, in parallel with it, is high. The individual currents developed in the inductance and capacitance at resonance can be very large but are out of phase with each other, resulting in the low combined current.

The above consideration leads to a unique solution for the resonant frequency. The resonant frequency may also be defined as that frequency at which the complex impedance passes through a minimum (in a series resonant circuit) or a maximum (parallel resonant circuit). In the series resonant case the solution is the same as above but in the case of parallel resonance a unique value is not necessarily found. The resonant frequency may have slightly different values that depend on the particular circuit parameter that is varied in order to achieve resonance. The differentials of Y with respect to $C, L,$ or ω may therefore lead to slightly different resonance conditions.

resonant line A ➤transmission line that exhibits ➤resonance at the operating frequency. ➤resonant frequency.

resonator A device or material that possesses a band-pass frequency-response characteristic, equivalent to an inductor-capacitor (LC) network. Some ceramic and crystalline materials possess such resonances with very high ➤Q factor, and can be used in filters and oscillators to determine the frequency behaviour.

restore ➤clear.

retentivity *Syn. for* remanence. ➤magnetic hysteresis.

return current *Syn.* reflected current. ➤reflection coefficient.

return-current coefficient ➤reflection coefficient.

return interval *Syn. for* flyback. ➤sawtooth wave; timebase.

return loss A figure of merit that indicates how closely a measured impedance matches a standard impedance, both in magnitude and ➤phase angle. For a perfect match the return loss would be infinite, indicating that all the incident power had been absorbed in the impedance with none of the energy being reflected; if the match was not perfect, some of the energy would be reflected.

return stroke ➤lightning stroke.

return trace *Syn. for* flyback. ➤timebase.

reverse active operation ➤bipolar junction transistor.

reverse bias *Syn.* reverse voltage. ➤reverse direction.

reverse conducting thyristor (RCT) ➤thyristor.

reverse current ➤reverse direction.

reverse direction The direction of operation of an electrical or electronic device in which the device exhibits the larger resistance. A voltage applied in the reverse direction is a *reverse bias* and the current flowing is the *reverse current.*
   A device such as a semiconductor ➤diode exhibits an extremely high reverse resistance and can therefore be used as a ➤rectifier or ➤switch. Such a device exhibits a very small *reverse saturation current* below ➤breakdown due to the movement of a few charge carriers across the p-n junction. The *reverse recovery time* of such a device, particularly a semiconductor diode, is the time interval between instantaneous switching from a forward bias to a reverse bias and the reverse current reaching the saturation value.

reverse recovery time ➤reverse direction.

reverse saturation current ➤p-n junction; reverse direction.

reverse voltage *Syn. for* reverse bias. ➤reverse direction.

reversible transducer ➤transducer.

RF (or **r.f.**) *Abbrev. for* radiofrequency.

RFC *Abbrev. for* radiofrequency choke.

RF heating ➤induction heating.

RFI *Abbrev. for* radiofrequency interference. ➤electromagnetic compatibility.

RHEED *Abbrev. for* reflection high-energy electron diffraction. ➤diffraction.

rhumbatron ➤cavity resonator.

ribbon microphone A ►microphone that consists of a very thin ribbon of aluminium alloy a few millimetres wide loosely fixed in a strong magnetic flux density parallel to the plane of the strip. A sound wave incident on the ribbon causes a pressure difference to be established between the front and back edges of the ribbon and it therefore experiences a force. The resultant motion causes a corresponding e.m.f. to be induced in the ribbon (►electromagnetic induction).

If the acoustic path difference across the ribbon is much smaller than a quarter wavelength, the pressure on the ribbon (and hence the e.m.f.) is proportional to the particle velocity and to the frequency. If the resonant frequency of the ribbon is made smaller than the frequency of the sound waves the frequency dependence becomes negligible and the induced e.m.f. is proportional to the particle velocity. The microphone has strong directional properties since sound waves that originate in the plane of the ribbon arrive at the front and back edges in phase, and therefore no resultant force is produced.

The ribbon microphone may be used to measure sound intensities. An alternating current of the same frequency as the sound wave is passed through the ribbon and the phase and amplitude are varied until the force due to the current just balances that due to the sound wave. The sound intensity is then calculated from the current amplitude and the magnetic field strength.

Richardson–Dushman equation The fundamental equation describing ►thermionic emission of electrons from the surface of a metal as a function of the temperature of the metal:

$$J = AT^2 \exp(-\Phi/kT)$$

where J is the emitted electron current, A is the *Richardson constant* (a function of the surface condition of the material), Φ is the work function of the metal (the height of the potential barrier to electron emission), k is the Boltzmann constant, and T the thermodynamic temperature.

ridged waveguide ►waveguide.

Rieke diagram A ►load impedance diagram plotted on polar coordinates: the radial coordinate represents the ►reflection coefficient measured in the transmission line that connects the active device to the load; the angular coordinate represents the angular distance of the minimum standing-wave voltage from a designated reference plane on the output terminal. Lines of constant power output, frequency, voltage, and efficiency may be drawn on the diagram.

Righi–Leduc effect If heat flows through a strip of conductor or semiconductor that is placed in a magnetic field, where the heat flow and magnetic field are at right angles, a temperature gradient is set up in the strip, orthogonal to both heat flow and magnetic field directions. This is the thermal analogue of the ►Hall effect.

right-hand rule (for a dynamo) ►Fleming's rules.

ringing 1. An unwanted low-frequency resonant tone that occurs in a radio receiver due to low-frequency oscillations produced in the receiver by the received radiofrequency waves. **2.** ➤pulse.

ring junction *Syns.* hybrid ring junction; rat-race. ➤hybrid junction.

ring modulator *Syn.* balanced modulator. A network of diodes or transistors in a ➤bridge arrangement to which a radiofrequency carrier and baseband signal are applied to produce a modulated RF signal. Alternative ring modulator circuits can be adapted from this basic circuit, using two nonlinear devices and a centre-tapped transformer to provide the full-wave switching characteristic.

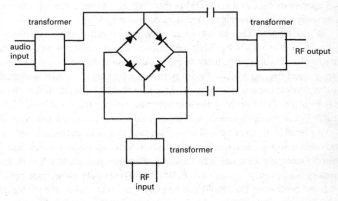

Ring modulator using four-diode bridge

ring network In digital communication networks, particularly ➤local area networks, a network that forms a closed loop. Each piece of equipment connected to the ring is connected in series and so makes contact with both halves of the loop.

ripple 1. An alternating-current component superimposed on a direct-current component resulting in variations in the instantaneous value of a unidirectional current or voltage. The term is particularly applied to the output of a ➤rectifier. The frequency of the a.c. component is the *ripple frequency;* for a full-wave rectifier it is twice the frequency of the input signal.

The magnitude of the ripple is given by the ratio of the root-mean-square value of the a.c. component to the mean value of the total and is usually expressed as a percentage. This is known as the *ripple factor.* Some form of ➤smoothing circuit or ➤regulator is normally used in order to reduce the amount of ripple present on the output of a rectifier, generator, etc. **2.** ➤pulse.

ripple factor ➤ripple.

ripple filter ➤smoothing circuit.

ripple frequency ➤ripple.

RISC *Abbrev. for* reduced instruction set computer. A computer with a relatively simple load-and-store ►instruction set and only register-to-register operate instructions. Typical RISC machines have large ►register sets, multiple functional units (such as addition, subtraction, and multiplication), and pipelined instruction and execution units in which each process can be performed in several steps.

rise time (of a pulse) ►pulse.

r.m.s. (or **RMS**) *Abbrev. for* root mean square.

ROM *Abbrev. for* read-only memory. A form of ►memory that retains information permanently and in which the stored information cannot be altered by a program or normal operation of a computer. ROM is therefore used to store control programs and commonly used ►microprograms in large computer systems.

ROM is fabricated as ►solid-state memory, and is usually formed by replacing storage capacitors of ►RAM by either open circuits or connections to earth. Information may be placed in the storage array during manufacture of the memory – so called *hard-wiring* – and the two possible binary states of each memory location are determined by the physical construction of the device. A form of ROM in which the information may be placed in the array by the user rather than during the initial manufacture is a *fusible-link ROM*. Each memory cell is provided with a fusible link to earth; information is placed in the array by applying a particular pattern of electrical impulses to the array that is strong enough to blow the fuses at locations where open circuits are required. Once the pattern has been formed it is retained permanently. The fusible-link memory is a form of *programmable ROM* (*PROM*) but can be programmed only once.

If the contents of the PROM can be erased (brought back to the unprogrammed state) by ultraviolet light, the chip is an erasable PROM, or *EPROM*. If the erasing process is electrical, the chip is an electrical-erasable PROM, or *EEPROM*.

root-mean-square (r.m.s.) value *Syns.* effective value; virtual value. The square root of the mean value of the squares of the instantaneous values of a periodically varying quantity averaged over one complete cycle. In the case of a sinusoidally varying quantity the r.m.s. value is equal to the peak value divided by $\sqrt{2}$.

rope ►confusion reflector.

rotator ►waveguide.

rotor ►motor.

routing In telephone or digital communication systems, the assigning of a communication path along which a call or message will pass from its source to its destination. ►►alternative routing.

R-S flip-flop ►flip-flop.

RS-232 interface A standard interface for computer and telecommunications systems. It was produced by the EIA (Electronics Industries Association).

RTL *Abbrev. for* resistor-transistor logic.

rumble Unwanted low-frequency noise heard on the loudspeaker of a high-fidelity record system. It is caused by mechanical vibrations of the record adding to the signal produced from the ➤pick-up cartridge.

run-length codes ➤digital codes.

Rutherford back scattering (RBS) The scattering through close to 180° of an ion when it impinges on an atom, the energy E of the scattered ion being given by

$$E = [(M - m)/(M + m)]^2 E_0$$

where E_0 is the initial energy of the ion and M and m are the masses of the atom and incident ion respectively. If the incident ion travels through a material before back scattering takes place, additional energy losses occur both incoming and outgoing. Such losses can be determined for different materials and different depths. Detection of the back-scattered ion and measurement of its energy can therefore determine both the mass and the depth of the scattering atom.

Use of a monoenergetic beam of ions and a high energy resolution detector can thus yield a picture of not only the impurity elements in a semiconductor crystal but also the perfection of the crystal structure, since back scattering also occurs from lattice defects. A depth resolution of approximately 10 nanometres is possible.

RZ PCM *Abbrev. for* return to zero pulse code modulation. ➤pulse modulation.

S

SAINT process *Short for* self-aligned implantation for n⁺ layer technology. A method of forming ►field-effect transistors for digital logic circuits on gallium arsenide wafers.

Sallen–Key filter Any of a range of ►biquad active second-order RC filters with a ►VCVS device being the single active element (also known as the *VCVS filter*), allowing low-pass, high-pass, and band-pass ►filters to be realized.

S-ALOHA *Short for* slotted ALOHA. ►ALOHA system.

SAM *Abbrev. for* scanning Auger microprobe. ►Auger electron spectroscopy.

sampling 1. A technique in which only some portions of an electrical signal are measured and are used to produce a set of discrete values that is representative of the information contained in the whole. In order that the output values represent the input signal without significant loss of information, the rate of sampling – the *sampling frequency* – of a periodic quantity must be at least twice the frequency of the highest component in the signal. The minimum sampling frequency is known as the ►Nyquist rate.

A sampling circuit is used to produce a set of discrete values representative of the instantaneous values of the input signal. The output may be in the form of a set of instantaneous values (*instantaneous sampling*) or in a coded form. The technique is widely used in ►analogue-to-digital converters, ►digital voltmeters, ►multiplex operation, ►pulse modulation, etc. **2.** A technique in which intermittent measurements of an electrical signal are made. The technique is used in ►feedback control systems where the controlled variable is sampled intermittently and a correction applied if necessary as a result of the instantaneous value (the *sample intelligence*). The technique is also used in radio-navigation systems when information from the navigation signal is extracted only when the sampling ►gate is activated by a selector pulse and therefore interrogates the signal and produces a corresponding output ►pulse or ►waveform. **3.** A system of quality control of mass-produced electronic components, circuits, devices, or other equipment. Random samples of the manufactured items are removed from the manufacturing point and tested exhaustively. The sampling process is usually carried out for each processing stage during manufacture.

sampling frequency ►sampling.

sampling period The reciprocal of the sampling frequency. ►sampling.

sampling synthesis A technique used in electronic musical instruments in which the sounds are sampled and stored digitally. ►►synthesis.

sapphire A synthetic form of aluminium oxide used as the substrate in microwave integrated circuits where low loss is paramount. ➤silicon-on-insulator.

sat *Short for* saturated mode.

satellite An artificial body that is launched from earth to orbit either the earth or another body of the solar system. There are two main classes: *information satellites* and ➤communications satellites.

Information satellites transmit signals containing many different types of information to earth. Typical uses include the provision of atmospheric and meteorological data, infrared, ultraviolet, gamma- and X-ray studies of celestial objects, surveys of the earth's shape, surface, and resources, and as navigational aids. Communications satellites receive radiofrequency signals from earth by means of highly directional antennas and return them to another earth location for purposes of long-distance telephony, TV broadcasting, etc.

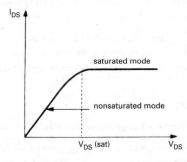

Saturated mode and nonsaturated mode of a FET

saturated mode The operation of a ➤field-effect transistor in the portion of its characteristic beyond the saturation voltage V_{DS} (sat) (see diagram), where V_{DS} is the drain voltage. The ➤drain current is independent of the drain voltage in this region. *Nonsaturated mode,* sometimes known as *triode-region operation,* is operation of the device in the portion of its characteristic below saturation.

saturated velocity The constant velocity of the charge carriers in a semiconductor at high electric fields: the velocity no longer follows an ohmic relation due to high energy loss mechanisms that the carriers experience at these high fields.

saturation 1. A condition where the output current of an electronic device is substantially constant and independent of voltage. In the case of a device such as a ➤field-effect transistor or thermionic valve, saturation is an inherent function of the device and produces the maximum current inherent to the device. In the case of a ➤bipolar junction transistor, saturation occurs when both junctions are forward biased. The collector current is limited by the circuit elements of the external circuit and changing these alters the magnitude of the ➤saturation current drawn from the device. **2.** (magnetic) The maximum possible degree of magnetization of a material; it is independent of the strength of the magnetic flux density applied to the material. All the domains in the material at saturation are assumed to be fully orientated with respect to the magnetic flux density (➤ferromagnetism).

saturation current 1. In field-effect transistors and thermionic valves, the current shown on the portion of the static characteristic where it is substantially constant and independent of voltage. Very little further increase of current with voltage occurs until ➤breakdown is reached. The value of the saturation current is a function of the device only. **2.** In bipolar junction transistors, the current flowing when both junctions are forward biased. This current is limited by the external circuit.

saturation resistance The resistance that occurs between the collector and emitter electrodes of a ➤bipolar junction transistor for a specified value of base current when the device is saturated (➤saturation) due to the external circuit.

saturation signal A signal, received by a ➤radar receiver, that has an amplitude larger than the ➤dynamic range of the receiver.

saturation voltage The residual voltage between the collector and emitter of a ➤bipolar junction transistor for a specified value of base current when the device is saturated (➤saturation) due to the external circuit. It is smaller than the operating voltage under nonsaturated conditions.

SAW *Abbrev. for* surface acoustic wave. ➤acoustic wave device.

SAW filter A ➤transversal filter that is realized using surface ➤acoustic wave structures.

sawing A method of separating individual ➤chips on a semiconductor ➤wafer preparatory to packaging, using a high-speed precision circular saw. The wafer moves on a table mounted below the blade, which is typically between 3 and 15 micrometres wide and coated with diamond. ➤➤scribing.

sawtooth oscillator A ➤relaxation oscillator that produces a ➤sawtooth waveform.

sawtooth pulse A ➤pulse that has a geometrical shape similar to one complete period of a ➤sawtooth waveform.

sawtooth waveform A periodic waveform whose amplitude varies approximately linearly between two values, the time taken in one direction, the *active interval,* being very much greater than the time taken in the other. The shorter period is termed the *flyback.* An ideal sawtooth waveform is linear with a sharp change of direction; in practice this is not achieved (see diagram) and the transition stage departs from linearity, often with a short unwanted *inactive interval* before the next cycle. Sawtooth waveforms are commonly produced by suitably designed ➤relaxation oscillators and are frequently used to provide a ➤timebase.

S band A band of microwave frequencies ranging from 2.00–4.00 gigahertz (IEEE designation). ➤frequency band.

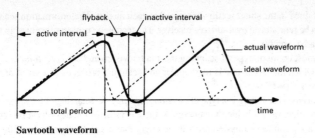

Sawtooth waveform

scaler A device, commonly used in counting circuits, that produces an output pulse whenever a designated number of input pulses have been received. The *scaling factor* is the number of input pulses per output pulse. Scaling factors of 10 or 2 are most commonly used: a *decade scaler* has a factor of 10 and a *binary scaler* has a factor of 2. The latter usually consists of a ➤flip-flop or ➤trigger circuit. Scaler circuits are most commonly used with radiation counters and in computers.

scaling factor ➤scaler.

scaling tube A ➤gas-discharge tube that contains several anodes (usually 10) and that is used in a ➤scaler circuit. An electrical signal input to the tube causes each successive anode to be activated in turn so that the glow discharge moves in position. The tube may be arranged so that a spot of light moves around the circumference or the glow may take up the shape of the digits 0–9 in turn.

scanner 1. *Syn.* rectilinear scanner. A device that visualizes the distribution of a radioactive compound in a particular system, usually the human body. The scanner consists of a ➤scintillation crystal, ➤photomultiplier tube, and amplifying circuits. The crystal is collimated, so that only the radiation from a small area is received at any given instant, and driven backwards and forwards to produce a rectilinear scan of the area investigated.

Double-headed scanners have been developed in which the output is the sum of the outputs of two crystals that simultaneously scan the area of interest. Anomalies arising from a source deep within a human body are minimized by this technique. **2.** ➤flying-spot scanner.

scanning 1. Causing one complete horizontal or vertical traverse of the spot of light on the screen of a ➤cathode-ray tube (CRT), i.e. one sweep of the screen, in response to a voltage generated by a ➤timebase circuit. If the deflection extends beyond the usable physical dimensions of the phosphor on the screen, *overscanning* occurs; if it is less than the usable size *underscanning* results. In the former case information can be lost from the ends of the scan; in the latter the image does not fill the screen.

A particular type of scanning sometimes used with a CRT is *spiral scanning*. In this case the electron beam is made to execute a spiral trace on the screen. **2.** Exploring a particular area, volume, or frequency range in a methodical manner in order to produce a variable electrical signal whose instantaneous values are a function of the information contained in the small section examined at each instant. The information scanned may then be reproduced by a suitable receiver. The technique is most often used in television, radar, and fax.

The most common type of scanning used in television and fax is *rectilinear scanning,* in which the target area is rectangular and scanning proceeds in a set of narrow parallel strips (➤television).

High-velocity scanning is a type of *electron scanning,* i.e. scanning of the target with a beam of electrons, in which the energy of the electrons in the beam is sufficient to produce a ➤secondary-emission ratio at the target that is greater than unity. If the energy of the electrons is less than the minimum velocity required to produce a secondary-emission ratio of unity, the scanning is termed *low-velocity scanning.*

Coarse scanning is frequently used in order to produce a 'rough' picture of the target before carrying out a more detailed investigation. In this case the size of the scanning spot is comparable to the image detail and is the diameter of the electron beam, light beam, or beam of radiowaves used to carry out the scan. Coarse scanning is most often used in radar systems.

Radar systems may also employ *circular scanning,* in which one complete scan is a horizontal rotation of the radar beam through 360°, or *conical scanning,* when the major lobe of the transmitted ➤antenna pattern generates a cone.

Most forms of scanning assume that the scanning speed is uniform throughout each scan. In some applications however it is useful to employ a variable speed of the scanning spot if one section of the scanned area is to be investigated in more (or less) detail than the rest. Some form of *distribution control* is used to achieve a variable speed distribution during the trace interval by controlling the output of the timebase circuit used.

scanning Auger microprobe (SAM) ➤Auger electron spectroscopy.

scanning electron microscope (SEM) ➤electron microscope.

scanning-transmission electron microscope (STEM) ➤electron microscope.

scanning-tunnelling microscope (STM) ➤electron microscope.

scanning yoke *Syn.* deflecting yoke. ➤cathode-ray tube.

scattering loss Energy loss from a beam of electromagnetic radiation because of deflection of the radiation by scattering. The individual particles or photons in the beam can interact with the nuclei or electrons present in the medium through which they are propagated, with photons of another field of radiation, or irregularities in a reflective surface.

scattering parameters *Syn.* s-parameters. Two-port parameters (➤network) that relate the incident and reflected ➤power waves of a network. If the incident and reflected power waves at the ports are denoted by a and b respectively, then for a two-port network the relationship is

$$b_1 = s_{11}a_1 + s_{12}a_2$$
$$b_2 = s_{21}a_1 + s_{22}a_2$$

or, in matrix form,

$$\begin{pmatrix} b_1 \\ b_2 \end{pmatrix} = \begin{bmatrix} s_{11} & s_{12} \\ s_{21} & s_{22} \end{bmatrix} \begin{pmatrix} a_1 \\ a_2 \end{pmatrix}$$

schematic *Informal* A circuit diagram.

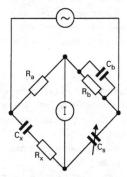

Schering bridge

Schering bridge A four-arm bridge for the measurement of capacitance (see diagram). At balance,

$$C_x = C_s R_b / R_a$$

$$R_x = R_a C_b / C_s$$

SCH laser *Short for* separate confinement heterostructure laser. ➤semiconductor laser.

Schmitt trigger A ➤bistable circuit in which the output voltage level is binary, is determined by the magnitude of the input signal, and is independent of the input signal waveform. The output level changes to the high level when the input signal exceeds a predetermined value. It falls to the low level when the input signal magnitude drops below a predetermined value. The circuit inevitably exhibits ➤hysteresis; the amount of hysteresis is determined by the components in the circuit and can be altered so that the desired switching values can be selected.

The Schmitt trigger can be used in binary ➤logic circuits in order to maintain the integrity of the logical one and zero levels. It can also be used with a variety of analogue waveforms as a level detector, i.e. it acts as a trigger for other circuits or devices when the magnitude of the input waveform exceeds or falls below the predetermined levels. The device can also generate a rectangular pulse train from a variety of input waveforms.

Schottky barrier A rectifying ➤metal-semiconductor contact. ➤Schottky effect; Schottky diode.

Schottky clamp A ►Schottky diode that is used to prevent the voltage at a particular point in a circuit from exceeding a predetermined value. The forward bias characteristic of a Schottky diode is such that in the conducting state the diode has an essentially constant small voltage drop across it. Carrier storage is negligible, leading to very fast switching between the 'on' and 'off' states of the diode. The most common application of Schottky clamps is in integrated ►logic circuits in which the bipolar transistors forming the logic gates are operated in ►saturation during part of the switching cycle. A Schottky diode connected across the base and collector prevents the collector-base voltage swinging too far in the forward direction and hence controls the depth of saturation of the transistor; the speed of operation of the gate is thus optimized.

Schottky diode *Syn.* metal-semiconductor diode. A rectifying diode formed from a junction between a metal and a semiconductor – a Schottky barrier (►Schottky effect). The term is reserved for Schottky barriers in which the energy gap and doping level of the semiconductor are such that minority carriers in the semiconductor do not contribute significantly to the current flowing in the diode.

If a forward bias is applied across the junction, majority carriers (electrons in an n-type semiconductor) that have energies greater than the Schottky barrier height can cross the barrier and current flows (►thermionic emission). Carriers may also cross the barrier by ►tunnelling if it is sufficiently narrow (►field emission). A mixture of these mechanisms often occurs and is known as *thermionic-field emission*, but thermionic emission is the dominant process in Schottky diodes. As the forward bias increases more hot carriers are present and the current increases rapidly. With sufficiently large voltages all the free carriers in the semiconductor can cross the barrier; the current can then only increase with increasing voltage as a result of acceleration of free majority carriers. The current-voltage characteristic is then linear in this region. Under reverse bias conditions the current falls to a small reverse saturation current. Since the minority carriers have a negligible contribution to the current in either the forward or reverse directions ►carrier storage at the junction is negligible and the diode has very fast switching speeds.

Schottky effect A reduction in the effective ►work function of a solid when an external accelerating electric field, *E,* is applied in a vacuum to the surface. In the case of a metal, image charges (►electric images) contribute to the effect, which is therefore sometimes termed *image-force lowering.* The external field lowers the potential energy of electrons outside the solid with a consequent distortion of the potential barrier at the surface. Electrons just inside the surface are liberated by surmounting the barrier (in contrast to the ►tunnel effect). A slight increase in electron emission from a thermionic cathode results from this effect.

A similar effect is observed when the metal surface is in contact with a semiconductor. Such a ►metal-semiconductor contact is termed a *Schottky barrier*; the energy levels associated with the junction are shown in the diagram. The magnitude of the lowering of the work function depends critically upon the surface state of the semiconductor but is usually less than the vacuum case described above. Schottky barriers are used

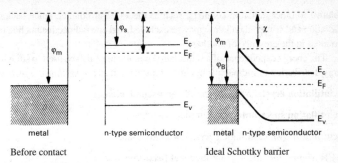

Energy diagrams before and after forming ideal Schottky barrier

to form ►Schottky diodes, in ►Schottky TTL and ►Schottky I²L logic circuits, and to form the gate electrode of one type of junction ►field-effect transistor.

Schottky-gate field-effect transistor ►MESFET.

Schottky I²L ►I²L.

Schottky noise ►shot noise.

Schottky photodiode ►photodiode.

Schottky TTL ►transistor-transistor logic.

Schrödinger (or **Schrödinger's**) **equation** The fundamental equation of ►quantum mechanics.

scintillation 1. A flash of light produced in certain materials (*scintillators*) when exposed to ionizing radiation. The frequency of the light emitted in the crystal is a function of the energy of the incident radiation. Each incident ionizing event produces one flash. **2.** A rapid fluctuation of the image of the target of a ►radar system about its mean position on the radar display. **3.** A small random fluctuation of the received signal in a radio transmission system about the mean signal value. This effect is analogous to the twinkling of light from a star and is caused by small variations in the density of the atmosphere.

scintillation counter A radiation ►counter that consists of a scintillator (►scintillation), ►photomultiplier, ►amplifier, and ►scaler and is used to measure the activity of a radioactive source. The scintillator crystal emits flashes of light of a characteristic frequency when exposed to gamma rays from a radioactive source. Each scintillation produces an output pulse from the photomultiplier. The count rate produced from the photomultiplier is measured and the activity of the source calculated.

The scintillation counter is energy-dependent since the frequency of the emitted light in dependent on the energy of the incident radiation. The energy of the electrons emitted from the ►photocathode is a function of the light frequency and hence pulses produced by radiation of energy other than that being considered may be excluded using a suitable ►discriminator circuit. The ►signal-to-noise ratio is therefore improved relative to other radiation counters, such as the ➤Geiger counter, since background radiation and scattered radiation may be excluded, and one radioactive nuclide may be counted in the presence of another.

The energy dependence of a scintillation counter may also be used to study the energy distribution of radioactive nuclides, thus using it as a *scintillation spectrometer.*

scintillation crystal *Syn. for* scintillator. ➤scintillation.

scintillation spectrometer ➤scintillation counter.

scintillator *Syn.* scintillation crystal. ➤scintillation.

SCR *Abbrev. for* silicon-controlled rectifier. ➤thyristor.

screen 1. The surface of a cathode-ray tube or other display device on which the visible pattern is produced. **2.** *Syn.* shield. A barrier or enclosure used to prevent or reduce the penetration of an electric or magnetic field into a particular region. ➤➤electric screening; magnetic screening.

screened pair ➤shielded pair.

screen grid ➤thermionic valve; tetrode.

screening test ➤life test.

scribing Scoring a ➤wafer of semiconductor with a precision diamond tool in order to separate individual ➤chips, each containing a circuit or component, preparatory to packaging. *Laser scribing* uses a laser to score the wafer rather than a mechanical tool. ➤➤sawing.

scribing channel A gap left between the areas on a semiconductor ➤wafer during the manufacture of several circuits or components on the wafer, in order to allow for ➤scribing into individual ➤chips.

s-domain circuit analysis A mathematical technique that simplifies the study of the transient behaviour of circuits by using the ➤Laplace transform, which results in a transformation into the *s-domain.* This technique avoids the formulation of differential equations in the ➤time domain in favour of much simpler algebraic manipulations in the s-domain. The function $F(s)$ given by the Laplace transform can be expressed as the ratio of two factored polynomials given by:

$$F(s) = K[(s + a_1)(s + a_2)...(s + a_n)]/[(s + b_1)(s + b_2)...(s + b_n)]$$

where $s = \sigma + j\omega$, the *complex frequency* or *complex operator.*

The roots of the denominator polynomial, $-b_1, -b_2,..., -b_n$, are called the *poles* of $F(s)$ and at these values of s, $F(s)$ tends to infinity. The roots of the numerator polynomial, $-a_1, -a_2,..., -a_n$, are called the *zeros* of $F(s)$ and at these values of s, $F(s)$ becomes zero. It is often convenient to visualize the poles and zeros of $F(s)$ as points on the ➤complex plane called the *s-plane.*

search coil *Syn. for* exploring coil. ➤flip-coil.

SECAM Acronym from *SE*quential *C*ouleur *À M*emoire. A ►line-sequential colour-television system developed and adopted in France. Cf. ►PAL.

second Symbol: s. The ►SI unit of time defined as the duration of 9 192 631 770 periods of the radiation corresponding to the transition between two hyperfine levels of the ground state of the caesium-133 atom.

secondary cell *Syns.* accumulator; storage cell. ►cell.

secondary electron An electron that is emitted from a material as a result of ►secondary emission.

secondary emission The emission of electrons from the surface of a material, usually a metal, as the result of bombardment by high-velocity electrons or positive ions. The total energy of the incident primary electrons is often sufficient to liberate several secondary electrons per incident particle. The *secondary emission ratio,* δ, is the number of secondary electrons emitted per incident particle. Secondary emission is mainly used in the ►electron multiplier.

secondary-emission ratio ►secondary emission.

secondary failure ►failure.

secondary-ion mass spectroscopy (SIMS) A method of analysing the surface of a material to detect impurities. An ion beam is used to sputter material in the form of secondary ions from the surface of a semiconductor. The secondary ions are electrostatically accelerated and then analysed using a mass spectrometer. The ion beam may be kept small and scanned across the sample and the term *ion microprobe* is sometimes used. Over 90% of the secondary ions are emitted from the two top atomic layers. A depth profile can be obtained by sputtering continuously in the vertical direction but the accuracy decreases with increasing depth. SIMS is a destructive analysis technique.

secondary radiator *Syn. for* passive antenna. ►directional antenna.

secondary service area ►service area.

secondary standard 1. A copy of a primary standard that has a known difference from the primary standard. **2.** A quantity that is accurately known in terms of a primary standard and that may be used as a unit. **3.** A measuring instrument that has been accurately calibrated in terms of a primary standard and that can be used for accurate measurements or calibration of other equipment.

➤➤primary standard.

secondary voltage 1. The voltage developed across the secondary (output) winding of a ►transformer. **2.** The voltage developed by a secondary ►cell.

secondary winding ►transformer.

secondary X-rays ►X-rays.

second breakdown A form of catastrophic destructive breakdown in ➤bipolar junction transistors. Local heating inside the transistor increases the intrinsic ➤carrier concentration, causing the local resistivity to fall. This enables a higher current density to flow, causing further heating. The current multiplication is rapid, leading to melting of the semiconductor and destruction of the transistor. This phenomenon occurs at high collector voltages and modest collector currents: it should not be confused with ➤avalanche breakdown – another operational limit in bipolar transistors – which is a voltage-induced effect.

second-order filter ➤order (of a filter).

second-order network A ➤network or ➤circuit that contains two energy-storage elements (such as capacitors or inductors) that cannot be reduced to a single equivalent element. ➤➤first-order network.

Seebeck effect ➤thermoelectric effects.

seed crystal A small single crystal that can act as a focus for crystallization, as from a supersaturated solution or a supercooled liquid. It is used to produce large single crystals of a ➤semiconductor during the manufacture of solid-state electronic components, circuits, or devices. ➤➤horizontal Bridgeman.

selective fading ➤fading.

selective interference Interference to radiofrequency signals that occurs only within a particular narrow band of signal frequencies.

selectivity The ability of a ➤radio receiver to discriminate against radiowaves that have a ➤carrier frequency different from that to which the receiver is tuned (➤tuned circuit). The selectivity is usually expressed on a graph that shows the power ratio E/E_o expressed in decibels against frequency, at specified values of the ➤modulation factor or index and modulation frequency; E_o is the output power at the resonant frequency, f_o, and E the output power at frequency f for the same value of input power as that producing E_o.

The graph indicates the factor by which the input power must be increased, as the carrier frequency is varied from the resonant frequency to which the receiver is tuned, in order to maintain a constant output. It is usual to render any ➤automatic gain control inoperative when selectivity curves are being obtained.

selenium Symbol: Se. A ➤semiconductor element, atomic number 34. In the form of its grey allotrope, selenium is markedly light-sensitive and is extensively used in photoconductive ➤photocells; it is also used as a donor dopant in ➤compound semiconductors.

selenium rectifier A ➤Schottky diode that consists of a selenium-iron junction and is used as a ➤rectifier. It is usual to construct a stack of such junctions in series.

self-aligned gate ➤MOS integrated circuit.

self-bias A required magnitude of bias developed from the main power supply at a point in a circuit by means of a ►dropping resistor rather than supplied by a separate battery.

self-capacitance The inherent distributed capacitance associated with an inductance coil or resistor. To a first approximation the self-capacitance of the coil or resistor may be represented as a single capacitance connected in parallel with it.

self-excited Denoting an ►oscillator that produces a build-up of output oscillations to a steady value upon application of power to the circuit. No separate input at the required output frequency is needed in order to produce the oscillations.

self-inductance ►inductance; electromagnetic induction.

self-quieting Reduced sensitivity in a radio receiver due to internally generated signals that capture the detector and thus inhibit the reception of the desired but weaker signal.

self-sustaining oscillations ►oscillation.

SEM *Abbrev. for* scanning electron microscope. ►electron microscope.

semianechoic chamber ►anechoic chamber.

semiconductor A (usually) crystalline solid that in pure form exhibits a ►conductivity midway between that of metals and insulators, typically 1–1000 siemens per metre. A pure, or *intrinsic* semiconductor has an electron ►energy band structure of the form shown schematically in the diagram. The ►valence band contains electrons that are used in bonding the atoms of the crystal together; at ►absolute zero of temperature ($T = 0$ K) this band is completely filled. The next highest allowed band of energies is the

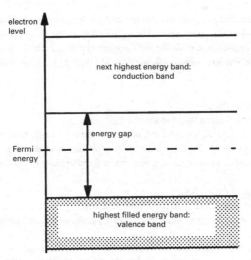

Energy band diagram of an intrinsic **semiconductor**

►conduction band, which in semiconductors is separated from the valence band by an energy band gap of about one electronvolt. At absolute zero the conduction band is completely empty, and the material behaves like an insulator.

At room temperature (about 300 K), some electrons in the valence band gain enough thermal energy from their surroundings to cross the band gap and appear in the conduction band. The two bands are now partially filled and the possibility of electrical conduction exists; conduction is by electrons in the conduction band and ►holes in the valence band. The conductivity is still low compared to metals because of the relatively small numbers of carriers involved in conduction. The ►number density of electrons (n) in the conduction band equals the number density of holes (p) in the valence band, which in turn equals the *intrinsic density* (n_i):

$$n = p = n_i$$

The ►Fermi energy level E_F in an intrinsic semiconductor lies halfway between the conduction and valence band edges. The following expressions for n and p also hold:

$$n = N_c \exp[-(E_c - E_F)/kT] = N_c \exp[-E_g/2kT]$$

$$p = N_v \exp[-(E_F - E_v)/kT] = N_v \exp[-E_g/2kT]$$

Hence,

$$n_i = \sqrt{\{(N_c N_v) \exp[-E_g/2kT]\}}$$

where N_c, N_v are the equivalent density of states in the conduction and valence bands, respectively, E_g is the energy gap, k the Boltzmann constant, and T the thermodynamic temperature. The intrinsic density, and hence conductivity, therefore has a strong temperature dependence.

The intrinsic conductivity is generally too low and too temperature dependent to be of practical use for electrical components. The ►carrier density can, however, be controlled by the careful introduction of *donor* and *acceptor impurities* into the semiconductor; this creates an *extrinsic semiconductor*. The ►impurities have the following specific properties.

A donor impurity

    replaces a host atom in the crystal;

    has one more valence electron than the atom it replaces;

    after bonding into the crystal, the extra valence electron can be easily lost;

    the free electron is in the conduction band, able to contribute to conduction;

    the donor atom becomes a positive ion, a +ve charge fixed in place in the crystal.

An acceptor impurity

    replaces a host atom in the crystal;

    has one fewer valence electrons than the atom it replaces;

    after bonding into the crystal, the vacant space is easily replaced by an electron from the valence band;

    this leaves a hole in the valence band, able to contribute to conduction;

    the acceptor atom becomes a negative ion, a –ve charge fixed in place in the crystal.

The addition of impurities to a semiconductor is known as ►doping. The impurity densities are chosen to be much greater than the intrinsic density, so that the temperature variation of carriers is overwhelmed, and the material conductivity is independent of temperature. The electron and hole densities in doped semiconductors are as follows:

In an *n-type semiconductor*:

    donor density N_D > acceptor density N_A >> intrinsic density n_i

    electron density $n = N_D - N_A$, so electrons are ►majority carriers,

    hole density $p = n_i^2/n$, so holes are ►minority carriers,

    the Fermi energy is in upper half of band gap.

In a *p-type semiconductor*:

    acceptor density N_A > donor density N_D >> intrinsic density n_i

    hole density $p = N_A - N_D$, so holes are majority carriers,

    electron density $n = n_i^2/p$, so electrons are minority carriers,

    the Fermi energy is in lower half of band gap.

At thermal equilibrium a dynamic equilibrium exists in a semiconductor. Mobile charges move around the crystal in a random manner due to scattering by the nuclei in the crystal lattice. The crystal retains overall charge neutrality and the number of charge carriers remains essentially constant. A continuous process of regeneration and recombination occurs, however, as thermally excited electrons enter the conduction band and other electrons return to the valence band and there combine with holes (►recombination processes).

If an electric field is applied to the semiconductor, charge carriers move under the influence of the field but still undergo scattering processes. The result of the field is to impose a drift in one direction onto the random motion of the carriers (►drift mobility; continuity equation).

Extrinsic semiconductors are used for making ►diodes, ►transistors, and ►integrated circuits (►p-n junction; metal-semiconductor contact). The most common semiconductor material is silicon, which is used for around 99% of commercial semiconductor products. ►Compound semiconductors, such as ►gallium arsenide, are used in specialist applications such as microwave devices and optoelectronics.

semiconductor device Any electronic circuit or device that depends for its operation on the flow of charge ►carriers within a ►semiconductor.

semiconductor diode A ►diode constructed from semiconducting material.

semiconductor laser *Syn.* diode laser. A ►laser that uses a ►p-n junction diode made from a ►direct-gap semiconductor material such as ►gallium arsenide, GaAs. When a p-n junction is operated in forward bias, majority carriers are injected across the junction and appear as excess minority carriers on the other side of the junction. In normal p-n junction operation these excess carriers would recombine with the majority carriers, and constitute the current flow across the junction. In a direct-gap semiconductor this ►recombination process can be *radiative*: the band-gap energy E_g that is released during recombination is given up as light, which has a wavelength $\lambda = hc/E_g$ and is thus ►monochromatic (h is the Planck constant and c the speed of light).

Stimulated emission of ►coherent radiation is arranged by cutting the semiconductor laser die so that the two end faces are parallel and reflective, and the radiated photons pass through the ►active region or optical cavity of the laser, causing further stimulated emission of photons. The light in the optical cavity is confined by the ►active area of the p-n junction contacts.

In a *heterostructure laser*, greater optical confinement is obtained by surrounding the active p-n junction region by semiconductor material of a slightly different dielectric constant and hence refractive index. This causes internal reflections of the photons at the interface between the active region and this confining layer, resulting in a higher photon density in the light beam and hence a higher intensity of light. An example of a heterostructure laser would use a GaAs p-n junction active region, sandwiched between layers of lattice-matched AlGaAs, which has a slightly higher dielectric constant. Variations on this theme can include further heterostructure layers to confine the laser active region and the optical cavity independently, a device known as a *separate confinement heterostructure* (SCH) *laser*.

In the above examples, reflective facets are used to provide the optical feedback necessary for stimulated emission to occur. In a *distributed feedback* (DFB) *laser*, the feedback is generated by corrugating the interface between the active and confining layers with a specific periodicity; this causes the emitted light to be scattered back into the optical cavity to interfere constructively with the light reflecting from the two facets. The optical feedback and hence the laser output is thereby increased.

semiconductor memory ►solid-state memory.

sensing element ►transducer.

sensitivity 1. In general, the change produced in the output of a device per unit change in the input. **2.** The magnitude of the change in the indicated value or deflection of a measuring instrument produced by a specified change in the measured quantity. It is usually quoted as the magnitude of the measured quantity required to produce full-scale deflection. **3.** The ability of a ►radio receiver to respond to weak input signals. It is the minimum signal input to the receiver that produces a particular output value under stated conditions, particularly a designated signal-to-noise ratio.

sensor ►transducer.

separate confinement heterostructure laser (SCH laser) ►semiconductor laser.

sequencer A device that enables ►MIDI data to be recorded, edited, stored, and replayed.

sequential circuit A ►logic circuit whose output depends on the inputs to the circuit and the current state of the circuit, i.e. it has a concept of history. The history is retained in a memory device within the sequential circuit. These memory devices are often ►flip-flops. ►►combinational logic.

sequential control Operation of a ►computer in which a sequence of instructions is produced and input to the computer during the solution of a problem.

sequential scanning ►television.

serial transfer Transfer of information along a single path in a computer or data processing system in which the characters move one after the other along the path.

serial transmission Transmission of units of information sequentially along a single line. ►►parallel transmission.

series Circuit elements are said to be *in series* if they are connected so that one current flows in turn through each of them (see diagram).

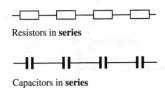

Resistors in **series**

Capacitors in **series**

For *n* resistors in series the total resistance, *R,* is given by:

$$R = r_1 + r_2 + \ldots r_n$$

where $r_1, r_2, \ldots r_n$ are the values of the individual resistors. The resistors thus behave collectively as one large resistance.

For *n* capacitors in series the total capacitance, *C,* is given by:

$$1/C = 1/c_1 + 1/c_2 + \ldots 1/c_n$$

where $c_1, c_2, \ldots c_n$ are the individual capacitances.

Machines, transformers, and cells are said to be in series when terminals of opposite polarity are connected together to form a chain. Several cells connected together in series add the values of the emfs and can therefore supply a larger voltage.
►►parallel.

series feedback *Syn. for* current feedback. ►feedback.

series-gated ECL ►emitter-coupled logic.

series network ►network.

series-parallel connection 1. An arrangement whereby electronic devices or circuits may be connected either in ►series or in ►parallel. **2.** A method of connecting the elements of a circuit or network containing resistors, inductors, and capacitors so that some are in ►series and some are in ►parallel with each other.

series resonant circuit *Syn.* acceptor. ►resonant circuit; resonant frequency.

series stabilization ►stabilization.

series supply A method of applying bias to an electrode of an active device so that the bias is applied across the same impedance in which the signal current flows (Fig. *a*).

Parallel supply is a method of applying the bias so that the bias is supplied across an impedance in ►parallel with that in which the signal current flows (Fig. *b*).

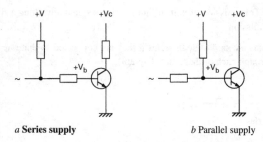

a **Series supply**   *b* Parallel supply

series transformer *Syn. for* current transformer. ►transformer.

service area The region covered by the useful range of a radiofrequency broadcast transmitter, such as a radio or television transmitter. The region is often represented pictorially on a *service-area diagram*.

The *primary service area* is the area within which satisfactory reception is possible, day or night, as a result of reception of the ►ground wave. The strength of the ground wave within this area is large compared to that of interference and ►indirect waves.

The *secondary service area* is the area within which satisfactory reception is possible using indirect waves. The strength of the ground wave within this area is attenuated to substantially less than the indirect waves.

The *mush area* is an area within which substantial ►fading or ►distortion of the received signal occurs. The unsatisfactory reception in this area results from interference either between waves from two or more synchronized transmitters or between direct and indirect waves from a single transmitter.

servomechanism ►control system.

settling time The time taken for the response of a ►second-order network to settle within a given voltage range.

set-up scale instrument ►suppressed-zero instrument.

shading 1. The generation of a nonuniform background level in the image produced by a television ►camera tube that was not present in the original scene. **2.** Compensation for the spurious signals generated by a television ►camera tube during the flyback interval.

shadow effect An effect observed when electromagnetic waves are transmitted, as during broadcast transmissions, due to the topography of the region between the transmitter and receiver. A loss of signal strength at the receiver is usually observed compared to the expected signal strength when transmission is over a uniformly flat region.

shadow mask ➤colour picture tube.

Shannon–Hartley theorem In communications theory, a theorem giving the relationship between the capacity of a channel when the channel bandwidth and noise are known. It states that the ➤channel capacity in bits/second is equal to

$$W \log_2(1 + P_S/P_N)$$

where W is the ➤bandwidth (in hertz) and P_S/P_N is the ➤signal-to-noise power ratio.

shaped-beam tube A ➤cathode-ray tube that is used to display characters or numerals. Suitable electric and magnetic fields are applied to the tube in order to produce a beam whose cross section is that of the desired character. All parts of the character are therefore shown on the screen simultaneously.

sheet resistance The resistance of a unit square of a thin-film material, such as a metal or thin layer of semiconductor, defined as $R_s = \rho/t$ where R_s is the sheet resistance, ρ the ➤resistivity, and t the thickness. A film of length L, width W, and thickness t therefore has a total resistance equal to $R_s(L/W)$. R_s has dimensions of resistance but is commonly given the unit 'ohms per square'.

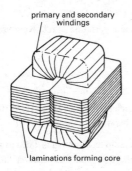

primary and secondary windings

laminations forming core

Single-phase **shell-type transformer**

shell-type transformer A ➤transformer in which most of the windings are enclosed by the ➤core (see diagram). The core is made from laminations and usually the windings are assembled and then the laminated core is built up around them. ➤➤core-type transformer.

SHF *Abbrev. for* superhigh frequency. ➤frequency band.

shield *Syn.* screen. A barrier or enclosure that prevents or reduces the penetration of an electric or magnetic field into a particular region.

shielded pair *Syn.* screened pair. A ➤transmission line that consists of two wires enclosed in a metal sheath. If the two wires are twisted around each other within the conducting sheath they form a *shielded twisted pair* (STP). The shield provides additional isolation from external sources of ➤electromagnetic interference. ➤➤pair.

shielding effectiveness (of a barrier or enclosure) The ratio of the field strength without the barrier or enclosure to the field strength when it is present. It may be specified for the electric or magnetic fields.

shift operator ➤z transform.

shift register A ➤digital circuit that can store a set of information in the form of pulses and displace it either to the left or right upon application of a *shift pulse*. If the information consists of the digits of a numerical expression, a shift of position to the left (or right) is equivalent to multiplying (or dividing) by a power of the base. Shift registers are extensively used in ➤computers and data processing systems as storage or delay elements. ➤➤delay line.

Shockley emitter resistance The dynamic resistance at the emitter terminal of a ➤bipolar junction transistor, given by

$$r_e = dV_{BE}/dI_E = (kT/e).(1/I_E)$$

Shockley equation *Syn.* ideal diode equation. ➤p-n junction.

short *Informal* A short circuit.

short circuit An accidental or deliberate electrical connection of relatively very low resistance between two points in a circuit.

short-circuit impedance ➤network.

short-time duty ➤duty.

short-wave Denoting a radiowave that has a wavelength in the range 10 to 100 metres, i.e. in the high-frequency band. ➤frequency band.

short-wave converter A ➤frequency changer in a radio receiver that is tuned to receive a broadcast transmission in the short-wave frequency band and that converts it to a frequency band within the dynamic range of a standard receiver.

shot noise *Syn.* Schottky noise. ➤Noise that occurs in any electrical device when electrons must overcome a potential barrier by means of their kinetic energy. As each electron crossing the barrier is a random event, the resulting current has a random nature. Shot noise occurs in junction devices, such as a ➤p-n junction diode or ➤bipolar junction transistor.

shunt 1. In general, *syn. for* parallel. **2.** *Syn.* instrument shunt. A resistor, usually of a relatively low value, that is connected in parallel with a measuring instrument, such as a galvanometer. Only a fraction of the current in the main circuit passes through the instrument so that the shunt increases the range of the instrument and also protects it from possible damage caused by current ►surges. ►►universal shunt.

shunt feedback *Syn. for* voltage feedback. ►feedback.

shunt network *Syn. for* parallel network. ►network.

shunt stabilization ►stabilization.

sideband ►carrier wave; amplitude modulation.

sideband splatter A form of distortion in transmitted signals, in particular amplitude-modulated signals, where the degree of modulation exceeds 100%. ►amplitude modulation.

side frequency ►carrier wave.

side lobe ►antenna pattern.

siemens Symbol: S. The ►SI unit of electrical ►conductance, ►susceptance, and ►admittance. An element possesses a conductance of one siemens if it has electrical ►resistance of one ohm. The siemens has replaced the ►mho.

Siemen's electrodynamometer An ►electrodynamometer that may be calibrated as an ammeter, voltmeter, or wattmeter. The signal to be measured produces an electromagnetic torque on the movable coil that in turn is balanced against the torque of a spiral spring connected to it by adjusting a calibrated torsion head attached to the spring. At the balance position the deflection of the movable coil is zero, and the value of the measured parameter is given by the setting of the torsion head.

signal A variable electrical parameter, such as voltage or current, that is used to convey information through an electronic circuit or system. The sequence of values of the parameter, recorded against time, represents the information. An *analogue signal* varies continuously in amplitude and time. The amplitude of a *digital signal* varies discretely: it will be at any one of a group of different levels, usually two, at any particular time.

In communications systems, signal transmission is either analogue or digital. In analogue transmission signals are transmitted in continuously variable form. In digital transmission there are two discrete signal levels representing a binary '1' and binary '0'. In *multilevel signalling systems* the number of signal levels is increased from two, thereby allowing more information to be transmitted in each 'bit'. ►►digital communications; digital codes.

signal flowgraph A graphical illustration of the relationships between two-port ►network parameters and the signals in the network. A signal flowgraph is a network of *directed branches* that are interconnected at *nodes*. The nodes have *node signals*, such as current, voltage, and ►power waves, and the branches have *branch transmittances* that specify the relationships between the signals at the source and sink nodes. The flow of signals in the graph is governed by the following basic rules: the node signal flows along the branch only in the direction of the arrow, and is multiplied by the transmittance of that branch; a node signal is equal to the algebraic sum of all signals entering that node; the signal at a node is applied to each of the outgoing branches from that node. ➤Mason's rules.

signal generator Any electronic circuit or device that produces a variable and controllable electrical parameter. The term is most commonly applied to a device that supplies a specified voltage of known variable amplitude, frequency, and waveform shape. A generator that produces ➤pulse waveforms is normally referred to as a ➤pulse generator, the term signal generator being reserved for a continuous-wave generator, particularly of sinusoidal and square waves.

signal level The magnitude of a ➤signal at a point in a transmission system.

signal processing The process of modifying one or more aspects of a ➤signal, often used to enhance the representation of features of particular interest. ➤digital signal processing.

signal-to-noise ratio At any point in an electronic circuit, device, or transmission system, the ratio of one parameter of a desired signal to the same or a corresponding parameter of the ➤noise. In broadcast communication the signal-to-noise ratio is often quoted in ➤decibels.

signal winding ➤transductor.

sign bit The bit that distinguishes positive from negative numbers in signed-magnitude, ➤one's-complement, and ➤two's-complement notation. The sign bit is 1 for positive, 0 for negative.

silent discharge An electrical discharge that occurs at high voltages and is inaudible to the human ear. Such a discharge involves a relatively high dissipation of energy and takes place most readily from a sharply pointed conductor.

silent zone *Syn.* skip zone. The portion of the skip area (➤skip distance) surrounding a particular transmitter that falls outside the range for ground-wave propagation of the transmitter. In practice there is a weak residual signal within the silent zone that results from scattering, localized reflection, or some abnormal means of propagation.

silica Symbol: SiO_2. An extremely abundant compound occurring in several different natural forms, the best known of which are probably quartz and common sand (in which the silica is discoloured by ferric oxides). Silica is important as a source of silicon for the manufacture of electronic components, devices, and integrated circuits, as a grown oxide for the ➤passivation of such equipment, and as natural ➤quartz, which has marked piezoelectric properties.

silica gel Deliquescent crystals consisting mainly of silica (SiO_2) that are used as a drying agent, particularly during dispatch and delivery of electronic and electrical equipment.

silicon Symbol: Si. A ➤semiconductor element, atomic number 14. It is very abundant in nature in the form of silicon dioxide (➤silica) and is the most widely used semiconductor in solid-state electronics. It is cheap and extremely versatile and rapidly replaced germanium except for a very few specialized applications. ➤➤amorphous silicon; polysilicon.

silicon-controlled rectifier (SCR) ➤thyristor.

silicon-gate technology ➤MOS integrated circuit.

silicon-on-insulator A silicon ➤integrated circuit technology using ➤MOSFETs that are made on an insulating layer. The insulator provides good isolation between the transistors for improved performance at high speeds and high frequencies, as parasitic coupling and hence ➤crosstalk between devices is reduced. The insulating layer can be provided by a nonconducting substrate upon which the silicon can be deposited, for example *silicon-on-sapphire*. Alternative techniques involve placing an insulating layer beneath the ➤active layer during the manufacture of the integrated circuit. A silicon dioxide layer can be created deep in the silicon by ➤ion implantation of oxygen; subsequent heat treatment produces a well-defined oxide layer.

silicon-on-sapphire ➤silicon-on-insulator.

silver Symbol: Ag. A metal, atomic number 47. The metal that best conducts electricity.

silver mica capacitor *Syn. for* bonded silvered mica capacitor. ➤mica capacitor.

simple magnet ➤permanent magnet.

simplex operation Operation of a communications channel in one direction only. ➤➤duplex operation.

SIMS *Abbrev. for* secondary-ion mass spectroscopy.

simulator A device, such as an analogue ➤computer, that imitates the behaviour of an actual physical system and can therefore be used to solve complex problems associated with the operation of the system. A simulator is usually fabricated from components that are easier, cheaper, or more convenient to manufacture than the system itself.

simultaneous equations A set of equations used in the solution of n linear algebraic equations in n unknowns, having the general form:

$$a_{11}x_1 + a_{12}x_2 + a_{13}x_3 + \ldots + a_{1n}x_n = b_1$$

$$a_{21}x_1 + a_{22}x_2 + a_{23}x_3 + \ldots + a_{2n}x_n = b_2$$

$$a_{31}x_1 + a_{32}x_2 + a_{33}x_3 + \ldots + a_{3n}x_n = b_3$$

$$\ldots\ldots$$

$$a_{n1}x_1 + a_{n2}x_2 + a_{n3}x_3 + \ldots + a_{nn}x_n = b_n$$

Cramer's rule states that the jth unknown is given directly by

$$x_j = \Delta_j/\Delta$$

where Δ is the determinant of the matrix of the coefficients a_{ij} ($i,j = 1,2,3,...,n$) and Δ_j is the determinant of the matrix obtained from the matrix of the coefficients a_{ij} by replacing the coefficients of the jth column with the coefficients b_j ($j = 1,2,3,...,n$). This assumes that the equations are linearly independent, i.e. that $\Delta \neq 0$.

The *Gaussian elimination* technique repeatedly combines different rows such that the original system of equations is transformed into a system of the type shown below. This is achieved by taking the first equation of the original system and multiplying it by a factor such that when added to the second equation the coefficient of x_1 is zero. This is repeated but with the third equation and so on until all the coefficients of x_1 become zero. With this new set of equations the technique is repeated so that the coefficients of x_2 become zero. This is repeated until x_n can be solved directly, i.e. $\alpha_{nn}x_n = \beta_n$. Then by a process of *back-substitution* all the other unknowns can be determined.

$$\alpha_{11}x_1 + \alpha_{12}x_2 + \alpha_{13}x_3 + \dots + \alpha_{1n}x_n = \beta_1$$

$$\alpha_{22}x_2 + \alpha_{23}x_3 + \dots + \alpha_{2n}x_n = \beta_2$$

$$\alpha_{33}x_3 + \dots + \alpha_{3n}x_n = \beta_3$$

$$\dots\dots$$

$$\alpha_{nn}x_n = \beta_n$$

Gaussian elimination and Cramer's rule are used extensively in circuit analysis, where the simultaneous equations are formulated using either node or loop analysis: x_1 through x_n are the unknown ➤node voltages (or ➤mesh currents), a_{ij} ($i,j = 1,2,3...,n$) are known ➤admittances (or ➤resistances), and b_1 through b_n are known source voltages (or currents).

sinc function A function of the form $(\sin x)/x$.

sine potentiometer ➤potentiometer.

sine wave ➤sinusoidal.

singing point ➤feedback.

single crystal A crystal in which the corresponding atomic planes are effectively parallel. This is demonstrated by the single spot diffraction pattern produced by such a crystal using a collimated beam of electromagnetic radiation. ➤➤mosaic crystal.

single-current system A telegraph system that uses a unidirectional electric current for the transmission of the signal. ➤➤double-current system.

single drift device ➤IMPATT diode.

single-ended Denoting a method of supplying an input signal or obtaining an output signal from a circuit in which one side of the input or output is connected to earth. A *single-ended amplifier* is one in which both the input and output are single-ended. A *double-ended* input (or output) is one in which neither side of the input (or output) is connected to earth and a differential signal is applied (or obtained).

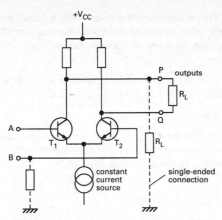

Single-ended and double-ended long-tailed pair

An example of a circuit that may be used with either single- or double-ended input or output is the simple ➤long-tailed pair (see diagram). When used as an ➤inverter, input B to the base of transistor T_2 is earthed to give single-ended input; the output is taken from output P across a load resistor, R_L, connected to earth.

The circuit may also be used as a ➤differential amplifier. The difference signal (double-ended) is applied across A and B and the output produced between P and Q. Single-ended output may also be obtained using either P or Q; with Q the output voltage developed is approximately half the value using P, assuming the components are all matched. Differential output may also be obtained using single-ended input.

single-ended amplifier ➤single-ended.

single-loop feedback *Syn. for* multistage feedback. ➤feedback.

single-phase system An electrical system or apparatus that has only one alternating voltage. ➤➤polyphase system.

single-shot multivibrator *Syn. for* one-shot multivibrator. ➤monostable.

single-shot trigger A ➤trigger circuit in which each triggering pulse initiates only one complete cycle of events in a driven circuit, the cycle ending in a stable condition.

single-sideband transmission (SSB) The transmission of only one of the two sidebands produced by ➤amplitude modulation of a ➤carrier wave. The carrier wave is usually suppressed at the transmitter (in addition to the suppressed sideband) and it is therefore necessary to reintroduce the carrier artificially at the receiver by means of a locally generated oscillation. The frequency of the original carrier must be reproduced in the local oscillation as nearly as possible but this requirement is not as stringent as in ➤double-sideband transmission.

The main advantages of SSB compared to transmission of the carrier and both side-bands are the reduction in transmitter power required for the transmission and the reduced ►bandwidth required for the transmission of signals within a designated ►frequency band.

single-tuned circuit A ►resonant circuit that may be represented as a single ►capacitance and a single ►inductance with the associated values of resistance.

sink 1. *Short for* heat sink. **2.** ►source.

sinusoidal Denoting a periodic quantity that has a waveform graphically identical in shape to a sine function; waveforms represented by the functions sin$x$ and cos$x$ would both be described as sinusoidal or as *sine waves* (see diagram).

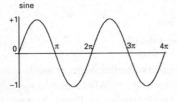

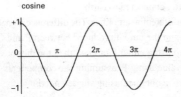

Sinusoidal waveforms

SIS *Short for* superconductor-insulator-superconductor mixer. A ►mixer that has a very nonlinear current-voltage characteristic at low voltages. It is used at low temperatures to minimize ►noise in very sensitive ►heterodyne receivers.

SI units The internationally agreed system of ►units intended for all scientific and technical purposes. The system is based on the ►MKS system and replaces the ►CGS and Imperial systems of units. Units in the SI system can be *base units, derived units,* or *dimensionless units.*

The base units are an arbitrarily defined set of dimensionally independent physical quantities. In any purely mechanical system of units only three base units – of mass, length, and time – are required. In a consistent electric and magnetic system four base units are needed. In the SI system there are seven base units: the ►metre, ►kilogram, ►second, ►ampere, ►kelvin, candela (the unit of luminous intensity), and mole (the unit of amount of substance). Each base unit has a special symbol.

The derived units are formed by the combination, by multiplication and/or division, of two or more base units without the use of numerical factors. The unit of speed, for example, is formed from the combination of one metre divided by one second; this can be expressed in terms of the symbols of the units: m s^{-1} or m/s. Some derived units have special names, such as coulomb, volt, hertz, or joule. The coulomb, which is the derived unit of charge, is formed from a combination of one ampere times one second (A s in symbol form). The named derived units have special symbols; for example, C is the symbol for coulomb. There are two named dimensionless units, the radian and steradian, which are the units of plane and solid angle, respectively.

A set of 14 prefixes, including micro- and kilo-, are used with the SI units to form decimal multiples and submultiples of the units. The symbol of a prefix can be combined with the symbol of the unit, as in mA (milliampere).

SI units and SI prefixes, together with their symbols, are given in Tables 6–9 in the back matter.

When considering electric and magnetic quantities a fourth term is required in addition to the fundamental units of mass, length, and time, for their complete definition. In the MKS system the fourth quantity is the permeability of free space, μ_0, which is defined as $4\pi \times 10^{-7}$ henry per metre. In the SI system it is the ampere that is the fundamental unit and μ_0 then has the value $4\pi \times 10^{-7}$ henry per metre.

size quantization ➤quantum dot; quantum wire.

skew The arrival of an electrical signal at two or more places in a circuit at significantly different times. In a ➤synchronous logic system, all components must be clocked by the same signal. As the clock signal propagates around the circuit, it should not be delayed or otherwise passed through logic gates. Such actions would produce skew in the clock signal – *clock skew* – and the different arrival times for the clock at different components in the circuit would cause or threaten malfunction of the circuit.

skin depth ➤skin effect.

skin effect A nonuniform distribution of current over the cross section of a conductor when carrying alternating current, with the greater current density located at the surface (or 'skin') of the conductor. The skin effect is caused by electromagnetic induction in the wire and increases in magnitude with increasing frequency. At sufficiently high frequencies the current is almost entirely confined to the surface of the conductor and results in a greater ➤I^2R loss than when the current is uniformly distributed. The *skin depth* of the conductor, d, is that distance from the surface that the current has decreased to 1/e of its surface value. It is given by

$$d = (2/\mu\omega\sigma)^{1/2}$$

for a metal, where μ is the magnetic permeability, ω the angular frequency, and σ the conductivity.

The ➤effective resistance of the conductor is therefore greater than the d.c. or ohmic resistance, when carrying alternating current, and for high-frequency applications the *high-frequency resistance* of a conductor can be substantially greater than the nominal d.c. value. The *surface resistance* of a metal, R_{surface}, is given by

$$R_{\text{surface}} = 1/\sigma d$$

A metallic conductor of thickness less than one skin depth will have a radiofrequency current distributed fairly uniformly throughout. The high-frequency resistance will be at a minimum in a conductor of a few skin depths cross section, with most of the current flow within one skin depth of the surface.

The skin effect may be minimized for high-frequency uses by employing hollow conductors or stranded conductors, such as ►Litzendraht wire. ·

skip distance The least distance, for a specified operating frequency, at which radiowaves are received in a given direction from the transmitter by reflection from the ionosphere. Reflection from the sporadic E-layer is customarily ignored. The area surrounding a particular transmitter swept out by a complete rotation of a radius vector equal in length to the skip distance is known as the *skip area*. This area is not necessarily circular as the skip distance is not always the same in all directions.

skip zone ►silent zone.

sky wave ►ionospheric wave.

slave circuit An electronic circuit that requires a ►trigger or a ►clock pulse from an external source (that may be common to several different slaves) in order to perform its cycle of operation. Examples include a clocked ►flip-flop and a *slave sweep,* i.e. a triggered timebase.

slew rate The rate at which the output from an electronic circuit or device can be driven from one limit to the other over the ►dynamic range.

slice *Syn.* wafer. A large single crystal of semiconductor material that is used as the ►substrate during the manufacture of a number of ►chips. Very large single crystals are grown and then sliced into wafers before processing.

slicer ►limiter.

slicing ►compact disc system.

slide wire A wire of uniform resistance provided with a sliding contact that can make a connection at any desired point along the length. Slide wires are used to provide a variable resistance, as a potentiometer, or to provide a desired resistance ratio (►Wheatstone bridge). An overall length of one metre is commonly chosen.

slope modulation *Syn. for* delta modulation. ►pulse modulation.

slope resistance The ratio of an incremental voltage change applied to a specified electrode in an electronic device to the corresponding current increment of that electrode, the voltages applied to all other electrodes being maintained constant at known magnitudes. The slope resistance is thus the differential resistance, or a.c. resistance, of the specified electrode. For example, the slope resistance, r_c, of the collector of a ►bipolar junction transistor is given by

$$r_c = \partial V_c / \partial I_c \qquad V_b, V_e = \text{constant}$$

where V_c is the collector voltage, I_c the collector current, and V_b and V_e the base and emitter voltages, respectively.

slot antenna *Syn. for* aperture antenna.

slotline ➤slotted line.

slot matrix tube ➤colour picture tube.

slotted line *Syn.* slotline. A transmission-line medium comprising a pair of conducting planes on the surface of a dielectric substrate, separated by a gap or 'slot' between the metal planes. The characteristic impedance is determined by the ratio of the slot width to the thickness of the dielectric substrate, and the dielectric constant of the substrate.

slotted waveguide ➤waveguide.

slow-break switch A manually operated switch whose speed of operation depends upon the speed at which the operating handle or lever is moved.

slow-wave structure ➤travelling-wave tube.

slug tuning *Syn.* permeability tuning. ➤tuned circuit.

small-outline package A very small volume package for an active device, for use in high-frequency circuits. The packages are usually made from ceramic materials, which have well-toleranced dielectric properties, designed to minimize ➤parasitic capacitance. The packages also have very short connecting leads to minimize the parasitic inductance. This enables the active devices to operate close to their maximum frequency of operation. Examples include small-outline diode, IC, and transistor packages – *SOD*, *SOIC*, and *SOT*, respectively.

small-scale integration (SSI) ➤integrated circuit.

small-signal Of or involving a signal whose amplitude is small enough so that the behaviour of the component or circuit can be considered to be linear; nonlinear effects can be ignored.

small-signal parameters ➤network; transistor parameters.

smearer ➤pulse.

S meter In radio receivers, the visual indication of the received signal strength.

Smith chart A graphical means of calculating the ➤impedances and ➤reflection coefficients along a ➤transmission line, developed by P. M. Smith in 1939. The chart incorporates the periodic variation of impedance and reflection coefficient with distance along the transmission line (see diagram overleaf), enabling rapid calculation of the impedances, admittances, reflection coefficients, ➤VSWR, etc.

smoothing choke ➤choke.

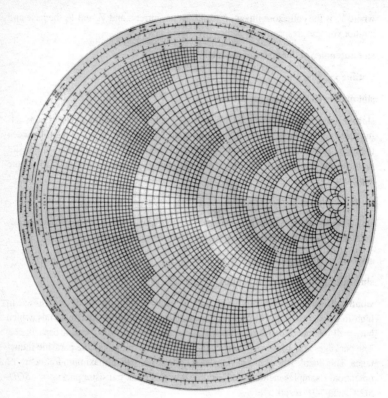

Smith's chart of impedances
reproduced by kind permission of
Tollit and Harvey Ltd, King's
Lynn, UK.

Smith chart of impedances

smoothing circuit *Syns.* ripple filter; rectifier filter. A circuit that is designed to reduce the amount of ➤ripple present in an essentially unidirectional current or voltage. A typical smoothing circuit consists of a low-pass ➤filter (see diagram) but a single inductance may be used.

snapback diode *Syn.* snap-off diode. ➤step-recovery diode.

snow An unwanted pattern appearing on the screen of a ➤television or ➤radar receiver that resembles falling snow. It may appear when the received signal is absent or transmitted at a lower power than is usual and is caused by noise associated with the circuits and equipment forming the receiver. In the case of a ➤colour picture tube, the snow is coloured.

SOD *Abbrev. for* small-outline diode package.

software The ►programs that can be run on a computer system, as opposed to the physical components (the ►hardware) of the system. A distinction can be drawn between *systems software*, which is an essential accompaniment to the hardware in providing an effective overall computer system, and *applications software*, which makes a direct contribution to meeting the needs of the computer user(s).

SOIC *Abbrev. for* small-outline integrated circuit package.

solar cell A ►photodiode whose spectral response has been optimized for daytime radiation from the sun. Solar cells are semiconductor devices that provide electrical energy from solar radiation by means of photogeneration of charge carriers at a ►p-n junction. Early solar cells used ►compound semiconductors such as cadmium sulphide, which has an ►energy band gap corresponding to visible light. More modern devices use silicon p-n junctions, where the band gap corresponds to ►infrared wavelengths, generating a greater electrical output. Compound-semiconductor solar cells based on indium phosphide are used in critical radiation-hard conditions; such devices produce the highest conversion efficiency, about 20%, but are much more costly than silicon devices.

For practical use, large numbers of solar cells can be assembled in flat arrays, known as *solar panels*. Solar panels are used mainly as a source of power in spacecraft.

solar panel ►solar cell.

solenoid A coil of wire that has a long axial length relative to its diameter. The coil is usually tubular in form and is used to produce a known magnetic flux density along its axis.

At a point on the axis inside the solenoid, ignoring any end effects, the magnitude of the magnetic flux density, B, is given by

$$B = \frac{1}{2}\mu_0 nI(\cos\theta_1 + \cos\theta_2)$$

where μ_0 is the magnetic constant, n the number of turns per unit length, I the current flowing through the solenoid, and θ_1 and θ_2 the semiangles subtended at the point by the ends.

A solenoid is often used to demonstrate ►electromagnetic induction and a bar or rod of iron that is free to move along the axis of the coil is usually provided for this purpose.

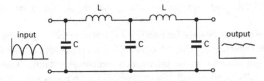

Simple low-pass filter as **smoothing circuit**

solid conductor A conductor that is composed of a single wire or uniform thin metal rather than being stranded or otherwise divided.

solid-state camera A ➤television camera in which the light-sensitive target area consists of an array of ➤charge-coupled devices (CCDs). Exposure to light energy results in the generation of electron-hole pairs in the semiconductor substrate (➤photoconductivity); the number of electron-hole pairs generated is a function of the light intensity. The majority carriers migrate into the bulk material and the minority carriers accumulate in the potential wells at the electrodes of the CCDs. The accumulated charge is transferred into other nonphotosensitive CCD storage sites during the retrace intervals of the television display system and is then transferred to the output device while further signals are being generated.

Two basic methods of obtaining the video signal are used. A *frame/field transfer device* is shown in Fig. *a*. A set of charge-transfer devices – the *integrating array* – is arranged in the vertical (field) direction and exposed to the optical image. During the vertical retrace interval of the television display system the charge pattern that has accumulated is clocked at high speed into the storage area. During the normal horizontal (line) blanking periods the pattern of charges in the storage area is moved downwards by one line into the bottom horizontal register and clocked horizontally to the output to form the video signal.

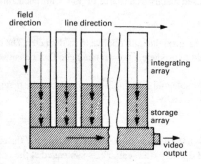

a Frame/field transfer **solid-state camera**

An *interline transfer device* is shown in Fig. *b*. In this type of solid-state camera the storage arrays are arranged alternately with the photosensitive arrays and are connected to them by a transfer gate. During the vertical retrace period the information is transferred horizontally from the photosensitive electrodes to the storage electrodes by means of the transfer gates and then read out line by line in a similar manner to the frame/field device.

Both types of device may be used for interlaced scanning (➤television). Interlacing is achieved in the frame/field device by adjusting the potentials applied to the gate electrodes of the integrating array so that information accumulating during the second field period is physically offset by half of a storage cell length compared to that accumulated during the first field period (Fig. *c*). The entire array is completely emptied

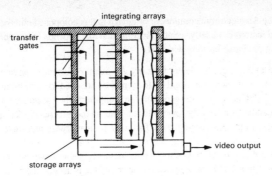

b Interline transfer device

of information during each field sweep and the integrating period is equal to the field period.

Interlacing in the interline transfer device is achieved by adjusting the potentials of the transfer gates so that the information in alternate sites (Fig. *b*) is transferred during each vertical retrace interval. The integrating time in each site is therefore for the full frame interval.

The frame/field transfer device requires only half the number of integrating sites required by the interline transfer device but the location of the storage area is such that additional area of semiconductor chip is required. The interline device has a more economical arrangement of electrodes but can be less sensitive than the frame/field device since the transfer gates and storage sites must be shielded from the optical image and twice as many storage sites are required. Integration for the full frame period can result in a slower response to fast moving objects; this can be overcome by reading out both sets of integration sites during each field period and pairing the outputs in a suitable manner.

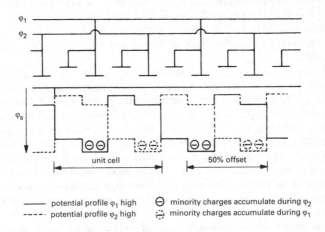

———— potential profile φ_1 high   ⊖ minority charges accumulate during φ_2
----- potential profile φ_2 high   ⊖ minority charges accumulate during φ_1

c Cross section of overlapping gate two-phase CCD array

solid-state device An electronic component or device that is composed chiefly or exclusively of solid materials, usually semiconducting, and that depends for its operation on the movement of charge carriers within it.

solid-state memory *Syn.* semiconductor memory. A ➤memory formed as an integrated circuit in a chip of semiconductor. Solid-state memories are used to store binary data patterns in digital electronic circuits, especially in computers; they are cheap, robust, compact, and operate on relatively low voltages. The memory capacity that can be stored on a single chip is increasing by a factor of four approximately every few years.

There are several different types of solid-state memory. One of the most important types is the read-write ➤RAM (random-access memory) that is used as the main working memory of computer systems and stores data and programs, which are accessed by the central processing unit of the system. Solid-state ➤ROM (read-only memory) is used for the permanent and semipermanent storage of information. ➤CCD memory, which is inherently slower than RAM and ROM, is used for applications that do not require such very high speeds.

Read-write RAM can be either *static* or *dynamic*. Static RAM (SRAM) is realized in either bipolar or MOS technology. Dynamic RAM (➤DRAM) is realized in MOS technology. The static memories have higher operating speeds than the dynamic memories but have a lower functional packing density, dissipate power continuously, and are more expensive. The bipolar memories have had faster operating speeds than the MOS memories but this is no longer always the case. The dynamic memories are slower than the static memories and require extra circuits for ➤refreshing the information but they have a larger density and dissipate power only when operating; the stand-by power dissipation is very low.

The basic very simple memory cell of dynamic RAM consists of a ➤MOS capacitor, in which the information is stored as electronic charges, and a ➤MOSFET transistor, which is used as a switch in order to connect the appropriate capacitor to the sense amplifier (➤DRAM). The interconnections between memory cells form the rows and columns of a rectangular matrix so that each memory cell has a unique address (➤➤RAM).

A cross section through a typical memory cell is shown in Fig. *a*. The MOS capacitor consists of a layer of highly doped ➤polysilicon that forms the upper plate (POLY 1). The lower plate is formed when a positive potential of +12 volts is applied to the polysilicon: an inversion layer is produced at the surface of the substrate and acts as the second plate. The MOSFET transistor is formed with a second polysilicon layer (POLY 2), which is the gate electrode. Overlapping the two layers of polysilicon allows the inversion layer of the capacitor to act as the source of the transistor. The bit or data line is formed from an n^+ diffusion into the substrate and also acts as the drain of the transistor. The thick oxide layer is produced using the ➤coplanar process and extra p-type ions are implanted below it, in order to provide isolation of the memory cells.

The memory cells are packed as closely as possible into the chip by suitable design of the arrangement of the cells and associated bit lines and the patterns of the vari-

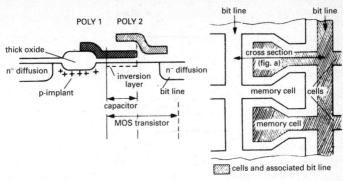

a Cross section of memory cell (not to scale)   *b* Schematic top section showing
                                                     memory cell packing

ous layers that make up the array. A small section is shown schematically in Fig. *b*;
Fig. *c* shows a plan of eight memory cells with parts of the various layers emphasized.

A particular storage location is selected by applying a high voltage level to the word
or address line, which is connected to the gate electrodes of all the transistors in the
column. This causes these switch transistors to be 'on'. Each memory cell is connected
to a different bit line, which in turn is connected to a sense amplifier. The stored charge
produces an appropriate output level on the chosen bit line by charge sharing. Built-

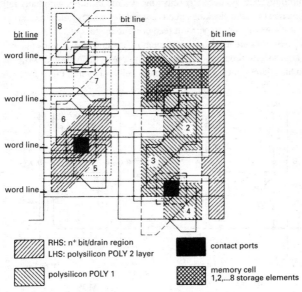

c Schematic top section of 8 bit section of MOSRAM memory

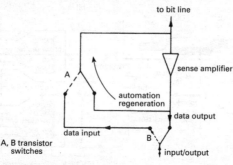

d Read-write circuit

in positive feedback is provided by a suitable arrangement of switch transistors (Fig. *d*) so that the condition of the selected storage capacitor is automatically regenerated following a read operation, i.e. logical 1 or logical 0. Leakage of charge occurs during the periods when the storage location is inactive and the data must therefore be regenerated periodically. This is achieved by periodically performing a read-write cycle automatically on every storage location. Information is input to the memory by switch transistors that produce the correct voltage levels for charge to be stored (or not stored) at a particular site on the memory.

A memory chip contains various logic circuits in addition to the memory cells and sense amplifiers: address decoders select the desired location; automatic refresh circuits regenerate the data; clocking circuits control the various functions and operate the switch transistors in order to perform read or write operations.

Static memories are used as ▶cache memory in computer systems that require extremely high speeds of operation. They usually consist of an array of ▶flip-flops connected to the address lines. A typical memory cell is shown in Fig. *e* for both bipolar

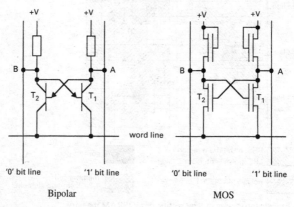

e Basic static memory cell

and MOS circuits. In the latter case extra MOSFETs may be used as the load resistors. A high logic level (logical 1) at points A causes the transistors T_2 to be 'on' and the voltage at points B therefore drops. This in turn causes transistors T_1 to be 'off', which maintains the high voltage level at points A and the circuit is latched in that state. A low logic level at points A causes the reverse situation and transistors T_1 are 'on'. Power is continuously dissipated since one or other of the transistors is always conducting. The data is read using sense amplifiers connected to the bit lines.

solid-state physics The branch of physics that studies the structure and properties of solids and any associated phenomena. Properties and associated phenomena that are dependent on the structure of the solid include electrical conductivity, semiconduction, superconductivity, photoconductivity, the photoelectric effect, and field emission.

solid-state valve ➤vacuum electronics.

sonar Acronym from *so*und *na*vigation and *r*anging. A method of detecting and locating underwater objects that operates on a similar principle to ➤radar but the transmitted pulse is a burst of sound energy, usually ultrasonic, rather than radiofrequency waves.

A sonar system that is used to measure the depth of the sea bed by projecting the pulse vertically downwards is termed an *echo-sounder.* Any sonar system that detects submarines is termed *asdic.* One type of asdic employs *echo-ranging,* in which the location of the target is deduced from the time difference between the two echoes from simultaneously transmitted sonic and ultrasonic pulses.

The specially designed electro-acoustic ➤transducer used to transmit the pulse in a sonar system is known as an *underwater sound projector*; the return echo is detected with a hydrophone.

sonde ➤radiosonde.

sonogram *Syn. for* spectrogram. ➤spectrograph.

sonograph ➤spectrograph.

SOT *Abbrev. for* small-outline transistor package.

sound carrier ➤television.

sound-level meter A device that gives a reading which relates to the perceived ➤loudness of a sound. This is achieved by frequency weighting a ➤sound pressure level measurement to compensate for the ear's sensitivity variation as a function of frequency.

sound pressure level (SPL) The ➤root-mean-square (r.m.s.) pressure of a sound wave at a particular instant (P_{actual}), usually expressed in ➤decibels (dBSPL) with respect to the average threshold of hearing at 1 kilohertz (P_{ref}) of 20 micropascal:

$$dBSPL = 20 \log_{10}(P_{actual}/P_{ref})$$

sound recording ➤recording of sound.

sound reproduction ➤reproduction of sound.

soundtrack ➤recording of sound.

sound wave ➤acoustic wave.

source 1. The point in a vector field at which lines of flux originate; an example is a positive electric charge in an electrostatic field. A point at which lines of flux terminate is a *sink*. **2.** Any device that produces electrical energy, e.g. a current source. **3.** The electrode in a ➤field-effect transistor that supplies charge ➤carriers (holes or electrons) to the interelectrode space.

source follower ➤emitter follower.

source impedance The ➤impedance presented to the input terminals of an electronic circuit or device by any source of electrical energy. The source impedance of an ideal voltage source is zero (i.e. $dV_s/dI_s = 0$) whereas that of an ideal current source is infinity (i.e. $dV_s/dI_s = \infty$).

space charge In any device, a charge density that is significantly different from zero in any given region. The region containing the space charge is a *space-charge region*. In a ➤semiconductor it is the region containing the depletion layers associated with the junction between two dissimilar conductivity types. In a ➤thermionic valve the space-charge region surrounds the cathode and contains electrons not immediately attracted to the anode. These two examples of space-charge regions can exist in the devices in equilibrium under conditions of zero applied bias; they constitute potential barriers that must be overcome, when bias is applied, before the device can conduct.

Space charge also causes divergence of a beam of electrons and ➤debunching in ➤velocity-modulated tubes. A radial field is often used to counteract the space-charge divergence of an electron beam and results in a cylindrical beam.

space-charge density The net electric charge per unit volume in a space-charge region (➤space charge).

space-charge limited region ➤thermionic valve.

space-charge region ➤space charge.

space diversity ➤diversity system.

space wave A radiowave that travels between a transmitting and a receiving antenna situated above the ground and that includes the ➤direct wave and the ➤ground-reflected wave. It is the component of the ➤ground wave that does not travel along the surface of the earth. If the two antennas are placed at a sufficient height above the ground the ➤surface wave is negligible and only the space wave needs to be considered.

s-parameters ➤scattering parameters.

spark A visible disruptive discharge of electricity between two points of high potential difference, preceded by ionization of the path. A sharp crackling noise occurs because of the rapid heating of the air through which the spark passes. The distance travelled is determined by the shape of the electrodes and the potential difference between them, and is not necessarily the shortest possible path.

Under specified conditions the distance between the electrodes is termed the *spark gap*. Specially designed electrodes are used to produce a spark over a given spark gap under particular conditions, as for ignition purposes in an internal-combustion engine. The insulation is self-restoring when the potential across the spark gap falls below that required to produce the spark. A spark is of much shorter duration than an ➤arc.

spark gap ➤spark.

spark interference ➤noise.

speaker *Short for* loudspeaker.

specific conductance *Obsolete syn. for* conductivity.

specific contact resistance ➤contact.

specific resistance *Obsolete syn. for* resistivity.

spectral analysis A ➤frequency domain representation of a signal in contrast to its ➤time domain representation, hence the conversion of a signal into its spectral characteristics (➤spectrum).

spectral characteristic A graph that shows the sensitivity or relative output as a function of frequency for any frequency-dependent device, circuit, or other equipment.

spectral response The range of frequencies that a given ➤transducer, circuit, or system will produce or react to when exposed to a wider range of input frequencies. The variation of amplitude or phase of the spectral response is generally expected here.

spectrogram ➤spectrograph.

spectrograph *Syn.* sonograph. In signal processing, a device that carries out a time, frequency, amplitude analysis of signals such as speech or music. The output from a spectrograph, known as a *spectrogram*, plots frequency on the Y-axis, time on the X-axis, and energy as the darkness or colour of marking. The trade-off between frequency and time accuracy is controlled by means of the analysis-filter bandwidth. This is often described as being 'wide' (for good time accuracy) or 'narrow' (for good frequency accuracy), these descriptions being with respect to the fundamental frequency of the signal under analysis. The output plot is known as a *wideband* or *narrowband spectrogram* respectively.

spectrum The range of possible frequencies that a particular (electrical) signal can have. For example, the audio spectrum is generally considered to extend from 20 hertz to 20 kilohertz, so a given audio signal will be found in this range, and a given instrument will have its own spectrum of frequencies or ➤spectral response within this range.

spectrum analyser An instrument that measures the frequency components contained within a given signal. Any complex periodic signal can be decomposed into a set of single-frequency sinusoidal components, as shown by ➤Fourier analysis; the spectrum analyser plots the amplitude versus frequency of each sine wave in the signal's ➤spectrum. The ➤time domain and ➤frequency domain representations of a simple signal are shown in the diagram. The spectrum analyser displays the frequency domain response.

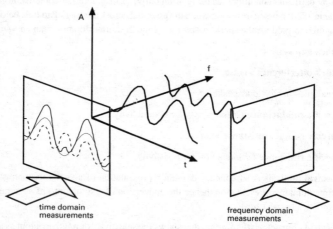

time domain
measurements

frequency domain
measurements

Relationship between time and frequency domains

Spectrum analysers use the heterodyne principle (➤heterodyne reception) to obtain the frequency domain response. Several mixing stages and intermediate frequencies are generally employed. The first local oscillator is a voltage-controlled oscillator driven by a sawtooth control voltage; this sweeps the oscillator across the frequency range desired for the frequency response measurement. The sawtooth control voltage also drives the x-display or timebase signal of the spectrum analyser CRT (➤cathode-ray tube) display, so the x-axis of the CRT corresponds to the frequency. When the first local oscillator is at such a point in its sweep that mixing with the incoming signal produces a response at the first intermediate frequency (➤mixer), then a signal appears at the appropriate location on the CRT display. Subsequent IF stages define the sensitivity and bandwidth of the displayed signal, the final stage defining the resolution of the analyser. This final filter stage can be varied over a wide range of bandwidths. The final displayed signal is a function of the resolution bandwidth, which in turn depends upon the swept frequency range and sweep time (period of the sawtooth control waveform). Modern spectrum analysers often use ➤digital filtering for the narrowest resolutions, as this can improve the speed of response at narrow bandwidths.

speech recognition device A device that produces orthographic text from the human ➤acoustic pressure waveform.

speed of light Symbol: *c*. The speed at which light and other ➤electromagnetic radiation travels in a vacuum. It is a universal constant with the defined value of 299 792 458 metres per second.

SPICE model ➤transistor parameters.

spike ➤pulse.

spin The intrinsic angular momentum possessed by all elementary particles, including the electron and proton. An atomic electron also possesses angular momentum as a result of its orbital motion. Spin is a quantized quantity, the *spin quantum number* (or *spin*), *s,* of particles having either a half integral or integral value. Both electron and proton have spin ½.

The magnetic fields produced by an atomic electron's spin and orbital motion interact in such a way that the electron can exist in either of two closely spaced energy levels. The spin in one level is +½ and in the other it is –½. Because of its spin an electron has an intrinsic ➤magnetic moment.

spiral inductor An inductor component in ➤monolithic microwave integrated circuits, made using the interconnect ➤metallization in the form of a straight-sided spiral with an ➤air bridge to connect to the centre of the spiral. Typical values of inductance range from 2 to 20 nanohenrys.

spiral scanning ➤scanning.

SPL *Abbrev. for* sound pressure level.

s-plane ➤s-domain circuit analysis.

spontaneous emission ➤laser.

spot 1. An area, such as on the screen of a cathode-ray tube, that is immediately affected by an electron beam impinging on or near to it. **2.** A local imperfection on the surface of an electrode.

spot speed 1. The product of the number of spots in the scanning line and the number of scanning lines per second in a television picture tube. **2.** The number of spots scanned or recorded per second in ➤facsimile transmission.

spreading 1. ➤Dispersion of a telecommunication signal, particularly along ➤optical fibres. **2.** Geometric divergence, usually of an airborne signal associated with path loss (➤Friis transmission equation).

spreading resistance When current flows through a small-area contact or ➤point contact into a conductor or semiconductor of much larger dimensions, the charge carriers will rapidly spread out from the contact to produce a uniform current density in the material. In the region of uniform current density the resistance of the material can be described by Ohm's law. In the region where the current flow is spreading from the point, the resistance is the spreading resistance. It is often included in the ➤contact resistance.

spread spectrum A form of digital modulation in which the frequency of the carrier for a modulated signal is rapidly switched between a number of predetermined possible frequencies (rather than a fixed allocation of frequencies). A number of transmitters may then share the same range of frequencies. ➤➤digital communications.

spurious rejection The ability of a radio receiver to reject signals that appear to be on the desired tuned channel, when this is not in fact the case. Spurious signals can arise in ►heterodyne receivers as a result of the mixing processes involved in the reception of signals.

sputter etching ►sputtering.

sputtering The action of dislodging particles from the surface of a solid by striking the surface with energetic heavy particles. The mechanism is one of momentum exchange, requiring the striking particles to be massive and travelling at high velocity. Uncharged inert particles, such as argon atoms, are commonly used. The argon atoms are accelerated towards the target from a glow discharge, being neutralized on the way. Accelerating voltages of a few hundred volts to several kilovolts are common. The process can be used for *sputter etching* features on semiconductor ►integrated circuits. The characteristics of the etching are almost uniform etching rate, independent of the sample material, and highly anisotropic features can be realized. The process can also be used for thin *film sputter deposition*, dislodging material from a target so that it is redeposited on a sample close by. This deposition process is very versatile, as alloys can be deposited without changing their composition in the process, and even dielectric and insulating layers can be deposited.

By using a reactive gas instead of inert argon, *reactive sputtering* is carried out, giving a wider range of etching and deposition processes for semiconductor device and integrated circuit manufacture.

square-law detector A ►detector that produces an output current that is proportional to the square of the input voltage. A ►diode detector is a square-law detector.

square wave A ►pulse train that consists of rectangular pulses the mark-space ratio of which is unity (see diagram).

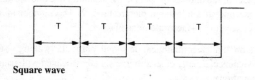

Square wave

square-wave response In general, the response of any electronic circuit or device to a square-wave input signal, given as the wave shape and peak-to-peak amplitude of the output signal.

In particular, it is the peak-to-peak amplitude obtained from a television ►camera tube in response to a test pattern of alternate black and white bars.

squegging oscillator A type of oscillator in which the main oscillations output from it alter the electrical conditions of the oscillator circuit so that the output amplitude periodically builds up to a peak value and then falls to zero. The ►blocking oscillator is a special type of squegging oscillator and is frequently used as a ►pulse generator for use, for example, in ►radar systems.

squelch circuit *Syn.* quieting circuit. A circuit used in a radio receiver that suppresses the audiofrequency output unless an input signal of a predetermined character is received. ➤automatic gain control.

SRAM *Abbrev. for* static RAM. ➤solid-state memory.

S-R flip-flop *Syn. for* R-S flip-flop. ➤flip-flop.

SSB *Abbrev. for* single sideband (transmission).

SSI *Abbrev. for* small-scale integration. ➤integrated circuit.

SS/TDMA *Abbrev. for* satellite-switched time-division multiple access. In satellite communications, the use of different antennas to direct a ➤TDMA signal to different geographic regions. A TDMA signal received by the satellite from a single ground station is cyclically connected to one of a number of different antennas, each directed to a different geographic region. The channels can be bidirectional, allowing the satellite to service the communication needs of a wider area than a single satellite.

stabilization The provision of negative ➤feedback in a circuit, such as an ➤amplifier, that contains inductance and capacitance and introduces substantial ➤gain between the input and the output, in order to provide overdamping (➤damped) in the circuit and thus prevent any oscillations occurring. *Shunt stabilization* is a type of stabilization in which the amplifier and the feedback circuit are connected in parallel. *Series stabilization* is provided by a feedback circuit connected in series with the amplifier. ➤compensator.

stable circuit A circuit that does not produce any unwanted oscillations over the entire dynamic operating range.

stable oscillations Oscillations that tend to decrease in amplitude with respect to time.

stage *Short for* amplifier stage.

staggered antenna *Syn. for* endfire array. ➤antenna array.

stagger tuned amplifier ➤tuned amplifier.

standard cell An electrolytic ➤cell that is used as a voltage reference standard. ➤Clark cell; Weston standard cell.

standardization 1. The process of relating a physical magnitude or the indicated value of a measuring instrument, such as a voltmeter, to the primary standard unit of that quantity. 2. A nationally or internationally agreed system for the standardization of electronic or electrical components or devices. If electronic devices, etc., are produced haphazardly by different manufacturers the ease of use of such components is severely limited. The convenience of the user is greatly increased when there are several sources from which devices may be chosen and when the devices are interchangeable. Interchangeability is greatly increased if the devices are produced only in certain physical sizes and in a range of predetermined values that are easily identifiable, usually by a ►colour code.

These values are the *preferred values* and are chosen by certain technical bodies although other values for special applications are not precluded. Broadcast transmissions and other means of international telecommunications are also greatly facilitated by international agreement on the frequency bands to be used and on harmonization of equipment between the sender and receiver.

The main international bodies that determine standards for telecommunications are the *International Telecommunication Union* (ITU) with its two associated international committees: the *International Telegraph and Telephone Consultative Committee* (CCITT) and the *International Radio Consultative Committee* (CCIR). The principal international standardizing bodies for electronic components and parts are the various technical committees of the *International Electrotechnical Commission* (IEC), with some broad areas defined by the *International Standards Organization* (ISO).

In the United Kingdom the body that determines national standards is the *British Standards Institute* (BSI), acting on advice from its various technical committees. The *Institution of Electrical Engineers* (IEE) is concerned with standardization for electronic equipment.

In the United States there are various bodies and committees that determine standards. These include the Electronic Industries Association (EIA), the Institute of Electrical and Electronics Engineers, Inc. (IEEE), the Joint Electron Device Engineering Council (JEDEC), the American National Standards Institute (ANSI), the National Institute of Standards and Technology (NIST), the National Aeronautical and Space Administration (NASA), the Federal Communications Committee (FCC), and the National Television System Committee (NTSC).

standing wave *Syn.* stationary wave. A wave that remains stationary, i.e. the displacement at any given point is always the same and a given displacement, such as that of a node, is not propagated along the wave. Standing waves result from the superimposition of two or more waves of the same period and usually occur when a wave is reflected totally or partially from a given barrier. ►travelling wave.

In ►distributed circuits any ►mismatch of impedances will cause reflections of the voltage or current waves at the mismatch. The superposition of the incident and reflected voltage or current waves will result in a standing wave. If the amplitudes of the incident and reflected waves are the same then the standing wave will have a maximum amplitude of twice the amplitude of the individual waves; it will also possess nulls – points of zero amplitude – where the two waves exactly cancel. These points are located at every half wavelength of the original waves. In general, the incident and reflected waves will have different amplitudes, as some of the incident energy will be absorbed at the mismatched impedance. In this case the amplitude of the standing wave will be less than the maximum value, and the minimum will be nonzero, as the waves will no longer cancel exactly. In this situation the *standing-wave ratio* is defined as the ratio of maximum to minimum values of the standing wave voltage or current, and it is a measure of the impedance mismatch. The *voltage standing-wave ratio* (*VSWR*) is equal to

$(1 + \Gamma)/(1 - \Gamma)$

where Γ is the ▶reflection coefficient of reflected to incident wave amplitudes due to the impedance mismatch.

standing-wave ratio ▶standing wave.

star circuit *Syns.* T (or tee) circuit; Y (or wye) circuit. A configuration of three impedances arranged as shown in the diagram.

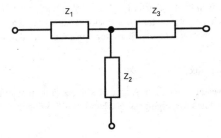

Star circuit

star-delta transformation *Syn.* tee-pi (T-π) transform. The transformation of the ▶star circuit configuration to the ▶delta circuit configuration, or vice versa, using the formulas given in the diagram.

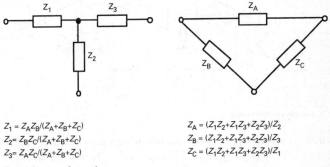

$Z_1 = Z_A Z_B/(Z_A + Z_B + Z_C)$

$Z_2 = Z_B Z_C/(Z_A + Z_B + Z_C)$

$Z_3 = Z_A Z_C/(Z_A + Z_B + Z_C)$

$Z_A = (Z_1 Z_2 + Z_1 Z_3 + Z_2 Z_3)/Z_2$

$Z_B = (Z_1 Z_2 + Z_1 Z_3 + Z_2 Z_3)/Z_3$

$Z_C = (Z_1 Z_2 + Z_1 Z_3 + Z_2 Z_3)/Z_1$

Star-delta transformation

star network In digital communication ▶networks, a configuration that consists of a single central hub node with various terminal nodes connected to the hub.

starter electrode An auxiliary electrode that is used in a glow-discharge tube (▶gas-discharge tube) in order to initiate the glow discharge; it is also used in an arc-discharge tube to initiate the arc discharge. A sufficiently high potential difference is applied between the starter electrode and the cathode and once the glow discharge is established the necessary voltage is applied to the anode. The maintaining voltage is considerably lower than the voltage required to initiate the discharge and this method of operation protects the main circuit from excessively high voltages. The starter electrode may also be used to maintain the discharge during a period when the main anode is 'off'.

starting current The magnitude of the current in an ►oscillator at which self-sustaining oscillations are initiated for specified load conditions.

start-oscillation current ►travelling-wave tube.

start-stop apparatus A device used in ►telegraphy in which each set of coded signal elements corresponding to a transmitted character is preceded by a start signal that activates the receiving apparatus in preparation for the character and is followed by a stop signal that brings the receiving apparatus to rest. Start-stop apparatus is used in ►telex systems. ➤teleprinter.

state diagram ►finite-state machine.

state variable filter *Syn.* universal active filter. A type of ►biquad filter that simultaneously provides second-order low-pass, high-pass, and band-pass filtering in one network (see diagram).

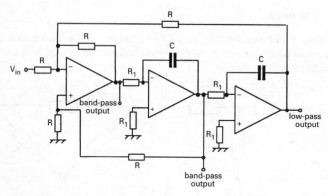

State variable filter

static 1. Unwanted random noises produced at the loudspeaker of a ►radio receiver. It is usually heard as an unpleasant crackling sound and is caused by atmospheric conditions, usually by the presence of static electricity in the air through which the radiowave is propagated. **2.** ►dynamic.

static characteristic ►characteristic.

static memory A ►solid-state memory that does not require refresh operations in order to retain the stored information. ➤dynamic memory.

static RAM ►solid-state memory.

stationary orbit ►geostationary earth orbit.

stationary state The state of an atom or other system that in the quantum theory or quantum mechanics is described by a given set of quantum numbers. Each of the various energy states that may be assumed by an atom is a stationary state of that atom.

stationary wave ➤standing wave.

stator ➤motor.

status register A ➤register in the ALU (arithmetic and logic unit) of a computer that contains ➤flags on the ALU status (e.g. carry, overflow, negative result, and zero result).

STD *Abbrev. for* subscriber trunk dialling. The name given to the UK's national subscriber telephone dialling system.

steady state A state reached by a system under steady operating conditions after any transient effects resulting from a change in the operating conditions have died away.

steepness factor A means of assessing the requirements of a ➤filter from the filter response specification. The steepness factor A_S for a low-pass filter response is given by the ratio f_S/f_c, where f_S is the frequency having the minimum required stop-band attenuation and f_c is the ➤cut-off frequency of the pass-band. Once the steepness factor is calculated from the required filter response, it is compared with published ➤normalized filter curves and a design is selected that meets or exceeds the requirements.

steerable antenna *Syn.* musa (acronym from *m*ultiple *u*nit *s*teerable *a*ntenna). A ➤directional antenna that consists of several fixed units in which the direction of maximum sensitivity (i.e. the direction of the major lobe of the ➤antenna pattern) can be altered by adjusting the phase relationship between the units from which it is formed.

step-down transformer ➤transformer.

step function ➤unit step function.

step-recovery diode *Syns.* snap-off diode; snapback diode. A p-n junction diode in which ➤carrier storage is the major factor contributing to the operation of the device. The diode is designed so that most of the injected minority carriers under forward bias are stored near the junction and are immediately available for conduction when reverse bias is applied. When the diode is switched from forward to reverse bias the diode conducts in the reverse direction for a short time interval then the current is abruptly cut off when all the stored charges have been dispelled. The diode therefore remains in a low-impedance state until the cut-off occurs. The reverse voltage then builds up rapidly at a rate determined by the reverse junction capacitance and the external circuit. The cut-off occurs in the range of picoseconds and results in a fast-rising voltage wavefront that is rich in harmonics. The diode is therefore used as a harmonic generator or as a pulse former.

Most step-recovery diodes are fabricated in silicon with relatively long minority-carrier lifetimes ranging from 0.5 to 5 microseconds. ➤fast-recovery diode.

step-stress life test A ➤life test consisting of several stress levels applied sequentially for periods of equal duration to one sample. During each period a stated stress level is applied and increased from one step to the next.

step-up transformer ➤transformer.

stereophonic sound reproduction ➤reproduction of sound; radio receiver.

stimulated emission ➤laser.

stochastic process Any process in which there is a random element.

Stokes' law ➤luminescence.

stop band ➤filter.

stopper 1. ➤parasitic oscillations. **2.** ➤channel stopper.

stopping potential ➤photoelectric effect.

storage battery A ➤battery that is formed from secondary ➤cells.

storage capacity *Syn. for* memory capacity. ➤memory.

storage cathode-ray tube ➤storage tube.

storage cell *Syn. for* secondary cell. ➤cell.

storage device Any device that holds information by a physical or chemical method. The term is particularly applied to a ➤memory in a computer, which holds information usually in binary form.

storage time 1. *Syn.* retention time. In any device that stores information, such as a ➤storage device or ➤storage tube, the maximum time that the information may be stored without significant loss of information. **2.** ➤carrier storage.

storage tube An electron tube that is used to store information for a determined and controllable time and from which the information may be extracted as required. Various principles are used to operate storage tubes and there are many different types of tube. The most common types of tube in general use are *charge-storage tubes* in which the information is stored as a pattern of electrostatic charges. The information may be extracted as a visual display, as in the storage cathode-ray tube, or as an electronic signal.

Charge-storage tubes contain a target plate on which the information is stored. The information to be stored is used to modulate the intensity of an electron beam and the beam is made to scan the target. ➤Secondary emission of electrons from the target plate occurs and an electrostatic charge pattern is left on the target. Each small area of the target that retains information and is distinguishable from neighbouring small areas forms a *storage element*. The number of secondary electrons emitted from each storage element is a function of the energy of the electron beam, the tube design, and the intensity of the beam. If the unmodulated beam produces a ➤secondary-emission ratio greater than unity a positive-charge image is produced; if it is less than unity a negative-charge image results. The information-modulated electron beam is called the *writing beam*; the rate at which information can be written to successive elements is the *writing speed*.

Storage cathode-ray tubes produce a visual display of controllable duration. The tube has two ►electron guns – the *writing gun* and the *flooding gun;* it also has a phosphor viewing screen and two fine mesh metal screens. One of the metal screens, the *storage screen,* is coated with a thin dielectric material to form the target and the other serves as an electron collector. A positive charge image is produced on the storage screen by scanning with a high-resolution intensity-modulated writing beam from the writing gun. It remains until it decays or is erased. Information is extracted by *flooding* the storage screen with an electron beam from the flooding gun. Each storage element effectively forms an elemental electron gun with each mesh hole forming a control element of one of the guns. The value of positive charge deposited at each aperture determines the amount of current (from the flooding beam) that can pass through to the phosphor viewing screen. At the viewing screen a light output is produced that is a function of the original information. The stored charges are erased by flooding the storage surface with low-velocity electrons, thereby depositing negative charge on each element, until the pattern is erased and the surface prepared for storing a new image.

Photoconductive storage tubes depend for their operation on ►photoconductivity or the related electron-bombardment conductivity, in which the conductivity of a material is temporarily increased when exposed to bombardment by light photons or electrons. The target consists of a back-plate electrode coated with a thin layer of photoconductive material. The information is deposited on the target either by exposing it to a light image or by scanning with a high-resolution intensity-modulated light beam or electron beam. The information is extracted by scanning the target with an unmodulated reading beam; the numbers of electrons, i.e. the signal intensity, reaching the back-plate electrode depends on the conductivity of each small element. The output is produced as a current in a load resistor in series with the back-plate electrode.

store ►memory.

STP *Abbrev. for* shielded twisted pair. ►shielded pair.

strain gauge An instrument that measures strain at the surface of a solid body by means of changes in the electrical properties of associated circuits. There are several types of strain gauge.

A *resistance strain gauge* consists of a fine wire attached to the surface. The length of the wire is altered by the strain, causing an associated variation in the resistance.

An *electromagnetic strain gauge* consists of a small soft-iron armature attached to the surface and free to move inside a fixed inductive coil. Movement of the surface alters the position of the armature causing an associated variation in the inductance of the coil.

A *variable capacitance strain gauge* consists of a parallel plate capacitor that has one fixed plate and the other attached to the surface. The strain causes the separation of the plates, and hence the capacitance, to be varied.

Piezoelectric or *magnetostriction strain gauges* have a suitable piezoelectric or magnetostrictive crystal attached to the surface. The strain is measured by the associated piezoelectric or magnetostrictive changes that it induces.

stray capacitance Any capacitance in an electronic circuit or device that is due to interconnections, electrodes, or the proximity of elements in the circuit and is additional to the intentional capacitance of the circuit or device. Stray capacitance is usually unwanted but can sometimes be utilized as part of the tuning of a ➤tuned circuit.

string electrometer An electrometer that consists of a fine metallized quartz fibre stretched between two parallel conducting plates. The plates are oppositely charged by the potential difference to be measured and the deflection of the quartz fibre is observed.

string galvanometer ➤Einthoven galvanometer.

stripline A generic term for a microwave circuit medium where the microwave signal flows as a wave along one or more conductors supported by a dielectric medium (which can be air). The stripline is a ➤distributed circuit, possessing reactance as a part of its structure. Practical examples of striplines include the ➤microstrip and ➤coplanar waveguide.

strong electrolyte ➤electrolyte.

stub A device used for ➤impedance matching of a ➤transmission line to a ➤load in microwave and ultrahigh-frequency applications. It consists of a short section of transmission line of similar properties to the line to be matched. The position and length of the stub is adjusted to give optimum energy transfer.

subcarrier A ➤carrier wave that is used to modulate a second different carrier.

subcarrier modulation ➤telegraphy.

subharmonic ➤harmonic.

subroutine ➤program.

subscriber's line In telephone or telecommunication systems, the communications line or channel that connects the user's equipment to the exchange.

subscriber station ➤telephony.

subsonic frequency A frequency of value less than the lower limit of the human hearing range, i.e. frequencies below about 20 hertz.

substation A complete assemblage of plant, equipment, and the necessary buildings at a place where electrical energy is received and where it may be either converted from alternating current to direct current, stepped up or stepped down by means of transformers, or used for control purposes. The substation usually receives power from one or more ➤power stations.

substrate A single body of material on or in which circuit elements or ➤integrated circuits are fabricated. The substrate may be passive, as with a printed circuit board, or active, as with a bulk semiconductor.

subsystem An interconnected set of related circuits (which may be integrated circuits) that form a subdivision of a piece of electrical equipment or an operational system and may be manufactured as an assembly or subassembly.

subtractive synthesis ➤synthesis.

summation instrument A single instrument that receives signals, such as current, power, or energy, from a number of separate circuits and measures the aggregate.

summing amplifier An ➤operational amplifier configuration (see diagram) that has a number of inputs, v_1, v_2, v_3, ..., v_n, and whose output is given by

$$v_o = -((R_f v_1)/R_1 + (R_f v_2)/R_2 + (R_f v_3)/R_3 + ... + (R_f v_n)/R_n)$$

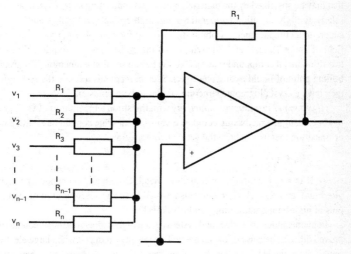

Summing amplifier

sum-of-products A ➤logic circuit in which input devices are AND gates and output devices are OR gates; it thus has the form AND-to-OR. This represents ➤Boolean algebra functions of the form

$$ABC'D + ABCD' + A'BC'D + AB'CD$$

expressing a sum of product terms, i.e. an OR of AND terms, containing complemented (negative) or uncomplemented variables (' after a variable denoting the complemented form).

superconductivity A phenomenon that occurs in certain metals and a large number of compounds and alloys when cooled to a temperature close to the absolute zero of thermodynamic temperature. At temperatures below a critical *transition temperature,* T_c, the electrical resistance of the material becomes vanishingly small and the material behaves as a perfect conductor. Currents induced in superconducting material have persisted for several years without significant decay.

The material also exhibits perfect ➤diamagnetism in weak magnetic fields: the ➤magnetic flux density inside the material is zero. If the material is in the form of a hollow cylinder, the magnetic flux density contained in the hollow region remains constant and trapped in the state existing at the transition temperature, while the flux density within the material becomes zero. These magnetic effects are termed *Meissner effects.* If the value of applied magnetic flux density rises to a value greater than a critical value, $\boldsymbol{B}_c$, the superconductivity is destroyed. The value of $\boldsymbol{B}_c$ – the *transition flux density* – is a function of the temperature of the material and its nature. A superconducting current in the material itself can produce an associated magnetic flux density greater than the critical value; there is therefore an upper limit to the current density that may be sustained by the material in the superconducting state. Certain alloys have relatively high transition temperatures and high critical field values and are used in superconducting magnets. Niobium-tin (Nb_3Sn) for example can produce a magnetic field of about 12 tesla at 4.2 kelvin, i.e. at the boiling point of liquid helium. Other transition metal compounds such as Nb_3Ge have transition temperatures at around the boiling point of liquid hydrogen (20 K). Such compounds fall into the Al5 series, i.e. they have a crystallographic structure similar to beta-tungsten.

High-temperature superconductivity with transition temperatures of 90 K or above has been demonstrated using complex ceramic oxides that contain rare earth elements or transition metals and have the general composition

$RBa_2Cu_3O_{1-7}$

where R is a rare earth ion or transition metal. Scandium, lanthanum, neodymium, ytterbium, and several other elements have all been successfully used to obtain samples of high-temperature superconductors (HTS).

Implementation of devices and systems using high-temperature superconductors has proved difficult because of the magnetic flux density produced when large electric currents flow in the HTS. Although a strong magnetic field destroys the superconductivity and a weak field cannot penetrate the material, at intermediate field strengths the field penetrates the HTS in thin tubes running across the material. In the core of each tube is a 'tornado' of electric current, and the tubes are thus known as *vortices*. Outside each vortex the material remains superconducting but within each vortex the superconductivity is destroyed. As the magnetic field-strength increases so do the number and density of the vortices. The current applied to the superconductor causes the vortices to move, which dissipates electrical energy, producing resistance. Vortices act like atoms and can form into solid or liquid states of *vortex matter* depending on the temperature and magnetic field strength. Locking vortices in place, for example by introducing crystal defects to trap them or by twisting vortices around one another, leads to HTS with better electrical properties.

One of the most successful theories of superconductivity at the lowest temperatures was given in 1957 by Bardeen, Cooper, and Schrieffer. This is the *BCS theory* in which electron pairs – *Cooper pairs* – can form in the presence of other electrons. The pairing results from interactions between the electrons and the quantized vibrations of the crystal lattice – phonons – and produces a highly ordered state with no dissipation of energy in the electron movement. The BCS theory, however, proposes a singlet pair state. At higher temperatures this is not sufficient to fully explain the phenomenon and various ideas have been proposed to explain the postulated attractive interactions between electrons and the configuration of the electron pairs. An understanding of the theoretical basis for superconductivity will be determined by careful analysis of data from experiments carried out on single crystal samples, and should help the search for materials with even higher critical temperatures.

The ►Josephson effect occurs when an extremely thin layer of insulating material is introduced into a superconductor. A current, below a certain critical value, can flow across the insulator in the absence of an applied voltage.

superconductor A material that exhibits ►superconductivity.

superheterodyne reception ►Heterodyne reception in which the incoming frequency is higher than the audiofrequency range.

superhigh frequency (SHF) ►frequency band.

Supermalloy *Trademark* An alloy of iron, nickel, and molybdenum that has a high magnetic permeability at low values of magnetic flux density and low hysteresis loss.

Supermendur *Trademark* A ferromagnetic material used as a magnetic ►core. It exhibits a substantially rectangular ►magnetic hysteresis curve.

superposition ►linear system.

super-regenerative reception A method of reception used for ultrahigh-frequency radiowaves in which the ►detector is a ►squegging oscillator. The frequency at which the oscillations are quenched – the *quench frequency* – is a function of the frequency of the received radiowaves. Very large values of amplification can be obtained using this method of reception as a result of the positive ►feedback employed in the detector. Compared however to ►heterodyne reception the ►selectivity is relatively poor.

supersensitive relay An electromechanical ►relay that operates with currents of less than about 250 microamps.

suppressed-carrier modulation A method of transmission of radiowaves in which the carrier component of the modulated wave is not transmitted. One or both of the sidebands only are transmitted. Suppressed-carrier transmission is used in ►single-sideband transmission and ►double-sideband transmission. It requires a local oscillator at the receiver that regenerates the carrier frequency and mixes it with the received signal in order to detect the modulating wave. This method of detection is termed *synchronous detection.*

suppressed-zero instrument *Syn.* set-up scale instrument. A measuring or recording instrument in which the zero position falls outside the ►dynamic range of the instrument; the moving part is not deflected until a predetermined value of the measured signal is reached.

suppressor ►interference.

suppressor grid ►thermionic valve; pentode.

surface acoustic wave ►acoustic wave device.

surface-channel FET ►field-effect transistor.

surface charge density ►charge density.

surface conductivity The reciprocal of ►surface resistivity.

surface leakage A leakage current that results from the flow of charge at the surface of a material rather than in the bulk material.

surface mount technology A means by which circuit components are mounted on one or both surfaces of a circuit board and are held in place by solder between pads on the board and the component terminations (rather than using their leads, passing through holes in the board, to secure the component). Because lead holes are unnecessary, components are smaller and the net result is a considerable increase in component density per unit area. Board assemblies made in this manner are known as *surface mount assemblies* and lend themselves to fully automated assembly processes; they are thus well suited to high-volume production so assembly costs can be reduced.

surface resistance ►skin effect.

surface resistivity The resistance between two opposite sides of a unit square of the surface of a material. The measured value can vary greatly depending on the method of measurement.

surface wave A ►radiowave that travels along the surface separating the transmitting and receiving antennas. The surface wave is affected by the properties of the ground along which it travels. ►►ground wave; space wave.

surge An abnormal transient electrical disturbance in a conductor. Surges are produced from many sources, such as a lightning stroke, sudden faults in electrical equipment or transmission lines, or switching operations.

susceptance Symbol: B; unit: siemens. The imaginary part of the ►admittance, Y, which is given by

$$Y = G + jB$$

where G is the ►conductance. For a circuit containing both resistance, R, and reactance, X, the susceptance is given by

$$B = -X/(R^2 + X^2)$$

susceptibility 1. *Syn.* magnetic susceptibility. Symbol: χ_m. A dimensionless quantity given by

$$\chi_m = \mu_r - 1$$

where μ_r is the relative ➤permeability of a material. Magnetic susceptibility describes the response of a material to a magnetic field, being the ratio of ➤magnetization to ➤magnetic field strength:

$$\chi_m = M/H$$

χ_m is a tensor when M is not parallel to H, otherwise it is a simple number. For crystalline material χ_m may depend on the direction of the field with respect to the crystal axes because of anisotropic effects. It has a wide range of values: diamagnetic materials have a negative value; paramagnetic materials (➤paramagnetism) have a small positive value; ferromagnetic materials can have a very large variable value (up to about one), which is dependent on the magnetic field strength (➤magnetic hysteresis).
2. *Syn.* electric susceptibility. Symbol: χ_e. A dimensionless quantity given by

$$\chi_e = \varepsilon_r - 1$$

where ε_r is the relative ➤permittivity. Electric susceptibility measures the ease of polarization of a dielectric and is given by the ratio

$$\chi_e = P/\varepsilon_0 E$$

where P is the ➤dielectric polarization, E the electric field strength, and ε_0 the permittivity of free space.

SVGA ➤VGA.

sweep ➤timebase.

sweep frequency ➤timebase.

sweep generator *Syn. for* timebase generator. ➤timebase.

sweep voltage The voltage output from an internal or external ➤timebase that when applied to the appropriate deflector plates or coils of a ➤cathode-ray tube causes a horizontal or vertical deflection of the electron beam.

swell pedal A foot pedal used with electronic musical instruments that enables the output volume or other features of the sound to be controlled, generally by means of ➤MIDI. The term originates from pipe organs.

swinging choke ➤choke.

switch 1. A device that opens or closes a circuit. **2.** A device that causes the operating conditions of a circuit to change between discrete specified levels. **3.** A device that selects from two or more components, parts, or circuits the desired element for a particular mode of operation.

In general, a switch may consist of a mechanical device, such as a ►circuit-breaker, or a solid-state device, such as a ►bipolar junction transistor, ►Schottky diode, or ►field-effect transistor.

switched capacitor filter A type of ►filter in ►integrated-circuit form that relies on a device consisting of a capacitor and a switch (see diagram). The switching between v_{in} and v_o results in the charging and discharging of the capacitor, which creates an average current flow between v_{in} and v_o dependent on the value of the capacitance and thus resembles the action of a resistor. In ►MOS technology it is difficult to fabricate the large values of resistance required by some filter networks. The switched capacitor filter enables the replacement of the resistive element of RC active filter designs, allowing them to be fabricated as an integrated circuit.

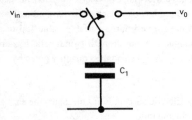

Switched capacitor filter

switched-mode power supply A power supply in which the incoming mains voltage is passed through a switch that is switched on and off at a very high frequency, which may be tens to thousands of kilohertz. This effectively produces a high-frequency waveform, which can then be transformed to lower voltage, rectified, and smoothed to produce a DC output voltage. The advantage over power supplies that operate at the mains frequency is that the ►transformers can be made much smaller, as the losses are much reduced at high frequency. The transformer cost is a significant fraction of the complete power supply, so using switched-mode techniques can offer considerable cost savings. Problems include high-frequency noise produced by the switching devices, which must be filtered.

switching system In telephone or telecommunication systems, that part of the system that diverts a call or digital message down a particular channel or physical line. An example of a switching system is a telephone exchange. ►►digital switching.

switching tube An arc-discharge tube that contains at least two anodes and is used as a switch. The current in the discharge flows to that anode held at the required voltage to sustain the arc; if the voltage is switched to a different anode the discharge path also moves. Switching tubes are used as transmit-receive switches and as scaling tubes.

syllable articulation score ►intelligibility.

symbols For symbols used in electronics, see the following tables in the back matter: Table 1 (graphical symbols), Table 4 (electric and magnetic quantities), Table 5 (fundamental constants), Tables 6–9 (SI units).

symmetric mode In two-wire mains plus earth transmission systems, symmetrical mode refers to signals that appear balanced, in antiphase, on the two mains lines; no signal component appears on the earth line. ➤asymmetric mode.

symmetric transducer ➤transducer.

symmetric two-port network ➤two-port network.

sync *Abbrev. for* synchronous or synchronizing signal or synchronism.

synchronism The relationship between two periodically varying quantities when they are in ➤phase.

synchronizing pulses ➤television.

synchronometer A device that counts the number of cycles of a periodically varying quantity that occur during a predetermined time interval.

synchronous *Syn.* clocked. Denoting any circuit, device, or system involved in computer control that is operated by means of clock pulses (➤clock). Sequential events take place at fixed times determined by the clock frequency. ➤asynchronous.

synchronous alternating-current generator *Syns.* alternator; synchronous generator. An alternating-current generator that consists of one or more coils that are made to rotate in the magnetic fields produced by several electromagnets excited by a direct-current source. The frequency, f, of the alternating currents and emfs induced in the coils is equal to the product of the speed, n_s, at which the coils rotate and the number of pairs of magnetic poles, p. This type of generator can operate and produce electrical power independently of any other source of alternating current and is the type most commonly employed in power stations.

synchronous clock A mains-operated electric clock in which the speed of the driving motor is a function of the mains frequency and therefore the time-keeping is controlled by the mains supply.

synchronous communications satellite ➤communications satellite.

synchronous computer A computer in which the timing of all operations is controlled by a ➤clock. Such a computer often employs *fixed-cycle operation* in which a fixed time is assigned in advance to each operation performed.

synchronous detection ➤suppressed-carrier transmission.

synchronous gate *Syn.* clocked gate. A ➤gate the output of which is synchronized to the input signal. The synchronizing signals may be derived from an independent clock so that the gate is operative during predetermined intervals of time; alternatively the input signal itself may be used as the trigger so that the gate operates only when an input signal is present.

synchronous generator ➤synchronous alternating-current generator.

synchronous logic A logic system that operates with *synchronous timing,* i.e. the timing of all the switching operations is controlled by ➤clock pulses. A logic system is said to be *asynchronous* when all the switching operations are triggered by a free-running signal so that successive stages or instructions are triggered by the completion of operation of the preceding stage.

Synchronous logic in general is slower and the timing more critical than asynchronous logic, but usually fewer and simpler circuits are required.

synchronous motor An alternating-current ➤motor in which a rotor comprising one or more pairs of fixed poles is made to rotate in the magnetic field produced by fixed coils in the stator excited by the alternating current. The magnetic field rotates at a constant angular velocity given by the frequency f of the exciting current and the number of coils n.

synchronous orbit ➤geosynchronous orbit.

synchronous timing ➤synchronous logic.

synchronous transmission In ➤digital communications, transmission of data where synchronization of a single clock source must be established between the end points of the communication link prior to data transmission. ➤➤asynchronous transmission.

synchrotron A cyclic particle ➤accelerator that is used to accelerate a beam of electrons – *electron synchrotron* – or protons – *proton synchrotron* – to very high energies. The particles travel along a circular evacuated tube of fixed radius: an applied magnetic flux density causes them to travel in the circular orbit. They are accelerated by a radiofrequency electric field applied across a gap in a metallic cavity inside the evacuated chamber. Acceleration also occurs as a result of electromagnetic induction within the acceleration chamber between the electric vector associated with the magnetic flux density and the beam current. In order to maintain the correct relationship between the magnetic flux density, the energy of the particles, and the radius of the chamber, the magnetic flux density is made to increase with time and the frequency of the applied r.f. field modulated. This prevents loss of synchronism due to the relativistic mass increase.

The particles are usually injected at high energies from a ➤linear accelerator in order to minimize the range of modulation of the r.f. field. An alternative method is to operate the machine without the r.f. field at lower energies; acceleration is then the result of the electromagnetic induction only.

sync separator ➤television.

synthesis In signal processing, the production of a signal by manipulating ➤time domain and/or ➤frequency domain elements. *Additive synthesis* is the generation of a complex signal by adding together individual sine-wave components. The resulting signal is usually a ➤periodic waveform and the sine-wave components are therefore harmonically related as integer multiples of the ➤fundamental frequency. *Subtractive synthesis* uses spectrally rich raw signals from which components are removed, usually by means of ➤filters. *FM synthesis* is a process that makes use of the ➤frequency modulation (FM) technique.

synthesizer 1. An electronic musical instrument, usually operated by keyboard and pedals, in which sounds are produced by, for example, additive, subtractive, or FM ➤synthesis, by ➤sampling acoustic waveforms, or by other modelling techniques. ➤➤attack decay sustain release. **2.** A laboratory instrument for generating single-frequency electrical signals. Frequency synthesizers are usually based upon ➤phase-lock loop techniques in either analogue or digital domains for precise definition of the desired output frequency.

systems software ➤software.

T

tandem A method of connecting two ►two-port networks so that the two output terminals of one network are connected to the input terminals of the other. ►►cascade.

tandem exchange ►telephony.

tangent galvanometer An obsolete moving-magnet instrument that has a small magnet in the centre of a large fixed coil. The magnet carries a long light nonmagnetic pointer that moves over a horizontal scale. The instrument must be set up correctly with the plane of the coil vertical and along the earth's magnetic meridian. The current is then proportional to the tangent of the angle of deflection.

tank network A ►resonant or tuned circuit, normally a parallel LC combination, whose centre frequency f_c, assuming small coil resistance, is approximated by the expression

$$f_c \approx 1/(2\pi\sqrt{(LC)})$$

tantalum Symbol: Ta. A metal, atomic number 73, that has an extremely high resistance to corrosion and is used for applications where this property is desirable.

tantalum capacitor ►electrolytic capacitor.

tap changer *Syn.* ratio adjuster. A device that alters the ratio of a ►transformer by selecting the desired ►tapping. One that is designed to be used only when the supply voltage is switched off is an *off-circuit tap changer.* If it is designed to operate when the supply voltage is switched on it is an *on-load tap changer,* although the transformer need not necessarily be on load for operation of the tap changer.

tape automated bonding A method used during the packaging of ►integrated circuits, usually when a large number of interconnections (over 100) is required between the chip and the ►leadframe. The leadframe is formed from plated copper on a strip of plastic (the tape), and extends to reach the ►bonding pads of the chip. Metal bumps are formed at the points to be connected to the bonding pads, and the pads themselves have metallic caps formed on them by ►sputtering. The chip is positioned on the tape and all the bonds formed simultaneously by thermal compression. The process is automated by producing the tape as a long strip of successive leadframes. The bonded assemblies are then encapsulated in moulded plastic and separated to produce the finished packages. The packages may have the form of ►pin grid arrays, ►leadless chip carriers, or ►dual in-line packages. ►►wire bonding.

tape recording ➤magnetic recording.

tapered window ➤windowing.

tape unit *Syn.* tape transport. ➤magnetic tape.

tapping A conductor, usually a wire, that makes an electrical connection with a point between the ends of a winding or coil. The number of turns included in the active portion of the coil can then be selected. More than one tapping may be made to a particular winding, such as that of a ➤transformer.

tap weight ➤digital filter.

target 1. ➤camera tube. **2.** *Syn.* anticathode. ➤X-ray tube. **3.** An object detected by a ➤radar or ➤sonar system.

target voltage The potential difference between the cathode and the signal electrode of a low-electron-velocity ➤camera tube. The minimum value that is required to produce a discernible video output is the target *cut-off voltage*.

Tchebyshev filter ➤Chebyshev filter.

T circuit ➤star circuit.

TDM *Abbrev. for* time-division multiplexing.

TDMA *Abbrev. for* time-division multiple access. ➤digital communications.

TDR *Abbrev. for* time domain reflectometer. ➤reflectometer.

TED *Abbrev. for* transferred-electron device.

tee circuit ➤star circuit.

tee-pi (T-π) transform ➤star-delta transformation.

Teflon *Trademark* Polytetrafluoroethylene (PTFE). An insulator that has an extemely high resistivity and is very resistant to moisture and temperature.

telecommunications The study and practice of the transfer of information by any electromagnetic means, such as wire or radiowaves.

telecommunication system The complete assembly of apparatus and circuits required to effect a desired transfer of information. Systems include television, radio, and telephony.

teleconference A communications system that allows a number of people at different geographic locations to hear, speak, see, and transmit digital information to each other. A *videoconference* is a conference where, traditionally, communication is limited to audio and visual, and an *audioconference* is a conference where communication is limited to audio only.

telegraphy Communication by means of a ➤telecommunication system that transmits documentary matter, such as written or printed matter or fixed images, and reproduces it at a distance. The matter is transmitted as a suitable signal code by means of wire or by radio (*radio telegraphy*). A telegraph network is a complete system of stations, installations, and communication channels that provides a telegraph service. ►Telex and ►fax are examples of telegraph services.

Telegraphy may be effected automatically or manually. A synchronous system is one in which the sending and receiving instruments operate at substantially the same frequency and are maintained with a desired phase relationship. Transmission may be in the form of a direct current applied to the line by the sending apparatus or a modulated carrier wave. Amplitude, frequency, or pulse-code modulation of the carrier can be used. *Subcarrier modulation* is a method that may be employed in radio telegraphy when a low-frequency carrier wave (the subcarrier) is frequency-modulated by the telegraph signal and then this modulated wave is used to modulate a second radiofrequency carrier wave.

telemeter ►telemetry.

telemetry Measurement at a distance. Data is transmitted over a particular telecommunication channel from the measuring point to the recording apparatus. A measuring instrument that measures a quantity and transmits the measured data as an electrical signal to a distant recording point is known as a *telemeter*. Space exploration and physiological monitoring in hospitals both require the use of telemetry.

telephone line ►telephony.

telephone set ►telephony.

telephone station ►telephony.

telephony Communication by means of a ►telecommunication system that is designed to transmit speech or sometimes data. A complete telephone system contains all the circuits, switching apparatus, and other equipment necessary to establish a communication channel between any two users connected to the main system. Communication between two points takes place along suitable cables (*telephone lines*) except where this is inappropriate; a particular access point may then be connected to the main system by means of a radio link (*radio telephone*), as in ship-to-shore telephony. ►cellular communications.

A *telephone set* is an assembly of apparatus that includes a suitable *handset* containing the transmitter and receiver, and usually a switch hook and the immediately associated wiring. A telephone set connected to a telephone system is a *telephone station;* one that has access to the public telephone network is a *subscriber station*. It is mainly large organizations that have a *private exchange* to interconnect telephone stations. Usually such an exchange is also connected to the public telephone system and is known as a *private branch exchange:* it is either a *PABX* (private automatic branch exchange) or *PMBX* (private manual branch exchange).

The public telephone system is organized in such a way as to facilitate the establishment of telephone channels. It consists of a large number of *local exchanges* that are switching stations where subscribers' lines terminate and where facilities exist for interconnection of local lines or for connection to the *trunk* (long-distance) network. Local exchanges are treated in a similar way to subscribers and are connected together through higher-order switching stations, termed *tandem exchanges.* These in turn are interconncted by means of toll centres, each serving a city, then group centres and zone centres so that a large region, such as the UK, is served economically. Usually connection is made automatically by suitable switching networks that are actuated by signals from the caller's telephone station. In the trunk network a system of repeaters is used that include one or more amplifiers to prevent loss of signal strength.

Communication between points in a telephone system may be made by means of a modulated carrier wave. If the carrier wave is suppressed when no audiofrequency signals are present the system is *quiescent-carrier telephony.* ►Voice-frequency telephony is also used. In the public telephone system voice-frequency telephony is employed for local connections between two lines in the same local exchange and carrier telephony for the trunk system. ►Multiplex operation allows the same telephone line to be used by several channels simultaneously.

The public telephone system is also used for electronic data transmission, such as ►electronic mail, ►telex, and ►fax. The data is transmitted in digital form using ►pulse code modulation. The telephone system is currently being converted to operate on a totally digital system. ►Analogue-to-digital converters are required to convert the analogue signals from telephones into digital form for transmission.

teleprinter A form of start-stop typewriter that comprises a *keyboard transmitter,* which converts keyboard information into electrical signals, and a *printing receiver,* which reverses the process. Teleprinters are used in ►telex systems and in some computing systems.

teletext An information service in which information can be displayed as pages of text on the screen of a commercial television receiver. The information is transmitted as part of the commercial television broadcast signal. It is in the form of pulse-code modulated signals that use a number of the unused lines in an ordinary television video signal transmitted during the normal vertical retrace period. Special decoding circuits are required for the extraction of the teletext signals from the normal television signals and for decoding them.

A typical teletext page consists of 24 rows with up to 40 characters in a row. A limited amount of colour information can be used and a flashing facility is also provided. The television systems transmit the coded lines during each field blanking period.

Teletext decoders contain facilities enabling the user to select the page required and to store and display the information. They also have facilities for inserting news flashes into the normal television picture, or to insert subtitles for people with impaired hearing. Some allow the full text to be superimposed on the normal picture or can store the required information for later display. It is also possible to display the current television programme in much reduced format as a small insert in the teletext display.

The two compatible television systems in current use in the UK are *Ceefax,* used by the British Broadcasting Corporation, and *Oracle,* used by the Independent Broadcasting Authority. ►►videotex.

television (TV) A telecommunication system in which both visual and aural information is transmitted for reproduction at a receiver. The basic elements of the system are as follows.

➤Television cameras and ➤microphones convert the original visual and aural information into electrical signals, i.e. into *video signals* and *audio signals* respectively.

Amplifiers and control and transmission circuits transmit the information along a suitable communication channel. Broadcast television uses a modulated radiofrequency ➤carrier wave; in digital television systems, the picture information is encoded into digital form at the transmitter and decoded at the receiver (➤digital codes).

A ➤television receiver detects the signals and produces an image on the screen of a specially designed ➤cathode-ray tube and a simultaneous sound output from a ➤loudspeaker.

The information on the target in the television ➤camera tube is extracted by ➤scanning and the spot on the screen of the receiver tube is scanned in synchronism with it to produce the final image. A process of rectilinear scanning is used in which the electron beam traverses the target area in both the horizontal and vertical directions. The horizontal direction is termed the *line* and the vertical direction the *field*. ➤Sawtooth waveforms are used to produce the deflections of the beam and in both the camera and receiver the flyback period is blanked out.

The *synchronizing pulses* synchronize the camera and the receiver and are transmitted during the flyback. The line synchronizing pulses are transmitted during the line flyback period and the field synchronizing pulses during the field flyback. A combination of overscanning (➤scanning) and blanking is often used in order to allow a sufficiently long interval for synchronization without loss of picture information. During the blanked interval the level of the signal is held at a reference value, termed the *blanking level,* that represents the blackest elements of the picture except for the synchronizing pulses. This allows easy recognition of the synchronizing pulses at the receiver. The interval immediately preceding the synchronizing pulse is termed the *front porch* and the interval immediately succeeding the sync pulse is the *back porch*. The synchronizing pulses are extracted from the video signal by means of a *sync separator* in the receiver.

For the maximum information to be obtained from the target area of the receiver the number of horizontal scans is made larger than the number of vertical scans so that as much of the target area is covered as possible. The number of lines traversed per second is the *line frequency;* the number of vertical scans per second is the *field frequency*. The scanning process is most important since imperfections in scanning or synchronization between transmitter and receiver can result in geometric distortion of the picture or in other faults, such as ➤pairing. A method of scanning that produces the entire picture in a single field is termed *sequential scanning*. The pattern of horizontal scanning lines is called the *raster*.

Most broadcast television systems use a system of *interlaced scanning*. In this system the lines of successive rasters are not superimposed on each other but are interlaced; two rasters constitute a complete picture or *frame*. The number of complete pictures per second is the *frame frequency,* which is half the number of rasters per second, i.e.

half the *field frequency.* The field frequency needs to be relatively slow to allow as many horizontal lines as possible but sufficiently fast to eliminate ►flicker. Various compromises are used. European television systems use a 50 hertz field frequency (25 Hz frame frequency) system with 625 lines per frame. American television uses a 60 Hz field frequency and 525 lines per frame.

Definition in television is a measure of the resolution of the system, which in turn depends on the number of lines per frame. High-definition systems have more lines. Some ►closed-circuit television systems use as many as 2000 lines per frame. The relationship between the total number of scanning lines per field and the corresponding bandwidth of the video signal is given by the *Kell factor.*

Positive or *negative transmission* may be employed for transmitting the video signals, positive transmission being most often used. In positive transmission an increase in amplitude or frequency above the reference *black level* of the received signal is proportional to the light intensity. The peak value of the video signal that corresponds to the lightest area of the picture is the *white peak.* In negative transmission the carrier wave value decreases below the black level in proportion to the light intensity.

The basic television system transmits images in black and white only (*monochrome television*). In colour television the broadcast signal is received on colour receivers. Monochrome receivers use the brightness information transmitted as part of the colour signal – the luminance signal – but the image produced is black and white (►colour television).

The transmitted signal contains both video and audio information. In analogue television systems the *picture carrier* is the ►carrier wave modulated by the video information. The audio signal modulates a second carrier wave, termed the *sound carrier.* The sound-carrier frequency differs from that of the picture carrier and is chosen so that both signals fall within a designated radiofrequency band but do not overlap each other.

television camera The device used in a ►television system to convert the optical images from a lens into electrical signals. The optical image formed by the lens system of the camera falls on to a photosensitive target. This is scanned, usually by a low-velocity electron beam and the resulting output is modulated with video information obtained from the target area. The output signal can be considered as an essentially constant d.c. signal with a superimposed a.c. signal. The amplitude variations of the latter correspond to the brightness, i.e. the *luminance,* of the target. The d.c. component is the value of the video signal that corresponds to the average luminance of the picture with respect to a fixed reference level.

The camera consists of three major parts housed in one container: optical lens system, ►camera tube, and preamplifier. The resulting output is further amplified and transmitted in the broadcasting network. Some cameras are self-contained, with the amplifier and transmitter in the same container. These cameras are usually employed in a closed-circuit system, or for special applications, such as an outside broadcast. Several types of camera tube have been developed; the major differences between the various types of tubes are in the composition of the photosensitive material used and the means of extracting the electrical information produced.

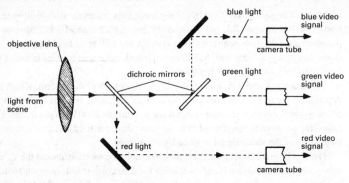

Colour-television camera

The camera used in ►colour television consists of three camera tubes each of which receives information that has been selectively filtered to provide it with light from a different portion of the visible spectrum. Light from the optical lens system is directed at an arrangement of ►dichroic mirrors each of which reflects one colour band and allows other frequencies to pass through. The original multicoloured signal is split into red, green, and blue components, and the video output from the three camera tubes represents the red, green, and blue components of the image (see diagram). The scanning systems in the three tubes are driven simultaneously by a master oscillator to ensure that the output of each tube corresponds to the same image point. The three outputs are then combined to provide brightness (luminance) and colour (chrominance) information (►►colour television).

television receiver A device that receives a television signal and effects the sound and vision reproduction of the original scene. The complete television receiver contains the following: detecting and amplifying circuits that extract the video and audio signals from the received signal; circuits that extract the synchronizing signals from the received signal and control the appearance of the image; a ►picture tube that reproduces the picture, and audio circuits that reproduce the transmitted sound. A desired broadcast channel may be selected using a ►tuner. A colour television receiver also contains extra decoding circuits that extract the chrominance (colour) information from the received signal. In digital television systems, the video information has been encoded into digital form at the transmitter and must be decoded at the receiver (►digital codes).

Various control circuits are provided in order to compensate for minor variations in the received signal level. *Automatic brightness control* is used to maintain automatically a constant value for the average luminance (brightness) level of the image. *Automatic contrast control* maintains a constant value between the black peak and the white peak of the image despite input variations. Various manual controls are also usually available:

horizontal hold and *vertical hold* are controls that adjust the starting points of the line sweep and field scan, respectively, relative to the television screen;

brightness and *contrast controls* adjust the average brightness and contrast of the image. They adjust the average luminance level and contrast range between the black and white peaks, respectively;

colour saturation control and *tint control* are provided in colour television receivers. The colour saturation alters the total intensity of all three electrons beams and makes the final image more coloured or less coloured. The tint control alters the relative intensities of the three electron beams to adjust the colour balance of the picture.

➤colour picture tube; colour television; television.

telex 1. A method of communicating written messages over the public telephone system, using teleprinters. **2.** A message so communicated.

temperature coefficient of resistance The incremental change in the resistance of any material as a result of a change in ➤thermodynamic temperature. In general conductors exhibit a positive coefficient of resistance; semiconductors and insulators have a negative coefficient of resistance.

In a conductor the distribution of electronic energy levels in the material is such that conduction levels are always available (➤energy bands). An increase in the thermodynamic temperature causes an increase in the vibration of nuclei in the crystal lattice. This leads to an increase in the amount of scattering of conduction electrons as they drift through the material and causes the resistance to increase.

In a semiconductor and an insulator the existence of a forbidden gap between the valence and conduction bands has the effect that as the temperature increases more charge carriers become available for conduction by crossing the forbidden gap. The resistance therefore decreases. The increased numbers of carriers more than offsets the effect of scattering by lattice nuclei.

For a given material at thermodynamic temperature T the resistance R_T is given by

$$R_T = R_0 + \alpha T + \beta T^2$$

where R_0 is the resistance at absolute zero ($T = 0$) and α and β are constants characteristic of the material. In general β is negligible and the temperature coefficient of resistance is given by α.

temperature saturation ➤thermionic valve.

TEM wave ➤mode.

tera- Symbol: T. A prefix to a unit, denoting a multiple of 10^{12} of that unit: one terahertz is 10^{12} hertz.

terminal 1. A device that provides ➤input/output facilities to a computer, often from a remote location. It may be used ➤interactively and usually contains a keyboard and/or ➤visual display unit. An *intelligent terminal* contains some local storage and processing ability and can perform simple tasks independently of the main computer. **2.** Any of the points at which interconnecting leads may be attached to an electronic circuit or device and at which signals may be input or output.

terminal impedance The complex ➤impedance at a pair of terminals of a ➤transmission line or other device under normal operating conditions but under ➤no-load conditions.

terminal repeater ➤repeater.

termination A load impedance connected across the output of a transmission line or transducer that completes the circuit while ensuring ➤impedance matching and preventing unwanted reflections.

tertiary winding An additional secondary winding on a ➤transformer. It can be used to supply a load when a different voltage is required from that of the main secondary or when a load must be kept electrically insulated from that of the normal secondary. It may also be used to interconnect supply systems that operate at different voltages.

tesla Symbol: T. The ➤SI unit of ➤magnetic flux density. One tesla is defined as one weber of magnetic flux per square metre of circuit area.

test pattern A chart used in television broadcasting that is transmitted at certain times when no programme is being transmitted. The pattern on the chart can be used for general testing purposes.

tetrode Any electronic device that has four electrodes. The term is most commonly applied to a ➤thermionic valve that contains a second grid. This auxiliary grid is usually a *screen grid* designed to decrease the anode-grid capacitance and hence to increase the resistance to high-frequency currents. It may also be used either to decrease the anode-cathode resistance or to modulate the main electron stream by injecting an alternating voltage.

TE wave *Syn.* H wave. ➤mode.

T flip-flop ➤flip-flop.

T-gate *Syn.* mushroom gate. A gate topology used in high-frequency FET technology. In such technology, the gate length must be made very short to minimize the transit time of carriers in the channel of the device, and thus maximize the speed of operation. The narrow gate can then present a high resistance to the input signal. The resistance can be reduced by producing a gate metal structure that is T-shaped or mushroom-shaped in cross section: this produces a short dimension at the base – the contact of the gate with the semiconductor – but a large area overall, reducing the resistance along the gate. This gate topology can be produced by ➤multilevel resist techniques to create the appropriate shape prior to metallization.

THD *Abbrev. for* total harmonic distortion. ➤harmonic distortion.

thermal battery ➤thermocouple.

thermal breakdown *Syn.* thermal runaway. ➤Breakdown of a reverse-biased ➤p-n junction caused by the generation of excess free charge carriers due to the cumulative interaction between increasing junction temperature and increasing power dissipation.

As the temperature is increased the effect is to reduce the voltage at which ►avalanche breakdown occurs.

thermal imaging *Syn.* thermography. Producing an image of an object by means of the infrared radiation emitted by it. A ►camera tube with a suitable lens system may be used to produce the image. Thermal imaging does not require any external source of illumination and is used to produce images in the dark, for example at night. It is also used for diagnostic purposes to discover any areas of the body that have an unusual temperature distribution.

thermal instrument *Syn.* electrothermal instrument. A measuring instrument that utilizes the heating effect of a current in a conductor. The conductor may be a hot wire, a ►bimetallic strip, or a ►thermocouple.

thermal noise *Syn.* Johnson noise. ►noise.

thermal resistance 1. The ratio of the difference in temperature between two specified points to the heat flow between these points, under thermal equilibrium conditions. **2.** In a semiconductor device, the ratio of the temperature difference between a region in the device and the ambient temperature to the power dissipation in the device. The thermal resistance of the device depends on the material used and the geometry of the device and affects the ease of cooling the device.

thermal runaway ►thermal breakdown.

thermionic cathode *Syn.* hot cathode. A ►cathode in which ►thermionic emission provides the source of electrons. A *directly heated cathode* is one that acts as its own source of heat: the cathode is in the form of a filament and the heater current is passed through it, superimposed on the normal cathode potential (usually earth potential). An *indirectly heated cathode* is one that has a separate heater. The heater is usually a coil of wire formed around the cathode and to which a current is applied.

The electron emission from the cathode may be increased by coating it with a thin layer of a suitable material. A *coated cathode* usually consists of a cylinder of platinum coated with an oxide of barium, strontium, or calcium. It operates at a lower temperature than the clean metal.

thermionic emission Electron emission from the surface of a solid that is a result of the temperature of the material. An electron can escape from the surface with zero kinetic energy if it has thermal energy just equal to the ►work function of the material (unlike ►photoemission). The numbers of electrons emitted increases sharply with temperature (►Richardson–Dushman equation). ►►Schottky effect.

thermionic-field emission ►Schottky diode.

thermionic valve *Syn.* vacuum tube. A multielectrode evacuated ►electron tube that contains a ►thermionic cathode as the source of electrons. Thermionic valves containing three or more electrodes are capable of voltage amplification: the current that flows through the valve between two electrodes, usually the anode and the cathode, is modulated by a voltage applied to one or more of the other electrodes. Thermionic valves have rectifying characteristics, i.e. current will flow in one direction only (the forward direction) when positive potential is applied to the anode.

The simplest type of thermionic valve is the *diode,* which has been most often used in rectifying circuits. Electrons are released from the cathode by thermionic emission. Under zero-bias conditions electrons released by the cathode form a ►space charge region in the vacuum surrounding the cathode and exist in dynamic equilibrium with the electrons being emitted. If a positive potential is applied to the anode, electrons are attracted across the tube to the anode and a current flows. The maximum available current, the ►saturation current, is given by

$$I_{sat} = AT^2 \exp(-B/T)$$

where A and B are constants and T is the thermodynamic temperature of the cathode. The current does not rise rapidly to the saturation value as the anode voltage is increased but is limited by the mutual repulsion of electrons in the interelectrode region. This is the *space-charge limited region* of the characteristic and the current obeys *Child's law* approximately where Child's law is given by

$$I = KV_a^{3/2}$$

where V_a is the anode voltage and K is a constant determined by the device geometry. Increasing the temperature of the cathode has very little effect on the current in this region of the characteristic curve and *temperature saturation* is said to occur. The motion of the electrons may be affected by the magnetic field associated with the current flowing in the heater and the electrons will be deflected from a linear path. This effect is called the *magnetron effect* and it contributes to the delay in reaching the saturation current.

Under conditions of reverse bias (►reverse direction) no current flows in the valve until the field across the valve is sufficient to cause ►field emission from the anode or ►arc formation; ►breakdown of the device will then occur. The characteristics of a simple diode (Fig. *a*) can be compared with the characteristics of a simple ►p-n junction diode (Fig. *b*), which is the solid-state analogue of the device.

The diode characteristic can be modified by interposing extra electrodes, called ►grids since they are usually in the form of a wire mesh, between the anode and the

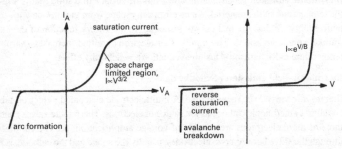

a Characteristic of a valve diode   *b* Characteristic of p-n junction diode

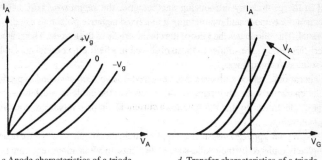

c Anode characteristics of a triode   d Transfer characteristics of a triode

cathode of a valve. The simplest such valve is the *triode* with only one extra electrode, a *control grid.* Application of a voltage to the grid affects the electric field at the cathode and hence the current flowing in the valve. A family of characteristics is generated for different values of grid voltage, similar in shape to the diode characteristic. The anode current at a given value of anode voltage is a function of grid voltage and amplification may be achieved by feeding a varying voltage to the grid; comparatively small changes of grid voltage cause large changes in the anode current. In normal operation the grid is held at a negative potential and therefore no current flows in the grid since no electrons are collected by it. Anode and transfer characteristics of a triode are shown in Figs. *c* and *d.* Triodes have been extensively used in ►amplifying and ►oscillatory circuits.

A disadvantage of the triode is the large grid-anode capacitance, which allows a.c. transmission, and extra electrodes have been added to reduce this effect. Such valves are called *screen grid valves,* the simplest of which are the ►tetrode and ►pentode. The tetrode has one extra grid electrode, the *screen grid,* placed between the control grid and the anode and held at a fixed positive potential. Some electrons will be collected by the screen grid, the number of electrons being a function of anode voltage. At high anode voltages the majority of electrons pass through the screen grid to the anode. An undesirable kink in the characteristics is therefore observed in a tetrode due to ►secondary emission of electrons from the anode, these secondary electrons being collected by the screen grid (Fig. *e*). Secondary electrons are prevented from reaching the screen

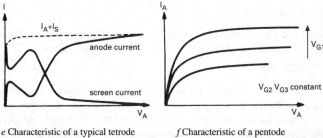

e Characteristic of a typical tetrode   f Characteristic of a pentode

grid in the pentode by introducing another grid, the *suppressor grid,* between the screen and the anode and maintaining it at a fixed negative potential, usually cathode potential. This eliminates the kink of the characteristic of the tetrode. The pentode characteristics (Fig. *f*) are similar to those observed in ►field-effect transistors, which are the solid-state analogues.

Thermionic valves with even more electrodes have been designed to produce particular characteristics. Multipurpose valves, such as the diode-triode, have an arrangement of electrodes so that the functions of several simpler valves are combined in a single envelope.

In everyday applications, such as amplification, thermionic valves have been almost completely replaced by their solid-state equivalents. In applications requiring high voltages and currents valves are still used but these are usually special-purpose valves as with the cathode-ray tube, magnetron, and klystron. For most applications solid-state devices, such as the p-n junction diode, bipolar junction transistor, and field-effect transistor, frequently in the form of integrated circuits, have the advantages of small physical size, cheapness, robustness, and safety as the power required is very much less than for valves.

thermistor A resistor, usually fabricated from semiconductor material, that has a large nonlinear negative ►temperature coefficient of resistance. The thermistor is usually shaped as a rod, bead, or disc and named accordingly. Applications include compensation for temperature variations in other components, use as a nonlinear circuit element, and for temperature and power measurements. A *thermistor bridge* is an arrangement of thermistors for measuring power.

thermocouple Two dissimilar metals joined at each end to form an electrical circuit. If the two junctions are maintained at different temperatures an electromotive force (e.m.f.) is developed between them as a result of the Seebeck effect (►thermoelectric effects). The e.m.f. is not affected by the presence of other metal junctions provided that they are all maintained at the same temperature. The thermocouple may therefore be connected to a suitable measuring instrument and used as a thermometer. It is convenient in use since it functions over a wide temperature range, it can be used to measure temperature at a very small area, and may be used remotely from the indicating instrument. The range of temperatures measured depends on the materials used.

In practice one junction is held at 0 °C and then the e.m.f. E is given by

$$E = \alpha T^2 + \beta T$$

where T is the temperature of the other junction – the 'hot' junction – and α and β are constants dependent on the metals used. At a temperature T_N where

$$T_N = -\beta/2\alpha,$$

E is a maximum. T_N is termed the *neutral temperature* and the use of the thermocouple is normally restricted to temperatures in the range $0 - T_N$ °C. Copper/constantan or iron/constantan thermocouples can be used up to 500 °C. Temperatures up to about 1500 °C may be measured using an platinum/platinum-rhodium alloy thermocouple and even higher temperatures may be measured by means of an iridium/iridium-rhodium alloy thermocouple.

Thermocouple instruments are measuring instruments that use a thermocouple to measure electric current, voltage, or power by means of the heating effect produced by the current in a metallic strip or wire. The current-carrying strip or wire plus the thermocouple are sealed into an evacuated container in order to minimize errors in the measurement due to conduction of heat by the air surrounding the conductor.

The sensitivity of a thermocouple instrument is increased by connecting several couples together in series to form a *thermopile*. An easily detectable output is produced by heat radiation impinging on the hot junctions. The *thermal battery* is a thermopile that is used to generate an e.m.f. when heat is supplied to the hot junctions.

thermocouple instrument ►thermocouple.

thermodynamic temperature *Syn.* absolute temperature. Symbol: T. Temperature that is measured as a function of the energy possessed by matter and as such is a physical quantity that can be expressed in units, termed ►kelvin. In the thermodynamic temperature scale changes of temperature are independent of the working substance used in a thermometer. The zero of the scale is ►absolute zero. The triple point of water is defined as 273.16 kelvin.

thermoelectric effects Phenomena that occur as a result of temperature differences in an electrical circuit.

The *Seebeck effect* is the development of an electromotive force between two junctions formed by joining two dissimilar metals if the two junctions are at different temperatures. The circuit constitutes a ►thermocouple. In general, the e.m.f. E is given by

$$E = a + b\theta + c\theta^2$$

where a, b, and c are constants and θ is the temperature difference between the junctions. If the colder junction is maintained at 0 °C then

$$E = \alpha T^2 + \beta T$$

where α and β are constants dependent on the metals used and T is the temperature of the hot junction. At temperatures below the neutral temperature (►thermocouple) if α is small (as is usually the case) then E is directly proportional to the temperature of the hot junction.

The *Peltier effect* is the converse of the Seebeck effect. If a direct current is passed round a circuit formed from two dissimilar metals or from a metal and a semiconductor, one junction gives off heat and is cooled and the other absorbs heat and becomes warm. The effect is reversible, i.e. if the current is reversed the cool junction becomes warm and the hot junction cools. Larger temperature differences are produced with metal-semiconductor junctions than with metal-metal junctions. A metal-n-type junction produces a temperature difference in the opposite sense to that of a metal-p-type junction for the same direction of current flow. A number of such junctions can be used to form a *Peltier element* (see diagram overleaf), which may be used as a heating or a cooling element.

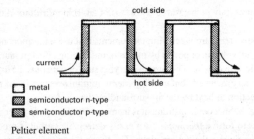

cold side

current

□ metal
▨ semiconductor n-type
▧ semiconductor p-type

hot side

Peltier element

In the *Kelvin effect* a temperature difference between different regions of a single metal causes an e.m.f. to be developed between them. Also a current that flows along a wire in which a temperature gradient exists causes heat to flow from one region of the wire to another. The direction of heat flow is a function of the particular metal used.

thermoelectric series An ordering of the metal elements so that if a ➤thermocouple is formed from two of them, the direction of current flow in the hotter junction is from the metal appearing earlier in the series to the other.

thermoelectron An electron emitted from the surface of a solid as a result of ➤thermionic emission.

thermography ➤thermal imaging.

thermojunction A junction of a ➤thermocouple.

thermoluminescence ➤luminescence.

thermopile ➤thermocouple.

thermostat An automatic temperature-control switch that is used in conjunction with heating systems, such as an immersion heater, to maintain the temperature of a given medium within predetermined limits. It contains a temperature-sensing device, such as a bimetallic strip, that is used to operate a ➤relay so that the source of heat is interrupted when the temperature reaches a predetermined value and is reconnected when the temperature falls to a lower value.

Thévenin's theorem A theorem that is used to simplify the analysis of resistive linear ➤networks. The theorem states that if two terminals (A, B) emerge from a network in order to connect to an external circuit, then as far as the external circuit is concerned the network behaves as a voltage generator. The e.m.f. of the voltage generator is equal to $V_{A,B}$, where $V_{A,B}$ is measured under open-circuit conditions, and it has an internal resistance given by

$$R_{A,B} = V_{A,B}/I_{A,B}$$

where $I_{A,B}$ is the short-circuit current.

Norton's theorem is an equivalent theorem to Thévenin's theorem and states that the network, under similar circumstances, can also be represented by a current generator shunted by an internal conductance. Proofs of both these theorems depend upon Ohm's law. Both theorems can also be applied to alternating-current linear networks but the resistance and conductance must be replaced by the complex impedance $Z_{A,B}$ or admittance $Y_{A,B}$, respectively.

thick-film circuit A circuit that is manufactured by thick-film techniques, such as silk screen printing, and usually contains only ➤passive components and interconnections. A thick film, up to about 20 micrometres thick and composed of a suitable glaze or cement, such as a ceramic/metal alloy, is deposited on a glass or ceramic substrate; the desired pattern for interconnections and passive components is then produced on it. Thick-film circuits may be used to form hybrid ➤integrated circuits: silicon chips containing ➤active components or devices are ➤wire bonded to the film circuit to form the completed circuit, which is then packaged.

thin-film circuit A circuit that is fabricated by thin-film techniques, such as ➤vacuum evaporation, and usually contains only ➤passive components and interconnections. A thin film only a few micrometres thick is deposited on a suitable glass or ceramic substrate and formed into the desired pattern for interconnections and components. Thin-film techniques have been used to form some ➤active components, such as the ➤thin-film transistor.

thin-film transistor (TFT) A ➤MOSFET that is fabricated using thin-film techniques on an insulating substrate rather than on a semiconductor chip. The insulating substrate reduces the bulk capacitance of the device and hence the operating speed can be increased. The technique was originally used for the fabrication of discrete cadmium sulphide transistors and the films could be deposited on the substrate in the order semiconductor, insulator, metal or vice versa. The technique is now used mainly for the construction of ➤silicon-on-insulator CMOS circuits.

Thomson bridge ➤Kelvin double bridge.

Thomson effect *Syn. for* Kelvin effect. ➤thermoelectric effects.

three-phase system ➤polyphase system.

three-phase transformer A ➤transformer that has three independent sets of windings, each usually with the same turns ratio, and is suitable for use with a three-phase (➤polyphase) mains input.

threshold The value at which a change in system state will occur. The term is usually applied to logic devices, and specifies the value of the inputs at which the output will change.

threshold frequency The frequency at which a particular phenomenon, such as the ➤photoelectric effect or ➤photoconductivity, just occurs or the frequency below which an electronic device, such as a high-pass ➤filter, does not operate. In the latter case the term *cut-off frequency* is also used.

threshold signal 1. ➤minimum discernible signal. **2.** ➤automatic control.

threshold voltage The voltage at which an electronic device first functions in a specific manner. In particular it is the voltage at which the conducting channel of a ➤MOSFET is just formed.

throat microphone ➤microphone.

through path ➤feedback control loop.

thyratron A ➤gas-filled tube containing three electrodes, the anode, cathode, and gate. The tube exhibits ➤bistable characteristics: a high impedance between anode and cathode – the 'off' state – and a low-impedance 'on' state. A voltage placed on the gate electrode initiates a discharge in the tube, switching from off to on. The current is limited by the external circuitry. The extinction of the discharge is effected by reducing the anode potential to a low value. The thyratron is now superseded by its solid-state analogue, the ➤thyristor.

thyristor *Syn.* p-n-p-n device. A general term for a family of ➤semiconductor devices that exhibit ➤bistable characteristics: they can be switched between a high-impedance low-current 'off' state and a low-impedance high-current 'on' state. The thyristor is the solid-state analogue of the ➤thyratron. It is a multilayer semiconductor device with a p-n-p-n structure; this is commonly analysed as a coupled pair of ➤bipolar junction transistors.

The diagram shows a thyristor with one gate contact; this device is usually known as a *silicon-controlled rectifier* (SCR). In the 'off' or *forward blocking* state, the outer pair of ➤p-n junctions is forward biased but the centre junction is reverse biased, and no current flows between anode and cathode. Injecting a gate current will cause a build up of charge in the bases of the two transistors until the centre junction becomes forward biased; at this point *forward breakover* or switching to the 'on' state occurs. Reducing the forward current below a *holding current* will switch off the thyristor. The *gate-turn-off thyristor* (GTO) can be turned on with a forward gate voltage and off with a reverse gate voltage, at any value of anode current. Applications include inverter circuits, chopper, and d.c. switching circuits.

A *reverse conducting thyristor* (RCT) is a three-terminal multilayer semiconductor device that exhibits similar behaviour to the conventional thyristor in the forward direction, but conducts a large reverse current. Applications include drivers for electroluminescent displays.

The *diac* (diode a.c. switch) and *triac* (triode a.c. switch) are bidirectional thyristors, and are useful in a.c. applications. The diac consists of a pair of ungated p-n-p-n structures connected in antiparallel. The electrical characteristics are those of a conventional thyristor (high-impedance 'off' state switching to low impedance 'on' state at a given breakover voltage), but operating in both forward and reverse directions. The triac is a more complex structure than the basic p-n-p-n thyristor, including additional n-type regions at the gate and anode regions. This enables the triac to switch currents in either direction under control of the gate voltage.

tie line In telephone or telecommunication systems, a private communications channel that links two or more geographically separate locations together. An example of a tie line is a permanent telephone line between two factories located a distance apart. The tie line cannot be used by outside parties so is more available.

tightly coupled multiprocessor ➤multiprocessor.

timbre The perceived difference between any two sounds that are presented in the same manner and have the same ➤pitch, ➤loudness, and duration.

timebase A voltage that is a predetermined function of time and that is used to deflect the electron beam of a ➤cathode-ray tube so that the luminous spot traverses the screen in a desired manner. One complete traverse of the screen, usually in a horizontal direction, is termed a *sweep*.

The most common type of timebase is one that produces a linear sweep: a sawtooth waveform is used to effect this. The circuit that produces the required voltage is a *timebase generator*; it may be free-running, in which a periodic sawtooth waveform is produced, or it may be clocked, when one sweep is produced on application of a trigger

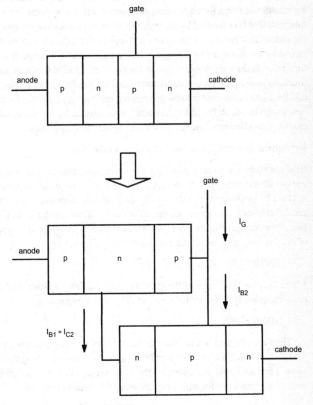

Thyristor with one gate contact: the SCR

pulse to the circuit. The period during which the spot returns to the starting point is the *flyback* and in many applications, such as in television receivers, the flyback is suppressed, i.e. no luminous spot is observed on the screen during the return interval. The *sweep frequency* is the repetition rate of the sweeps across the screen.

A *Miller sweep generator* is a timebase generator that contains a ►Miller integrator in the circuit in order to improve the linearity of the sweep. In some applications it is necessary to produce a timebase in which the electron beam moves at a faster speed during part of the sweep, termed an *expanded sweep*. An *expanded-sweep generator* is one in which the incremental output voltage with respect to time is greater during a portion of the sweep in order to produce such an expanded sweep. A *delayed sweep* is a sweep produced by a synchronous timebase generator in which a predetermined delay time is introduced between application of the trigger pulse and commencement of the sweep on the screen.

Timebases are used to control the spot on the screen in many applications. ►Cathode-ray oscilloscopes usually contain a circuit that generates a free-running sawtooth waveform with an adjustable sweep frequency. ►Radar systems use a synchronized timebase that is controlled by the transmitter so that each sweep is synchronized with the transmitted pulses and the return echo appears at a distance along the trace determined by the distance of the target from the transmitter. ►Television systems employ timebases in the camera tubes and in the receivers to scan the lines and frames. The timebase in the receiver is usually controlled by synchronizing pulses in order to retain the correct relationship to the transmitter; alternatively a *flywheel timebase* is sometimes used in which the frame frequency is controlled by the electrical inertia of the circuit. This eliminates the need for frame-synchronizing pulses.

timebase generator *Syn.* sweep generator. ►timebase.

time constant The time required for a unidirectional electrical quantity, such as voltage or current, to decrease to $1/e$ (approximately 0.368) of its initial value or to increase to $(1 - 1/e)$ (approximately 0.632) of its final value in response to a change in the electrical conditions in an electronic circuit or device. Thus at any instant following initiation of the change in electrical equilibrium, such as switching a d.c. supply voltage on or off, the instantaneous change of a quantity, V, is given by

$$dV/dt = (V_f - V)/\tau$$

where V_f is the final value of the quantity and τ the time constant. In the case of a decreasing quantity V_f is usually zero and the expression becomes

$$dV/dt = -V/\tau$$

The time constant is a measure of the speed of operation of any circuit or device; circuits that contain capacitance or inductance can have very long time constants (of about a few seconds). For example, the time constant of a circuit containing a resistance of R ohms in series with a capacitance of C farads is given by

$$\tau = RC \text{ seconds}$$

One that contains a resistance, R, in series with an inductance of L henries has a time constant given by

$$\tau = L/R \text{ seconds}$$

time delay ►time lag.

time discriminator A circuit in which the magnitude of the output is a function of the time interval between two input pulses and the sense of the output is determined by the order in which the inputs are received.

time-division multiple access (TDMA) ►digital communications.

time-division multiplexing (TDM) In communication systems, the division of a single channel into time slots which are allocated to different users or input signals. The time slots can be allocated to each input signal in turn, an approach called ►polling, or to each as it has information to transmit. On reception, each carrier can be distinguished from the others by its time position within the received signal, thereby allowing the original signal to be demodulated. Time-division multiplexing is extensively used in local area networks and satellite communication systems.

time-division switching In communication systems, the transmission of multiple signals down a single channel by allocating them specific time slots. At the receiving end a synchronized switch is used to divert each time slot to a separate output signal channel. In this way the original signals can be recovered.

time domain A term used to refer to a situation in which an electrical signal is represented as a function of time, for example representing how a voltage varies with time. An ►oscilloscope displays signals in the time domain. ►►spectrum analyser.

time domain reflectometer (TDR) ►reflectometer.

time lag *Syn.* time delay. The time that elapses between operation of a circuit-breaker, relay, or similar apparatus and the response of the current in the main circuit. A time lag may be deliberately introduced into a particular system. A definite time lag is a predetermined interval, sometimes adjustable, that is independent of the magnitude of the electrical quantity causing the operation. An inverse time lag is a delay time that is an inverse function of that quantity. ►►overcurrent release.

time sharing A form of ►time-division multiplexing by which a number of users may communicate directly with a computer by means of a number of individual terminals. The speed of operation of the machine is such that each user receives the impression of being the sole user of it. ►►interactive.

time to flashover ►impulse voltage.

time to puncture ►impulse voltage.

time to trip The natural ►delay, apart from any intentional delay, between the instant at which a predetermined signal is applied to a switch, circuit-breaker, or other similar device and the instant at which the device operates.

timing diagram A diagram that shows the relative timing of a number of input and/or output lines in a ►logic circuit. Timing diagrams are used particularly when connecting devices together that require some cooperation, such as memory devices and a microprocessor.

tin Symbol: Sn. A metal, atomic number 50, that has good resistance to corrosion and alloys readily with other metals, such as copper. It is used as a component in many solders, and is extensively used to make permanent ohmic contacts in electronic circuits, devices, or other equipment.

tint control ►television receiver.

T-junction *Syn.* hybrid T-junction. ►hybrid junction.

TM wave *Syn.* E wave. ►mode.

token ring In digital communications, a means of connecting computers or digital equipment together such that one can communicate with any other. The communication is achieved by passing *tokens* between the pieces of equipment. A token is a special bit pattern that is unique to the network. A token ring is a ►ring network architecture.

tolerance The maximum permissible error or variation permitted in the electrical properties or physical dimensions of any component or device.

tone control A device that alters the relative frequency response of an audiofrequency amplifier used in the reproduction of sound.

tone jamming In radar systems, the transmission of ►continuous-wave signals in an attempt to prevent a receiver from receiving a signal.

topside ionosphere ►ionosphere.

total capacitance The capacitance between one conductor in a system and all the other conductors in the system when electrically connected together.

total emission The peak value of the current that may be obtained by ►thermionic emission from a cathode under normal heating conditions. The anode of the valve containing the cathode, together with all other electrodes, must be raised to sufficiently high potentials so that saturation is ensured.

total harmonic distortion (THD) ►harmonic distortion.

touch screen A computer screen which, when touched, sends the position of that point on the screen to the computer; it is thus a type of input device.

touch-sensitive keyboard ►velocity-sensitive keyboard.

Townsend avalanche ➤avalanche.

Townsend discharge ➤gas-discharge tube.

trace The image traced out by the luminous spot on the screen of a cathode-ray tube.

trace interval *Syn. for* active interval. ➤sawtooth waveform.

track 1. The portion of a moving storage medium, such as magnetic tape or disk, that is accessible to a given reading device in a computer. **2.** An interconnecting metal path on an integrated circuit.

tracking 1. Maintaining a prescribed relationship between an electrical parameter of one electronic circuit or device and the same or a different parameter of a second device, suitably arranged, when both are subjected to the same stimulus. In particular it is the maintaining of a predetermined frequency relation in ➤ganged circuits, especially a constant difference between the ➤resonant frequencies of two ganged ➤tuned circuits. **2.** Maintaining a radar or radio beam set on a target while its position is determined. **3.** The formation of unwanted conducting paths on the surface of a dielectric or insulator when it is subjected to a high electric field. Such paths often result from carbonization on the surface. **4.** Maintaining the stylus of a ➤pick-up accurately within the grooves of a gramophone record. **5.** Maintaining the spot of light from the laser accurately on the track of a ➤compact disc.

tracking generator A signal source providing a swept-frequency output (➤timebase) that tracks the internal swept-frequency local-oscillator signal in a ➤spectrum analyser, enabling ➤frequency-response measurements to be made on two-port circuits using the spectrum analyser. The tracking generator is driven by the same timebase as the local oscillator so that the frequency sweeps are synchronized, and the output frequency of the tracking generator is offset from the local oscillator by the value of the first intermediate frequency, thus producing the desired input frequency.

traffic intensity ➤network traffic measurement.

trailing edge (of a pulse) ➤pulse.

transadmittance The ratio of output current of a circuit or network to its input voltage, when either or both may be complex quantities.

transadmittance amplifier An ➤amplifier that provides an output current for an input voltage; the gain is measured in siemens. This type of amplifier has a high input impedance and a high output impedance. It is often referred to as a *transconductance amplifier*.

transceiver A piece of equipment, a circuit, or integrated circuit that contains both a ➤receiver and a ➤transmitter.

transconductance Symbol: g_m. The ratio of output current of a circuit or network to its input voltage, when both are real. This is also the small-signal gain of a transistor:

$$g_m = \partial I_{out} / \partial V_{in}$$

where the output current is the drain or collector current, and the input voltage is the gate-source or base-emitter voltage, for ►field-effect transistors and ►bipolar junction transistors, respectively. The symbol g_m is derived from the earlier term for transconductance, *mutual conductance*.

transconductance amplifier A voltage-controlled current source (VCCS). ►dependent sources. ►►transadmittance amplifier.

transducer *Syn.* sensor. Any device that converts a nonelectrical parameter, e.g. sound, pressure, or light, into electrical signals or vice versa. The variations in the electrical signal parameter are a function of the input parameter. Transducers are used in a wide range of measuring instruments and have a variety of uses in the electroacoustic field. Gramophone pick-ups, microphones, and loudspeakers are all *electroacoustic transducers.* The term is also applied to a device in which both the input and output are electrical signals. Such a device is known as an *electric transducer.*

The physical quantity measured by the transducer is the *measurand.* The portion of the transducer in which the output originates is the *transduction element.* The device in the transducer that responds directly to the measurand is the *sensing element* and the upper and lower limits of the measurand value for which the transducer provides a useful output is the ►dynamic range.

Several basic transduction elements can be used in transducers for different measurands. They include capacitive, electromagnetic, electromechanical, inductive, photoconductive, photovoltaic, and piezoelectric elements. Most transducers require external electrical excitation for their operation; exceptions are self-excited transducers, such as piezoelectric crystals and photovoltaic and electromagnetic types.

Most transducers provide analogue output, i.e. the output is a continuous function of the measurand, but some provide digital output in the form of discrete values. Most transducers are *linear transducers,* i.e. they are designed to provide an output that is a linear function of the measurand since this allows easier data handling. If the measurand varies over a stated frequency range the output of the transducer varies with frequency. The frequency response of the transducer is the change with frequency of the output to measurand amplitude ratio. The portion of the response curve over which attenuation of the measurand is significant is the ►rejection band of the transducer.

Like many networks transducers may be considered as ►two-port devices but one pair of terminals is not necessarily electrical. A *symmetric transducer* is one in which the input terminals and output terminals may each be simultaneously reversed without affecting the operation of the device; otherwise it is *asymmetric.* A transducer that operates in one direction only is a *unilateral transducer*; otherwise it is *bilateral.* If the energy loss in a bilateral transducer is the same for both directions of operation, it is a *reversible transducer.*

An *active transducer* is one that introduces gain, i.e. it derives energy from a source that is independent of the input signal energy. The transducer gain is defined as the ratio of the energy delivered to a suitable predetermined load to the available power at the input. In the case of a *passive transducer,* in which no gain is introduced, the loss is defined as the ratio of the available power at the input to the power delivered to the load under specified operating conditions.

transduction element ➤transducer.

transductor Acronym from *trans*fer in*ductor*. A device that consists of a magnetic core carrying several windings. The state of magnetic flux density in the core is controlled by a fixed alternating current in one of the windings. This current is sufficiently large to cause saturation of the core. Small variations in the current of one of the other windings – the *signal winding* – then cause large variations in the power in another circuit coupled by another winding – the *power winding*. The device thus operates by *magnetic modulation* and is used in control circuits, such as lighting circuits, and, particularly, in aircraft.

Variations of the current in the signal winding are caused by the control circuit and must be slow relative to the frequency of the alternating current in the supply circuit; frequencies up to about 2 kilohertz are possible and therefore control of signals in the lower audiofrequency range is available. The controlled signal may be output directly from the power winding.

transfer characteristic 1. ➤characteristic. **2.** The relation between the degree of illumination of a television camera tube and the corresponding output current under specified conditions.

transfer constant *Short for* image transfer constant.

transfer current 1. The current applied to the ➤starter electrode of a glow-discharge tube (➤gas-discharge tube) in order to initiate a glow discharge across the main gap. **2.** The current at one electrode of a ➤gas-filled tube that causes ➤gas breakdown to occur at a different electrode.

transfer function A function describing how a network or circuit relates the output signal to its corresponding input signal for both ➤magnitude response and ➤phase response. The transfer function defines the operation of the network or circuit in a mathematical form.

transfer layer ➤multilevel resist.

transfer length ➤contact.

transfer parameter The tangent at any point on a given transfer ➤characteristic of any network, amplifying circuit, transducer, or device. At the point the transfer parameter is the incremental change in the electrical output quantity to the incremental change in the input quantity. The most common transfer parameter used is the ➤transconductance, the transfer characteristic being the output current, I_{out}, versus the input voltage, V_{in}. The ➤transresistance is the parameter derived from the characteristic showing the active output voltage, V_{out}, versus the input current, I_{in}; other transfer parameters, such as the ➤transimpedance, are derived by plotting the appropriate transfer characteristics.

transferred-electron device (TED) *Syn.* Gunn diode. A negative-resistance microwave oscillator that operates by means of the ►Gunn effect. It is a diode formed from a sample of low-resistivity n-type gallium arsenide that produces coherent microwave oscillations when a large electric field is applied across it. The most fundamental form of operation of the device is *transmit time mode* in which the output frequency of the oscillations is determined by the time taken for a domain of high-energy low-mobility carriers to drift across the length of the semiconductor (►Gunn effect). This mode has several disadvantages, including fixed frequency and low efficiency.

Delayed domain mode of operation is a more efficient and useful operating mode. The diode is connected to an external tuned circuit that determines the operating frequency of the diode. The field across the diode is biased by a radiofrequency (r.f.) field so that the threshold voltage is only exceeded during the positive cycles of the r.f. fields. Domains can only form during the positive cycles and therefore the output current pulses are forced to occur at an externally determined frequency.

Limited space-charge accumulation (*LSA*) *mode* is an operating mode that produces an output frequency much greater than the transit time frequency. The frequency of the r.f. bias, determined by the external tuned circuit, is high and therefore stable domains do not have time to form. The field within the device remains above the threshold value for most of the operating cycle because stable domains are not present, and the electrons experience a negative reistance as they flow across the diode.

transformer An apparatus that has no moving parts and that transforms electrical energy at one alternating voltage into electrical energy at another usually different alternating voltage without change of frequency. It depends for its action upon mutual induction (►electromagnetic induction) and consists essentially of two electric circuits coupled together magnetically. The usual construction is of two coils (or windings) with a magnetic core suitably arranged between them. One of these circuits, called the *primary,* receives energy from an a.c. supply at one voltage; the other circuit, called the *secondary,* delivers energy to the load, usually at a different voltage.

The general symbol for a transformer and the circuit diagram for a typical transformer feeding a load impedance, Z_2, are shown in the diagram. In the case of an *ideal transformer* there is complete ►coupling between the primary and secondary windings and therefore

$$M^2 = L_1 L_2$$

where M is the mutual inductance and L_1 and L_2 are the self-inductances of the two coils. The self-inductances are related to the squares of the numbers of turns, n_1, n_2, of the two coils by the equation

$$L_1 L_2 = (n_1/n_2)^2 = n^2$$

where n is the ►turns ratio of the given transformer.

In the ideal case energy dissipation in the core can be ignored and it can be shown that

$$Z_p = V_1/I_1 = Z_1 + Z_2/n^2$$

where Z_p is the impedance of the primary circuit and Z_2/n^2 represents the effect of the secondary circuit on the primary and is known as the *reflected impedance*. The current in the secondary circuit is I_2 and the impedance is Z_s. They are given by

$$I_2 = -nV_1/Z_s$$

$$Z_s = Z_2 + Z_1n^2$$

From these equations it can be seen that

$$Z_s = n^2Z_p \qquad \text{(impedance transformation)}$$

$$V_2 = -nV_1 \qquad \text{(voltage transformation)}$$

$$I_2 = I_1/n \qquad \text{(current transformation)}$$

In practice complete coupling is not achieved and the mutual inductance is given by

$$M = k(L_1L_2)^{1/2}$$

and

$$V_2 \approx -knV_1$$

where k is the coupling coefficient.

Suitable design of the device can optimize the coupling and minimize the energy dissipation: values of k almost equal to unity may be achieved. Energy dissipation is kept to a minimum by winding the coils around laminated cores; the magnetic circuit of the cores is completed by forming them as a yoke. The two coils are sometimes interwound but insulation can be a problem where high voltages are used and they are then wound side by side. Uniformity of the magnetic flux density within the coils can be maintained by using an extra core (a *limb*) that surrounds the coils.

The property of voltage transformation is used in the *voltage transformer*. This can be used as an ➤instrument transformer in order to measure voltages. The primary winding is connected in parallel with the main circuit and the secondary winding connected to a suitable measuring instrument. The voltage transformer may also be used as a *power transformer* in order to supply electrical power at a predetermined voltage to a

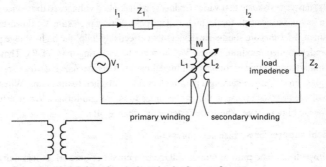

General symbol (left) and equivalent circuit of a **transformer**

circuit. The primary winding is usually connected to the electrical mains supply. The secondary winding consists either of several windings or of one winding with several ➤tappings in order to supply different magnitudes of voltage or to supply more than one circuit. The transformer is described as *step-up* or *step-down* according to whether the secondary voltage is respectively greater or less than the primary voltage.

The *current transformer* utilizes the current transformation property and is most often used as an ➤instrument transformer. The primary winding is connected in series with the main circuit and the secondary to the appropriate measuring instrument. Current transformers are also used to operate protective ➤relays in order to prevent the current in the main circuit rising above a predetermined value.

➤autotransformer; core-type transformer; shell-type transformer; variocoupler.

transformer ratio ➤turns ratio.

transient A phenomenon, such as damped oscillations or a voltage or current surge, that occurs in an electrical system following a sudden change in the dynamic conditions of the system and that is usually relatively short-lived. A transient may be caused by the application of an impulse voltage or current to the system or by the application or removal of a driving force. The nature of the transient is a function of the system itself but the magnitude depends on the magnitude of the impulse or the driving force.

The *transient response* of an electronic device, such as an amplifier, is the change in output that occurs as a result of a specific sudden change in the input.

transient response ➤transient.

transient suppression A means of removing the high-voltage noise that can occur when switching on mains-level voltages; for example, thermostats in domestic central heating and refrigerators, fluorescent light starters, etc., if unsuppressed, can generate bursts of high-voltage peaks of up to a kilovolt for a time duration of a few microseconds. *Transient suppressors* remove the high-voltage content, and often also slow down the rise time of the voltage peaks so that they are no longer a problem.

The following are examples of transient suppressors. *Voltage-dependent resistors* (VDRs) have a resistance that varies (reduces) with applied voltage, so the device provides a low-resistance path for the high-voltage peak thereby protecting the downstream equipment. *Varistors* are solid-state devices that effectively 'clip' the high-voltage peak to a predetermined maximum level. Varistors are faster acting than VDRs. They are often manufactured under commercial model names. *Gas discharge devices* are gas-filled glass envelopes containing a pair of electrodes, attached to the mains. When the high-voltage peaks are present, the gas becomes ➤ionized and forms a low-resistance path for the high voltage. These are faster acting than VDRs.

transient suppressor ➤transient suppression.

transimpedance The ratio of output voltage of a circuit or network to its input current, when either or both may be complex qualities.

transimpedance amplifier An ►amplifier that provides an output voltage for an input current; the gain is measured in ohms. This type of amplifier has a low input impedance and a low output impedance.

transistor A multielectrode ►semiconductor device in which the current flowing between two specified electrodes is controlled or modulated by the voltage applied at the third (control) electrode. The term transistor was originally derived from the phrase *trans*fer re*sistor*, as the resistance of the output electrodes was controlled by the input circuit (transferred). Transistors fall into two major classes: the ►bipolar junction transistor (BJT) and the ►field-effect transistor (FET).

The bipolar junction transistor was derived from the ►point-contact transistor, which was invented at Bell Telephone Laboratories in 1947 by Bardeen, Brattain, and Shockley. The BJT comprises two ►p-n junctions placed back-to-back in close proximity, creating three regions in the device – *emitter*, *base*, and *collector*. It utilizes the flow of both ►electrons and ►holes across the junctions for its electrical behaviour. The current flow through the emitter and collector electrodes is controlled by the voltage across the base–emitter p-n junction.

In the case of the field-effect transistor, the current flow inside the transistor between *source* and *drain* electrodes is controlled by the internal electric field, which is a function of the voltage applied at the third terminal, the *gate*. The junction ►field effect transistor (JFET) was invented at Bell Laboratories by Shockley in 1952. In this device the current flow is in the bulk of the semiconductor, along a *channel* whose cross-sectional area is controlled by the width of the ►depletion region associated with a reverse-biased p-n junction formed by the gate and channel semiconductors. The current flow is thus in the region of the channel that is free of the gate field.

In a *surface field-effect transistor*, the current flow is along a surface of the semiconductor that is separated from the gate electrode by an insulating region. Such a device is known as an insulated-gate ►field-effect transistor, exemplified by the ►MOSFET. The surface field-effect transistor was first proposed in the 1930s. It was only realized practically in 1960, when device technology had improved sufficiently to reduce the density of surface defects to a low enough level to enable control of the surface potential by an applied voltage.

transistor parameters The electrical behaviour of ►transistors can be described more or less accurately by a number of equations relating the movement of the ►electrons in the ►semiconductor due to the applied potentials, and the resulting current flows. While this is useful for understanding transistor physics it is impractical for circuit design. Electrical models of the transistor behaviour have been devised that can be used in straightforward circuit analysis and design: these are *equivalent circuit models*, and the ►parameters of these models are the transistor parameters.

For circuit analysis by hand, the equivalent circuits are relatively simple. Generally they are divided into 'DC' models, for determining the steady bias conditions or ►operating point of the circuit, and small-signal models for determining the circuit response to an applied AC signal. Examples of such models for a ►bipolar junction transistor are the *Ebers–Moll model*, which is a DC model, and the *hybrid-π model*, which

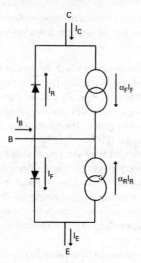

a Ebers–Moll DC model for a bipolar junction transistor

is a small-signal model (Figs. *a*, *b*). The transistor parameters in each model are the values for the equivalent-circuit components. For example, the parameters of the Ebers–Moll model are the current transfer ratios, α_F and α_R, and the saturation currents of the ►p-n junctions formed by the emitter-base and the collector-base junctions; the parameters of the hybrid-π model are the resistor and the capacitor values, and the transconductance.

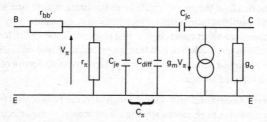

b Hybrid-π small-signal model for a bipolar transistor

Some small-signal models are derived from the two-port analysis of the transistor behaviour (►network). For example, the *hybrid parameter model* for a bipolar transistor in ►common-emitter configuration results in terms such as the input resistance h_{ie}, the output admittance h_{oe}, and the forward-current gain h_{fe} (gain parameter on the current-controlled current source at the output port). This hybrid parameter model is relatively archaic and should not be confused with the hybrid-π model of the bipolar transistor. The term h_{fe} is often found in transistor data sheets, and can be identified with the forward-current gain β.

In practical devices the equivalent-circuit models can also include components to model the packages that the devices are placed in. This is particularly important at high frequencies where the packages can contribute significant capacitance and inductance that can strongly affect the overall electrical behaviour of the transistor.

When computer-aided design is used in the circuit analysis and design, more sophisticated equivalent-circuit models can be employed, making use of the computational power available to perform many more calculations than would be possible by

Model parameter		Default	Unit	Scaling
IS	saturation current	1E-16	amp	area
BF	ideal maximum forward current gain	100		
NF	forward current ideality factor	1		
VAF	forward Early voltage	∞	volt	
IKF	corner for BF high-current roll-off	∞	amp	area
ISE	base-emitter leakage satn. current	0	amp	area
NE	base-emitter leakage ideality factor	1.5		
BR	ideal maximum reverse current gain	1		
NR	reverse current ideality factor	1		
VAR	reverse Early voltage	∞	volt	
IKR	corner for BR high-current roll-off	∞	amp	area
ISC	base-collector leakage satn. current	0	amp	area
NC	base-collector leakage ideality factor	2		
RB	zero-bias (maximum) base resistance	0	ohm	1/area
RBM	minimum base resistance	RB	ohm	1/area
RE	emitter ohmic resitance	0	ohm	1/area
RC	collector ohmic resistance	0	ohm	1/area
CJE	base-emitter zero-bias junction cap.	0	farad	area
VJE	base-emitter built-in voltage	0.75	volt	
MJE	base-emitter p-n doping grading	0.33		
CJC	base-collector zero-bias junction cap.	0	farad	area
VJC	base-collector built-in voltage	0.75	volt	
MJC	base-collector p-n doping grading	0.33		
XCJC	fraction of C_{bc} connected internal to R_b	1		
CJS	collector-substrate zero-bias junction cap.	0	farad	area
VJS	collector-substrate built-in voltage	0.75	volt	
MJS	collector-substrate p-n doping grading	0		
FC	forward-bias junction cap. coefficient	0.5		
TF	forward transit time	0	sec	
TR	reverse transit time	0	sec	
EG	energy band gap	1.11	eV	
KF	flicker noise coefficient	0		
AF	flicker noise exponent	1		

SPICE parameters for bipolar transistor (subset)

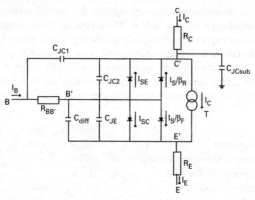

c SPICE model for bipolar transistor

hand. The *SPICE model* for a bipolar transistor is shown in Fig. *c*, and the associated parameter list is given in the table. Various levels of sophistication can be included in such models.

Similar model approaches are employed for junction ►field-effect transistors and ►MOSFETs.

transistor-transistor logic (TTL) A family of high-speed integrated ►logic circuits in which the input is through a multiemitter ►bipolar junction transistor; usually the output stage is ►push-pull. ►Diode-transistor logic operates in a similar manner but the input is through a number of ►diodes. A typical circuit (a three input NAND circuit) is shown in Fig. *a*. If the input levels are all high, the emitter-base junctions of the input transistor T_1 will all be reverse biased: current through the base will flow across the forward-biased collector junction to the phase-splitting transistor T_2, which will switch on. Current flows through T_2 to T_4 and turns on T_4. T_3 will remain off as current is shunted away from the base and the output voltage will be low. If any one or more of the inputs is low, the emitter-base junction of T_1 will be forward biased and current flows out of the base through the emitter of T_1. The current is therefore diverted away from T_2, and T_2 and T_4 will be turned off. The current through R_2 flows into the base of T_3 and T_3 is switched on and the output voltage is high. The output voltage will change rapidly when the input conditions change since the transistors drive the level in both directions.

The biggest limitation in speed is caused by the ►delay time due to hole storage in the saturated output transistor T_4. The speed may be improved by adding a ►Schottky diode with low diode forward voltage across the base-collector junction of T_4. This circuit is called *Schottky TTL* and part of the circuit is shown in Fig. *b*. This diode prevents T_4 saturating. Hole storage therefore does not occur in the collector-base junction and since it does not occur in Schottky diodes, T_4 will be turned off very rapidly when the base current is cut off.

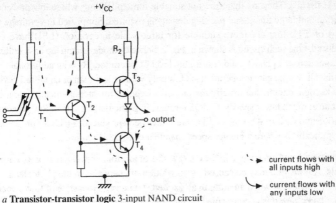

a **Transistor-transistor logic** 3-input NAND circuit

- - - ► current flows with all inputs high
- ╲ current flows with any inputs low

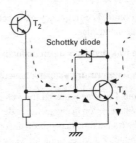

b Schottky **transistor-transistor logic** circuit

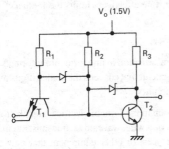

c Low-voltage **transistor-transistor logic**

TTL is one of the most widely used types of integrated logic circuit for high-speed applications and, together with ►emitter-coupled logic (ECL), tends to be regarded as a standard against which all other logic circuits are judged. TTL is also characterized by medium power dissipation and ►fan-out and good immunity to ►noise. TTL circuits may have high operating speeds but at the expense of power dissipation, since the higher the speed the greater the power consumed. They are also medium-scale integration

(MSI) circuits. They are therefore not suitable for applications where low power dissipation and large functional packing density is required. Simplified low-voltage versions of TTL that are more suitable for large-scale integration (LSI) have been produced. One such circuit is shown in Fig. *c*. Schottky diodes clamp the base-collector voltages across T_1 and T_2 and control the level of saturation. R_2 is an additional resistor that allows the gate to operate from a supply voltage less than or equal to 1.5 volts. ►MOS logic circuits are usually the circuits of choice for such applications but they operate at much lower speeds. CMOS circuits, in particular, have very low power dissipation and CMOS versions of TTL gates have been produced. ►I^2L circuits are used for applications where a greater speed than that of MOS is required.

transition A sudden change in the energy state of an atom between two of its ►energy levels. A transition may be caused by absorption of energy, as in the photoelectric effect, or may occur when an atom in an excited state reverts to a state of lower energy. In the latter case it is accompanied by the emission of a photon of energy. Both these effects are utilized in the ►laser.

transition band The frequency band between an electronic ►filter's pass-band edge and the stop-band edge.

transition flux density ►superconductivity.

transition temperature ►superconductivity.

transit time The time required for a charge ►carrier to travel directly from one designated point or region in an electronic device to another, under specified operating conditions. The transit time depends on the operating conditions, the device geometry, and in semiconductor devices upon the velocity of the carriers within the semiconductor; in the absence of ►carrier storage the transit time effectively limits the frequency at which a given device may be operated.

transit time mode ►transferred electron device.

translator A device or circuit that converts information in one form into information in another form. In telephony a translator is a circuit that converts the dialled digits into information appropriate to the call routing circuitry.

transmission The act of conveying information in the form of electrical signals from one designated location to another by means of a wire, waveguide, transmission line, or radio channel and using any circuits, devices, or other equipment that may be necessary.

transmission electron microscope ►electron microscope.

transmission gain ►transmission loss.

transmission level The ratio of the electrical power at any point in a ►telecommunication system to the power at some arbitrary reference point, usually the sending point in a two-wire system.

transmission line 1. *Syn.* power line. An electric line, such as an overhead wire, that is used to convey electrical power from a power station or substation to other stations or substations. **2.** An electric ➤cable or ➤waveguide that conveys electrical signals from one point to another in a ➤telecommunication system and that forms a continuous path between the two points. **3.** *Syn.* feeder. The one or more conductors that connect an antenna to a transmitter or receiver and that are substantially nonradiative.

Any of the above transmission lines is described as smooth or uniform if its electrical parameters are distributed uniformly along its length. A *balanced line* is a transmission line that has conductors of the same type, equal values of resistance per unit length, and equal impedances from each conductor to earth and to other electrical circuits. The transmission characteristics of a particular transmission line are usually frequency-dependent and *loading* is often used to improve the transmission characteristic throughout a given frequency band. Loading is the addition of inductance to a line: in *coil loading* inductance coils (*loading coils*) are placed in series with the line conductors at regular intervals; in *continuous loading* a continuous layer of magnetic material is wrapped along each conductor.

The presence of a discontinuity in a transmission line can cause some signal energy to be reflected back along the line. The presence of such line reflections can lead to the production of ➤standing waves in the transmission line. A transmission line that has no line reflections and therefore no standing waves is a *nonresonant line*.

The most efficient transfer of signal power from a transmission line to the ➤load occurs when the load is matched to the transmission line, i.e. the load impedance, Z_L, is equal to the *characteristic impedance, Z_0*, of the transmission line. Z_0 is given by the ratio V/I at every point along a transmission line when no standing waves are present in the line. V and I are the complex voltage and current at each point along the line.

transmission line matrix (TLM) A method of computing wave propagation similar to the ➤finite difference time domain method. The region of interest is broken up into a collection of small cubic elements, each of which contains information about the fields and material parameters at that point. However, instead of a calculation directly on the field components, an analogue to transmission lines is invoked where the fields are translated into voltages and currents and the material parameters are translated into capacitances and inductances. A three-dimensional network of transmission lines is then modelled using these parameters, with signals being applied and propagated at successive small time steps. When the calculation is terminated, the voltages and currents are then translated back to fields where desired.

transmission loss The reduction in power between any two points 1 and 2 in a telecommunication system given by the ratio P_2/P_1, where P_1 and P_2 are the powers at points 1 and 2 respectively and 2 is further away from the source of power than 1. For a given transmission path at a designated frequency the transmission loss is given by the ratio of the available power at the input of a receiver to the power transmitted from the output of the transmitter.

In any system where ➤gain is introduced to the signal, the *transmission gain* is given by the ratio of the powers P_2/P_1. This represents the increase in power between the points 1 and 2 where 2 is again further from the source of power than 1.

transmission mode ➤mode.

transmission primary Any of the set of three primary colours used to provide the chrominance signals in a ➤colour-television system. The colours chosen are red, green, and blue.

transmit-receive switch (TR switch) A switch used in a ➤radar system that has a common antenna for both transmission and reception. The TR switch automatically decouples the receiver from the antenna during the transmitting period. It is commonly used in conjunction with an *anti-transmit-receive switch* (ATR switch) that automatically decouples the transmitter from the antenna during the receiving period; extra protection is thus provided for the transmitter from return echos received by the antenna.

A common type of transmit-receive switch is a combination of a ➤gas-discharge tube and a ➤cavity resonator. The cavity resonator is used to connect the antenna to the transmitter (or receiver). When a discharge is established in the gas-discharge tube, i.e. when the tube is fired, the electrical conditions of the cavity resonator become such that resonance is not possible and the transmitter (or receiver) is therefore disconnected. A *keep-alive circuit* is frequently incorporated in the switch to maintain the discharge in the tube during periods when it is not operative.

transmittance ➤signal flowgraph.

transmitted-carrier transmission A telecommunication system that uses ➤amplitude modulation of a carrier wave in which the carrier is transmitted. ➤➤suppressed-carrier transmission.

transmitted-reference spread spectrum ➤digital communications.

transmitter The device, circuit, or apparatus used in a ➤telecommunication system to transmit an electrical signal to the receiving part of the system.

transmitting antenna ➤antenna.

transponder A combined transmitter and receiver system that automatically transmits a signal when a predetermined ➤trigger is received by it. The trigger, which is often in the form of a ➤pulse, is known as the *interrogating signal*. The minimum amplitude of the trigger signal that initiates a response from the transponder is the *trigger level*.

transposed transmission line A transmission line in which the conductors of which it is composed are interchanged in position at regular intervals of distance, in order to reduce the electric and magnetic coupling between the line and other lines.

transresistance The ratio of output voltage of a circuit or network to its input current, when both are real. ➤➤dependent sources.

transversal filter *Syn.* finite-impulse response (FIR) filter. A series of cascaded time-delay elements that are sampled or 'tapped' along the length of the delay line. The sample points may have different gains associated with them. The delayed and sampled signals are then summed to provide the output signal; the ►transfer function can be tailored to produce a filter characteristic, as shown in the diagram. Transversal filters can be realized using ►charge-coupled devices, ►bucket-brigade devices, and ►acoustic-wave structures. ►►digital filter.

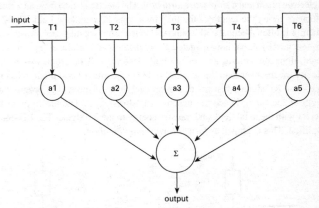

Transversal filter and the resulting sampled time function yielding the signal envelope

transverse wave A ►wave in which the displacement at each point along the wave is in a direction perpendicular to the direction of propagation. An electromagnetic wave is an example of a transverse wave. ►►mode.

trapezium distortion ►distortion.

trapping recombination ►recombination processes.

travelling wave An electromagnetic wave that is propagated along and is guided by a ►transmission line. In the case of a hypothetical lossless line of uniform cross section and infinite length, a sinusoidal a.c. supply at one end of the line (the sending end) causes electrical energy to be transmitted along the line with instantaneous values of current and voltage at any given point varying sinusoidally. If the line is situated in a medium of relative permittivity ε_r and relative permeability μ_r, the sine waves are propagated along the line with a velocity v given by

$$v = c/\sqrt{(\varepsilon_r\mu_r)}$$

where c is the speed of light. In the case of a lossless line of finite length terminated with a load impedance equal to the characteristic impedance of the line the above equation holds. In practice dissipation in the line causes both a reduction in the velocity and ►attenuation. Nonsinusoidal travelling waves, such as an ►impulse or ►surge voltage, may also be propagated.

Impedance discontinuities at any point on the line result in partial reflection of the initial wave, which then consists of two waves: the *reflected wave,* which travels in a backward direction towards the sending end, and the transmitted wave, which travels forward towards the receiving (output) end of the line.

travelling-wave amplifier ➤distributed amplifier.

travelling-wave tube (TWT) A microwave linear-beam vacuum tube that consists of an electron beam and a *slow-wave structure*. The electron beam from an ➤electron gun is focused along the length of the tube by a constant magnetic field. The slow-wave structure is often a helical coil wound round the tube (see diagram). An applied radio-frequency (RF) signal is propagated along the coil and produces an electric field directed along the central axis of the coil. This axial field progresses with a phase velocity much lower than the speed of light (in the ratio of coil pitch to its circumference – the RF signal must travel around each loop of the coil to progress along the length of the tube), hence the description 'slow-wave'. The electrons in the beam interact with the axial field, and transfer energy to the RF signal. The RF signal is thus amplified. This is known as *forward-wave amplification*.

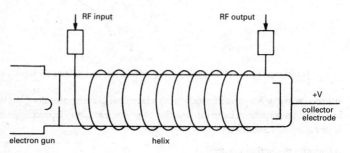

Travelling-wave tube

The *backward-wave oscillator* is a form of linear-beam travelling-wave tube in which the optimum transfer of power produces RF waves in the backward direction, i.e. the group velocity and phase velocity differ by 180°. A beam current of sufficient magnitude interacts with the slow-wave structure to produce RF oscillations that are delivered as microwave power at the electron-gun end of the structure. The minimum beam current required to produce oscillations is the *start-oscillation current*. At currents below this value the tube may be used as an amplifier by supplying an RF-wave input at the collector end of the structure. The interaction efficiency of this type of tube is increased by using an electron beam with a hollow cross section. This is achieved by magnetically confining the electron flow from the cathode.

trellis code ➤digital codes.

triac *Short for* triode a.c. switch. ➤thyristor.

triboelectricity ➤frictional electricity.

triboluminescence ➤luminescence.

trickle charge A small continuous charge applied to a battery in order to maintain it in a fully charged condition during storage. The charging current is maintained at a value that just compensates for internal dissipation due to local action within the battery.

trigger Any stimulus that initiates operation of an electronic circuit or device. Also the act of initiating operation in the circuit or device. In general the response to a trigger continues after the cessation of the stimulus. A circuit, such as a flip-flop or multivibrator, that is used to trigger other circuits is known as a *trigger circuit.* The output signal is often in the form of a *trigger pulse,* such as a ➤clock pulse (➤edge triggering).

trilevel resist ➤multilevel resist.

trimmer *Syn.* trimming capacitor. A relatively small variable capacitor used in ➤parallel with a large fixed capacitor in order to adjust the total capacitance of the combination over a limited range of values.

trimming capacitor ➤trimmer.

Trinitron *Trademark. See* colour picture tube.

triode Any electronic device with three electrodes, such as a ➤bipolar junction transistor, a ➤field-effect transistor, or, in particular, a ➤thermionic valve that has three electrodes.

triode-hexode A multiple (or multielectrode) ➤thermionic valve that contains both a triode and a hexode within the same envelope. The triode-hexode is most often used as a frequency changer: it acts as a combined oscillator (triode portion) and ➤mixer (hexode portion).

triode region ➤field-effect transistor; saturated mode.

trip coil ➤tripping device.

tripler ➤frequency multiplier.

tripping device A device that normally constrains a ➤circuit-breaker in the 'on' position until actuated, when the circuit-breaker is allowed to break the circuit. Manual operation is common in many types of tripping device. Other types are operated electromagnetically. A typical example is the *trip coil,* which consists of a coil that controls a movable plunger or armature; the plunger controls the action of the circuit-breaker.

tristate logic gate A logic device that has three possible output states: '1', '0', and 'high impedance' (rather than the normal two states '1' and '0'). The high-impedance state isolates the device electrically from the rest of a circuit. Tristate logic gates allow many devices to be connected onto the same data lines, such as data and address buses. However, only one device is 'connected' at any one time, all others being in their high-impedance state and thus electrically disconnected.

tropospheric ducting The phenomenon of long-range ►radiowave propagation caused by downward refraction of the waves in the troposphere – the earth's lowest atmospheric layer – and subsequent reflection at the earth's surface. The refraction and reflection continues and the waves appear to be conveyed through a duct between the maximum height of the bending and the earth's surface. The refraction is caused by the continuous variation in the atmospheric ►dielectric constant. ►►tropospheric scattering.

tropospheric scattering The phenomenon whereby ►radiowaves and low-frequency ►microwaves are scattered by sharp variations in the atmospheric ►dielectric constant that occur in regions of the troposphere (the earth's lowest atmospheric layer). These variations can account for beyond-the-horizon transmissions in telecommunication systems. ►►tropospheric ducting.

TR switch *Abbrev. for* transmit-receive switch.

truncated test ►life test.

trunk ►trunk feeder.

trunk feeder *Syns.* interconnecting feeder; interconnector; trunk; trunk main. A ►transmission line that is used to interconnect two electric ►power stations or two electric power distribution networks.

trunk main ►trunk feeder.

trunk telephony ►telephony.

truth table A table used in formal logic that lists the truth or falsity of the outcome when a logical operator, such as 'and' or 'or', is applied to combinations of logical statements. The truth table has been adapted to describe the operation of ►logic circuits by listing the outputs of a binary logic gate, such as a ►NAND circuit or ►flip-flop, for all possible combinations of inputs. The 'true' state corresponds to the voltage level representing a logical 1 and 'false' to logical 0.

T-section ►two-port network; filter.

TTL *Abbrev. for* transistor-transistor logic.

tube *Short for* electron tube, especially a vacuum tube.

tunable magnetron ►magnetron.

tuned amplifier An amplifier that is designed to operate only with a very narrow range of input frequencies. A typical frequency response curve is shown in Fig. *a*. The response may be modified to give an essentially flat response to the frequencies in the pass band by using stagger tuning. A *stagger tuned amplifier* consists of two or more ►amplifier stages each of which is tuned to slightly different frequencies, f_1 and f_2. The overall response is essentially flat within the pass band (Fig. *b*), the centre of the peak occurring at the frequency corresponding to

$$|f_1 - f_2|/2$$

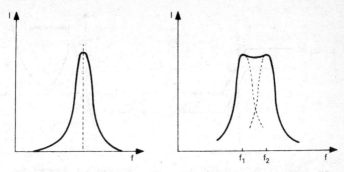

a Frequency response of a **tuned amplifier**   *b* Response of a stagger **tuned amplifier**

tuned circuit A ►resonant circuit that contains some form of *tuning* so that the natural ►resonant frequency of the circuit may be varied. The resonance condition for ►forced oscillations can thus be changed. A tuned circuit is used to select the frequency of resonance of a variety of frequency-selective devices, such as audiofrequency amplifiers. Tuning may be carried out by adjusting the value of the capacitance (*capacitive tuning*), or the inductance (*inductive tuning*), or both. Inductive tuning is commonly achieved by altering the position of a suitably shaped piece of soft ferromagnetic material (a slug) relative to a coil in the circuit (*slug tuning*).

tuner 1. A device, such as a variable capacitor or inductor, that is used to alter the resonant frequency of a ►tuned circuit. **2.** The first stage of a ►radio or ►television receiver that is used to select a particular broadcast channel. It contains either a ►tuned circuit in which the resonant frequency may be altered to accept the desired channel frequency or a set of ►resonant circuits each of which is tuned to an individual channel and may be selected (►turret tuner).

tungsten Symbol: W. A heavy metal, atomic number 74, that has an extremely high melting point and is extensively used to form lamp filaments. It has also been widely used to form ►thermionic cathodes. It is resistant to ►electromigration and is used as an interconnection on some IC processes.

tuning ►tuned circuit; electronic tuning.

tuning screw ►waveguide.

tunnel diode *Syn.* Esaki diode. A p-n junction diode that has extremely high doping levels on each side of the junction, i.e. it is doped into ►degeneracy. The energy bands are shown in Fig. *a* for zero applied bias. At such high doping levels tunnelling of electrons (►tunnel effect) across the junction can occur in the forward direction (positive applied voltage on the p-region). As the positive bias is increased the height of the potential barrier at the junction is decreased and the width increased; the diode therefore exhibits a negative resistance portion of the characteristic as the tunnel effect contributes progressively less towards the conductance. A minimum current is reached at the

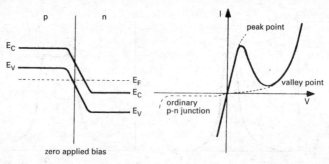

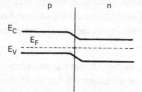

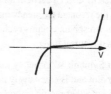

a Energy bands of **tunnel diode**   *b* Characteristic of **tunnel diode**

valley point when the tunnel effect ceases and for voltages above this point the diode behaves as a normal p-n junction diode. Tunnelling also occurs in the ►reverse direction in a similar manner to that in the ►Zener diode but the effective Zener breakdown voltage can be considered to occur at a small positive value of voltage, termed the *peak point*. A typical tunnel diode characteristic is shown in Fig. *b*.

The *backward diode* is similar to the tunnel diode but the doping levels are slightly lower so that the semiconductor regions are not quite degenerate (Fig. *c*). The tunnel effect occurs readily for small values of reverse bias but the negative-resistance portion of the characteristic disappears. The current flowing in the reverse direction is larger than in the forward direction (Fig. *d*).

c Energy bands of backward diode   *d* Characteristic of backward diode

The backward diode can be used for rectification of small signals, when the conventional forward-direction positive bias on the p-region becomes the reverse direction for this device. ►Carrier storage at the junction does not occur and the backward diode therefore has a high speed of response and may be used at microwave frequencies. Very little variation of the current-voltage characteristics occurs with temperature or incident radiation. Unfortunately the high doping levels required for both tunnel diodes and backward diodes make reliable manufacture difficult.

tunnel effect (or **tunnelling**) The crossing of a potential barrier by a particle that does not have sufficient kinetic energy to surmount it. The effect is explained by wave mechanics. Each particle has an associated wave function that describes the probability of finding the particle at a particular point in space. As a particle approaches the potential barrier the wave function is considered to extend inside the region of the potential barrier. Provided that the barrier is not infinitely thick a small but finite probability exists that the particle will appear on the other side of the barrier. As the thickness of the barrier decreases the probability of tunnelling through it increases.

The tunnel effect is the basis of ►field emission and the ►tunnel diode. ►►quantum mechanics.

Turing machine A mathematical model of a device that changes its internal state and reads from, writes on, and moves a potentially infinite tape, all in accordance with its present state, thereby constituting a model for computer-like behaviour. The model was published by Alan Turing in 1936.

turns ratio *Syn.* transformer ratio. Symbol: *n*. The ratio of the number of turns, n_2, active in the secondary circuit of a transformer to the number of turns, n_1, in the primary winding. In the case of a simple transformer with no ►tappings n_2 is the number of turns in the secondary winding. In the case of a current transformer the ratio of the currents is the inverse of the turns ratio, i.e.

$$I_{out}/I_{in} = 1/n$$

►►transformer.

turret tuner A tuning device used in a ►television or ►radio receiver. It contains a set of resonant circuits each tuned to the frequency of one of the separate broadcast channels. One or more manually operated switches, termed *band switches,* allow the particular circuit corresponding to the desired channel to be selected by the user.

tweeter A physically small loudspeaker that reproduces sounds of relatively high frequency, for example, frequencies from five kilohertz upwards. In a high-fidelity sound-reproduction system a tweeter and a ►woofer are used together.

twin cable ►paired cable.

twin-T network *Syn.* parallel-T network. ►two-port network.

twisted pair ►pair.

two-phase system *Syn.* quarter-phase system. ►polyphase system.

two-port analysis ►network.

two-port network *Syns.* four-terminal network; quadripole. A ►network that has only four terminals, i.e. a pair of input terminals (the input ►port) and a pair of output terminals (the output port). The behaviour of a two-port network is usually described by the impedances presented at its terminals at specified frequencies. If the electrical properties are unchanged when the input and output ports are reversed the two-port network is *symmetric*; otherwise it is *asymmetric*. The network is described as *balanced* if its electrical properties are unchanged when both the input and output ports are interchanged simultaneously; otherwise it is *unbalanced*.

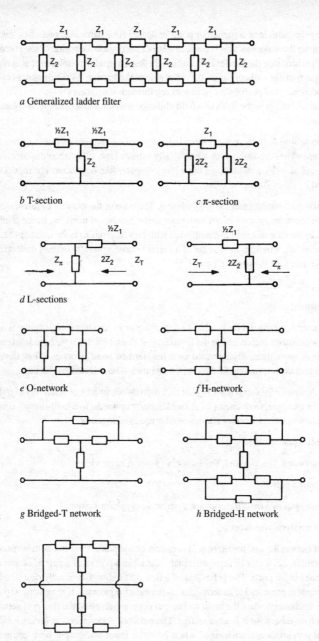

a Generalized ladder filter

b T-section

c π-section

d L-sections

e O-network

f H-network

g Bridged-T network

h Bridged-H network

i Twin-T network

A very common arrangement, used particularly for ►attenuators and ►filters, is a ladder network consisting of a number of series and shunt impedances (Fig. *a*). This arrangement may be broken down for analysis into identical sections each with the same characteristic impedance. In order to avoid power dissipation in the load by reflections, a ladder network must be terminated by an impedance equal in value to the ►iterative impedance of the sections. The ladder filter shown in Fig. *a* may be analysed as a series of *T-sections* (Fig. *b*) terminated in the iterative impedance Z_T. The same network may be considered as a series of π-*sections* (Fig. *c*.) that must be terminated in the iterative impedance Z_π. Comparison of the two shows that the half-section or *L-section* (Fig. *d*) acts as an impedance transformer from Z_π to Z_T. Such half-sections may be used to match a two-port network to a load; they are especially important when composite networks are being designed. It can be shown that if the impedances used in the ladder are Z_1 and Z_2, as shown in Fig. *b*, then

$$Z_T Z_\pi = Z_1 Z_2$$

In general Z_1 and Z_2 and hence Z_T and Z_π will be dependent on frequency. If the product $Z_T Z_\pi$ is substantially independent of frequency the network is a *constant-R network*.

Other arrangements of elements used to form ladder filters or attenuators are the *O-network*, *H-network*, *bridged-T*, *bridged-H*, and *twin-T networks*, which are shown in Figs. *e, f, g, h,* and *i*.

two-port parameters ►network.

two's-complement notation An integer representation in ►binary notation in which negative numbers are obtained from positive numbers, and vice versa, by complementing all the bits of a number – converting 1s to 0s and vice versa – and adding 1 (discarding the carry); for example, +3 in 4-bit two's complement is 0011_2 and –3 in 4-bit two's complement is $1100_2 + 1 = 1101_2$.

two-tone modulation ►amplitude modulation; frequency modulation.

two-wire circuit A circuit that consists of two conductors insulated from each other and that provides simultaneously a two-way communication channel in the same frequency band between two points in a ►telecommunication system. The circuit may be a ►phantom circuit in which case it is formed from two groups of conductors. A circuit that operates in the same manner but that is not limited to only two conductors (or groups of conductors) is termed a *two-wire type circuit*. ►transmission line; four-wire circuit.

TWT *Abbrev. for* travelling-wave tube.

U

UART *Abbrev. for* universal asynchronous receiver/transmitter. An input/output device that sends and receives information in bit-serial fashion.

UHF *Abbrev. for* ultrahigh frequency. ►frequency band.

UJT *Abbrev. for* unijunction transistor.

u-law In digital communications, the coding law used by the standard 24-channel ►pulse code modulation systems.

ultrahigh frequency (UHF) ►frequency band.

ultrasonic communication Underwater communication at ultrasonic frequencies using suitably modified ►sonar.

ultrasonic delay line An acoustic ►delay line that uses acoustic waves of ultrasonic frequencies as the delay element.

ultrasonics The study and application of sound frequencies that lie above the upper limit of the human hearing range, i.e. frequencies above about 20 kilohertz.

Ultrasonic waves may be generated electronically by *magnetostriction generators* or *piezoelectric generators.* In the former case a high-frequency electric oscillation is applied to a material that exhibits ►magnetostriction so as to produce ultrasonic waves. In the latter the ►piezoelectric effect transforms the electric oscillations into ultrasonic waves.

Ultrasonic waves may be detected using the ►hot-wire microphone. The only electrical receiver that is sensitive to ultrasonic waves is the ►piezoelectric quartz crystal. When excited by an ultrasonic wave that is propagated in a suitable direction and corresponds in frequency to the natural frequency of the receiving crystal, mechanical vibrations are set up in the crystal and produce varying electric potentials. Since two crystals of the same natural frequency are very difficult to obtain, the receiving crystal must be electrically tuned to the frequency of the received wave. A type of detector due to Pierce uses the same piezoelectric crystal as both transmitter and receiver.

Ultrasonic waves have many applications. In electronics the main applications include degassing of melts during the manufacture of electronic components and circuits, cleaning of electronic components, and ►echo sounding and underwater communication. Ultrasonics is used in medicine for imaging of, for example, a brain tumour or a foetus.

ultraviolet photoelectron spectroscopy (UPS) ►photoelectron spectroscopy.

ultraviolet radiation Electromagnetic radiation lying between light and X-rays on the electromagnetic spectrum. Radiation of frequency close to that of light is *near ultraviolet* radiation; that at the high-frequency end of the range is *far ultraviolet*.

unabsorbed field strength The field strength of a radiowave that would exist at the receiving point in the absence of any ➤absorption between the transmitting and receiving antennas.

unbalanced two-port network ➤two-port network.

unclocked flip-flop ➤flip-flop.

underbunching ➤velocity modulation.

undercoupling ➤coupling.

undercurrent release A switch, circuit-breaker, or other tripping device that operates when the current in a circuit falls below a predetermined value. A current that causes the release to operate is termed an *undercurrent*. ➤➤overcurrent release.

underdamping ➤damped.

underscanning ➤scanning.

undershoot An initial transient response to a change in an input signal that precedes the desired response and is in the opposite sense. ➤➤overshoot.

uniconductor waveguide ➤waveguide.

unidirectional current ➤current.

unidirectional microphone ➤cardioid microphone.

unidirectional transducer *Syn. for* unilateral transducer. ➤transducer.

uniform cable A cable with constant electrical characteristics along its length.

uniform line A transmission line with electrical properties that are substantially identical along its length.

uniform waveguide A waveguide that has constant electrical and physical characteristics along its axial length.

uniform window ➤windowing.

unijunction transistor (UJT) A ➤bipolar junction transistor that has three terminals and one junction (➤Fig. *a*). The device consists of a bar of lightly doped (high-resistivity) ➤semiconductor, usually n-type, with an opposite-polarity region of highly doped (low-resistivity) material located near the centre of the bar. Ohmic contacts are formed to each end of the bar (base 1 and base 2) and to the central region (emitter). Originally the emitter region was alloyed into the bulk material but planar diffused or planar epitaxial structures are now produced.

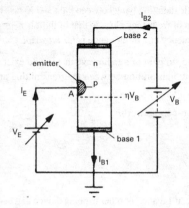

a **Unijunction transistor**

Under normal operating conditions base 1 is earthed and a positive bias, V_B, applied to base 2. A point A on the least positive side of the emitter junction is considered. The voltage on the n side of the junction at point A is given by ηV_B, where η is the *intrinsic stand-off ratio*:

$$\eta = R_{B1}/R_{BB}$$

R_{B1} is the resistance between point A and base 1 and R_{BB} the resistance between base 1 and base 2.

If the voltage, V_E, applied to the emitter is less than ηV_B the junction is ►reverse biased and only a small reverse saturation current flows. If V_E is increased above ηV_B the junction becomes ►forward biased at point A and holes are injected into the bar. The electric field within the bar causes the holes to move towards base 1 and increase the conductivity in the region between point A and base 1. Point A becomes less positive and therefore more of the junction becomes forward biased; this causes the emit-

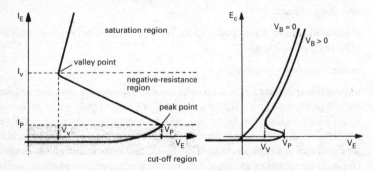

b Part of emitter I-V characteristic   *c* Emitter I-V characteristics

ter current, I_E, to increase rapidly. As I_E increases the increased conductivity causes the emitter voltage to drop and the device exhibits a negative-resistance portion of the current-voltage characteristic (Fig. *b*).

The emitter voltage at which the device starts to conduct, V_p, is known as the *peak point*. At voltage V_v, known as the *valley point,* the device ceases to exhibit negative resistance. The switching time between the peak point and the valley point depends on the device geometry and the biasing voltage, V_B, applied to base 2. It has been found to be proportional to the distance between point A and base 1.

If base 2 is open circuit, the I-V curve is essentially that of a simple ➤p-n junction (Fig. *c*). As V_B is increased the peak point, V_p, and the current, I_v, of the valley point also increase. The characteristics show little temperature dependence.

The most common application of the unijunction transistor is in ➤relaxation oscillator circuits.

unilateral network ➤network.

unilateral transducer *Syn.* unidirectional transducer. ➤transducer.

uninterrupted duty ➤duty.

unipotential cathode An indirectly heated ➤thermionic cathode.

unit A precisely specified quantity, such as the ampere or second, in terms of which the magnitude of other physical quantities of the same kind can be stated:

physical quantity = numerical value × unit

current = n amperes

charge = n amperes per second = n coulombs

In a purely mechanical system the units can be defined in terms of three base units of mass, length, and time: other units are derived by multiplying and/or dividing these base units. In any system concerned with electric and magnetic properties a fourth base unit is necessary to make the system totally self-consistent. In the past, various choices of this quantity and its value led to several different systems of units, including the ➤CGS and ➤MKS systems.

The system of ➤SI units was formulated, by international agreement, in order to produce a coherent set of units that can be used in every field of science and technology in a consistent manner. There are seven base SI units from which almost every other unit can be derived. ➤➤Tables 4–9 in the back matter.

unit circle A circle of unit radius centred at the origin of the z-plane. ➤z transform.

unit delta function ➤Dirac delta function.

unit sample A sampled waveform, usually denoted $\delta[n]$, that consists of an isolated sample with an amplitude of unity and all other samples with amplitudes of zero:

$\delta[n] = 0, \qquad n \neq 0$

$\delta[n] = 1, \qquad n = 0$

unit-step function An electrical signal that is zero for all times before a predetermined instant and unity for times following that instant.

universal active filter ►state variable filter.

universal motor An electric ►motor that may be used with a DC or AC power supply.

universal shunt A galvanometer ►shunt that is tapped so that it can pass designated fractions (0.1, 0.01, etc.) of the main current and can thus be used with galvanometers of widely varying internal resistance. An example is the *Ayrton shunt,* which is a relatively large resistance with suitable tappings.

univibrator ►monostable.

unstable oscillation ►oscillation.

up conversion Mixing of a (radio) signal with a local-oscillator signal to result in a new signal at a higher frequency than the originally received radiofrequency. ►heterodyne reception. ►►down conversion.

uplink The radio-communications path from a fixed transmitting site to a moving receiving device. ►►downlink.

upper sideband ►carrier wave.

UPS *Abbrev. for* ultraviolet photoelectron spectroscopy. ►photoelectron spectroscopy.

upsampling The process of increasing the sampling rate of a sampled signal.

useful life ►failure rate.

UTP *Abbrev. for* unshielded twisted pair. ►pair.

V

vacancy A site in a crystal lattice that is not occupied by an atomic nucleus.

vacuum evaporation A technique used during the manufacture of electronic circuits and components in order to produce a coating or thin film of one solid material on the clean surface of another, in particular a metal on a semiconductor. The solid to be deposited is heated in a vacuum in the presence of the cool substrate material. Atoms evaporated from the material (either in solid or liquid form) suffer few collisions from the residual low-pressure gas and travel directly to the substrate. They condense on the substrate surface and form a thin film. ➤thin-film circuit.

vacuum microelectronics A branch of ➤microelectronics in which the ➤active devices use the principles of ➤vacuum tubes for their operation: they are known as *solid-state valves*. The ➤cathode is a ➤field emitter constructed from a fine point of ➤semiconductor material, usually silicon. The ➤anode can be a metallic contact that collects the electrons emitted from the field-emission cathode. The flight of the electrons between the electrodes can be modified or modulated by interspersing other electrodes or grids to control the anode current. The complete structure may be only a few micrometres to a few millimetres in size, much smaller than conventional tubes; it is thus able to operate at very high frequencies, including microwave frequencies of around 10 gigahertz. These devices can be made using the standard processing techniques of silicon planar technology, and are therefore less expensive and more reliable than their glass tube counterparts.

vacuum tube An ➤active device in which conduction between two electrodes takes place in an envelope that is sealed and evacuated to a sufficiently low pressure that its electrical characteristics are independent of any residual gas. Tubes frequently contain additional electrodes to modify the electrical behaviour. The term vacuum tube, originally American, is applied in particular to the ➤thermionic valve. Such devices, largely displaced by their solid-state equivalents, are still used for example for high-frequency high-power applications such as radio communications.

valence band The band of energies in a solid in which the electrons cannot move freely but are bound to individual atoms. Mobile ➤holes are formed in the valence band. ➤conduction band; energy bands.

valence electrons Electrons that occupy the outermost (highest energy) energy levels of an atom and are involved in chemical and physical changes, such as bonding.

valency The number of hydrogen (or equivalent) atoms with which an atom will combine or that it will displace.

valley ➤pulse.

valley point 1. ➤unijunction transistor. **2.** ➤tunnel diode.

valve *Syn.* electron tube. An ➤active device in which two or more electrodes are enclosed in an envelope, usually of glass, one of the electrodes acting as a primary source of electrons. The electrons are most often provided by thermionic emission (in a ➤thermionic valve) and the device may be either evacuated (vacuum tube) or gas-filled (gas-filled tube). The name derives from the rectifying properties of the devices, i.e. current flows in one direction only. The word valve is becoming obsolete and is being replaced by electron tube.

valve diode *Syn.* vacuum diode. ➤thermionic valve.

valve voltmeter A type of ➤voltmeter that has now generally been superseded by the ➤digital voltmeter. It consists of an ➤amplifier of extremely high ➤input impedance and containing one or more ➤thermionic valves, with a measuring instrument in the output circuit. Direct voltage or alternating voltage may be measured with this instrument.

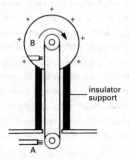

Van de Graaff generator

Van de Graaff generator An electrostatic ➤voltage generator that can produce potentials of millions of volts. An electric charge from an external source is applied to a continuous insulated belt at points A (see diagram). The belt travels vertically up into a large hollow metallic sphere and the charge is collected at points B. The charge resides on the exterior of the sphere and the possible voltage generated is limited only by leakage.

vapour phase epitaxy (VPE) The most common method of ➤epitaxy. The material to be deposited on the substrate is heated to gaseous form in a deposition furnace that also contains the substrate material. The substrate is held at a temperature just below the solidification point. As the gas molecules reach the substrate they are deposited on the surface, replicating the substrate crystal structure. The conditions in the deposition furnace can be adjusted with respect to temperature and pressure to allow particular desired combinations of substrate and deposited material to be produced. ➤chemical vapour deposition.

vapour plating A technique for producing a thin film of one solid material on the clean surface of another solid. A compound of the material to be deposited is vaporized and thermally decomposed in the presence of the substrate material. Atoms from the vapour condense on the surface to be plated and form a thin film on it.

var The unit of reactive ➤power of an alternating current.

varactor Acronym from *var*iable re*actor*. A ➤p-n junction diode or ➤Schottky diode operated with reverse bias so that it behaves as a voltage-dependent capacitor. The ➤depletion layer at the junction acts as the dielectric and the n- and p-regions form the plates (see diagram). A diode intended for use as a varactor generally has a particular impurity profile designed to give an unusually large variation of junction capacitance and to minimize the series resistance.

In a diode with an ➤abrupt junction the voltage dependence is given by

$$C \propto V^{-1/2}$$

where C is the capacitance and V the reverse voltage. ➤hyperabrupt varactor.

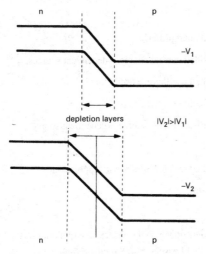

Energy bands in a **varactor**

varactor tuning Capacitive tuning (➤tuned circuit) that is employed in ➤receivers, such as television receivers, in which the variable capacitance element is provided by a ➤varactor.

variable capacitance gauge ➤strain gauge.

variable impedances ➤Impedances, such as capacitors, that are adjustable so as to present variable values of the impedance. The method of adjustment is usually provided by a movable contact or by physical adjustment of the size of the device.

variable inductance gauge *Syn. for* electromagnetic strain gauge. ➤strain gauge.

variable resistance gauge *Syn. for* resistance strain gauge. ➤strain gauge.

Variac *Trademark* An ➤autotransformer used in laboratories to provide a continuously variable a.c. voltage supply.

variation ➤deviation.

variocoupler A ➤transformer used in radio circuits that is constructed in such a way that the mutual inductance between the windings can be varied while the self-inductance of each winding remains substantially constant.

varistor *Short for* variable resistor. A two-terminal device whose resistance varies with the magnitude of the applied voltage, in particular increasing rapidly at high voltages. Such a device can be used to suppress large voltage spikes, although the switching time is not rapid. ➤transient suppressor.

varying duty ➤duty.

V-beam radar ➤radar.

VCA *Abbrev. for* voltage-controlled amplifier.

VCCS *Abbrev. for* voltage-controlled current source. ➤dependent sources.

VCO *Abbrev. for* voltage-controlled oscillator. ➤oscillator.

VCR *Abbrev. for* video cassette recorder. ➤videotape.

VCVS *Abbrev. for* voltage-controlled voltage source. ➤dependent sources.

VCVS filter ➤Sallen–Key filter.

VDR *Abbrev. for* voltage-dependent resistor. ➤transient suppression.

VDU *Abbrev. for* visual display unit.

vector field ➤field.

velocity microphone A ➤microphone in which the electrical output is a function of the particle velocity of the detected sound waves. ➤ribbon microphone.

velocity-modulated tube ➤velocity modulation.

velocity modulation A process that introduces a radiofrequency (r.f.) component into an electron stream and thereby modulates the velocities of the electrons in the beam. Individual electrons will be either accelerated or retarded by the radiofrequency signal depending on the relative phase of the r.f. component at the point of interaction with the electrons. The velocity modulation therefore causes *bunching* of the electron beam as it travels down the electron tube since the faster electrons catch up with preceding slower ones (and conversely).

The amount of bunching of an electron stream varies with the distance travelled from the point of first interaction with the r.f. field. *Ideal bunching* is the production of small sharply defined bunches of electrons with no electrons in the regions between bunches. In practice however this is not achieved. *Optimum bunching* occurs at a particular distance down the tube for any given tube and is the condition when the minimum size of bunches containing the maximum possible numbers of electrons is achieved. *Underbunching* is the condition of less than optimum bunching. If the electron beam is allowed to travel beyond the point of optimum bunching faster electrons begin to leave the slower ones behind. This condition is termed *overbunching*. If the electron-beam direction is reversed by a reflector during the transit interval the resulting bunching is known as *reflex bunching*.

Velocity-modulated tubes are used as microwave oscillators and amplifiers. They include the ►klystron and the ►travelling-wave tube. The r.f. field can be made to interact with the electron stream in one sharply defined region, such as a ►cavity resonator; the region is known as the *buncher*. The electron beam is then allowed to travel through a field-free drift space. This method is employed in the klystron. An alternative method is used in travelling-wave tubes: the r.f. field is propagated along the length of the electron tube and interaction between the electron beam and the field is a continuous two-way process.

velocity-sensitive keyboard *Syn.* touch-sensitive keyboard. A keyboard on an electronic musical instrument where the velocity at which keys are depressed and released can be used to control ►synthesis parameters by means of ►MIDI.

vertical blanking ►blanking.

vertical FET (VFET) A FET (►field-effect transistor) in which the ►source is formed on the back of a semiconductor wafer and current flows vertically through the device rather than horizontally as in the more usual structure. The cross section of a VFET is shown in Fig. *a*. A variation of the VFET is the *permeable base transistor* in which an epitaxial layer is formed over the ►gate electrodes before the ►drain is formed (Fig. *b*).

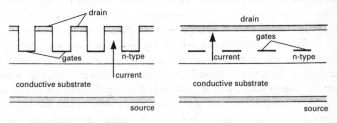

a Cross section of **vertical FET**   *b* Cross section of permeable base transistor

vertical hold ➤television receiver.

vertical sync *Syn for.* vertical hold. ➤television receiver.

very high frequency (VHF) ➤frequency band.

very large scale integration ➤VLSI.

very low frequency (VLF) ➤frequency band.

vestigial sideband ➤carrier wave.

vestigial-sideband transmission A form of ➤suppressed-carrier transmission in which one sideband and the corresponding vestigial sideband (➤carrier wave) are transmitted. ➤double sideband transmission

VFET *Short for* vertical FET (field-effect transistor).

VGA *Abbrev. for* video graphics array. An IBM standard video display that has a resolution of 480×640 pixels with 16 colours or 320×200 pixels with 256 colours. *SVGA* (*superVGA*) is a version of VGA that has a resolution of 800×600 pixels with 16 colours; the term SVGA is often used to describe systems with higher spatial and colour resolutions as well.

V-groove technique An etching technique used to produce a very precise edge in a silicon crystal. The technique is used to produce mesa layers or during the manufacture of ➤VMOS circuits.

A silicon crystal with (100) orientation is etched from the surface by a suitable etch, such as potassium hydroxide, that will not etch perpendicularly to the (lll) plane. The etching process therefore follows the direction of the (lll) crystal planes and stops at any point where the (lll) planes intersect. This results in a very precise V-groove in the material. The depth of the groove depends on the size of the original opening in the surface oxide layer and etching to a precise predetermined depth is therefore possible with this technique.

VHDL A ➤hardware description language (HDL) developed for very high speed integrated circuits. It is a language for describing digital electronic systems in an unambiguous way, and is designed to fill a number of needs in the design process. Firstly, it allows description of the structure of a system – how it is decomposed into subsystems and how those subsystems are interconnected. Secondly, it allows the specification of the function of a system using familiar programming language forms. Thirdly, as a result, it allows the design of a system to be simulated before being manufactured, so that designers can quickly compare alternatives and test for correctness without the delay and expense of hardware prototyping. Fourthly, it allows the detailed structure of a design to be synthesized from a more abstract specification, allowing designers to concentrate on more strategic design decisions and reducing time to market.

VHF *Abbrev. for* very high frequency. ➤frequency band.

via A plated hole used to provide low-inductance connections on an ➤integrated circuit. A via may be formed through the dielectric layer on the top of a wafer to interconnect two metallization patterns. A via may also be formed though the back of a wafer containing FETs where a low-inductance earth is required; this is especially useful for microwave or power FETs, and metal plating is used on the back of the wafer to provide a heat sink and earth plane simultaneously (see diagram).

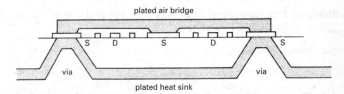

Vias formed through the back of a wafer

vibration galvanometer ➤galvanometer.

vibrator A device that produces an alternating current by periodically interrupting or reversing a continuous steady current from a direct-current source. The vibrator consists of an electromagnetic ➤relay with a vibrating armature that alternately makes and breaks one or more pairs of contacts.

The most common application is in a power-supply unit in which a high-voltage direct current must be produced from a low-voltage d.c. source, such as a battery. The vibrator produces a low-voltage periodically varying current that is transformed (➤transformer) into a high-voltage a.c. supply and then rectified to produce a high-voltage d.c. supply. A ➤rectifier circuit may be used to produce the direct-current output or the vibrator itself can be used for this purpose. In the latter case the vibrator (termed a *synchronous vibrator*) is fitted with an extra pair of contacts that are used to reverse the connections to the secondary winding of the transformer in synchronism with the reversal of the current in the primary winding so that the output from the transformer secondary circuit is a direct current.

video 1. A video recorder that is used with domestic television receivers. ➤videotape. **2.** A cassette of videotape on which a film, TV programme, etc. has been recorded.

video amplifier ➤video frequency.

video camera A ➤television camera that is designed for use with ➤videotape rather than direct broadcasting. The term most commonly applies to cameras used to prerecord outside broadcasts or for small portable domestic systems.

video cassette recorder (VCR) ➤videotape.

videoconference ➤teleconference.

video frequency The frequency of any component of the output signal from a ➤television camera. Video frequencies are within the range 10 hertz to 2 megahertz. An amplifier that is designed to operate with video-frequency signals is termed a *video amplifier.*

video IF system ➤intercarrier system.

video mapping A technique used in the display of a ➤radar system in which a chart or map of the area covered by the radar is superimposed electronically on the radar display.

videophone A telephone receiver that can receive and display visible images simultaneously with the telephone signals.

video recorder ➤videotape.

video signal *Syn.* picture signal. ➤camera tube; television.

videotape A form of magnetic tape that is suitable for use with a ➤television camera. Simultaneous recording of the video signal from the camera and the audio signal from the microphone system is carried out on separate tracks on the tape. The signals may later be output directly to the modulating circuits of the transmission system. Many television programmes are recorded on videotape prior to transmission rather than being broadcast live.

A form of videotape recorder is available for use with domestic television receivers. Usually referred to as a *video recorder* or *video cassette recorder* (VCR), or simple as a *video*, it can be used either to record a received broadcast on videotape or to replay a prerecorded videotape directly into the television receiver. The videotape is contained within a closed *video cassette.*

videotex An information service in which information from one or more sources may be transmitted to users over telephone lines. Communication with the information source is made using the keyboard of a computer, which is connected to the telephone through a modem. The information is transmitted in the form of pulse-code modulated signals. ➤teletext.

vidicon A low-electron-velocity photoconductive ➤camera tube that is widely used in closed-circuit television and as an outside broadcast camera since it is smaller, simpler, and cheaper than the image orthicon.

The photosensitive target area of the vidicon consists of a transparent conducting film placed on the inner surface of the thin glass faceplate (Fig. *a*). A thin layer of photoconductive material (➤photoconductivity) is deposited on the conducting layer and a fine mesh grid placed in proximity to the photoconductive layer. The conducting layer acts as the *signal electrode* from which the output is obtained and is held at a positive potential. The photoconductor can be considered as an array of leaky capacitors with one plate electrically connected to the signal electrode and the other floating except when subjected to an electron beam (Fig. *b*). The surface of the photoconductive layer is charged to cathode potential by a beam from the electron gun and each elemental capacitor therefore becomes charged.

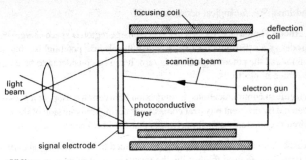

a **Vidicon**

The optical image is produced on the target electrode, and the value of each effective leakage resistor is determined by the intensity of illumination of the corresponding element on the target area (➤photoconductivity). During the frame period, each elemental capacitor discharges by an amount that depends on the value of the corresponding leakage resistor; a positive potential pattern thus appears on the electron-gun side of the target, corresponding to the optical image.

The target is scanned with a low-velocity electron beam and each elemental capacitor is again charged to cathode potential by electrons from the electron beam. As the target is scanned current flows in the signal electrode; the magnitude of the current developed is a function of the charge on the target area and hence of the illumination from the optical-lens system.

In the *plumbicon,* a development of the vidicon tube, the target elements may be considered as semiconductor current sources controlled by light energy. When not illuminated, the target elements are essentially reverse-biased Schottky diodes with a very low reverse saturation current. The principle advantages of these tubes compared to the vidicon are the low dark current and good sensitivity and light-transfer characteristics.

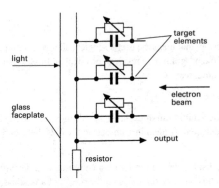

b Photosensitive target area of **vidicon**

virtual address *Syn. for* logical address.

virtual cathode The surface, located in a space-charge region (➤space charge) between the electrodes of a ➤thermionic valve, at which the electric potential is a mathematical minimum and the potential gradient is zero. It can be considered to behave as if it were the source of electrons.

virtual circuit In communications, a connection between two users of a network that can be realized by different physical connections during transmission of the message. Virtual circuits commonly exist in networks that use ➤packet switching.

virtual earth A node in a circuit that can be assumed to be (virtually) at earth potential. The term is often used to describe the inverting input of an ➤operational amplifier connected in the inverting amplifier configuration (see diagram). The output voltage of the amplifier is given by

$$v_o = A(v_2 - v_1) \text{ and hence } v_2 - v_1 = v_o/A$$

If A is large then $v_2 - v_1 \approx 0$, i.e. virtually zero.

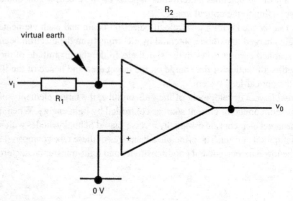

Virtual earth

virtual memory A memory system that maps the user's ➤logical addresses into physical ➤addresses; hence parts of the program may reside at physical addresses that bear no relationship to the logical addresses.

virtual value ➤root-mean-square value.

visibility factor The ratio of the ➤minimum discernible signal (mds) of a ➤television or ➤radar receiver that can be detected by ideal instruments to the mds that can be detected by a human operator.

visual display unit (VDU) A display device used with a computer that displays information in the form of characters and line drawings on the screen of a ➤cathode-ray tube. A VDU is most often associated with a ➤keyboard and/or ➤touch screen and/or

➤light pen, which allow the information in the display to be altered or new data to be input, and a printer for a permanent record of the display. ➤terminal.

VLF *Abbrev. for* very low frequency. ➤frequency band.

VLSI *Abbrev. for* very large scale integration, describing all ➤integrated circuits that are larger than a few thousand gates: this includes most ➤microprocessor, ➤memory, and ➤digital signal processing ICs currently available.

VMOS MOS circuits or transistors that are fabricated using the ➤V-groove technique. Regions of the required conductivity type are formed in (100). orientation silicon crystals using a combination of planar diffusion and epitaxy. V-groove etching is then performed, and the gate electrode formed in the groove, in order to produce vertical ➤MOSFET transistors of the same structure as ➤DMOS. This technique is used to produce MOS devices of very precise channel length; as with DMOS the channel length is determined by the diffusion rather than by ➤photolithography. VMOS devices also contain an n⁻ drift region (➤DMOS) in order to prevent ➤punch-through. The length of the drift region determines the breakdown voltage of the device, as with DMOS devices, and the structure shown in Fig. *a* is used to form discrete high-power MOSFETs.

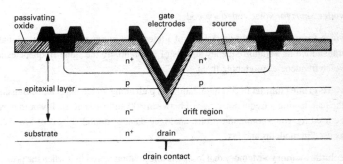

a **VMOS** power transistor

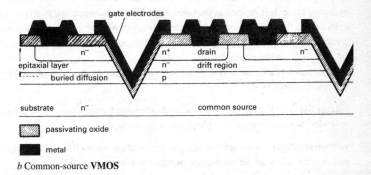

b Common-source **VMOS**

Operating voltages of about 300 volts can be achieved using a drift length of about 25 micrometres although voltages of about 100 V are more usual. The 'on' impedance can be kept low (of the order of a few ohms) using several long grooves on each transistor chip.

The technique is also used to produce an array of transistors for applications, such as ►ROM (read-only memory), that are used with a common source (Fig. *b*). In this case the device is physically inverted and the groove must be taken right through to the substrate. The drift region can therefore only be about two micrometres in length in order to accommodate the gate electrodes. An inverted VMOS circuit is only suitable for low-power high-speed applications up to microwave frequencies. It has however an increased functional packing density compared to DMOS.

vocoder *Short for* voice coder. A device that enables a direct spectral representation of a speech signal to be obtained by finding the energy in each filter band of a bank of band-pass ►filters using a ►rectifier and ►smoothing circuit. This representation is coded and transmitted along with the excitation signal, which is either *voiced* (periodic), in which case a voicing flag is set and its fundamental frequency is coded, or *voiceless* (nonperiodic) when the voiced flag is unset.

voder *Short for* voice coder. ►vocoder.

voice-frequency telephony A form of ►telephony in which the electric signals are transmitted as ►audiofrequency signals of substantially the same frequencies as the voice frequencies producing them.

voice-grade channel In communications, a path along which data can be transmitted, the path having a frequency range of between 300 and 3400 hertz. It would normally be used for voice communications, analogue data, or digital signals of low data rate as used in modems and fax.

volatile memory ►Memory that loses the information stored in it when the power supply is switched off.

volt Symbol: V. The ►SI unit of ►electric potential, ►potential difference, and ►electromotive force. It is defined as the potential difference between two points on a conductor when the current flowing is one ampere and the power dissipated between the points is one watt. In practice volts are measured by comparison with a ►Weston standard cell using a ►potentiometer.

This unit replaced the *international volt* (V_{int}) as the standard unit of voltage: one V_{int} equals 1.000 34 V.

voltage Symbol: *V*; unit: volt. The potential difference between two points in a circuit or device. ►active current (or voltage); reactive voltage.

voltage amplifier An ►amplifier that provides voltage ►gain. It has a high input impedance and a low output impedance. ►dependent sources.

voltage between lines *Syns.* line voltage; voltage between phases. The voltage between the two lines of a single-phase electrical power system or between any two lines of a symmetrical three-phase system.

In the case of a symmetrical six-phase power system the six lines can be considered as arranged around the periphery of a regular hexagon in correct order of phase sequence. Then the voltage between any two consecutive lines is the *hexagon voltage*. The *delta voltage* is the voltage between alternate lines and the *diametrical voltage* is the voltage between lines arranged opposite to each other on the hexagon.

voltage-controlled amplifier (VCA) An ➤amplifier whose ➤gain is controlled by a voltage.

voltage-controlled current source (VCCS) ➤dependent sources.

voltage-controlled voltage source (VCVS) ➤dependent sources.

voltage divider ➤potential divider.

voltage doubler An arrangement of two ➤rectifiers that produces an output voltage amplitude twice that of a single rectifier. In a typical circuit using ➤diode rectifiers (see diagram) each rectifier separately rectifies alternate half cycles of the input alternating voltage and the two outputs are then summed.

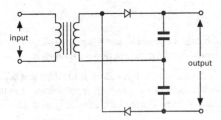

Voltage doubler

voltage drop The voltage between any two specified points of an electrical conductor, such as the terminals of a circuit element or component, due to the flow of current between them.

In the case of direct current the voltage drop is equal to the product of the current in amperes and the resistance in ohms. In the case of alternating current the product of the current in amperes and the resistance in ohms is the *resistance drop,* which is in ➤phase with the current. The product of the current and the reactance gives the *reactance drop,* which is in ➤quadrature with the current.

voltage feedback *Syn.* shunt feedback. ➤feedback.

voltage gain ➤gain.

voltage generator Any source of electrical power that is used to supply voltage to a circuit or device. ➤constant-voltage source.

voltage jump An abrupt unwanted change or discontinuity in the operating voltage of any electronic circuit or device, particularly a glow-discharge tube (►gas-discharge tube).

voltage level The ratio of the voltage at a point in a transmission system to an arbitrary voltage reference. The arbitrary reference point is specified by the CCITT (►standardization) as a point of ►zero level, and expressed in dBm0, i.e. decibels measured with reference to the zero relative level. Absolute measurements of received signal magnitudes or noise levels at various receivers may be compared by converting them to the zero relative values in order to assess the performance of the receiving circuits.

voltage-mode circuits ►current-mode circuits.

voltage multiplier An arrangement of ►rectifiers that produces an output voltage amplitude that is an integral multiple of that of a single rectifier, i.e. of the peak value of applied alternating voltage. ►►voltage doubler.

voltage reference diode ►diode forward voltage.

voltage regulator ►regulator.

voltage-regulator diode ►Zener diode.

voltage relay ►relay.

voltage selector ►limiter.

voltage source Ideally, a two-terminal circuit element where the voltage between the terminals is constant and independent of the current flowing through it.

voltage stabilizer A circuit or device that produces an output voltage that is substantially constant and independent of variations either in the input voltage or in the load current, i.e. it acts as a ►constant-voltage source. Such a device is most often used as a voltage ►regulator.

An early form of voltage stabilizer was based on the ►gas-discharge tube. Such circuits have now largely been superseded by circuits based on solid-state devices, such

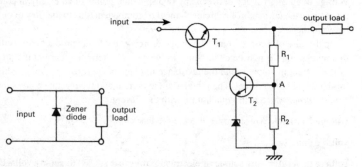

a Zener diode stabilizer circuit   *b* Series stabilizer circuit

as the ►Zener diode (Fig. *a*) or the bipolar junction transistor (Fig. *b*). In the series stabilization circuit shown in Fig. *b* the load impedance is in series with the circuit.

voltage transformer *Syn.* potential transformer. ►transformer; instrument transformer.

volt-ampere Symbol: VA. The ►SI unit of apparent ►power, defined as the product of the ►root-mean-square values of voltage and current (as measured by a voltmeter and an ammeter) in an alternating-current circuit. ►power; active volt-amperes; reactive volt-amperes; var.

volt efficiency ►cell.

voltmeter A device that measures voltage. Voltmeters in common use include ►digital voltmeters, ►cathode-ray oscilloscopes, and d.c. instruments (such as permanent-magnet ►moving-coil instruments).

In order to provide the minimum disturbance in the circuit containing the voltage to be measured, voltmeters are required to pass very little current and therefore require a very high input impedance. Digital voltmeters, cathode-ray oscilloscopes, and the now little used ►valve voltmeter comply with this requirement. A large series resistance however is required in the case of the moving-coil voltmeter and the *electrostatic voltmeter* in order to increase their input impedances. The electrostatic voltmeter is a voltmeter based upon the principle of operation of a ►quadrant electrometer or other type of electrometer.

volt per metre The ►SI unit of ►electric field strength.

volume The magnitude of the complex ►audiofrequency signals in an audiofrequency transmission system.

volume charge density ►charge density.

volume compressor A device that automatically reduces the range of amplitude variations of an ►audiofrequency signal in a transmission system. It operates by decreasing the amplification of the signal when it has a value greater than a predetermined amplitude and increasing the amplification when the signal amplitude is less than a second predetermined value. A *volume expander* is a device that produces the opposite effect, i.e. it automatically extends the range of amplitude variations of the transmitted audiofrequency signal.

A suitably designed expander used at one point of a transmission system can be made to compensate for the effect of a compressor in another part of the system and thus restore the original audiofrequency signal. A compressor and expander used together in this manner are termed a *compandor.*

In the ►recording of sound a compressor may be used to reduce the volume range of the signals recorded on a gramophone record or on a film track where the dynamic range that can be recorded is less than the dynamic range of the sound to be recorded. The sound-reproducing apparatus will then include an expander to compensate.

A telecommunication system, such as a radio-telephone system, often employs a compandor to improve the ►signal-to-noise ratio of the system; the compressor is used at the transmitter and the expander at the receiver. The relative increase in the smaller transmitted signals reduces the effect of noise on these signals.

volume control ➤automatic gain control; gain control.

volume expander ➤volume compressor.

volume lifetime *Syn. for* bulk lifetime. ➤semiconductor.

volume limiter An amplitude ➤limiter that operates on an audiofrequency signal.

volume resistivity ➤resistivity.

volumetric radar ➤radar.

von Hann window ➤windowing.

von Neumann machine Any ➤computer that consists of a single control unit connecting a memory to a processing unit: instructions and data are fetched one at a time from the memory and fed to the processing unit under control of the control unit. The processing speed of the entire machine is limited by the rate at which instructions and data can be transferred from memory to processing unit.

VPE *Abbrev. for* vapour phase epitaxy. ➤➤chemical vapour deposition.

VSWR *Abbrev. for* voltage standing-wave ratio. ➤standing wave.

W

wafer *Syn.* slice. A large single crystal of ►semiconductor material that is used as the substrate during the manufacture of a number of ►chips. Very large single crystals are grown and then sliced into wafers before processing.

Wagner earth connection A method of connection used with an alternating-current ►bridge circuit that minimizes the ►admittance to earth of the bridge. The earth connection is formed by means of the centre tapping of a three-terminal adjustable resistor, R, in ►parallel with the a.c. supply (see diagram). The input of the a.c. supply is usually transformer-coupled. If frequencies greater than the audiofrequency range are involved an adjustable capacitor, C, is also used as shown. The indicating instrument, I, is connected across the bridge with a two-position switch. The bridge is balanced with the switch in position B and then rebalanced, using R and C, in position G. The original balance is then rechecked.

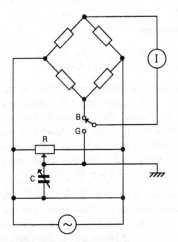

Wagner earth connection

wait state A dummy bus cycle that a computer's CPU executes when an attached device (or memory) cannot deliver the requested item of data within the required amount of time; for example, a 1-wait-state machine inserts one wait state for each memory access if the processor's clock speed is too fast for the attached memory chips. The term wait state generally applies to microprocessors.

walk-out A phenomenon observed in semiconductor devices that are subjected to repeated ➤avalanche breakdown. It is a progressive increase in the value of the avalanche-breakdown voltage of the device and results from the injection of ➤hot electrons into the surface passivating oxide layer. The change in the breakdown characteristics are caused by the change in the surface field due to the injected electrons.

WAN *Abbrev. for* wide area network.

warble The periodic variation, several times a second, of a frequency between two limits that are usually small compared to the nominal frequency. A *warble tone generator* consists of an oscillator whose output frequency is varied using a small variable capacitor in the tuned circuit. A warble tone is often used in a reverberation chamber in order to produce a uniform sound field containing no standing waves.

warm-up The period after a particular electronic device, circuit, or apparatus has been switched on during which it reaches a state of thermal equilibrium with its surroundings. A circuit may not be fully operational until after the warm-up period, as the electrical characteristics of its components tend to drift towards their steady value during this period. The warm-up includes the time that any individual device contained in the apparatus under consideration takes to heat up to a steady temperature.

watt Symbol: W. The ➤SI unit of ➤power. It is defined as the power resulting when one joule of energy is dissipated in one second. In an electric circuit one watt is given by the product of one ampere and one volt.

watt-hour One thousandth of a ➤kilowatt-hour.

watt-hour meter *Syn.* integrating wattmeter. An ➤integrator that measures and records electrical energy in watt-hours or more usually in ➤kilowatt-hours.

wattmeter An instrument that measures electric power and is calibrated in watts, multiples of a watt, or submultiples of a watt. The most commonly used type of wattmeter is the electrodynamic wattmeter (➤electrodynamometer). In circuits that have substantially constant currents and voltages an induction wattmeter (➤induction instrument) may be used. For standardization and calibration purposes an *electrostatic wattmeter* is used. This consists of a ➤quadrant electrometer that is arranged in a suitable circuit containing noninductive resistors so that it measures power directly.

wave A periodic disturbance, either continuous or transient, that is propagated through a medium or through space and in which the displacement from a mean value is a function of time or position or both. Sound waves, water waves, and mechanical waves involve small displacements of particles in the medium; these displacements return to zero after the disturbance has passed. With electromagnetic waves (➤electromagnetic radiation) it is changes in the intensities of the associated magnetic and electric fields that represent the disturbance and a medium is not required for propagation of the wave.

The instantaneous values of the periodically varying quantity plotted against time gives a graphical representation of the wave that is known as the *waveform*. If the waveform is sinusoidal in shape it is usually described as undistorted; a nonsinusoidal waveform is distorted. The *wavefront* is the imaginary surface over which the displacements are all of the same ►phase. The *amplitude* of the wave is the peak value of the displacements relative to the equilibrium state or to some arbitrary reference level, usually zero. If a wave suffers ►attenuation the amplitude of successive periods is continuously reduced (➤➤propagation coefficient).

The *wavelength* is the distance between two displacements of the same phase along the direction of propagation. If v is the velocity of the wave and λ its wavelength, then the frequency of vibration, f, is given by

$$f = v/\lambda$$

The frequency is the reciprocal of the ►period, T, of the wave. The frequency (or wavelength) of electromagnetic radiation is commonly used to describe particular regions of the ►electromagnetic spectrum, such as the visible or radio regions (➤➤frequency band).

An alternating current propagated through a long chain ►network or ►filter behaves as if it were a wave. Elementary particles, such as electrons, have associated wavelike characteristics. ➤➤Doppler effect.

wave analyser A device, such as a ►spectrum analyser or frequency analyser, that resolves a complex waveform into its fundamental and harmonic components.

wave duct ►waveguide.

wave equation ►quantum mechanics.

waveform ►wave.

wavefront 1. *Syn.* wave surface. ►wave. **2.** ►impulse voltage.

wavefunction ►quantum mechanics.

waveguide A ►transmission line that consists of a suitably shaped hollow conductor, which may be filled with a dielectric material, and that is used to guide ultrahigh-frequency electromagnetic waves propagated along its length. The transmitted wave is reflected by the internal walls of the waveguide and the resulting distribution within the guide of the lines of electric and magnetic flux associated with the wave is the transmission ►mode. At any instant the phase and amplitude of the wave are given by the appropriate ►propagation coefficient.

For any given transmission mode there is a lower limit to the frequency that may be propagated through the waveguide. This *cut-off frequency* occurs when the complex propagation coefficient becomes real. The electromagnetic wave is then attenuated exponentially and soon becomes substantially zero. (A short length of waveguide used below cut-off may sometimes be employed as a known ►attenuator.) The waveguide therefore acts effectively as a high-pass ►filter. In an ideal (lossless) waveguide, above the cut-off frequency, the propagation coefficient is purely imaginary and no attenuation occurs. In practice however some attenuation of the wave always occurs due to energy dissipation in the guide.

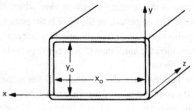

a Rectangular **waveguide**

The most common shapes of waveguide are rectangular (Fig. *a*) and cylindrical; the most common dielectric is air. A cylindrical waveguide is sometimes known as a *wave duct;* one that contains a solid rod of dielectric is a *uniconductor waveguide.* If a wide range of frequencies is to be transmitted a *ridged waveguide* (Fig. *b*) may be used. The presence of the ridges extends the possible range of frequencies that may be propagated in a particular transmission mode but the attenuation is greater than in the equivalent rectangular waveguide.

Single ridged   Double ridged

b Cross section of ridged **waveguides**

Electromagnetic waves may be excited in a waveguide by the electric and magnetic fields associated with electromagnetic waves present in another device, such as a ►cavity resonator or ►microwave tube. A suitable *mount* is used to connect the waveguide to the source of radiation in order to achieve the optimum transfer of energy and to excite the desired transmission mode. Energy may also be extracted from a waveguide in a similar manner.

Energy may also be transferred using either a probe to which a voltage is applied or a coil that carries an electric current. The alignment of the exciting probe or coil depends on the desired transmission mode. The probe must be aligned along the direction of lines of electric flux in order to excite the electric field; the coil is aligned so as to excite the desired magnetic field. Energy may be similarly extracted from the waveguide. Probes are also used to examine the distribution of the fields within the waveguide.

A *slotted waveguide* is one that is provided with nonradiative slots at intervals along its length that allow the insertion of a probe. The presence of the slots affects the distribution of the fields inside the guide, as can junctions (flanges) between sections of a waveguide. Junctions and slots in a guide contribute towards the distributed capacitance of the device and also contribute to the energy dissipation in the guide. A waveguide in which no reflected waves occur at any of the transverse sections is known as a *matched waveguide*.

Bends in a waveguide are usually made smoothly in order to prevent unwanted reflections. An *E-bend* is a smooth bend in a waveguide in which the direction of polarization is parallel to the axis (i.e. a horizontally polarized TM wave). An *H-bend* is a smooth bend in a guide in which the direction of polarization is perpendicular to the axis (i.e. a vertically polarized TE wave). If the direction of polarization has to be changed a special joint, known as a *rotator*, is used. In a rectangular guide twisting of the guide structure can be used to rotate the plane of polarization.

It is possible to produce local concentrations of electric or magnetic energy within a waveguide by inserting suitably shaped metallic or dielectric pieces. These act essentially as lumped inductances or capacitances. Capacitive elements can be formed from screws; inductive elements can be formed from diaphragms or posts. Either a *tuning screw* or a *waveguide plunger* can be used in order to change the impedance of a waveguide. This may be necessary when the impedances of different sections of a waveguide are slightly different. Coupling between two waveguides of different impedances is effected using a suitable *waveguide transformer*, such as a ➤quarter-wavelength line, for ➤impedance matching. Unwanted frequencies or modes of transmission are eliminated using a waveguide filter. An *iris* consists of a shaped inductive diaphragm and capacitive screw inserted into a guide to act as a reactance, susceptance, or waveguide filter.

Waveguides are widely used for the transmission of electromagnetic waves. Electromagnetic energy may also be stored by using a section of waveguide short-circuited at each end to form a ➤cavity resonator. A section of coaxial transmission line may also be used. If the length of the section of waveguide or transmission line is h, resonance occurs when

$$h = n(\lambda_g/2)$$

where n is an integer and

$$\lambda_g = \lambda[\varepsilon - (\lambda/\lambda_c)^2]^{1/2}$$

λ is the wavelength in free space, λ_c the wavelength of the waveguide cut-off and ε the relative dielectric constant of the waveguide dielectric. A cavity resonator formed in this way is termed either a *waveguide resonator* or a *coaxial resonator*.

waveguide plunger ➤waveguide.

waveguide resonator ➤waveguide.

waveguide transformer ➤waveguide.

wave impedance The ratio of the electric to magnetic fields in an electromagnetic wave.

wavelength ➤wave.

wavelength constant *Syn. for* phase-change coefficient. ➤propagation coefficient.

wave mechanics ➤quantum mechanics.

wavemeter An apparatus that measures the frequency or wavelength of a ➤radiowave. It consists essentially of a capacitively tuned circuit together with a current-detecting instrument. The variable capacitor is calibrated directly to read frequency or wavelength. A current maximum is obtained when the resonant frequency of the tuned circuit corresponds to the frequency of the received radiowave.

wave surface *Syn. for* wavefront. ➤wave.

wavetail ➤impulse voltage.

wavetrain A succession of ➤waves, particularly a small group of waves of limited duration.

wavetrap A tuned circuit, usually a parallel ➤resonant circuit, that is included in a ➤radio receiver in order to reduce interference from unwanted radiowaves of the particular frequency to which the circuit is tuned.

wave vector ➤momentum space.

weak electrolyte ➤electrolyte.

wear-out failure ➤failure.

wear-out failure period ➤failure rate.

weber Symbol: Wb. The ➤SI unit of ➤magnetic flux. One weber is the magnetic flux that, linking a circuit of one turn, produces an electromotive force of one volt when it is reduced to zero at a uniform rate in one second.

Weiss constant *Syn.* paramagnetic Curie temperature. ➤paramagnetism.

Weston standard cell *Syn.* cadmium cell. A cell that has a substantially constant terminal voltage and is used as a reference standard for electromotive force. This cell is constructed in an H-shaped glass vessel. The positive electrode is mercury and the negative electrode is a cadmium and mercury amalgam with a saturated cadmium sulphate solution as the electrolyte. The e.m.f. developed by the cell at 20 °C is 1.018 58 volts. The cell has a very low temperature coefficient of e.m.f. ➤➤Clark cell.

wet cell ➤cell.

wet etching ➤etching.

Wheatstone bridge A four-arm ➤bridge used for measuring resistance. Each arm contains a resistance (see diagram), with the unknown and reference resistances, R_1 and R_2, being connected at a point. At balance, when a null response is obtained from the indicating instrument,

$$R_1/R_2 = R_3/R_4$$

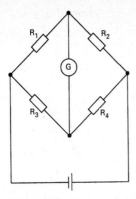

Wheatstone bridge

The two arms R_3 and R_4 are known as the ratio arms and may take the form of a wire of uniform resistance, which is tapped by a sliding contact. R_3 and R_4 are proportional to the lengths of wire, l_1 and l_2, on each side of the contact, so that

$R_1/R_2 = l_1/l_2$

white compression Compression (➤volume compressor) applied to the components of a ➤television video signal that correspond to the light areas of the picture.

white noise ➤Noise that is uniform in energy over equal intervals of frequency.

white peak *Syn.* picture white. ➤television.

white recording ➤recording of sound.

wide area network (WAN) A collection of computers whose interconnection network spans large distances, often by using telephone lines or microwave communications.

wideband *Syn.* broadband. Denoting a system or circuit that is operational over a frequency range which is large compared to its ➤centre frequency. A common measure of wideband is where the bandwidth is greater than a few percent of the centre frequency.

wideband FSK ➤frequency modulation.

Wiedemann effect ➤magnetostriction.

Wien bridge A four-arm bridge that is used for the measurement either of capacitance or frequency. A typical network is shown in the diagram. At balance, when a null response is obtained from the indicating instrument,

$C_x/C_s = (R_b/R_a) - (R_s/R_x)$

$C_s C_x = 1/\omega^2 R_s R_x$

where ω is the angular frequency. For the measurement of a frequency f, it is convenient to make $C_s = C_x$, $R_s = R_x$, and $R_b = 2R_a$ so that $f = (2\pi CR)^{-1}$.

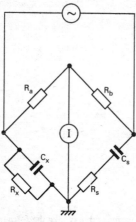

Wien bridge

Wilson effect Electrical polarization of an insulator when it is moved through a region containing a magnetic flux. The motion induces a potential difference across the material that results in its polarization since the insulating properties inhibit the creation of an electric current.

Wimshurst machine An early electrostatic ➤generator.

winding A complete group of insulated conductors in an electrical machine, transformer, or other equipment that is designed either to produce a magnetic field or to be acted upon by a magnetic field. It may consist of a number of separate suitably shaped conductors, electrically connected, or a single conductor shaped to form a number of loops or turns.

window 1. The thin sheet of material, usually mica, that covers the end of a ➤Geiger counter and through which the radiation enters the instrument. **2.** ➤radio window. **3.** ➤confusion reflector. **4.** ➤windowing.

windowing The tapering of a sampled signal prior to a transformation being applied in order to reduce the effect of any discontinuities at the edges. This is achieved in practice by multiplying the portion of the time domain signal to be transformed by a *window function*, which is equivalent to ➤convolution in the frequency domain. When, for example, a Fourier transform is being applied, the choice of a suitable window is made as a compromise between keeping local spreading of individual spectral components to a minimum as well as keeping spectral spreading elsewhere low. A number of window functions have been proposed, which are defined for

$$-\tfrac{1}{2}(N-1) \leq n \leq \tfrac{1}{2}(N-1)$$

and zero elsewhere (see table).

Blackman window:	$w_B[n] = 0.42 + 0.5 \cos(2\pi n/[N-1]) + 0.08 \cos(4\pi n/[N-1])$
Bartlett or *triangular window:*	$w_T[n] = 1 - 2|n|/[N-1]$
Hamming window:	$w_H[n] = 0.54 + 0.46 \cos(2\pi n/[N-1])$
Hanning or *raised cosine window:*	$w_C[n] = 0.5 + 0.5 \cos(2\pi n/[N-1])$
Kaiser window:	$w_K[n] = \dfrac{I_0(\alpha\sqrt{\{1 - (2n/[N-1])^2\}})}{I_0(\alpha)}$
	where I_0 is a Bessel function and α controls the degree of edge taper
rectangular or *uniform window:*	$w_R[n] = 1$
tapered window:	$w_t[n] = 1$ for $-\frac{1}{2}(N-1) < n < \frac{1}{2}(N-1)$
	$w_t[n] = 0.5$ for $n = -\frac{1}{2}(N-1)$ and $n = \frac{1}{2}(N-1)$
von Hann window:	$w_V[n] = 0.5 + 0.5 \cos(2\pi n/[N+1])$

Window functions

wire bonding A method used during the packaging of ➤integrated circuits to connect the chip to the ➤leadframe. The leadframe is formed by stamping the interconnection pattern required from a strip of thin copper sheeting. The chip is positioned in a small depression in the centre and the leads connected to the ➤bonding pads of the chip using individual wire bonds. In the case of packages with ceramic casings, the leadframe is mounted in the package before the chip is bonded in place. For plastic packages, the chip is bonded to the leadframe and then encapsulated in moulded plastic. The finished package may be in the form of a ➤pin grid array, ➤leadless chip carrier, or ➤dual in-line package. ➤➤tape-automated bonding.

wireless 1. A means of electrical communications between systems or parts of a system by means of radio, infrared, or acoustic waves, without using connecting wiring. Examples include personal wireless communications by mobile phone (➤cellular communications) and wireless ➤local area networks (WLANs) in which infrared or microwaves are used as a carrier between the network outlet and end-user device. **2.** A radio receiver.

wire-wound resistor ➤resistor.

wiring diagram A diagram of a piece of electronic equipment showing the interconnections between assemblies and subassemblies. Such a diagram is particularly useful for the maintenance or repair of such equipment.

wobbulator A ➤signal generator in which an automatic periodic variation is applied to the output frequency so that it varies over a predetermined range of values. It can be used to investigate the frequency response of an electronic circuit or device or as a test instrument for a ➤tuned circuit.

Wollaston wire Extremely fine platinum wire that is used for ►electroscope wires, microfuses, and ►hot-wire instruments. It is produced by coating platinum wire with a sheath of silver, drawing them together into a relatively fine uniform diameter wire, and then dissolving the silver with a suitable acid. Diameters down to about one micrometre may be produced by this method.

woofer A relatively large ►loudspeaker that is used to reproduce sounds of relatively low frequency in a high-fidelity sound reproduction system. ►►tweeter.

word A string of ►bits that stores a unit of information in a ►computer. The length of the word depends on the particular machine. Typical word lengths contain 32, 64, or 128 bits.

word-addressable ►address.

word line ►RAM.

work function Symbol: Φ; unit: electronvolt. The difference in energy between the Fermi level (►energy bands) of a solid and the energy of free space outside the solid, i.e. the vacuum level. It is the minimum energy required to liberate an electron from the solid at absolute zero temperature.

In a metal the ►image potential that an electron just outside the surface would experience also contributes to the value of the work function. In a semiconductor the work function is greater than the ►electron affinity, χ (see diagram). ►►photoelectric effect.

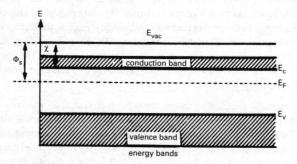

Work function and electron affinity of a semiconductor

workstation 1. Traditionally, a computer or data terminal used by one person for computational work, including computationally intensive ►modelling, and high-quality graphics. Workstations are common in design and drawing offices where they are used for specific computer-aided design tasks, such as finite-element analysis, three-dimensional drawings, etc. **2.** In production, a place of work equipped with all the necessary tools and parts to assemble or produce part of a larger product or the final assembly of a product.

wound core ►core.

wow Low-frequency (up to about 10 hertz) audible periodic variations in the pitch of the sound output in a sound-reproduction system. Wow is an unwanted form of ➤frequency modulation and is usually caused by a nonuniform rate of reproduction of the original sound, such as by nonuniform tape speed or nonuniform rotation of a gramophone turntable. ➤➤flutter.

wrist strap ➤electromagnetic compatibility.

write To enter information into a storage element of a ➤memory.

wye circuit ➤star circuit.

X

X.21 ►X series.

X.25 ►X series.

X-axis The horizontal axis on a graphical plot such as on the screen of a cathode-ray tube or spectrum analyser.

X band A band of microwave frequencies ranging from 8.00–12.00 gigahertz (IEEE designation). ►frequency band.

XGA *Abbrev. for* extended graphics array. A colour video standard, introduced by IBM in 1991, that supports a resolution of 1024×768 ►pixels with 256 colours or 640×480 pixels with 65 536 colours.

X-guide A ►transmission line that is used for the propagation of surface waves and consists of a length of dielectric material with an X-shaped cross section.

XPS *Abbrev. for* X-ray photoelectron spectroscopy. ►photoelectron spectroscopy.

X-ray crystallography ►X-rays.

X-ray fluorescence (XRF) ►electron microprobe.

X-ray lithography A method of ►lithography that uses X-rays rather than light beams. The basic process uses soft X-rays to expose appropriate resists. Because of the low wavelength (0.2–1.0 nanometre), diffraction effects are virtually nonexistent, and back scattering or reflection from the substrate material is almost nonexistent. Masks can be placed proximal to the substrate rather than in contact with it, and most dust is transparent to X-rays and therefore does not affect the process. X-ray lithography offers the advantages of excellent resolution combined with a large depth of field, vertical-walled patterns, and simplicity of the system – no complex lenses, mirrors, or electron-beam lenses are required. Disadvantages have arisen in the development of suitable X-ray sources, masks, and alignment procedures.

The optimum X-ray source is the ►synchrotron, which produces highly collimated intense beams of X-rays; these machines, however, are scarce and expensive. Masks must be opaque to X-rays and therefore must be made from high atomic number materials such as gold. The resulting masks are very fragile and require careful handling. X-ray resists also present problems. With the exception of synchrotron-produced X-rays, most other sources of X-rays do not produce a collimated intense beam and few suitable resist materials exist with sufficient sensitivity, particularly positive resists.

►Multilevel resist techniques must be used, where the pattern is produced in the very thin top layer of resist and transferred to the lower layers by dry ►etching.

X-ray photoelectron spectroscopy (XPS) ►photoelectron spectroscopy.

X-rays Electromagnetic radiation of frequencies ranging approximately between those of ultraviolet radiation and gamma rays. They are produced when matter is bombarded with sufficiently energetic electrons and were first observed by Roentgen in 1895. X-rays can be produced as a result of electron transitions from higher to lower energy levels within an atom. X-rays resulting from transitions have a frequency that is characteristic of the material and are therefore termed *characteristic X-rays*. X-rays are also produced during the rapid decelerations of electrons as they approach the nucleus. These X-rays, a form of ►bremsstrahlung radiation, have a relatively wide frequency range and are known as *continuous X-rays*. X-rays of relatively low energy are termed *soft X-rays*; those at the high-energy portion of the frequency spectrum are *hard X-rays*.

X-rays can be reflected, refracted, and polarized and also exhibit interference and diffraction. They interact with matter to produce relatively high energy electrons: the mechanism is the same as in the ►photoelectric effect. These electrons are usually of sufficiently high energy to ionize a gas or to produce *secondary X-rays* from the matter. The ionization of a gas can be utilized, in the ►ionization chamber, to measure the intensity of an X-ray beam.

X-rays are widely used in *radiography* to investigate materials that are opaque to visible light but relatively transparent to X-rays. Radiography is used to detect flaws in structures and for diagnostic examinations. Sufficiently high intensities of X-rays of higher energies damage human tissue and X-rays are used in *radiotherapy* for the therapeutic destruction of diseased tissue. X-rays are also used in *X-ray crystallography* to investigate and determine crystal structures by using the crystal as a three-dimensional ►diffraction grating. ►X-ray lithography systems are being developed.

X-ray topography ►diffraction.

X-ray tube An ►electron tube that produces X-rays. Early X-ray tubes were gas-filled tubes but only electron beams of relatively low energy could be produced because of disruptive discharges occurring through the gas. Modern X-ray tubes are invariably hard tubes (i.e. evacuated to a high vacuum) developed from the Coolidge type of tube; they are more stable than the earlier gas-filled sort.

A beam of electrons, produced by ►thermionic emission from the cathode, is accelerated and focused on to a *target electrode*. The target electrode is usually the anode of the tube. The X-ray spectrum produced from the target contains frequencies that are characteristic of the target material and in addition contains a continuous range of frequencies of which the short-wave (high-energy) limit is determined by the energy of the electrons and hence by the accelerating voltage across the tube. The current in a vacuum tube can be controlled by varying the temperature of the cathode and hence the numbers of electrons emitted from it.

In order to produce high-energy X-rays the voltage across the tube must be very large: values of greater than one megavolt have been used. These extremely high voltages are usually supplied by a step-up transformer: if the secondary winding is connected across the tube it can act as its own rectifier. If this mode of operation is not suitable, for example when a continuous steady output is required, separate rectifying and smoothing circuits are necessary.

When operated at such energies cooling of the target is important. The extremely small size of the focal spot causes the generation of localized high temperature. The target is therefore constructed of a metal, such as tungsten, that has an extremely high melting point. The rest of the target electrode is usually made from a metal, such as copper, that is a good conductor of heat. A *rotating-anode tube* contains an anode that rotates about a centre that is not the focal spot of the electron beam; the same part of the target is thus not always bombarded. A rotating hollow cylinder with water cooling can also be used.

XRF *Abbrev. for* X-ray fluorescence. ➤electron microprobe.

X series A series of standards established by the CCITT covering the communications between digital equipment, such as computers and modems, and digital networks. The X.25 standard addresses the interface between a computer or data terminal and a digital network and is the most common standard in the series. The X.21 standard addresses the interface to modems. Other standards within the series address how blocks of data are formatted, how public data networks should be numbered, and the open systems interconnection (OSI) model.

X-Y plotter An instrument that produces a graph showing the relationship between two varying signals. One signal causes the pen to move in the direction of the ➤X-axis and the other independently causes it to move in the direction of the ➤Y-axis. The corresponding instantaneous values of the two signals are plotted on the graph.

Y

Yagi antenna A sharply directional ➤antenna array from which most antennas used for ➤television have been developed. The active part of the Yagi consists of one or two ➤dipoles together with a parallel reflector antenna and a set of parallel directors (➤directional antenna). The directors are relatively closely spaced, being from 0.15 to 0.25 of a wavelength apart. When the antenna is used for transmission, the directors absorb energy from the back lobe of the dipole ➤antenna pattern and rereflect it in the forward direction; the major lobe is thus reinforced at the expense of the back lobe. When used for reception the inverse process occurs causing the signal to be focused on the dipole.

Y-axis The vertical axis on a graphical plot such as on the screen of a cathode-ray tube or spectrum analyser.

Y circuit ➤star circuit.

YIG *Abbrev. for* yttrium-iron-garnet. A ➤ferrite that is widely used for microwave applications. The magnetic properties are altered by the amount of trace elements in the material. The most common trace elements are calcium, vanadium, and bismuth. Calcium-germanium YIGs may be easily grown on a substrate. This material is used as a thin magnetic film on a nonmagnetic substrate, such as ➤G^3, for the production of solid-state magnetic circuits.

yoke A piece of ferromagnetic material that is used to connect permanently two or more magnetic ➤cores and thus complete a magnetic circuit without surrounding it by a winding of any kind.

y parameters ➤network.

Z

Zener breakdown A type of ►breakdown observed in a reverse-biased ►p-n junction that has very high ►doping concentration on both sides of the junction. At sufficiently high doping levels, the ►tunnel effect is the dominant mechanism under reverse bias. The built-in field (►p-n junction) is high and the depletion layer narrow as a result of the high level of doping. The application of a small reverse voltage (of up to about six volts) is sufficient to cause electrons to tunnel directly from the valence band to the conduction band (►energy bands). At the Zener breakdown voltage a sharp increase in the reverse current is obtained. Above the breakdown value the voltage across the diode remains substantially constant. This process is reversible since the dielectric properties of the material remain unchanged. Unlike ►avalanche breakdown no multiplication of charge carriers occurs.

Avalanche breakdown is the dominant breakdown mechanism in most semiconductor devices except those with high doping concentrations. Zener breakdown is utilized in ►Zener diodes.

Zener diode *Syn.* voltage-regulator diode. A ►p-n junction ►diode that has sufficiently high doping concentrations on each side of the junction for ►Zener breakdown to occur. The diode therefore has a well-defined reverse ►breakdown voltage (of the order of a few volts only) and can be used as a voltage regulator. Unlike ►tunnel diodes, the doping level is not high enough for the semiconductor to become ►degenerate and the diode behaves like a normal p-n junction in the forward direction.

The term is also applied to less highly doped p-n junction diodes that have higher breakdown voltages (up to 200 volts) and undergo ►avalanche breakdown. Most so-called Zener diodes are in fact avalanche breakdown diodes. True Zener diodes have a low value of reverse breakdown voltage.

zero error ►index error.

zero IF ►direct conversion receiver.

zero level An arbitrary reference level used in telecommunication systems to compare the relative intensities of transmitted signals or of ►noise.

zero potential ►earth potential.

zeros (and poles) ►s-domain circuit analysis.

zinc Symbol: Zn. A metal, atomic number 30, that is used as an electrode in some electrolytic ►cells.

zirconium Symbol: Zr. A metal, atomic number 40, that can be used as a ►getter in hard vacuum electron tubes and for coating tungsten field emitters to reduce their work function and improve efficiency.

z-modulation (of a cathode-ray tube) ►intensity modulation.

zone refining A method of redistributing impurities within a solid material, such as a semiconductor, by melting parts of the material and causing the molten zones to move along the sample. Impurities travel along the sample in a direction determined by their effect on the freezing point of the material. If the impurity depresses the freezing point it will travel in the same direction as the molten zone; if it raises the freezing point it will travel in the opposite direction. *Zone levelling* is the use of zone refining in order to distribute impurities evenly throughout the bulk of the material. *Zone purification*

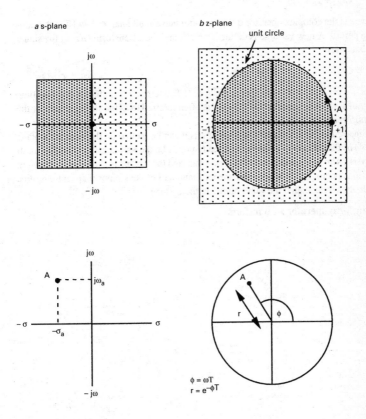

z transform: relationship between (a) s-plane and (b) z-plane

is the application of zone refining in order to reduce the concentration of an impurity in a material.

In zone refining the sample of the material is moved slowly past a heater so that the molten zone effectively moves along the length of the bar. Melting is achieved by induction heating, electron bombardment, or the heating effect of a current through a resistance coil. Zone purification carried out using a large number of heaters can reduce the impurity concentration to about one part in 10^{10}.

z parameters ➤network.

z-plane ➤z transform.

z transform An extension to the ➤Fourier and ➤Laplace transforms for sampled-data signals. The Laplace transform $G(s)$ of a sampled-data signal $x(t)$ is given by:

$$G(s) = x_0 + x_1 e^{-sT} + x_2 e^{-s2T} + x_3 e^{-s3T} + \ldots$$

where s is the complex operator $\sigma + j\omega$ (➤s-domain circuit analysis) and T is the sampling period. A new variable is defined, $z = e^{sT}$, and the z transform $G(z)$ of the signal $x(t)$ becomes:

$$G(z) = x_0 + x_1 z^{-1} + x_2 z^{-2} + x_3 z^{-3} + \ldots$$

Since $z = e^{sT}$ and $s = \sigma + j\omega$, then $z = e^{\sigma T} e^{j\omega T}$. The term $e^{-j\omega T}$ implies a constant time delay of T seconds, and for this reason z is often referred to as the *shift operator* or the *z transform operator.*

Since z has both real and imaginary parts it can be plotted on the ➤complex plane known as the *z-plane* in a similar fashion to points on the ➤s-plane (➤s-domain circuit analysis). The relationship between the s-plane and the z-plane is shown in the diagram (previous page). It can be seen that the left-hand half of the s-plane maps into the area inside the circle, called the ➤unit circle, in the z-plane.

z transform operator ➤z transform.

Table 1

Table 1   Graphical Symbols – mainly solid-state components also some electron tubes; IEC approved symbols are indicated O, popular usage is indicated □

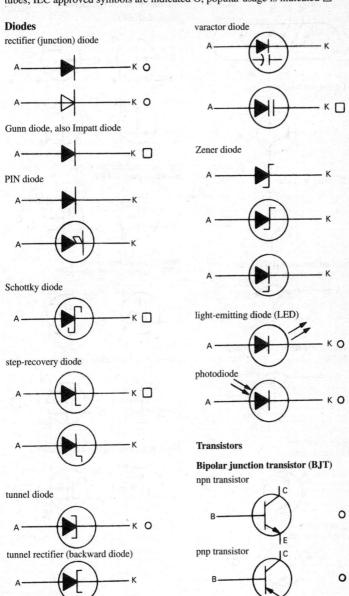

Diodes

rectifier (junction) diode

Gunn diode, also Impatt diode

PIN diode

Schottky diode

step-recovery diode

tunnel diode

tunnel rectifier (backward diode)

varactor diode

Zener diode

light-emitting diode (LED)

photodiode

Transistors

Bipolar junction transistor (BJT)

npn transistor

pnp transistor

Table 1 638

multiple-emitter npn-transistor

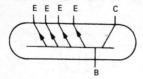

npn Darlington transistor

npn Schottky transistor

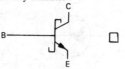

unijunction transistor (UJT) with
p-type base

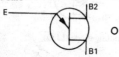

unijunction transistor (UJT) with
n-type base

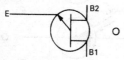

npn phototransistor,
no base connection

npn phototransistor,
with base connection

Field-effect transistor (FET)
n-channel   p-channel

junction (JFET)

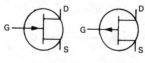

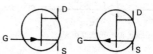

three-terminal depletion-type
insulated-gate (IGFET)

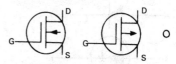

three-terminal depletion-type
IGFET, substrate tied to source

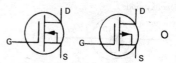

four-terminal depletion-type IGFET

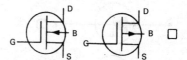

four-terminal enhancement-type IGFET

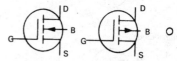

Table 1

neon lamp, a.c. type

fluorescent lamp, two-terminal

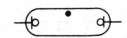

Crystals
piezoelectric crystal

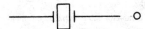

Circuit protectors
fuse

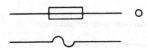

circuit-breaker

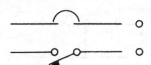

Audio devices
loudspeaker

microphone

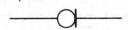

Amplifiers
single-ended amplifier

differential amplifier (or comparator)

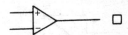

Norton (current) amplifier

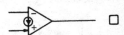

Measuring instruments

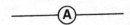

type indicated by letter in circle
A   ammeter
G   galvanometer
V   voltmeter

Filters
general

band-pass filter

band-reject filter

low-pass filter

Table 1   640

high-pass filter

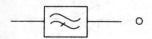

frequency changer

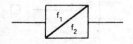

matched phase shifter

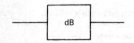

attenuator

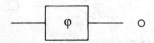

Antennas
receiving

transmitting

earth (ground)

chassis or frame connection

Interconnections
arrow indicates direction of signal flow;
other characteristics may be indicated

conducting path

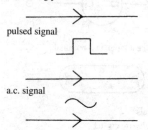

pulsed signal

a.c. signal

Cables
two-conductor cable with earthed shield

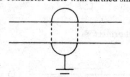

coaxial cable with earthed shield

twisted pair

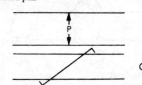

Waveguides
circular waveguide

rectangular waveguide

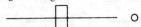

flexible waveguide

twisted waveguide

Table 1

Thyristors
four-layer (pnpn, Schottky) diode

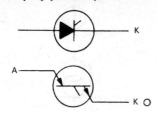

silicon-controlled rectifier (SCR)

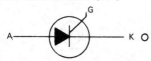

silicon-controlled switch (SCS)

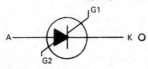

triac (gated bidirectional switch)

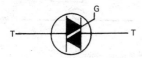

Circuit Components
Impedences

Resistors
fixed value resistor

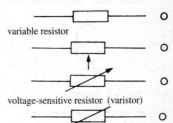

variable resistor

voltage-sensitive resistor (varistor)

Capacitors
fixed-value capacitors

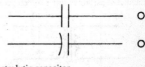

electrolytic capacitor

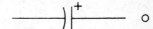

variable capacitor

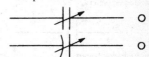

Inductors
fixed-value inductor

fixed-value inductor
with magnetic core

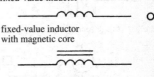

variable inductor

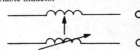

Table 1

642

Transformers

transformer, air cored

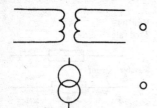 o

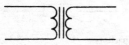 o

transformer with magnetic core

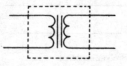

shielded transformer with magnetic core

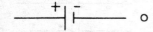

Batteries and sources

single-cell battery

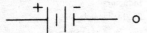 o

multiple-cell battery

 o

constant-voltage source

□

constant current source

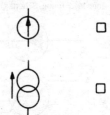 □

□

a.c. oscillator source

Lamps

incandescent lamp

signal lamp

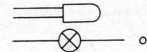 o

flashing signal lamp

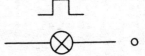 o

neon lamp, d.c. type

 o

Table 1

Contacts

fixed relay contact

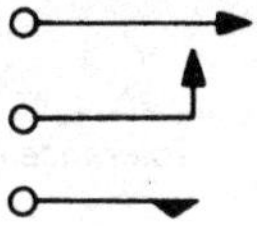

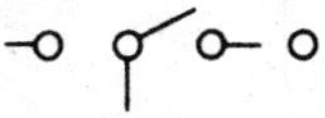

double-pole double-throw (dpdt)

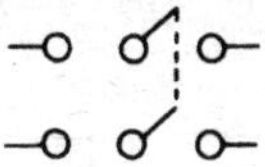

push-button switch, normally open

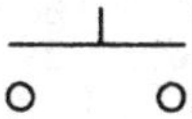

push-botton switch, normally closed

multiposition switch

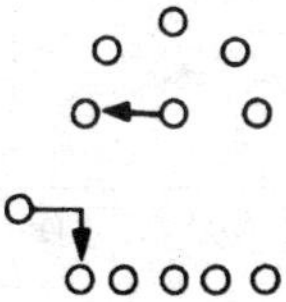

fixed switch contact

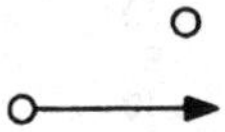

moving contact, locking

moving contact, nonlocking

closed contact

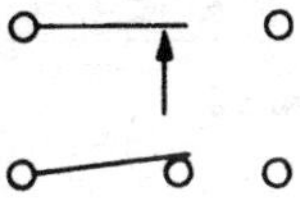

open contact

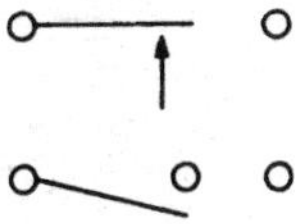

Switches

single-pole single-throw (spst)

single-pole double-throw (spdt)

Relay contact arrangements
(heavy arrow indicates direction of operation)

form A, spst normally open (make)

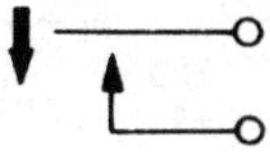

form B, spst normally closed (break)

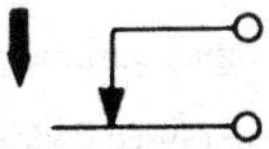

form C, spdt (break, make)

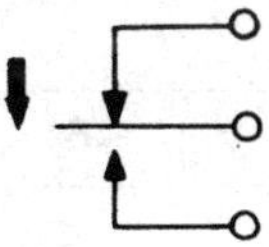

form D, spdt (make, break)

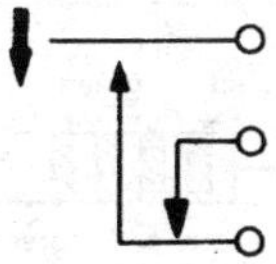

Electron tubes

evacuated envelope

gas-filled envelope

Electrodes

anode

main

intermediate

cathode

indirectly heated

directly heated

liquid electrode

photoelectric

cold cathode

grid

trigger electrodes

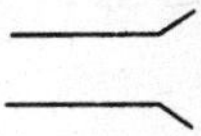

Table 2   644

Table 2 Colour Codes: indications on components of their value, tolerance, voltage rating, etc., by means of circumferential coloured bands. Axial lead components: to determine value, etc., start with band nearest end; band may cover endcap.

Standard colour coding

Colour	Digit	Tolerance (%)	Colour	Digit	Tolerance (%)
black	0	±20	blue	6	±6
brown	1	±1	violet	7	±12.5
red	2	±2	grey	8	±30
orange	3	±3	white	9	±10
yellow	4	*	gold		±5
green	5	±5	silver		±10

* guaranteed minimum value: variation of 0 to +100%

Resistors Values given in ohms; code of four or five bands

Band	A	B	C	D	E
significance of 4 bands	1st digit	2nd digit	additional zeros	tolerance	
5 bands	1st digit	2nd digit	3rd digit	additional zeros	tolerance

example:

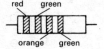

$$R = 25\ 000 \text{ ohms} \pm 5\%$$

Ceramic capacitors Values given in picofarads; code of five or six bands (endcap covered) or five bands (endcap uncovered)

Table 2 Colour Codes *(continued)*

Band	Significance of 5 band endcap covered	Significance of 6 band endcap covered	Significance of 5 uniform bands
A	temperature coefficient	↑ temperature coefficient	1st digit
B	1st digit	↓	2nd digit
C	2nd digit	1st digit	additional zeros
D	additional zeros	2nd digit	tolerance
E	tolerance	additional zeros	working voltage (digit × 100V)
F		tolerance	

Miniature foil capacitors Values coded in picofarad; multiply value by 10^{-6} to give capacitance in microfarad; code of five adjacent bands

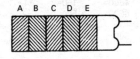

Band	A	B	C	D	E
Significance	1st digit	2nd digit	additional zeros	tolerance	working voltage (digit × 100)

Tantalum bead capacitors Values in microfarads

Region	A	B	*C	*D
Significance	1st digit	2nd digit	additional zeros	working voltage

*see table on following page

Table 2   646

Table 2 Colour Codes *(continued)*

Special colour coding for tantalum bead capacitors

Colour	Multiplier	Voltage	Colour	Multiplier	Voltage
black	× 1μF	10	blue		20
brown	× 10		grey	× 0.01	25
red	× 100		white	× 0.1	3
yellow		6.3	pink		35
green		16			

Table 3 Properties of Important Semiconductors

Semiconductor	Type	Energy gap at 300K (eV)	Drift mobility at 300 K (cm^2 V^{-1} s^{-1})		Dielectric constant
			electrons	holes	
silicon (Si)	element	1.09	1500	600	11.8
germanium (Ge)		0.66	3900	1900	16
selenium (Se)		2.3	0.005	0.15	6.6
gallium arsenide (GaAs)	III-V compound	1.43	8600–11 000	3000	10.9
indium phosphide (InP)		1.29	4800–6800	150–200	14
cadmium sulphide (CdS)	II-VI compound	2.42	300	50	10

Table 4 Electric and Magnetic Quantities

Quantity	Symbol	SI unit	SI symbol
electric current	I	ampere	A
electric charge, quantity of electricity	Q	coulomb	C
charge density, volume density of charge	ρ	coulomb per cubic metre	C/m^3
surface charge density	σ	coulomb per square metre	C/m^2
electric field strength	$\boldsymbol{E}$	volt per metre	V/m
electric potential	V, ϕ	volt	V
potential difference	U	volt	V
electromotive force	E	volt	V
electric displacement	$\boldsymbol{D}$	coulomb per square metre	C/m^2
electric flux	Ψ	coulomb	C
capacitance	C	farad	F
permittivity	ε	farad per metre	F/m
permittivity of free space	ε_o	farad per metre	F/m
relative permittivity	ε_r		
electric susceptibility	χ_e		

Table 4 Electric and Magnetic Quantities *(continued)*

Quantity	Symbol	SI unit	SI symbol
dielectric polarization	$\boldsymbol{P}$	coulomb per square metre	C/m^2
electric dipole moment	$\boldsymbol{p}$, $(\boldsymbol{p_e})$	coulomb metre	C m
permanent dipole moment of a molecule	p, μ	coulomb metre	C m
induced dipole moment of a molecule	p, p_i	coulomb metre	C m
electric polarizability of a molecule	α	coulomb metre squared per volt	$C\ m^2/V$
electric current density	$\boldsymbol{j}$, $\boldsymbol{J}$	ampere per square metre	A/m^2
magnetic field strength	$\boldsymbol{H}$	ampere per metre	A/m
magnetic potential difference	U_m	ampere	A
magnetomotive force	F, F_m	ampere	A
magnetic flux density	$\boldsymbol{B}$	tesla	T
magnetic flux	Φ	weber	Wb
self inductance	L	henry	H
mutual inductance	$\boldsymbol{M}$, $\boldsymbol{L_{12}}$	henry	H
coupling coefficient	k		
leakage coefficient	σ		
permeability	μ	henry per metre	H/m
permeability of free space	μ_o	henry per metre	H/m
relative permeability	μ_r		
magnetic susceptibility	χ_m		
magnetic moment	$\boldsymbol{m}$	ampere metre squared	$A\ m^2$
magnetization	$\boldsymbol{M}$	ampere per metre	A/m
magnetic polarization	B_i	tesla	T
speed of light in a vacuum	c	metre per second	m/s
resistance	R	ohm	Ω
resistivity	ρ	ohm metre	Ω m
conductance	G	siemens	S
conductivity	σ	siemens per metre	S/m
reluctance	R, R_m	reciprocal henry	H^{-1}
permeance	Λ	henry	H
phase displacement	ϕ		
number of turns on winding	N		
number of phases	m		
number of pairs of poles	p		
impedance	Z	ohm	Ω
modulus of impedance	$|Z|$	ohm	Ω
reactance	X	ohm	Ω
quality factor	Q		
admittance	Y	siemens	S
modulus of admittance	$|Y|$	siemens	S
susceptance	B	siemens	S
active power	P	watt	W
apparent power	S, (P_s)	watt	W (= V A)
reactive power	Q, (P_q)	watt	W
Faraday constant	F	coulomb per mole	C/mol
wavelength	λ	metre	m
frequency	υ, f	hertz	Hz

Table 4   648

Table 4 Electric and Magnetic Quantities (continued)

Quantity	Symbol	SI unit	SI symbol
angular frequency	ω	hertz	Hz
period	T	second	s
relaxation time	τ	second	s
thermodynamic temperature	T	kelvin	K
energy	E	joule	J

Table 5 Fundamental Constants

Constant	Symbol	Value
speed of light	c	$2.997\ 924\ 58 \times 10^8$ m s^{-1}
permeability of free space	μ_0	$4\pi \times 10^{-7} = 1.256\ 637\ 0614 \times 10^{-6}$ H m^{-1}
permittivity of free space	$\varepsilon_0 = \mu_0^{-1} c^{-2}$	$8.854\ 187\ 817 \times 10^{-12}$ F m^{-1}
charge of electron or proton	e	$\pm 1.602\ 177\ 33 \times 10^{-19}$ C
rest mass of electron	m_e	$9.109\ 39 \times 10^{-31}$ kg
rest mass of proton	m_p	$1.672\ 62 \times 10^{-27}$ kg
rest mass of neutron	m_n	$1.674\ 92 \times 10^{-27}$ kg
electron charge-to-mass ratio	e/m	$1.758\ 820 \times 10^{11}$ C kg^{-1}
electron radius	r_e	$2.817\ 939 \times 10^{-15}$ m
Planck constant	h	$6.626\ 075 \times 10^{-34}$ J s
Boltzmann constant	k	$1.380\ 658 \times 10^{-23}$ J K^{-1}
Faraday constant	F	$9.648\ 531 \times 10^4$ C mol^{-1}

Table 6 Base SI Units

Physical quantity	Name	Symbol
length	metre	m
mass	kilogram	kg
time	second	s
electric current	ampere	A
thermodynamic temperature	kelvin	K
amount of substance	mole	mol
luminous intensity	candela	cd

Table 7 Derived SI Units with Special Names

Physical quantity	Name	Symbol	Dimensions of unit derived	base
frequency	hertz	Hz		s^{-1}
energy	joule	J		$kg\ m^2\ s^{-2}$
force	newton	N	$J\ m^{-1}$	$kg\ m\ s^{-2}$
power	watt	W	$J\ s^{-1}$	$kg\ m^2\ s^{-3}$
pressure	pascal	Pa	$N\ m^{-2}$	$kg\ m^{-1}\ s^{-2}$
electric charge	coulomb	C		$A\ s$
potential difference	volt	V	$J\ C^{-1}$	$kg\ m^2\ s^{-3}\ A^{-1}$
resistance	ohm	Ω	$V\ A^{-1}$	$kg\ m^2\ s^{-3}\ A^{-2}$
conductance	siemens	S	Ω^{-1}	$s^3\ A^2\ kg^{-1}\ m^{-2}$
capacitance	farad	F	$C\ V^{-1}$	$s^4\ A^2\ kg^{-1}\ m^{-2}$
magnetic flux	weber	Wb	$V\ s$	$kg\ m^2\ s^{-2}\ A^{-1}$
inductance	henry	H	$Wb\ A^{-1}$	$kg\ m^2\ s^{-2}\ A^{-2}$
magnetic flux density	tesla	T	$Wb\ m^{-2}$	$kg\ s^{-2}\ A^{-1}$

Table 8 Dimensionless SI Units with Special Names

Physical quantity	Name	Symbol
plane angle	radian	rad
solid angle	steradian	sr

Table 9 Prefixes used with SI Units

Factor	Name of prefix	Symbol	Factor	Name of prefix	Symbol
10^{-1}	deci-	d	10	deca-	da
10^{-2}	centi-	c	10^2	hecto-	h
10^{-3}	milli-	m	10^3	kilo-	k
10^{-6}	micro-	μ	10^6	mega-	M
10^{-9}	nano-	n	10^9	giga-	G
10^{-12}	pico-	p	10^{12}	tera-	T
10^{-15}	femto-	f	10^{15}	peta-	P
10^{-18}	atto-	a	10^{18}	exa-	E

Table 10   650

Table 10 Electromagnetic Spectrum

wavelength (m)		frequency (Hz)
3×10^7		10
3×10^6	audiofrequency range	10^2
3×10^5		10^3
3×10^4	VLF	10^4
3×10^3	LF	10^5
3×10^2	MF	10^6
30	HF	10^7
3	radio / VHF	10^8
3×10^{-1}	UHF	10^9
3×10^{-2}	SHF	10^{10}
3×10^{-3}	EHF	10^{11}
3×10^{-4}		10^{12}
3×10^{-5}	infrared	10^{13}
3×10^{-6}		10^{14}
3×10^{-7}	visible	10^{15}
3×10^{-8}		10^{16}
3×10^{-9}	ultraviolet	10^{17}
3×10^{-10}		10^{18}
3×10^{-11}	X-rays	10^{19}
3×10^{-12}		10^{20}
3×10^{-13}	gamma rays	10^{21}

Table 11 Periodic Table of the Elements - giving group, atomic number, and chemical symbol

Group	1	2	3	4	5	6	7	8	9	10	11	12	13	14	15	16	17	18
Period 1	1 H																	2 He
2	3 Li	4 Be											5 B	6 C	7 N	8 O	9 F	10 Ne
3	11 Na	12 Mg											13 Al	14 Si	15 P	16 S	17 Cl	18 Ar
4	19 K	20 Ca	21 Sc	22 Ti	23 V	24 Cr	25 Mn	26 Fe	27 Co	28 Ni	29 Cu	30 Zn	31 Ga	32 Ge	33 As	34 Se	35 Br	36 Kr
5	37 Rb	38 Sr	39 Y	40 Zr	41 Nb	42 Mo	43 Tc	44 Ru	45 Rh	46 Pd	47 Ag	48 Cd	49 In	50 Sn	51 Sb	52 Te	53 I	54 Xe
6	55 Cs	56 Ba	57-71 La-Lu	72 Hf	73 Ta	74 W	75 Re	76 Os	77 Ir	78 Pt	79 Au	80 Hg	81 Tl	82 Pb	83 Bi	84 Po	85 At	86 Rn
7	87 Fr	88 Ra	89-103 Ac-Lr															

	3	4	5	6	7	8	9	10	11	12	13	14	15	16	17
6 Lanthanides	57 La	58 Ce	59 Pr	60 Nd	61 Pm	62 Sm	63 Eu	64 Gd	65 Tb	66 Dy	67 Ho	68 Er	69 Tm	70 Yb	71 Lu
7 Actinides	89 Ac	90 Th	91 Pa	92 U	93 Np	94 Pu	95 Am	96 Cm	97 Bk	98 Cf	99 Es	100 Fm	101 Md	102 No	103 Lr

The above is the modern recommended form of the table using 18 groups.
Older group designations are shown below.

Modern form	1	2	3	4	5	6	7	8	9	10	11	12	13	14	15	16	17	18
European convention	IA	IIA	IIIA	IVA	VA	VIA	VIIA		VIII (or VIIIA)		IB	IIB	IIIB	IVB	VB	VIB	VIIB	0 (or VIIIB)
N. American convention	IA	IIA	IIIB	IVB	VB	VIB	VIIB		VIII (or VIIIB)		IB	IIB	IIIA	IVA	VA	VIA	VIIA	VIIIA (or 0)

Table 12   652

Table 12 Electron Arrangement in Atomic Shells of neutral atoms in their lowest energy state

Electron arrangement

Atomic number	Element	K	L		M			N				O					P				Q	
		1s	2s	2p	3s	3p	3d	4s	4p	4d	4f	5s	5p	5d	5f	5g	6s	6p	6d	6f	7s	7p
1	H	1																				
2	He	2																				
3	Li	2	1																			
4	Be	2	2																			
5	B	2	2	1																		
6	C	2	2	2																		
7	N	2	2	3																		
8	O	2	2	4																		
9	F	2	2	5																		
10	Ne	2	2	6																		
11	Na	2	2	6	1																	
12	Mg	2	2	6	2																	
13	Al	2	2	6	2	1																
14	Si	2	2	6	2	2																
15	P	2	2	6	2	3																
16	S	2	2	6	2	4																
17	Cl	2	2	6	2	5																
18	Ar	2	2	6	2	6																
19	K	2	2	6	2	6		1														
20	Ca	2	2	6	2	6		2														
21	Sc	2	2	6	2	6	1	2														
22	Ti	2	2	6	2	6	2	2														
23	V	2	2	6	2	6	3	2														
24	Cr	2	2	6	2	6	5	1														
25	Mn	2	2	6	2	6	5	2														
26	Fe	2	2	6	2	6	6	2														
27	Co	2	2	6	2	6	7	2														
28	Ni	2	2	6	2	6	8	2														
29	Cu	2	2	6	2	6	10	1														
30	Zn	2	2	6	2	6	10	2														
31	Ga	2	2	6	2	6	10	2	1													
32	Ge	2	2	6	2	6	10	2	2													
33	As	2	2	6	2	6	10	2	3													
34	Se	2	2	6	2	6	10	2	4													
35	Br	2	2	6	2	6	10	2	5													
36	Kr	2	2	6	2	6	10	2	6													
37	Rb	2	2	6	2	6	10	2	6			1										
38	Sr	2	2	6	2	6	10	2	6			2										
39	Y	2	2	6	2	6	10	2	6	1		2										
40	Zr	2	2	6	2	6	10	2	6	2		2										
41	Nb	2	2	6	2	6	10	2	6	4		1										
42	Mo	2	2	6	2	6	10	2	6	5		1										
43	Tc	2	2	6	2	6	10	2	6	5		2										
44	Ru	2	2	6	2	6	10	2	6	7		1										
45	Rh	2	2	6	2	6	10	2	6	8		1										
46	Pd	2	2	6	2	6	10	2	6	10												
47	Ag	2	2	6	2	6	10	2	6	10		1										
48	Cd	2	2	6	2	6	10	2	6	10		2										
49	In	2	2	6	2	6	10	2	6	10		2	1									
50	Sn	2	2	6	2	6	10	2	6	10		2	2									
51	Sb	2	2	6	2	6	10	2	6	10		2	3									
52	Te	2	2	6	2	6	10	2	6	10		2	4									
53	I	2	2	6	2	6	10	2	6	10		2	5									
54	Xe	2	2	6	2	6	10	2	6	10		2	6									
55	Cs	2	2	6	2	6	10	2	6	10		2	6				1					
56	Ba	2	2	6	2	6	10	2	6	10		2	6				2					
57	La	2	2	6	2	6	10	2	6	10		2	6	1			2					
58	Ce	2	2	6	2	6	10	2	6	10	2	2	6				2					
59	Pr	2	2	6	2	6	10	2	6	10	3	2	6				2					
60	Nd	2	2	6	2	6	10	2	6	10	4	2	6				2					
61	Pm	2	2	6	2	6	10	2	6	10	5	2	6				2					
62	Sm	2	2	6	2	6	10	2	6	10	6	2	6				2					
63	Eu	2	2	6	2	6	10	2	6	10	7	2	6				2					
64	Gd	2	2	6	2	6	10	2	6	10	7	2	6	1			2					
65	Tb	2	2	6	2	6	10	2	6	10	9	2	6				2					
66	Dy	2	2	6	2	6	10	2	6	10	10	2	6				2					
67	Ho	2	2	6	2	6	10	2	6	10	11	2	6				2					
68	Er	2	2	6	2	6	10	2	6	10	12	2	6				2					
69	Tm	2	2	6	2	6	10	2	6	10	13	2	6				2					
70	Yb	2	2	6	2	6	10	2	6	10	14	2	6				2					
71	Lu	2	2	6	2	6	10	2	6	10	14	2	6	1			2					
72	Hf	2	2	6	2	6	10	2	6	10	14	2	6	2			2					
73	Ta	2	2	6	2	6	10	2	6	10	14	2	6	3			2					
74	W	2	2	6	2	6	10	2	6	10	14	2	6	4			2					
75	Re	2	2	6	2	6	10	2	6	10	14	2	6	5			2					
76	Os	2	2	6	2	6	10	2	6	10	14	2	6	6			2					
77	Ir	2	2	6	2	6	10	2	6	10	14	2	6	9								
78	Pt	2	2	6	2	6	10	2	6	10	14	2	6	9			1					
79	Au	2	2	6	2	6	10	2	6	10	14	2	6	10			1					
80	Hg	2	2	6	2	6	10	2	6	10	14	2	6	10			2					
81	Tl	2	2	6	2	6	10	2	6	10	14	2	6	10			2	1				
82	Pb	2	2	6	2	6	10	2	6	10	14	2	6	10			2	2				
83	Bi	2	2	6	2	6	10	2	6	10	14	2	6	10			2	3				
84	Po	2	2	6	2	6	10	2	6	10	14	2	6	10			2	4				
85	At	2	2	6	2	6	10	2	6	10	14	2	6	10			2	5				
86	Rn	2	2	6	2	6	10	2	6	10	14	2	6	10			2	6				
87	Fr	2	2	6	2	6	10	2	6	10	14	2	6	10			2	6			1	
88	Ra	2	2	6	2	6	10	2	6	10	14	2	6	10			2	6			2	
89	Ac	2	2	6	2	6	10	2	6	10	14	2	6	10			2	6	1		2	
90	Th	2	2	6	2	6	10	2	6	10	14	2	6	10			2	6	2		2	
91	Pa	2	2	6	2	6	10	2	6	10	14	2	6	10	2		2	6	1		2	
92	U	2	2	6	2	6	10	2	6	10	14	2	6	10	3		2	6	1		2	

Table 13 Electrochemical Series – giving reduction potentials relative to the standard hydrogen electrode

Element	Reaction	Potential volts	Element	Reaction	Potential volts
Li	$Li^{+}+e^{c}\,Li$	−3.045	Ga	$Ga^{3+}+3e^{-}\rightleftharpoons Ga$	−0.56
Rb	$Rb^{+}+e^{-}\rightleftharpoons Rb$	−2.925	S	$S+2e^{-}\rightleftharpoons S^{2-}$	−0.508
K	$K^{+}+e^{-}\rightleftharpoons K$	−2.924	Fe	$Fe^{2+}+2e^{-}\rightleftharpoons Fe$	−0.409
Cs	$Cs^{+}+e^{-}\rightleftharpoons Cs$	−2.923	Cd	$Cd^{2+}+2e^{-}\rightleftharpoons Cd$	−0.4026
Ba	$Ba^{2+}+2e^{-}\rightleftharpoons Ba$	−2.90	In	$In^{3+}+3e^{-}\rightleftharpoons In$	−0.338
Sr	$Sr^{2+}+2e^{-}\rightleftharpoons Sr$	−2.89	Tl	$Tl^{+}+e^{-}\rightleftharpoons Tl$	−0.3363
Ca	$Ca^{2+}+2e^{-}\rightleftharpoons Ca$	−2.76	Co	$Co^{2+}+2e^{-}\rightleftharpoons Co$	−0.28
Na	$Na^{+}+e^{-}\rightleftharpoons Na$	−2.7109	Ni	$Ni^{2+}+2e^{-}\rightleftharpoons Ni$	−0.23
Mg	$Mg^{2+}+2e^{-}\rightleftharpoons Mg$	−2.375	Sn	$Sn^{2+}+2e^{-}\rightleftharpoons Sn$	−0.1364
Y	$Y^{3+}+3e^{-}\rightleftharpoons Y$	−2.37	Pb	$Pb^{2+}+2e^{-}\rightleftharpoons Pb$	−0.1263
La	$La^{3+}+3e^{-}\rightleftharpoons La$	−2.37	Fe	$Fe^{3+}+3e^{-}\rightleftharpoons Fe$	−0.036
Ce	$Ce^{3+}+3e^{-}\rightleftharpoons Ce$	−2.335	D_2	$D^{+}+e^{-}\rightleftharpoons \frac{1}{2}D_2$	−0.0034
Nd	$Nd^{3+}+3e^{-}\rightleftharpoons Nd$	−2.246	H_2	$H^{+}+e^{-}\rightleftharpoons \frac{1}{2}H_2$	0.000
H	$\frac{1}{2}H_2+e^{-}\rightleftharpoons H^{-}$	−2.23	Re	$Re^{3+}+3e^{-}\rightleftharpoons Re$	+0.3
Sc	$Sc^{3+}+3e^{-}\rightleftharpoons Sc$	−2.08	Cu	$Cu^{2+}+2e^{-}\rightleftharpoons Cu$	+0.340
Th	$Th^{4+}+4e^{-}\rightleftharpoons Th$	−1.90		$Cu^{+}+e^{-}\rightleftharpoons Cu$	+0.522
Np	$Np^{3+}+3e^{-}\rightleftharpoons Np$	−1.9	I_2	$I_2+2e^{-}\rightleftharpoons 2I^{-}$	+0.535
U	$U^{3+}+3e^{-}\rightleftharpoons U$	−1.80	Hg	$Hg_2^{2+}+2e^{-}\rightleftharpoons 2Hg$	+0.7961
Al	$Al^{3+}+3e^{-}\rightleftharpoons Al$	−1.706	Ag	$Ag^{+}+e^{-}\rightleftharpoons Ag$	+0.7996
Be	$Be^{2+}+2e^{-}\rightleftharpoons Be$	−1.70	Pd	$Pd^{2+}+2e^{-}\rightleftharpoons Pd$	+0.83
Ti	$Ti^{2+}+2e^{-}\rightleftharpoons Ti$	−1.63	Br_2	$Br_2(1)+2e^{-}\rightleftharpoons 2Br^{-}$	+1.065
V	$V^{2+}+2e^{-}\rightleftharpoons V$	−1.2	Pt	$Pt^{2+}+2e^{-}\rightleftharpoons Pt$	+1.2
Mn	$Mn^{2+}+2e^{-}\rightleftharpoons Mn$	−1.029	Cl_2	$Cl_2+2e^{-}\rightleftharpoons 2Cl^{-}$	+1.358
Te	$Te+2e^{-}\rightleftharpoons Te^{2-}$	−0.92	Au	$Au^{3+}+3e^{-}\rightleftharpoons Au$	+1.42
Se	$Se+2e^{-}\rightleftharpoons Se^{2-}$	−0.78		$Au^{+}+e^{-}\rightleftharpoons Au$	+1.68
Zn	$Zn^{2+}+2e^{-}\rightleftharpoons Zn$	−0.7628	F_2	$\frac{1}{2}F_2+e^{-}\rightleftharpoons F^{-}$	+2.85
Cr	$Cr^{3+}+3e^{-}\rightleftharpoons Cr$	−0.74			

Table 14   654

Table 14 Major Discoveries and Inventions in Electricity and Electronics

1745	capacitor: Leyden jar	P. van Musschenbroek
1747–48	positive and negative electricity postulated	B. Franklin
1767	inverse square law: postulated	J. Priestley
	demonstrated 1785	C. A. de Coulomb
1800	electric battery: voltaic pile	A. Volta
1808	atomic theory	J. Dalton
1820	electromagnetism	H. Oersted
1820–21	Ampère's laws	A. M. Ampère
1821	thermoelectricity	J. Seebeck
1823	electromagnet: first made	W. Sturgeon
	improved 1831	J. Henry
1827	Ohm's law	G. S. Ohm
1831	electromagnetic induction	M. Faraday
1831	transformer	M. Faraday
1832	self-induction	J. Henry
1833–34	analytical engine conceived	C. Babbage
1834	electrolysis laws	M. Faraday
1845	Kirchhoff's laws	G. Kirchhoff
1847	magnetostriction	J. Joule
1856	gas-discharge tube: low pressure	H. Geissler
1860	microphone: diaphragm	J. P. Reis
	carbon-granule 1878	D. Hughes
1864	Maxwell's equations	J. C. Maxwell
1876	telephone	A. G. Bell
1876–79	cathode-rays studied	W. Crookes
1877	gramophone	T. A. Edison
1879	Hall effect	E. Hall
1880	piezoelectricity	P. Curie
1887–88	electromagnetic waves (radio) first produced	H. Hertz
1887	antenna	H. Hertz
1887	photoelectric effect: observed	H. Hertz
	explained 1905	A. Einstein
1896	radio telegraphy: short distance	G. Marconi
	transatlantic 1901	G. Marconi
1897	electron discovered, proposed as	J. J. Thomson
	constituent of all matter	
1897	cathode-ray oscilloscope	F. Braun
1898	magnetic recording: wire	V. Poulsen
	tape 1927	US patent
1900	quantum theory	M. Planck
1901–02	ionosphere: postulated	A. Kennelly;
		O. Heaviside
	demonstrated 1924	E. Appleton
1904	thermionic valve: diode	J. A. Fleming
	triode 1906	L. de Forest
1906	radio broadcasting: first successful	R. Fessenden
	commercial 1919–20	UK, US
1911	superconductivity: observed	H. Kamerlingh Onnes
	BCS theory 1957	J. Bardeen, L. Cooper &
		J. Schrieffer
1911	atomic nucleus	E. Rutherford
1913	Bohr atom	A. Bohr

Table 14 Major Discoveries and Inventions in Electricity and Electronics
(continued)

1923	television: electronic system (iconoscope)	V. Zworykin
	mechanical system 1926	J. L. Baird
	electronic system 1927	P. Farnsworth
	colour demonstrated 1928	J. L. Baird
	image orthicon 1946	A. Rose, P. Heimer & H. Law
1924–25	radar: first experiments	E. Appleton; et al
	developed 1935–45	R. Watson-Watt
1924–25	wave mechanics	E. Schrödinger; L. de Broglie; W. Heisenberg; et al
1926	Yagi antenna	H. Yagi
1929–35	TV broadcasting: mechanical scanning	BBC
	electronic system	BBC
	colour, NTSC 1953	US
1933	FM radio: patented	E. Armstrong
1936	waveguide	various
1937	pulse code modulation: invented	A. Reeves
	developed 1940s	
1937	klystron	R. & S. Varian
1939	magnetron	J. Randall & H. Boot
1939–46	computer: using vacuum tubes	UK; US
	transistors 1956	
	integrated circuits 1968	
1947	transistor: point contact, demonstrated	J. Bardeen, W. Brattain
	(announced 1948)	& W. Shockley
	bipolar junction 1948	W. Shockley
1948	single crystal growth: germanium	G. Teal & J. Little
	silicon 1952	G. Teal & E. Buehler
1952	field-effect transistor: JFET	W. Shockley
	MOSFET 1958/60	various
1955	optical fibre: demonstrated	N. Kapany
	long-distance transmission 1966	C. Kao & G. Hockham
	commercial use 1977	UK
1958	integrated circuit: prototype	J. Kilby; R. Noyce
	using planar process 1959	J. Hoerni & R. Noyce
	MOS technology 1970	
1960	laser	T. Maiman
1960	communications satellite: passive, Echo	US
	active, Telstar 1962	US
	GEOS, Early Bird 1965	Intelsat
1969	computer internet: ARPANET	US research groups
	generally available as Internet 1984	
1971	microprocessor: Intel 4004	T. Hoff
	8-bit 8008 1972	Intel
1975	personal computer: kit form, Altair 8800	E. Roberts
	Apple II 1977	S. Jobs & S. Wozniak
	IBM PC 1981	IBM
1979	telephony: 1st digital exchange	UK
1982	compact disc: audio player	CBS & Sony; Philips
	CD-ROM for data storage 1984	Sony & Philips
1986	superconductivity: high-temperature	A. Müller & G. Bednorz

Table 15   656

Table 15 The Greek Alphabet

Letters		Name	Letters		Name
A	α	alpha	N	ν	nu
B	β	beta	Ξ	ξ	xi
Γ	γ	gamma	O	o	omikron
Δ	δ	delta	Π	π	pi
E	ε	epsilon	P	ρ	rho
Z	ζ	zeta	Σ	σ	sigma
H	η	eta	T	τ	tau
Θ	θ	theta	Y	υ	upsilon
I	ι	iota	Φ	φ	phi
K	κ	kappa	X	χ	chi
Λ	λ	lambda	Ψ	ψ	psi
M	μ	mu	Ω	ω	omega

READ MORE IN PENGUIN

In every corner of the world, on every subject under the sun, Penguin represents quality and variety – the very best in publishing today.

For complete information about books available from Penguin – including Puffins, Penguin Classics and Arkana – and how to order them, write to us at the appropriate address below. Please note that for copyright reasons the selection of books varies from country to country.

In the United Kingdom: Please write to *Dept. EP, Penguin Books Ltd, Bath Road, Harmondsworth, West Drayton, Middlesex UB7 0DA*

In the United States: Please write to *Consumer Sales, Penguin Putnam Inc., P.O. Box 999, Dept. 17109, Bergenfield, New Jersey 07621-0120*. VISA and MasterCard holders call 1-800-253-6476 to order Penguin titles

In Canada: Please write to *Penguin Books Canada Ltd, 10 Alcorn Avenue, Suite 300, Toronto, Ontario M4V 3B2*

In Australia: Please write to *Penguin Books Australia Ltd, P.O. Box 257, Ringwood, Victoria 3134*

In New Zealand: Please write to *Penguin Books (NZ) Ltd, Private Bag 102902, North Shore Mail Centre, Auckland 10*

In India: Please write to *Penguin Books India Pvt Ltd, 210 Chiranjiv Tower, 43 Nehru Place, New Delhi 110 019*

In the Netherlands: Please write to *Penguin Books Netherlands bv, Postbus 3507, NL-1001 AH Amsterdam*

In Germany: Please write to *Penguin Books Deutschland GmbH, Metzlerstrasse 26, 60594 Frankfurt am Main*

In Spain: Please write to *Penguin Books S. A., Bravo Murillo 19, 1° B, 28015 Madrid*

In Italy: Please write to *Penguin Italia s.r.l., Via Benedetto Croce 2, 20094 Corsico, Milano*

In France: Please write to *Penguin France, Le Carré Wilson, 62 rue Benjamin Baillaud, 31500 Toulouse*

In Japan: Please write to *Penguin Books Japan Ltd, Kaneko Building, 2-3-25 Koraku, Bunkyo-Ku, Tokyo 112*

In South Africa: Please write to *Penguin Books South Africa (Pty) Ltd, Private Bag X14, Parkview, 2122 Johannesburg*

READ MORE IN PENGUIN

SCIENCE AND MATHEMATICS

Six Easy Pieces Richard P. Feynman

Drawn from his celebrated and landmark text *Lectures on Physics*, this collection of essays introduces the essentials of physics to the general reader. 'If one book was all that could be passed on to the next generation of scientists it would undoubtedly have to be *Six Easy Pieces*' John Gribbin, *New Scientist*

A Mathematician Reads the Newspapers John Allen Paulos

In this book, John Allen Paulos continues his liberating campaign against mathematical illiteracy. 'Mathematics is all around you. And it's a great defence against the sharks, cowboys and liars who want your vote, your money or your life' Ian Stewart

Dinosaur in a Haystack Stephen Jay Gould

'Today we have many outstanding science writers ... but, whether he is writing about pandas or Jurassic Park, none grabs you so powerfully and personally as Stephen Jay Gould ... he is not merely a pleasure but an education and a chronicler of the times' *Observer*

Does God Play Dice? Ian Stewart

As Ian Stewart shows in this stimulating and accessible account, the key to this unpredictable world can be found in the concept of chaos, one of the most exciting breakthroughs in recent decades. 'A fine introduction to a complex subject' *Daily Telegraph*

About Time Paul Davies

'With his usual clarity and flair, Davies argues that time in the twentieth century is Einstein's time and sets out on a fascinating discussion of why Einstein's can't be the last word on the subject' *Independent on Sunday*

READ MORE IN PENGUIN

SCIENCE AND MATHEMATICS

In Search of SUSY   John Gribbin

Many physicists believe that we are on the verge of developing a complete 'theory of everything' which can reduce the four basic forces of nature – gravity, electromagnetism, the strong and weak nuclear forces – to a single superforce. At its heart is the principle of super-symmetry (SUSY).

Fermat's Last Theorem   Amir D. Aczel

Here, weaving together history and science, Amir D. Aczel offers a thrilling, step-by-step account of the search for the mathematicians' Holy Grail. 'Mr Aczel has written a tale of buried treasure ... This is a captivating volume' *The New York Times*

Insanely Great   Steven Levy

It was Apple's co-founder Steve Jobs who referred to the Mac as 'insanely great'. He was absolutely right: the machine that revolutionized the world of personal computing was and is great – yet the machinations behind its inception were nothing short of insane. 'A delightful and timely book' *The New York Times Book Review*

The Artful Universe   John D. Barrow

This thought-provoking investigation illustrates some unexpected links between art and science. 'Full of good things ... In what is probably the most novel part of the book, Barrow analyses music from a mathematical perspective ... an excellent writer' *New Scientist*

The Jungles of Randomness   Ivars Peterson

Taking us on a fascinating journey into the ambiguities and un-certainties of randomness, Ivars Peterson explores the complex interplay of order and disorder, giving us a new understanding of nature and human activity.

READ MORE IN PENGUIN

REFERENCE

The Penguin Dictionary of the Third Reich
James Taylor and Warren Shaw

This dictionary provides a full background to the rise of Nazism and the role of Germany in the Second World War. Among the areas covered are the major figures from Nazi politics, arts and industry, the German Resistance, the politics of race and the Nuremberg trials.

The Penguin Biographical Dictionary of Women

This stimulating, informative and entirely new Penguin dictionary of women from all over the world, through the ages, contains over 1,600 clear and concise biographies on major figures from politicians, saints and scientists to poets, film stars and writers.

Roget's Thesaurus of English Words and Phrases
Edited by Betty Kirkpatrick

This new edition of Roget's classic work, now brought up to date for the nineties, will increase anyone's command of the English language. Fully cross-referenced, it includes synonyms of every kind (formal or colloquial, idiomatic and figurative) for almost 900 headings. It is a must for writers and utterly fascinating for any English speaker.

The Penguin Dictionary of International Relations
Graham Evans and Jeffrey Newnham

International relations have undergone a revolution since the end of the Cold War. This new world disorder is fully reflected in this new Penguin dictionary, which is extensively cross-referenced with a select bibliography to aid further study.

The Penguin Guide to Synonyms and Related Words
S. I. Hayakawa

'More helpful than a thesaurus, more humane than a dictionary, the *Guide to Synonyms and Related Words* maps linguistic boundaries with precision, sensitivity to nuance and, on occasion, dry wit' *The Times Literary Supplement*

READ MORE IN PENGUIN

REFERENCE

The Penguin Dictionary of Troublesome Words   Bill Bryson

Why should you avoid discussing the *weather conditions*? Can a married woman be celibate? Why is it eccentric to talk about the aroma of a cowshed? A straightforward guide to the pitfalls and hotly disputed issues in standard written English.

Swearing   Geoffrey Hughes

'A deliciously filthy trawl among taboo words across the ages and the globe' Valentine Cunningham, *Observer*, Books of the Year. 'Erudite and entertaining' Penelope Lively, *Daily Telegraph*, Books of the Year.

Medicines: A Guide for Everybody   Peter Parish

Now in its seventh edition and completely revised and updated, this bestselling guide is written in ordinary language for the ordinary reader yet will prove indispensable to anyone involved in health care: nurses, pharmacists, opticians, social workers and doctors.

Media Law   Geoffrey Robertson QC and Andrew Nichol

Crisp and authoritative surveys explain the up-to-date position on defamation, obscenity, official secrecy, copyright and confidentiality, contempt of court, the protection of privacy and much more.

The Penguin Careers Guide
Anna Alston and Anne Daniel; Consultant Editor: Ruth Miller

As the concept of a 'job for life' wanes, this guide encourages you to think broadly about occupational areas as well as describing day-to-day work and detailing the latest developments and qualifications such as NVQs. Special features include possibilities for working part-time and job-sharing, returning to work after a break and an assessment of the current position of women.

READ MORE IN PENGUIN

DICTIONARIES

Abbreviations
Ancient History
Archaeology
Architecture
Art and Artists
Astronomy
Biographical Dictionary of
  Women
Biology
Botany
Building
Business
Challenging Words
Chemistry
Civil Engineering
Classical Mythology
Computers
Contemporary American History
Curious and Interesting Geometry
Curious and Interesting Numbers
Curious and Interesting Words
Design and Designers
Economics
Eighteenth-Century History
Electronics
English and European History
English Idioms
Foreign Terms and Phrases
French
Geography
Geology
German
Historical Slang
Human Geography
Information Technology

International Finance
International Relations
Literary Terms and Literary
  Theory
Mathematics
Modern History 1789–1945
Modern Quotations
Music
Musical Performers
Nineteenth-Century World
  History
Philosophy
Physical Geography
Physics
Politics
Proverbs
Psychology
Quotations
Quotations from Shakespeare
Religions
Rhyming Dictionary
Russian
Saints
Science
Sociology
Spanish
Surnames
Symbols
Synonyms and Antonyms
Telecommunications
Theatre
The Third Reich
Third World Terms
Troublesome Words
Twentieth-Century History
Twentieth-Century Quotations